SHAKESPEARE
and
OPERA

SHAKESPEARE & OPERA

Gary Schmidgall

New York   Oxford
OXFORD UNIVERSITY PRESS
1990

Oxford University Press

Oxford   New York   Toronto
Delhi   Bombay   Calcutta   Madras   Karachi
Petaling Jaya   Singapore   Hong Kong   Tokyo
Nairobi   Dar es Salaam   Cape Town
Melbourne   Auckland

and associated companies in
Berlin   Ibadan

Copyright © 1990 by Oxford University Press, Inc.

Published by Oxford University Press, Inc.,
200 Madison Avenue, New York, New York 10016

Library of Congress Cataloging-in-Publication Data
Schmidgall, Gary, 1945–
Shakespeare and opera / Gary Schmidgall.
p.   cm.
Includes bibliographical references.
ISBN 0–19–506450–X
1. Opera.   2. Shakespeare, William, 1564–1616.
3. Music in literature.
4. Music—Philosophy and aesthetics.
I. Title.
ML3858.S373   1990
782.1—dc20   90–6861   MN

1 3 5 7 9 8 6 4 2
Printed in the United States of America
on acid-free paper

For Michael Plummer

Contents

OPERAS FROM SHAKESPEARE

ACKNOWLEDGMENT

I am grateful to the American Council of Learned Societies for a fellowship that supported the final stages of research on this book.

NOTE ON THE TEXT

The sources for all quotations are given in the endnotes, pages 335–383. Those notes which contain further important or relevant information are denoted in the text by an asterisk (*). Except where noted, all quotations and line references from Shakespeare are made from *The Riverside Shakespeare,* edited by G. Blakemore Evans and others (Houghton Mifflin, 1974). Unless otherwise indicated, all translations are my own.

INTRODUCTION
The Great Shakespearean Orange

Is there an art form that receives more satirical or downright derisive publicity than opera? Here is one tirade that touches most of the usual bases: we find, says the author, "nothing but death made sensational, despair made stage-sublime, sex made romantic, and barrenness covered by sentimentality and . . . plenty of bogus characterization. . . . At the great emotional climaxes we find passages which [rely] on symmetry of melody and impressiveness of march to redeem a poverty of meaning." The reader will nod familiarly at these clichés of opprobrium. But I hope the reader will also be a little discomfited to learn that they were aimed (by George Bernard Shaw) not at opera but at the works of William Shakespeare.

For several years I have taught and written about the subjects of Shakespeare and opera, often separately but sometimes in tandem. Over these years my conviction that Shakespearean and operatic drama are profoundly alike has experienced a crescendo worthy of a Rossini finale. This conviction has led me, in my English courses, to become something of a clandestine propagandist for opera. The "cover" is perfect and irresistible: teacher of Shakespeare. For if ever there was an author famous for being famous it is, as W. H. Auden described him, the Top Bard. To advance through life in an English-speaking culture without at least a smattering of Shakespeare is possible, of course, but not without loss of caste and some burden of guilt. *Shakespearean* is a privileged epithet even on the world's literary stage. Ivan Turgenev wrote to Gustave Flaubert about Émile Zola, "I can well believe that he has never read Shakespeare. It is a sort of deformity of mind he will never get rid of." And Hector Berlioz wrote of a close friend: "to have reached the age of forty-five or fifty and not to know *Hamlet*—it's like having lived all one's years in a coal mine!" The attention of my students enforced by the deep harmony of *Shakespearean* and eager to erase their "deformity of mind," they have been very vulnerable to offhand interpolations, during class discussion, of such terms as *tempo, rhythm, counterpoint, reprise, duet,* and *aria.* When the syllabus brings forth *Othello,* I never fail to surprise unwary scorners of opera with Shaw's typically bold and witty assertion: "The truth is that instead of *Otello* being an Italian opera written in the style of Shakespear, *Othello* is a play written by Shakespear in the style of Italian opera . . . [T]he plot is pure farce plot: that is to say, it is supported on an artificially manufactured and desperately precarious trick with a handkerchief which a chance word might upset at any moment. With such a libretto, Verdi was quite at home: his success with it proves, not that he could occupy Shakespear's plane, but that Shakespear could on occasion occupy his." And my students have been fair game for some carefully laid snares like the one in my first paragraph.

I think I have ridden my operatic hobbyhorse amiably and not too insistently. My students, for their part, have accepted these seeming digressions with good grace and some awareness of how baldly I am seeking to raise the cachet of *operatic* by association with the all-conquering splendor of *Shakespearean*. I have made shameless pedagogical use of the association, too: opera talk almost inevitably tickles the comic Muse, and Thalia is a very useful colleague in the classroom. But there are serious as well as amusing aspects of my variations on the theme of synonymy between *operatic* and *Shakespearean,* and these will be my principal quarry in the following excursions. While bowing to none in my enjoyment of the unintended humor derived from opera (and opera singers and performances), I think the epithet *operatic* deserves to be rescued from all the tongue-in-cheek, elbow-in-the-ribs condescension with which it is usually uttered. No path for such a rescue seems more inviting or promising than that of stressing at length and in detail the shared characteristics of Shakespearean and operatic drama.

That *operatic* deserves such solicitude becomes apparent if we sample contemporary appearances of the word, which the *Oxford English Dictionary* says was first used in the mid-eighteenth century. It is likely to crop up nowadays in strange but unfailingly colorful circumstances:

My grandmother, in her own way, shines like a beacon down the stormy American past. She was a bootlegger in a little county up in the state of Washington. She was also a handsome woman, close to six feet tall, who carried 190 pounds in the grand operatic manner.

Joseph Pintauro is one playwright who will never be accused of being too subtle. . . . [His plays offer] blunt, melodramatic scenarios that get right to the heart of the matter . . . flagrantly provocative . . . visions [that] seem raw and strident. . . . Mr. Pintauro's writing is impatiently, crudely operatic.

[The] battle between leather-jacketed hoodlums and T-shirt–clad fathers, a battle fought on one side with chains and on the other with baseball bats, hoes and garbage-pail lids—is rendered by Ms. McDermott with an almost operatic grandeur.

On the field, managers went to operatic extremes to show up umpires, their tempers bursting like sewer pipes.

[Rebecca West's] extravagance of feeling was certainly operatic in the sense that it was orchestrated and, in fact, led to disturbing, and even fantastic, illusions.

Enter a deliciously described world of sharply painted, dramatically costumed heroes and heroines posing, with many a spectacular gesture and eloquent aria, in magnificent landscapes maintained by invisible hands as a kind of huge stage set. This operatic Europe, like opera itself, would call us into largeness.

You get the idea. The word appears, and one can sense the writer's lips curling, however slightly. *Operatic* means "not normal": extravagant, perhaps crazy, but not dangerously so; not given to mincing or minuets; unafraid of being perceived by normal folks as vulgar or outlandish. Operatic characters would be hard to live with in real life. Auden ventured that the great ones like Norma, Lucia, Tristan, and Brünnhilde "would all be bores, even Don Giovanni." In real life most of them would tend to loom, as if wearing those high-heeled buskins that Greek tragedians put on to give themselves greater stage presence. *Operatic* suggests mold-breakers.

Such figures make an easy target for humor, and much of this humor comes from their staunchest devotees. The novelist John Gardner, who cut an operatic swath through life himself, believed that opera was the most important influence upon him as a youth, yet he still thought of the operatic stage as offering "a grand cartoon—Wagner's mountainscapes and gnomes, Mozart's crazies, Humperdinck's angels, the weirdness and clowning that show up everywhere from *La Bohème* to *The Tales of Hoffmann*." Pertinent to Gardner's view is the fact that the Japanese word for opera comes from the verb *kabuku,* which means "to be far out, offbeat." Opera's penchant for the character who is "far out" is vividly suggested by what happened to the main character in several Sir Walter Scott novels, so favored by nineteenth-century librettists. Julian Budden writes that "with rare exceptions the central figure of a Scott canvas is usually too normal and well-behaved for operatic treatment," and so he is often entirely removed from the libretto plot.

Operatic characters, in short, are very . . . Shakespearean. Name three "normal" characters of any significance in his plays, if you can. The only normal person in *Hamlet* is Fortinbras. Remember Fortinbras? Characters who are normal or mundane in their speech, thought, and actions rarely commandeer Shakespeare's stage—and then, as with Polonius, only when they are screwed up to parodic pitch. The dull-witted and hence dull-tongued folks like Constable Dull in *Love's Labour's Lost* have trouble getting a word in edgewise. Virtually all of the memorable characters, both large and small, beggar credulity in a way that each of the authors quoted above was striving to suggest with the word *operatic*. My point, of course, is that each author might have chosen *Shakespearean* with similar justness. "To be far out, offbeat"—often noisily and elaborately so—is to be Shakespearean. It takes some courage in an audience and much indeed from an actor or actress—shall I say an operatic courage?—to embrace the vivid, voluble, incredible elements that are dominant in most of the major roles.

Shaw made this point when he launched himself *con brio* at an actress who thought she could play Lady Macbeth and remain "ladylike." As often, he relied on musical analogies: "Neither [Shakespeare] nor any other musician ever wrote music without *fortissimi* and thundering ones too. It is only your second-rate people who write whole movements for muted strings and never let the trombones and the big drum go. It is not by tootling to him *con sordino* that Lady

Macbeth makes Macbeth say 'Bring forth men children only.' She lashes him into murder." John Updike writes acutely that opera calls us "into largeness," and Shaw is here making the same point in his very practical stage-director's way. He is calling Stella Patrick Campbell to a largeness of imagination . . . *and* vocalization. He ends his lecture by reminding her how outrageous Lady Macbeth is to the notions of "normal" and "credible": "If you want to know the truth about Lady Macbeth's character, she hasn't one. There never was no such person. She says things that will set people's imagination to work if she says them in the right way: that is all." As we shall see on several occasions, such is the supreme task of the operatic artist as well. I anticipate here a point that will be pursued later by way of giving a flavor of the unexpected juxtapositions in the following rather picaresque journey. I cannot but expect that Shakespeareans and opera aficionados will find some of the material drawn from their respective bailiwicks to be familiar, but I am hopeful that the unusual context in which much of it is placed will, as Shaw phrased it, set the imagination to work.

√ "A private railroad car is *not* an acquired taste," Mrs. August Belmont once observed; "one takes to it immediately." My experience with initiates to Shakespeare and opera has led me to believe that, in general, these art forms fall into the same category as private railroad cars. The magical poise of sensibility being just right, one takes to them immediately. I am inclined to think that the happy crisis is a matter of luck, momentary openness of the senses, and serendipity. That my own fourteen-year-old virgin palate was, against all reasonable expectation, whetted by Richard Wagner's *Die Meistersinger* (one of the longest operas on record) in Los Angeles' cavernous Shrine Auditorium, in the company of a relative who knew nothing about opera, underscores my point. (My Shakespearean initiation was more auspicious: *A Midsummer Night's Dream* with Bert Lahr as Bottom, in the company of the favorite uncle who was to introduce me seriously to opera some years later.) The effect of first contacts is crucial and telling. In class it quickly becomes clear who has taken to Shakespeare: if a student bridles impatiently at the artificiality and verbal complexities of a Shakespearean sonnet, it is the sure hint of a semester's uphill struggle. Similarly, I have been in the company of several opera debutants; the reactions they express during the first intermission will usually tell the tale. (There are happy exceptions. I recall a student, in a dormitory over which I had some cultural charge, whose athletic devotions doomed my efforts to interest him in opera. Several years later I was flabbergasted to meet him again at a performance of, incredibly, *Parsifal* . . . as an enthusiastic standee, no less.)

Which brings me to the brutal corollary for Mrs. Belmont's witticism: there are people in the world who have no palate for private railroad cars, Shakespeare, or opera. They may instead prefer Porsches, Harley-Davidson motorcycles, Ernest Hemingway's novels and Neil Simon's plays, the latest smash Broadway musical, or minimalist "performance pieces." Such preferences ought to occasion no snobbish condescension. "Chacun à son goût," says Prince Orlofsky in *Die Fledermaus*. It is rather an occasion for regret and even sympathy that, because he is the Top Bard, Shakespeare will always attract a healthy percentage of students in an introductory class who do not have and will never acquire a taste for his plays. They will leave his "great feast of languages" with a mere "alms-basket of words" (phrases from *Love's Labour's Lost*). Similar regret and sympathy must be extended to the many people in a typical opera audience who have yielded to the *force majeure* of a parent, spouse, lover, or irresistible friend and not because they have acquired the taste for opera. Like Edgar Guest, they "cannot pass the citadels / To tread the path where music dwells." Yet there they sit, their eyes—and their mind's eye—glazed with boredom.

These outcasts from opera and outcasts from Shakespeare are legion. The former tend to be a more complacent, feistier lot and can fend for themselves well enough. The latter excite a little more compassion; exposed to Bardolatry from childhood, their exile is borne more wistfully. Indeed, Guest's poem "The World of Music" captures well the regret of those who feel a little ashamed that Shakespeare's "music of men's lives" does not move them:

> I know it is a realm ablaze
> With beauty I shall never see.
> I'd like to walk its shores alone
> And feel my soul with rapture swept,
> But I was fashioned deaf to tone,
> So at the outer wall I'm kept.
> Friends freely pass within the gate
> And leave me in the cold to wait.

Outcasts will appear only peripherally in what follows, but they will be helpful. Knowing what has driven them away can help us understand more vividly and precisely the "strong toil of grace" that enchants and captivates the admirers of Shakespeare and opera, as Antony was captivated by Cleopatra. And it is worth adding in defense of these outcasts that, however quickly one acquires a taste for Shakespearean and operatic theater, refining and educating the palate to a full appreciation can cost much in energy and time. These worlds present an intimidatingly large vocabulary of sounds and words (Shakespeare's is by far the record vocabulary in world literature). Almost nothing in opera and in Shakespeare is as simple or forthright as at first appears. To enjoy them, it helps to bring a taste for ambiguity, for complexity, for intricate thought and structure.

A few warnings now about the organization and purposes of this book. The reader who comes looking for theoretical formulations and analysis will be disappointed. If any theory is dominant here, it is the shamelessly opportunistic pragmatism of the two creators who will be most prominent in these pages, two men of the theater rather than of theory: Shakespeare and Giuseppe Verdi. Shakespeare has been well described as possessing "a supremely wandering mind" and as "always wantoning on the verge of anarchy." He was not of a discursive, rule-mongering temper. He did not—as did his rival, Ben Jonson—seek to mold his audience or "the theater" to his own particular notions of morality and order. His one fit of prescription—Hamlet's admonitions to the itinerant players—

scarcely rises to the status of theory. These admonitions are as enigmatic as the *Mona Lisa* when it comes to judging precisely how natural or artificial Hamlet/ Shakespeare desired acting to be. Shaw observed that "no man ever really thinks about a thing until it is his job," and the plays of Shakespeare reveal everywhere the thought of an actor, a producing co-owner, an author, and (presumably) a director. Hence their efficiency and effectiveness, often when he is blithely trampling on some supposed "rule" of playwriting. "Utility is Shakespeare's sole test," wrote Harley Granville-Barker. "He will employ any sort of device, however old and worn, if he can make it dramatically useful."

Verdi was of the same bent. In 1871 the Italian Minister of Education asked him to assist in improving the nation's conservatories—domain of the musicologists. He declined, but offered instead to direct his "concern to a surer, more useful and more practical purpose: the theater." A true Shakespearean, Verdi goaded his librettist by saying, "When the situation demands it, one must have no scruples," and he echoes Granville-Barker's conclusion about the playwright when, while working on *Otello*, he threw up his hands at the current wars of musical (especially Wagnerian) theory: "In a word: melody, harmony, coloratura, declamation, instrumentation, local color . . . all these are only means. Make good music with these means, and I will accept everything, and every genre."

Among such company—for whom, in Alexander Pope's famous phrase, "license is a rule"—the desire to discover or extrapolate theory is risky and unprofitable. One can so easily settle into the rule-mongering of a Polonius or Beckmesser. Shakespeare's plays show him generally of Ogden Nash's mind: "Take this as a rule of thumb / Too clever is dumb." This, in less compact form, is precisely Berowne's point in *Love's Labour's Lost*:

> So study evermore is overshot:
> While it doth study to have what it would,
> It doth forget to do the thing it should;
> And when it hath the thing it hunteth most,
> 'Tis won as towns with fire—so won, so lost.   (1.1.142–146)

Then there is Apemantus' terrible put-down in *Timon of Athens*, which lies in wait for any author who, while spinning theoretical arabesques, loses his subject . . . or his reader: "So thou apprehend'st it, take it for thy labor" (1.1.209). And so I have striven here for a more relaxed, serendipitous passage through the colorful and exotic terrain: as the medieval poet John Gower said of his *Confessio Amantis*, "I undertook to make a book which stands between earnest and game."

Nor will I attempt to canvass exhaustively opera's Shakespearean repertory. That would, alas, be a glum business. As an Italian essayist wrote in 1879, apropos rumors of a Verdian *Otello* and the apparent demise of a Verdian *King Lear*, "Shakespeare has always exercised a kind of fascination over composers, but when they try to measure up to him they all, even the most illustrious, remain quite inferior." Sad, but (almost) true. Watching operas tortured, as they often are, on the Procrustean bed of Shakespeare's genius naturally tempts one into an incensed frame of mind. It certainly did for Paul Dukas, who raged thus on behalf of all the fed-up: "Only a few of Shakespeare's plays have escaped being mutilated, transformed, perverted, and abused with a view to musical adaptation. Never has a man of genius been so scandalously exploited by the official purveyors of opera in our theaters, and no one ever lived upon whose glorious brow has fallen such an avalanche of dull verse united with so many sacrilegious semiquavers."

Obviously, that way dyspepsia lies, as Winton Dean has sanely suggested in his perceptive, witty monograph "Shakespeare and Opera." Simply judging how far an opera takes the measure of its source play, Dean warns, "planes away the contours of history by burying much interesting material and diminishing the perspective in which the masterpieces appear." Dean might have added, too, that a jealous preoccupation with the literary "merit" of a source can easily lead one to miss the distinctive *raisons d'être,* the discrete (but seldom discreet) joys of opera itself. One can, in other words, become a stick like Harriet Herriton, in E. M. Forster's *Where Angels Fear to Tread,* at a provincial Italian performance of *Lucia di Lammermoor*: "Harriet, like M. Bovary on a more famous occasion, was trying to follow the plot. Occasionally she nudged her companions, and asked them what had become of Walter Scott. She looked round grimly. The audience sounded drunk, and even Caroline, who never took a drop, was swaying oddly. Violent waves of excitement, all arising from very little, went sweeping around the theater." As the performance reaches its climax, Harriet can take no more: "Call this classical! . . . It's not even respectable! Philip! take me out at once."

Variations on "Call this Shakespeare!"—uttered in just Miss Herriton's primly outraged tone of voice—have echoed through the last three centuries. Perhaps they have echoed so often that an unnecessarily negative atmosphere has come to envelop all musical adaptations of the Bard. A musicologist mounting a "defence" of Henry Purcell's *Fairy Queen* (based on *A Midsummer Night's Dream*) a few years ago understandably objected to what he saw as "an orthodoxy of distaste" for Shakespearean operas. This is not to say, though, that many a brutal dismissal has been unjust. Indeed, there is much cause for mirth in the numerous spirited, acerbic reactions to Shakespearean plays that have been traumatized for music's sake. Verdi, for instance, heard of a Parisian *Hamlet* and wrote to a friend there, "Is it true that two ballets have been interpolated? *Hamlet* and dance tunes!! What a cacophony! Poor Shakespeare!" In a few months he had read the libretto and reported, "It is impossible to have done worse. Poor Shakespeare! how they have screwed you up! There is nothing but the scene between Hamlet and the Queen, which is well-managed, theatrical, and suitable for music. The rest . . . *Amen.*" After Shaw visited Ambroise Thomas' opera he coolly suggested: "The title of the work ought to be changed. Since St. Ambroise Thomas has honestly done his very best not to remind us of Shakespear, why should the subject be dragged in by calling the people in the libretto Ophelia, Hamlet, Laertes, and so on?" Berlioz said his happiest Italian memories were of Florence, but they evidently did not include attending Vincenzo Bellini's *Capuleti e i Montecchi* at the city's Teatro della Pergola: "Bitter disappointment! The opera contained no ball at the Capulets, no Mercutio, no garrulous nurse, no grave and tranquil hermit, no balcony scene, no sublime soliloquy for Juliet, no duet in the cell between the banished Romeo and the disconsolate friar, no Shakespeare, no nothing." And

Berlioz added of Bellini, "That little fool has apparently not been afraid that Shakespeare's shade might come and haunt him in his sleep." Claude Debussy, to take another example, attended a *Midsummer Night's Dream* opera and expressed his distaste with this cryptic warning: "One should never fall in love with fairies. Except when one is very young, and in books generally bound in red." Much more recently, Andrew Porter attended Berlioz' own *Béatrice et Bénédict* and

✓ sensibly observed that it is "*Much Ado* without the ado." Less often, attempts are made to *épater les Bardolâtres* with similar invidious comparisons. Not long ago a critic made bold to suggest that Verdi's *Falstaff* is "more truly Shakespearean than its Shakespearean source." Another commentator saw this and raised the ante: "Verdi and Boito have in *Falstaff* improved, even redeemed, Shakespeare."

These skirmishes along the border between legitimate and musical drama are often exhilarating, and one is sometimes inclined to enjoy, without necessarily sharing, the furious desire to protect (or detonate) Shakespeare's reputation. On the other hand, one must chasten the gallant protectors and point out how insolently and lavishly rewritten, rearranged, cut, bowdlerized, and upholstered the plays themselves have been over the last three centuries. When Ralph Vaughan Williams felt obliged to defend the "impertinence" of offering the public another Falstaffian opera, *Sir John in Love,* he said simply, "With regard to Shakespeare my only excuse can be that he is fair game, like the Bible." And fair game he has truly been, beginning with the Restoration. An image in the prologue to the Sir William D'Avenant–John Dryden version of *The Tempest* (1670) candidly tells all:

> As when a Tree's cut down a secret root
> Lives underground and thence new Branches shoot
> So, from old Shakespeare's honour'd dust, this day
> Springs up and buds a new reviving Play.

Not a few branches trimmed or lopped, mind you, but the tree itself cut down! Most notorious of the Restoration transformations was Nahum Tate's 1681 *King Lear.* One can tell from the dedication that there will be mayhem: "I found the whole [play] a Heap of Jewels, unstrung and unpolish't; yet so dazling in their Disorder, that I soon perceiv'd I had seiz'd a Treasure." Very few marauding opera librettists have matched Tate's boldness in seizing the Shakespearean booty. He dispensed with the Fool, invented a love affair between Edgar and Cordelia, and "made the Tale conclude in a Success to the innocent distrest Persons." Thus, Lear, Kent, and Gloucester retire to a "cool Cell" for "calm Reflections on [their] Fortunes past." Tate explains, "Otherwise I must have incumbred the Stage with dead Bodies, which Conduct makes many Tragedies conclude with unseasonable jests." And he adds this monumentally fatuous bit of self-congratulation: "Neither is it so Trivial an Undertaking to make a Tragedy end happily, for 'tis more difficult to Save than 'tis to Kill." Tate's *Lear* held the English stage *almost 150 years,* until 1823, when Edmund Kean restored the tragic ending. The Fool had to wait until 1838 to be reinstated by William Charles Macready.

The eighteenth century was dominated by Colley Cibber's versions of eighteen of the plays, all involving cut speeches and scenes, altered poetry, and additions. With plays, like operas, becoming vehicles for star performers, many changes were designed to throw a title role into relief. Thus, Cibber's *Richard III* was cut by 1500 lines and five Richard-less scenes. The crookback's lines rose from thirty to forty percent of the whole, and he received *seven* new soliloquies. This makes the solicitude of Handel, Mozart, Rossini, and other composers for their divas and "divos" seem almost modest by comparison. Cibber's version of *Richard III* (Shakespeare's most popular play in the eighteenth and nineteenth centuries) appeared in 1700 and held the stage well into this century. A thoroughly Cibberitic *Richard III* apparently played in Boston as late as 1930, and Olivier's movie version owes much to Cibber.

In the nineteenth century advancing stage technology made more elaborate sets and lighting effects possible; Victorian Shakespeare productions rose to an average of ten or twelve scene changes. As these thickly upholstered productions became ever more lugubrious, the only salvation was to cut. The result, as described in 1930 by Harley Granville-Barker, the prominent English actor, director, and playwright, sounds not a little like the stately stop-and-go pace of nineteenth-century opera performances: "Modern producers . . . bring down the curtain upon a display of virtuosity in [Juliet's] 'potion scene,' long drawn out, worried to bits, and leave us to recover till they are ready with Romeo in Mantua." What vanished during the scene change was simply 175 lines of the play.

The history of Shakespeare on the stage until well into this century, then, has been one of huge liberties taken. Though some of these liberties now seem offensive or ludicrous, all have behind them a dramaturgical, commercial, or artistic rationale—and usually a rationale peculiar to a particular time, place, and troupe of executants. It seems fruitful to approach such liberties with a more open and enquiring attitude . . . and less judgmental, proprietary fervor. H. N. Hudson may have felt better after penning his warning in 1848, "Withered be the hand, palzied be the arm, that ever dares to touch one of Shakespeare's plays," but it was a lonely and quixotic voice then as now among those who put the plays on stage. Hudson's menacing purism, anyway, is thoroughly un-Shakespearean. That the plays were originally cut for performance is most notably shown in the shorter acting version of *Hamlet* in the quarto editions (the First Folio properly includes the full 3,900 mind-numbing lines Shakespeare wrote for his *Hamlet* project). The very first contemporary reference to the playwright accuses him of plagiarism ("an upstart Crow, beautified with our feathers"), and he did turn out to be the supreme thieving magpie of the stage. Not once did he bother to invent his own plot rather than purloin one—the only possible exception being *A Midsummer Night's Dream*.* Everything in his own literary and theatrical heritage, it seems, was fair game to him, so it is some justice that his works have in turn proved fair game in every artistic medium imaginable.

My point, obviously, is that harsh judges of upstart composers who beautify their operas with Shakespeare's feathers should perhaps be a little more bashful. Lord Byron attended a performance of Gioachino Rossini's *Otello* and reported, "They have been crucifying *Othello* into an opera: the music good, but lugubrious; but as for the words, all the real scenes with Iago cut out, and the greatest nonsense inserted; the handkerchief turned into a *billet-doux.*" Points well taken . . . and yet! We might be a little loath to ridicule Rossini for presenting his *Otello* in 1816 without that handkerchief but *with* one of Italian opera's first tragic endings, while, *in Shakespeare's native land,* a *King Lear* with a gloriously cheery denouement still held the stage. No, a rather more *laissez-faire* attitude is in order, for Shakespeare's plays, like most successful operas, are the products not of theory but of exigencies opportunistically embraced. T. S. Eliot—highest of highbrows—felt that "Elizabethan art . . . is impure art," and he was dead right as far as Shakespeare is concerned. His art was impure simply because the theater is inevitably an impure site. The life of the theater and humanity itself is—in the phrasing of that most characteristic of all Shakespearean insights—"of a mingled yarn, good and ill together."* Artists who would work in pure yarn are bound to feel restive and at odds with the theater; they very rarely succeed on their own terms. One who did—Wagner, a worker in pure yarn if ever there was one— finally had to invent and build a theater of his own in order to find artistic happiness. A life spent in the theater is far more likely to leave one a realist, perhaps even a cynic, than an idealist. After twenty-five years before the public and while revising *Macbeth,* Verdi wrote, "I know the world in general but the theater in particular, so I am surprised by neither the small nor the great perfidies which can be committed there." Is it imaginable that Shakespeare left for Stratford with a rosier view of the theater?

I am arguing here rather as the garrulous tippler Toby Belch does when he asks the puritanical steward Malvolio his withering question in *Twelfth Night:* "Dost thou think because thou art virtuous there shall be no more cakes and ale?" Every Shakespearean opera, in a sense, represents a collision between virtue (fidelity to the original play) and cakes and ale (the pure pleasure of musical sounds allied with words, with passion). Shakespeare's resolution of the conflict was, as usual for him, a compromise: he leaves the audience in the middle, wishing finally to be neither a Malvolio nor a Belch. Likewise, it is impossible to imagine Shakespeare succeeding as opera solely by virtuous fidelity to his text or by filling up on cakes and ale. Compromise is at the heart of the process. There is virtue *in* cakes and ale, and vice versa.

This is not to say, however, that it is idle or foolish to contemplate an ideal that might govern the translation of Shakespeare into opera. Perhaps the most charm-ing expression of such an ideal appears in a letter Arrigo Boito wrote to Verdi, telling of his troubles setting to work on the *Falstaff* libretto. It is especially worth including here at the outset, since the ideal issued in that rarest of events, a Shakespearean transformation full to the brim with cakes, ale, *and* virtue:

In the first few days I was in despair. To sketch the characters with a few strokes, to set the intrigue in motion, to extract all of the juice from that great Shakespearean orange [*quelle enorme melarancia Shakespeariana*] without letting the useless pips slip into the little glass, to write with color, clarity, and brevity, to delineate the musical plan of the scenario so that an organic unity results that is a *piece of music* and yet at the same time is not one, to make the joyous comedy live from top to bottom, to make it live

with a natural and engaging gaiety, is difficult, difficult, difficult, and yet one must ✓
make it all seem easy, easy, easy. *Corragio e avanti.*

'Tis a consummation devoutly to be wished, and it is obviously all to the good to
approach Shakespeare with such a noble standard as a guide. But there are hints in
Boito's phrasing that we must set our sights a bit lower. A "useless pip" to a
composer may be a very important part of the original play. Othello's superb and
crucial final speech, "Soft you, a word or two," for instance, was cut by Verdi
himself from Boito's draft libretto. And Boito's paradox of the "piece of music"
that both exists and vanishes is eloquently daunting. Perhaps more appropriate to
our subject is the note of compromise and common sense (the Shakespearean
note) struck by Berlioz in a review of Bellini's *Romeo and Juliet* opera: "It is not
possible to transform any sort of play into opera without modifying it, disturbing
it, corrupting it more or less. I know this. But there are many intelligent ways to
prosecute this task of profanation [*ce travail profanateur*] that is imposed by
musical exigencies."*

Finally, I should warn that certain time-honored topics associated with any
discussion of literature and music will here be of only passing concern. Many
students of Shakespeare and opera are not of a compromising sensibility. Their
taste runs instead to dialectics and the pleasures of controversy, and they
sometimes do not mind setting artistic media at loggerheads in order to generate
excitement. We have them to thank for the hoary debates as to whether
Shakespeare is pre-eminently a poet or a dramatist and whether it is *prima la
musica, e dopo le parole,* or vice versa. Not long ago Peter Conrad displayed this
penchant for sowing the seeds, as it were, of aesthetic civil war. On the subject of
nineteenth-century Shakespearean opera he wrote, "The fate of Shakespeare in
the romantic period . . . offers a further demonstration of the enmity between
music and drama. Music's ambition is to render drama obsolete." And on his
book's last page he summarizes, "Words and music are enemies, perpetually
jealous of one another, because each longs for the privileges of the other while
resenting intrusions on its own territory. . . . Words and music are united in
antagonism. Opera is the continuation of their warfare by other means." Belligerent notions!*

My experience, though, is that those who are most likely to indulge in rigid
dialectics of this kind are often situated farthest from the supposed war zone—in a
comfortable chair in a book-lined study. The more practically involved in the
actual performance of plays and operas, the more likely is the observer to develop
more diplomatic, pacific notions about words and music, poetry and drama.
Shakespeare, I am convinced, would have been either mystified or bemused at a
debate about whether he was a poet *or* a playwright. He was both at the same
time: the dedication and prefatory poems for the first collection of his plays refer
to him as a poet six times. It should be small surprise that the sane attitude toward
this trumped-up debate has been expressed by a scholar keenly interested in how
to act Shakespeare and put his plays on the stage: "it is a mistake to speak as if we
have a choice between a literary Shakespeare and a theatrical one. The choice is

rather between some of Shakespeare and more of him." Thus, a contemporary description of him as a "scenicke Poet" is the most accurate of all. Needless to add, the same un-tendentious attitude should prevail when we ask, Is such-and-such an opera composer a musician or a dramatist?

As to *prima la musica* . . . it is true that Verdi told his first Lady Macbeth that he wished "the performers to serve the poet better than the composer" and his first Macbeth that he would "never stop urging" him "to study the dramatic situation and the words; the music will come by itself." But I think this was by way of emphatically redressing a heavy bias, on the Italian stage of the 1840s, in favor of cakes and ale. Verdi had little taste for theoretical fancy dancing, and I imagine he would have been puzzled (in his later years, probably even annoyed) by Conrad's eagerness for aesthetic warfare. Perhaps he would even have made bold to point out that, in reality, there is no question of *prima* and *dopo:* in opera words and notes go arm in arm, just like Sir John and Ford as they exit from the Garter Inn at the end of *Falstaff*'s third scene.

Wesley Balk, another thoroughly practical man of the theater, has taken precisely Conrad's conceit, but he handles it with a lighter and ultimately more pacific twist. He describes two neighboring kingdoms: Musiconia, whose motto is "In Score We Trust," and Theatrylvania, whose motto is "Honor That Which Communicates." The former, Balk says, is a "law and order" society (i.e., virtuous), the latter is "permissive" (cakes and ale). For Balk, however, the buffer state called Operania is not the site for Conradian warfare, but for the diplomatic integration of words and music, without which no opera performance can finally succeed. In the real theatrical world it is usually the artful and practical compromisers, the peacemakers, who have the good fortune to survive long enough to flourish on their own terms, as Verdi or Wagner did. That the pacific rather than belligerent key will dominate in the following pages can perhaps be underlined by noting the conclusion of a Renaissance sonnet "In Praise of Musique and Poetrie": "One God is God of Both."

Style

1

A Passage to Opera

Wherein lies the synonymy of *Shakespearean* and *operatic?* The close examination of a rather typical speech from one of the plays will, I think, help to open discussion of this large question. It will also produce something of a pastiche overture that gathers together several themes that will be reprised in subsequent chapters. Hoping for a fresh response to Shakespeare's methods from an operatic perspective, I have deliberately avoided a familiar passage or one from a play famously transformed into opera. Instead, I have chosen an excerpt from one of his least often performed tragedies, *Titus Andronicus. Titus,* incidentally, is one of the few Shakespearean plays that have so far failed to attract an opera composer. (The others, aside from *Two Gentlemen of Verona,* are all chronicle plays: *King John, Henry VI, Richard II,* and *Henry V.*) Though, Lord knows, *Titus* bids fair to be the most grandly operatic play of the entire canon, its plot making the gaudy action of *Il Trovatore* seem almost Ibsenesque by comparison. The pace of events is reckless, and a baker's dozen of characters are murdered under various appalling circumstances by the end.

The passage I have chosen occurs at about the midpoint of the action and can be briefly introduced. "How rapidly falls the avalanche of destiny," sings the chorus at one point in Amilcare Ponchielli's lurid opera, *La Gioconda.* That, in few words, is the premise of the *Titus* plot. We first meet the title character as a triumphant Roman general, "Patron of virtue, Rome's best champion," but he is soon hemmed in with horrors. He has seen his ravished daughter, Lavinia, with her hands chopped off and her tongue cut out; and two of his sons, wrongly accused of murder, have just been led off to execution. Only moments before the following dialogue takes place, Titus allows his own hand to be chopped off as "ransom for their fault." Now Marcus, Titus' brother and the play's voice of patience and common sense, tries to console him. Here is the passage, arranged typographically to throw certain operatic parallels into relief:

> Marcus: O brother speak with possibility,
> And do not break into these deep extremes.
> Titus: Is not my sorrow deep, having no bottom?
> Then be my passions bottomless with them.　　(*Recitative*)
> Marcus: But yet let reason govern thy lament.　　5
> Titus: If there were reason for these miseries,
> Then into limits could I bind my woes.
>
> When heaven doth weep, doth not the earth o'erflow?
> If the winds rage, doth not the sea wax mad,
> Threat'ning the welkin with his big-swoll'n face?　　10
> And wilt thou have a reason for this coil?
> I am the sea. Hark how her sighs do flow:　　√(*Cavatina or slow aria*)
> She is the weeping welkin, I the earth:
> Then must my sea be moved with her sighs,
> Then must my earth with her continual tears,　　15
> Become a deluge: overflow'd and drown'd:
>
> For why, my bowels cannot hide her woes,
> But like a drunkard must I vomit them:　　√(*Cabaletta or fast aria*)
> Then give me leave, for losers will have leave,
> To ease their stomachs with their bitter tongues.　　20
> √ (*Enter a messenger with two heads and a hand.*)

"Losers will have leave" . . . numerous memorable moments in opera are generated when heroes and heroines become losers and take their elaborately vocal "leave" as Titus does: the Countess with "Dove sono" in *Figaro,* Otello in his final "O gloria," Orfeo's "Che faro senz' Euridice," and Elektra's first monologue, to think of a few of the most famous. The structure of this dramatic climax is also reminiscent of many an operatic set piece. One could practically insert after line seven the ubiquitous "ascolta!" or "udite!" ("listen!") that precede so many arias in Italian opera. As everywhere in opera, a major character has here been maneuvered into an emotionally volatile situation; it is ignited during a preliminary conversation that finds him reacting contrarily to a confidant; and, finally, because the reaction is so elaborate and extensive, the stage clock must stop until further action—the entrance of the messenger—leads on to a new crisis. One might even urge more specific parallels from operatic dramaturgy and cast the exchange between the brothers as recitative, the first two lines of Titus' long speech as a stormy arioso, then the next nine lines as a leisurely, expansive two-part aria in a slow tempo (the imagery is complex and impossible to grasp if reeled off hastily), and the last four lines of the passage, where Titus' imagery turns more blunt and corrosive, as an emotional, fast-tempo *cabaletta.*

Quintessentially operatic, too, is the *kind* of character Titus is. Inevitably, Shakespearean and operatic dramaturgy, as we shall see, is perfectly "irrational" or "unreal." Its reason for being, one might even venture, lies in the experience and expressiveness of characters who, like Titus, are not inclined to let "reason govern" their emotions. It goes against their grain to bind their emotions "into limits." The Shakespearean list of *dramatis personae* is thus rent by a huge dichotomy epitomized in Marcus and Titus—a dichotomy between all those characters who coolly "speak with possibility" and all those who memorably allow their feelings and the words that convey them to "break into . . . deep extremes." The former urge reticence, safety, patience (Balthasar's "I do beseech you, sir, have patience" to Romeo), and conservative husbandry of resources (Polonius' "neither a borrower nor a lender be"). The latter class of characters is willing to put considerable emotive energies at risk. This characteristic adventurousness tends to separate Shakespeare's principals from his comprimarios.

This dichotomy is most notably expressed by Hamlet, when he praises his levelheaded, stoic friend Horatio "As one in suff'ring all that suffers nothing" and as one whose "blood and judgment are so well co-meddled." And Hamlet then adds these famous lines:

> Give me that man
> That is not passion's slave, and I will wear him
> In my heart's core, ay, in my heart of heart,
> As I do thee.

The irony, of course, is that Hamlet himself is often and to his eternal fame a slave of passion. The Elizabethan stage was a poor place to display a cool head. Those who do act coolly there usually seem the poorer for it. Cleopatra, toward her end, asks of Antony, "Think you there was, or might be, such a man / As this I dreamt of?" And Dolabella, speaking like Marcus "with possibility," answers: "Gentle madam, no." Desdemona amazedly asks Emilia if "there be women do abuse their husbands," and Emilia, firmly resident in the real world, tells her: "There be some such, no question." But good advice and down-to-earth views, like good behavior, are not very interesting, especially on the stage. As in the case of Cleopatra, Desdemona, and Titus and their sensible confidants, our (and Shakespeare's) interest lies with those who are constitutionally unable to "speak with possibility."

Auden could have been describing Shakespeare's daring heroes and heroines when he made a very similar generalization in his essay "Some Reflections on Music and Opera." Opera, he wrote, "is an imitation of human willfulness; it is rooted in the fact that we not only have feelings but insist upon having them at whatever cost to ourselves. . . . The quality common to all great operatic roles, e.g. Don Giovanni, Lucia, Tristan, Isolde, Brünnhilde, is that each of them is a passionate and willful state of being." The motto of such willful characters, one might say, is uttered by Beethoven's Leonore: "Ich folg' dem innern Triebe" ("I follow the inner force"). They are virtually bound to ignore the voice of reason, so the exchange between Marcus and Titus is replayed time and again on the opera stage. When we first meet Leonora in *Il Trovatore,* her attendant Ines is warning, "You cherish a dangerous passion!" This serves only to ignite Leonora's passionnate "Tacea la notte placida," describing her meeting with her mysterious troubador. This in turn elicits further sensible advice from Ines: "I am afraid. . . . I am doubtful. . . . Try to forget him. Yield to a friend's counsel." Leonora cuts her off, and, after a browbeating rejection of this advice in two lines (compare

Titus' lines 6–7), launches into the "deep extreme" of "Di tale amor," with its high-flying, *brillante* fireworks. Perhaps the most gargantuan exaltation of heroic commitment to recklessness in the entire repertory occurs in *Turandot,* where fully 239 pages of the 384-page score are devoted mainly to successive warnings from the Chinese people, Timur, Liù, then Ping, Pang, and Pong, and finally the Emperor and Turandot herself that Calaf should not attempt to solve the three riddles. Of course, to no avail: *un*common sense, as so often in Shakespeare and opera, prevails and generates the central theatrical excitement.

There are many who like to say, as Bernard Shaw did, "I have long since learnt to leave my common sense at the door when I go to the Opera." Shaw happened to be fuming at Thomas' *Hamlet* opera when he said this, but in the present context it is possible to accept his point in a less satiric way. For we have seen that common sense—that is, mundane, unextraordinary, commonplace sense—is a safe passport to the periphery and comprimario status on Shakespeare's stage. In the audience it is an almost certain passport to boredom or frustrated antipathy. Thus, those who attend a performance of Shakespeare without checking their common sense at the door may find themselves similarly distanced from the heart of the action. The reader who soberly lets "reason govern" his responses to the passage from *Titus* can come to but one conclusion: it is an absurdity, and a prolix one at that. No actual person, fallen into such "bottomless" miseries, could muster such polished, complex, image-laden, and resounding utterance. Titus literally does not speak with possibility, nor did Shakespeare write with it. Tested by common sense or the norms of real-life speech, the passage is psychologically implausible and poetically overripe . . . the product of conspiracy between a self-indulgent hero, a self-indulgent playwright, and probably a self-indulgent actor.

More than any other English-speaking writer, Shaw loved to rattle John Bull's teacup with acerbities of this kind. He professed to find Elizabethan literature thin on substance, and on one occasion described a contemporary of his own as "being what I call an Elizabethan, by which I mean a man with an extraordinary and imposing power of saying things, and with nothing whatever to say."* But, as often as the thoroughly politicized Shaw rampaged about Shakespeare's apolitical lethargy, he seldom failed to remind us that his plays are essentially *aural* and (insofar as music implies the arrangement of sounds) *musical* entities. Tested by Shaw's paradoxically passionate common sense, Shakespeare seldom escaped unscathed; but, since Shaw brought a keen musician's ear to the plays, he managed finally to love Shakespeare with a love not too far this side of idolatry. Musical metaphors were one of his primary means of critical expression when he came to the plays. Of the role of Othello he observed: "Tested by the brain, it is ridiculous; tested by the ear, it is sublime. [The actor] must have the orchestral quality about him. . . . It is no use to *speak* 'Farewell the tranquil mind,' for the more intelligently and reasonably it is spoken the more absurd it is. It must affect us as [Otello's aria] 'Ora [e] per sempre addio, sante memorie' affects us when sung by Tamagno." Sir Johnston Forbes-Robertson's noble Moor, Shaw mercilessly suggested, was not in Tamagno's league: "For the most part one has to listen to the music of Shakespear . . . as one might listen to a symphony of

Beethoven's with all the parts played on the bones, the big drums, and the Jew's harp." And in a column titled "Poor Shakespear!" Shaw announced with typical verve, "The ear is the sure clue to him: only a musician can understand the play of feeling . . . it is the score and not the libretto that keeps the work alive and fresh; and this is why only musical critics should be allowed to meddle with Shakespear—especially early Shakespear."

Titus being decidedly early Shakespeare, let us meddle as musical critics might with the nine lines (8–16) that are the heart of our passage. First, though, it is worth remarking on an operatic aspect of the playwright's expansion of his simple simile: raging storm = extreme emotion. For the richness of these lines is conceptual as well as aural. In fact, Shakespeare develops his imagery in a way very similar to that of the "simile aria" common in baroque opera librettos. Patrick Smith, in his study of libretto poetry, describes the standard form of the simile aria in this way: "It consisted of two parts, or two thoughts, connected as similes, as oppositions, or as a syllogism, often using images from nature (animals, flowers, or the weather—particularly the weather at sea)."* Shakespeare does precisely this for Titus' speech. The conceit is elaborated in the first four lines: rain = tears, winds = Lavinia's sighs, flood = excess of woe, rough sea = grief-swollen face. Then it is applied to the situation in the next five lines. The effect of the whole is of a steadily building crescendo that arrives at the emphatic, climactic phrase "overflow'd and drown'd." Note, too, that Shakespeare—having with these nine lines halted his action as surely as "Vissi d'arte" halts the flow of *Tosca*'s second act—carefully brackets them with two *woe*'s at lines 7 and 17, as if to emphasize that the elaborate conceit is an "extra" or parenthetical part of the action.

But now, to use Shaw's distinction, let us briefly consider the "score" rather than the "libretto" of Titus' nine-line aria. Shakespeare wrote in common or duple time and in five-measure units—otherwise known as iambic pentameter. Like all composers, he achieved this underlying rhythmic norm only to depart significantly and affectingly from it. The English director John Barton, in *Playing Shakespeare,* emphasizes that off-the-beat stresses are "very, very important because they help make a text vigorous and tough rather than smooth and bland . . . make us sit up because they break the norm and give us a jolt." Several accentual abnormalities in the present passage demonstrate Barton's point. Line 1 establishes the norm with five perfect iambs (a lightly stressed syllable followed by a heavy one: ˘ /), but then strong stresses begin to cluster together ("wińds ráge," "séa wáx mád," "bíg-swóll'n fáce"). Five of the next seven lines commence abnormally on the heavy stress of a trochee (the reverse of an iamb: / ˘). The effect, as Barton suggests, is to invigorate and toughen Titus' declamation. Composers give important words emphasis by assigning them longer note-values, and Shakespeare several times approximates this method by deployment of the strong stress. Likewise, metaphorical equivalents are thrown into relief through strong stress (*weep* = *rage, o'erflow* = *mad*). And almost every line ends with purposeful stress on a telling one-syllable word.

Strong stresses are several times used in conjunction with the auricular figures (as rhetoricians call them) of assonance and alliteration (*w*eeping *w*elkin, *w*ax m*a*d, con*t*inual *t*ears). And almost every line achieves a feeling of balance by the placement of a pause (caesura) or the subtle counterpoise of consonants: *winds/ wax* in line 9, *thou/this* in line 11, *seas/sighs* in line 15, *deluge/drown'd* in line 16. Composers are always achieving similar effects by emphasizing verbal relationships through corresponding notes, note values, or melodic placement. Reiteration is also a common element in Shakespearean syntax, and it can be found in the present passage, too (*doth not* and *Then must my* are repeated).

Calling Titus' speech an aria is scarcely farfetched. It is a highly artificial expression, with the three successive rhetorical questions of the first part and the massed equivalents of the second part all integrated into a crescendo that must call upon an actor's full vocal range. As with any effective operatic aria, tiny local effects constantly challenge interpretation, and yet an organic, unified whole must emerge in the end. Which brings me to another important operatic aspect of the passage: in it, not only does a fallen Roman general experience an emotional crisis, but an actor also confronts an artistic challenge. Titus' speech is also a "performance" by the actor because its lines are the product of detailed and scrupulously calculated architecture. It is impossible to experience an operatic aria or a Shakespearean set speech—small (like Titus') or large (like Hamlet's "O what a rogue and peasant slave am I!" or Shylock's "Hath not a Jew eyes?")—without an awareness that both a fictional character and an artist are, as it were, performing together. In opera and in Shakespeare, it is impossible to display compelling emotions on stage without at the same time displaying technical virtuosity. It is thus no coincidence that the performance of an aria or a set speech can usually be appraised according to the same technical desiderata: breadth of range, suppleness of phrasing, justness of emphasis, keenness of diction, mastery of tonal nuance, security of projection, coherence of the whole, and so on.

The central and overwhelming problem of Shakespearean and operatic performance of such artificial set pieces as Titus' nine lines, then, is coping with this dual awareness. "The problem," as John Barton says of Shakespearean set speeches and soliloquies, "is how to do justice to the rhetoric and yet to keep it human." Singers confront the same problem in an even more exaggerated form when they cope with the "rhetoric" of musical notes. Barton turns to Titus' speech and finds in it a common dilemma of Shakespearean (and operatic) interpretation: "The obvious problem of such a speech is that in real life the emotion of the speaker would be much greater than his power to articulate that emotion in words. So it's very understandable if an actor confronted with such a speech recoils from the literary content, tries to express the emotion to the full, and so leaves the words to take care of themselves." Finally, though, Barton concludes that this approach has the unfortunate effect of encouraging the audience to stop listening to or following the words. He concludes on a Shakespearean note of compromise: "I suspect that the only way to do the speech, and to make an audience both listen to and feel with it, is for Titus to . . . use the words to try to *cope with* as much as to *express* the emotion."

Reconciling the emotions themselves and the technique necessary to project them in a theater or opera house requires great stamina and concentration. As the soprano Grace Bumbry simply puts it, "You've got to keep your wits about you." The *Titus* passage typifies the challenge on a modest scale . . . the scale, say, of such short arias as Cavaradossi's "Recondita armonia" or Falstaff's "Va, vecchio John." Proportionately more daunting is the challenge of sustaining the more elaborate Shakespearean *scenas* like Mercutio's Queen Mab *jeu d'esprit,* Gaunt's "sceptred isle" extravaganza, or Antony's oration to the Roman plebes . . . the equivalents of, say, Norma's "Casta diva," Aïda's "O patria mia!," or Butterfly's "Un bel dì." The primrose path to tedium in both genres is the loss of sustained, specific concentration. Barton's observation that "it's so easy to play a kind of *summary* of a speech and not discover it line by line" applies emphatically to performance of arias in the opera house. Sadly but perhaps inevitably, in both theaters the Gielguds, Callases, Ashcrofts, Vickerses, Söderströms, Oliviers, McKellens, Rysaneks, Denches, Stratases, Schwarzkopfs, and other superlative artists who discover their roles line by line are bound to be a treasured minority and the legion of "summarizers" the vast majority.

2

Things Like Truths

Unreal": the favorite epithet of satirists and despisers of opera. Seldom is it aimed at Shakespeare, however, though it is scarcely "real" when a character speaks in ornate iambic pentameter. "Real" and "natural" are, however, relative terms, and few would quarrel with the suggestion that spoken drama is more real than opera. Few, too, would quarrel with the conclusion that Richard Wagner drew from his experience of "the *Romeo and Juliet* symphony of our friend Berlioz":

> The Dramatist must lay hands on quite other means than the Musician; he stands much nearer to the life of everyday, and is intelligible solely when the idea he presents us is clothed in an Action whose various component "moments" so closely resemble some incident of that life that each spectator fancies he is also living through it. The Musician, on the contrary, looks quite away from the incidents of ordinary life, entirely upheaves its details and its accidentals, and sublimates whatever lies within it to its quintessence of emotional content—to which alone can Music give a voice, and Music only.

Wagner's formulation of the dramatist's style is surely more apt to the dramaturgy of Ibsen, Chekhov, and their twentieth-century followers than to Shakespeare. The musician's style as Wagner formulates it is *precisely* Shakespeare's style: it is his habit to look "quite away" from the incidents of ordinary life, to "entirely upheave" the mundane details, and to aim for a "quintessence of emotional content." This is the path to unreality or non-naturalism, and the following pages will insist on this shared unreality of our two dramatic styles. These pages will also offer various ways one might mount an apologia for this unreality.

A critic recently described the "new naturalism" that Shakespeare's style must have represented when it was introduced in London. But surely it is possible to call him naturalistic only by comparison with what had preceded this style. He was bound to seem more natural to audiences accustomed to Christopher

Marlowe's bombast, as Marlowe no doubt seemed more natural than his predecessor John Lyly, who wrote in a clotted, maddeningly balanced rhetorical style. And they all must have appeared infinitely more naturalistic than the old morality-play cycles. From our own perspective, though, Shakespeare has thoroughly outgrown the "naturalistic" epithet. "All stage dialogue except the modern," Eric Bentley summarizes, "is highly rhetorical; that is to say, it is raised far above the colloquial by deliberate artifice . . . it is 'rhetorical' in the popular sense—high-flown, florid, oratorical." Shakespeare's dramaturgy and expressive style are rhetorical in this sense. That is why trying to deliver his high-flying cadences in a realistic mode is bound to prove self-defeating. One commentator, in an essay, "Acting Shakespeare Now," has acidly noted, for instance, that "American Shakespeare (from the British standpoint) is necessarily vulgar and sometimes downright stupid, since the actors treat the stage personalities as though they were indeed 'real' people."

The alternative to being "real" is, to many in a modern audience, ham acting. With Shakespeare, however, this is the direction in which to err. The director Margaret Webster wrote in *Shakespeare Without Tears,* "The fear of being 'ham' is the bogey of the modern actor, trained to a tradition of Anglo-Saxon self-consciousness coupled with a reticent 'reserve' which is all too often barren of anything to reserve." Bentley wittily seconds her point in his own attack on "a narrow and philistine Naturalism" in certain actors and directors: "That we are all ham actors in our dreams means that melodramatic acting, with its large gestures and grimaces and its declamatory style of speech, is not an exaggeration of our dreams but a duplication of them. In that respect, melodrama is the Naturalism of dream life." It is but a small step to Prospero's "revels" speech, so often read as an allusion to Shakespeare's own characters: "We are such stuff as dreams are made on. . . ."

Theatrical "truth," as Prospero suggests, is *sui generis.* Especially with Shakespeare's drama, it cannot be grasped through the colloquial and prosaic means of naturalism. Donald Sinden implied as much when he spoke candidly of his attempts to get right one of Malvolio's exits in *Twelfth Night:* "I never really succeeded in bringing it off theatrically, even if I did [succeed] truthfully." That the "theatrical" can be antithetical to (and must take priority over) the merely "truthful" is in fact urged by Ben Kingsley as he addresses the main challenge of acting Shakespeare: "I find more and more when I am on stage that naturalistic acting—that is, totally reported nature—is inappropriate. . . . [The actor] must remember that his landscape as an actor, the play itself, is a compressed, organized, condensed version of the truth." Peggy Ashcroft, when asked what she felt was the main objective in playing Shakespeare, makes a corollary point. She says the goal is simply the "truth of reality . . . [the] truth of poetry, which is a little bit of super-reality."

The point being made here by all these actors has often been expressed before. William Hazlitt did so when he called the stage "an epitome . . . of the world, with the dull part left out." Oscar Wilde broadened the generalization: "No great artist sees things as they really are." "Truth, *cher ami,* is a colossal bore," says a character in Albert Camus' *The Fall.* Especially pertinent to Kingsley's mention of

condensation is Aldous Huxley's eloquent essay, "Tragedy and the Whole Truth."
The reader will see that this passage urges us again in the direction of opera:

> Tragedy is something that is separated out from the Whole Truth, distilled from it, so
> to speak, as an essence is distilled from the living flower. Tragedy is chemically pure.
> Hence its power to act quickly and intensely on our feelings. . . . From the reading
> or the hearing of a tragedy we rise with the feeling that
>
> > Our friends are exultations, agonies,
> > And love, and man's unconquerable mind.

Neither Shakespeare nor opera traffics in the Whole Truth of naturalism, with all
the dull and inert parts left in. These are dramaturgies of distillation and elevation.
Paul Valéry speaks for them and makes Huxley's point, coincidentally, by also
referring to the perfumer's art: "If a perfume manufacturer were to adapt the
'naturalistic' aesthetic, what kind of scents would he bottle?"

This simplifying impetus is summarized differently in a passage written by
Friedrich Nietzsche, who thought often about opera and was, successively, the
most perfect and imperfect of Wagnerites. Nietzsche is speaking here of the
creative process in general, but the relevance of his point to Shakespearean and
operatic methods will be clear: "What is essential . . . is the feeling of increased
strength and fullness. Out of this feeling one lends to things, one *forces* them to
accept from us, one violates them—this process is called *idealizing*. Let us get rid
of a prejudice here: idealizing does not consist, as is commonly thought, in
subtracting or discounting the petty and inconsequential. What is decisive rather
is a tremendous drive to bring out the main features so that the others disappear in
the process." Idealizing, in other words, must inevitably violate "reality." The
effort of bringing the "main features" of experience into full prominence requires
tremendous expressive "drive"—drive that iambic pentameter and music are
particularly able to generate.

Huxley's "distillation" and Nietzsche's "idealizing" are adroit ways of speaking
about splendid forms of falsehood. Ben Jonson described the situation succinctly
when he wrote that "poet never credit gained / By writing truths, but things like
truths well feigned." The deceptions of art have, of course, been exposed and
defended in many brilliant ways. No one did so more simply or elegantly than
Erasmus in *The Praise of Folly:* "To be deceived, they say, is miserable. Quite the
contrary—not to be deceived is the most miserable of all. For nothing could be
further from the truth than the notion that man's happiness resides in things as
they actually are . . . the human mind is so constituted that it is far more taken
with appearances than reality." And there is perhaps no more piercing defense of
the falsehood of the stage in particular than St. Augustine's:

> On the stage [the actor] Roscius was a false Hecuba by choice, a true man by nature;
> but by that choice also a true tragic actor because he fulfilled his purpose, yet a false
> Priam because he imitated Priam but was not he. And now from this comes something
> amazing, which however no one doubts . . . that all these things are true in some
> respects for the very reason that they are false in some respects, and that only the fact

that they are false in one sense helps them toward their truth. Hence they cannot in any way arrive where they would be or should be if they shrink from being false. . . . So if the fact that they are false in one respect helps certain things to be true in another respect, why do we fear falseness so much and seek truth as such a great good?

Or consider Ludovico Ariosto's *Orlando Furioso,* a huge epic poem written with Shakespearean exuberance, humor, and psychological finesse that proved a favorite quarry for librettists (and was a favorite of Verdi's, too). During the "Age of Reason," however, it required a defense in terms rather similar to those of Erasmus and Augustine: "Critics may talk what they will of Truth and Nature, and abuse the Italian poets, as they will, for transgressing both in their incredible fictions. But believe it, my friend, these fictions with which they have studied to delude the world are of that kind of creditable deceits, of which a wise antient pronounces with assurance, 'That they who deceive are honester than they who do not deceive; and they, who are deceived, wiser than they who are not deceived.'"

These defenses mounted by Erasmus, Augustine, and Richard Hurd are pertinent here because the falsehoods of Shakespeare and opera are so extravagant and shameless. George Puttenham, in the most important poetry treatise of Shakespeare's day, wrote at length of how important to poets are the rhetorical figures of "false semblance and dissimulation." Among these is hyperbole, which Puttenham describes thus: "when we speake in the superlative and beyond the limites of credit, that is by the figure which the Greeks call *Hyperbole* . . . I for his immoderate excesse calle him the over reacher . . . or *lowd lyar.*" Shakespeare's is an essentially hyperbolical art. His memorable characters, being passion's slaves, tend to speak not "with possibility" but "beyond the limites of credit."* They are loud liars.   LOUD LIARS

These characters, magnificently free from the inhibitions of normal or prosaic utterance, obviously share strong similarities with the great operatic characters. Shaw, writing of himself in the third person, credited both Shakespeare and opera for his own ability to create such characters for the stage: "Handel's and Mozart's symmetry of design formed his classical taste; and from Rossini and Verdi as from Shakespear he acquired the grandiose and declamatory elements hidden beneath his superficially colloquial and 'natural' dialogue." And Shaw also declared Shakespearean his power to endow his characters with "freedom from inhibitions, which in real life would make them monsters of genius. It is the power to do this that differentiates me (or Shakespear) from a gramophone and a camera."* It is perhaps not surprising that Verdi also found occasion to refer to Shakespeare and photography when he insisted on the unreality of staged "life" and the folly of seeking after that will-o'-the-wisp of totally reported nature: "To copy reality is a good thing, but to invent reality is better, much better. A contradiction may seem to exist in these three words *to invent reality,* but just ask Papa Shakespeare [*ma domandatelo al Papa*]. . . . Copying reality is a nice business, but it is photography, not painting."

A more famous and eloquent defense of unreality on stage was evoked from

Verdi, aptly enough, by consideration of one of the most artificial and arbitrary conventions of nineteenth-century opera, the *cabaletta*. In *Opera as Drama*, Joseph Kerman described it as "one of the worst lyric conventions of early nineteenth-century opera . . . a fast, vehement aria or duet of extremely crude form and sentiment; it always came after a slower, quieter piece . . . and served to provide a rousing curtain." Among the most famous *cabalettas* are Manrico's "Di quella pira" or Violetta's "Sempre libera," which Kerman calls "coarse in sentiment." Kerman further chides the "crude" simplification and stereotyping, the "elementary boisterousness" involved in deploying the *cabaletta*. In an odd but telling way, several of Kerman's remarks could do service in describing the Shakespearean soliloquy, which so often functions as a rousing exit aria and which has been nominated a crude theatrical device by many a commentator. "Elementary boisterousness" is just about the perfect phrase for Iago's big soliloquies in *Othello*—and many others in the canon.

The Wagnerian revolution forced the *cabaletta* onto the ropes, and Verdi himself used it less and less in his later operas. (Shakespeare's reliance on the soliloquy also declined steadily over time.) But even as late as 1880, Verdi was willing to defend this convention, accused of crudity on all sides, when theatrical logic required it. Looking at the second act of his 1857 *Simon Boccanegra* with a view to revising it, Verdi wrote to his publisher, Giulio Ricordi:

> From a musical point of view, the *cavatina* for the woman could be retained, the duet with the tenor, and the other between father and daughter, despite the *cabalettas!* (Open, O earth!) In any case, I don't have such a horror of *cabalettas,* and if a young composer were to appear tomorrow who could write any as worthwhile as, for example, [Bellini's] "Meco tu vieni, O misera" or "Ah perché non posso odiarti," I would listen to them with all my heart and renounce the harmonic sophistries, all the affectations of our learned orchestrators. Ah, progress, science, realism! Alas, alas! Be as real as you like, but . . . Shakespeare was a realist, though he did not know it. He was an inspired realist; we are planning and calculating realists. So taken all in all, system for system, the *cabalettas* are still better. The beauty of it is that, in the fury of progress, art is turning backwards. Art which lacks spontaneity, naturalness and simplicity is no longer art.

It is ironic to find the "crude" *cabaletta* convention praised for its theatrical effect of "naturalism," but this helps to underscore both the arbitrariness and conventionality of Shakespearean and operatic drama. The point is not that the *cabaletta* and soliloquy are crude or "elementary" devices, but that—when evoked by the right dramatic situation and when brilliantly composed and sung in performance—they are capable of producing very potent effects on an audience. "Have no doubt, I do not abhor *cabalettas,*" Verdi wrote Antonio Ghislanzoni, his *Aïda* librettist, "but I want a subject and a pretext for them."

The question, then, was not *whether* but *where* to employ the *cabaletta*. Ghislanzoni wrote some lines for one just after Aïda's father brutally forces her to betray Radames in act 3, and Verdi sensibly resisted: "I have always been of the opinion that *cabalettas* should be used when the situation demands them. . . . In the duet between father and daughter it seems especially out of place.

Aïda is in such a state of fear and moral depression that she cannot and must not sing a *cabaletta."* Verdi could vigorously defend the device, but he could also goad his librettist, with acerbic emphasis, for being afraid of two things, *"bold, theatrical strokes* and of *not writing cabalettas!"* A deeply pragmatic, utilitarian artist (like Shakespeare), Verdi was perfectly happy to have it both ways. He was willing to pay realism lip-service as "a nice business" ("una bella cosa"), but he knew that "the deepest veracity of all"—to borrow a phrase from Walt Whitman—was to be won by the "inspired" realism, the loud lying of a Shakespeare. A dramatic style that could embrace the soliloquy could also happily embrace the *cabaletta*.

The German composer Hans Werner Henze might have had Shakespeare in mind when he wrote that in opera "the improbable is transformed into the probable, fiction into highest truth, artificiality into nature." Indeed, the *tema con variazioni* of these last pages has been the shared quintessence of unreality in Shakespearean and operatic drama, or, rather, how often an apologia for the dramaturgy of one can prove quite serviceable for the other. I have saved for last the longest and most incisive variation on my theme. It comes from an essay by Richard Strauss's principal librettist, Hugo von Hofmannsthal, and will function as the resounding coda that every theme-and-variations requires. In it Hofmannsthal describes the gestation of their opera *Die ägyptische Helena,* which premiered in 1928. He reconstructs, elaborately and with obviously fictional precision, a conversation he had had several years earlier in Strauss's opera-house office. Toward the end of this conversation comes the crucial exchange in which Hofmannsthal rejects the idea of collaborating on a realistic "psychological conversation-piece" in naturalistic, contemporary dialogue. Shakespeare, as it happens, provides the starting point for the librettist's eloquent apologia for all non-naturalistic, presentational—as opposed to illusionistic or representational—stage actions:

Hofmannsthal: I don't like it when drama moves on the dialectical plane. I mistrust dialogue full of artistic purpose as a vehicle for the dramatic. I shrink from words; they ruin what is best. . . .

Strauss: But the writer has nothing else with which to bring his characters into existence but what he has them say. For you, then, words are what notes are for me and colors are for a painter.

Hofmannsthal: Certainly. Yes, words. But not expeditious, ingeniously reasoned out speech—not what is called "the art of dialogue" or "psychological" dialogue and which from Hebbel to Ibsen and even beyond has seemed to be held in such high esteem . . . and which is present in Euripides and even in Shaw. . . .

Strauss: And this is so with Shakespeare?

Hofmannsthal: Oh, not in the least! With him the word is always expression, never "communication." In this sense Shakespeare has written genuine operas [*lauter Opern*]; he is very close to Aeschylus and miles away from Euripides. Hasn't it ever struck you that in life nothing is really decided by speech. One is never so solitary, so convinced of the

insolubility of a situation as after one has tried to resolve it through
speech. The falsifying power of speech goes so far that it not only
distorts the character of the speaker but finally dissolves it
. . . forces the "I" out of existence. I maintain that a writer has the
choice of creating speech, or characters!

Strauss: That's too paradoxical for me! The writer really has no other means
but speech.

Hofmannsthal: He does have others—the most secret, most precious, least well-
known: what is purely efficacious [*einzig wirksamen*]. The writer will
be capable of anything who renounces the notion that his characters
must authenticate their existence by direct "communication."

Strauss: What kind of artistic means is this?

Hofmannsthal: He can, in his construction, *convey* rather than *communicate* his
action. He can make something live in the listener without his
suspecting how this has happened to him. He can make one feel how
the seemingly simple is pieced together, how closely gathered together is that which is widely scattered. He can show how a woman
becomes a goddess, how some one who is dead emerges from a living
being . . . he can make that which is silent resound, give sudden
presence to what is off in the distance. He can cause his characters to
grow out of themselves and into giants, because mortals do that in
certain rare moments. In dialogue that is fashioned to be "natural"
there is no room for all this. The "natural" is the projection of
ungraspable life into a very arbitrarily chosen context. The utmost of
our human nature—embracing time and space with cosmic
animation—will not be captured by naturalness.

Strauss: What kind of artistic means are these? Can't you define them for me?

Hofmannsthal: The way I handle the action, weave the themes, allow what is latent to
stir to consciousness, cause what is remembered to vanish . . .
through the similarity of characters, through analogies of the situation . . . through inflection, which often says more than words.

Strauss: But these are all mine . . . the methods of the composer!

Hofmannsthal: And the methods of the lyrical drama—they are the same
. . . These huge inner spaces, these raging inner tensions, this Here
and Elsewhere, all of which is the signature of our life . . . it is not
possible to gather up all this in prosaic [*bürgerlichen*] dialogue. Let us
make mythological operas—that's the truest of all genres. You can
believe me.

3

Worlds of Sound

Behavioral psychologists, especially those concerned with learning processes, have in recent years developed categories to describe an individual's dominant mode of sensual response. Thus, one might venture that a ballet audience will contain a high proportion of "kinaesthetic mode" dominants, an art gallery will be populated by "visual mode" dominants, and a symphony concert will largely attract "aural mode" dominants. These modes not only affect powers of appreciation but are thought to play an important part in the emergence of creative and executive talent. Applying and refining this theory in his study, *Performing Power,* Wesley Balk has amusingly suggested that the perfection of the facial/emotional mode dominant actor will be a Ronald Reagan, while perfection of the aural/vocal mode dominant actor will be a Sir John Gielgud. Indeed, the director Peter Brook, in his analysis of Gielgud's career, described just such an outstanding performer without the psychologists' jargon: "His tongue, his vocal cords, his feeling for rhythm compose an instrument that he has consciously developed all through his career in a running analogy with his life. . . . His art has always been more vocal than physical: at some early stage in his career he decided that for himself the body was a less supple instrument than the head. He thus jettisoned part of an actor's possible equipment but made true alchemy with the rest."

Since, as Shaw asserted, the ear is the sure clue to Shakespeare, this was the obvious repertory a Gielgud could excel in. Shakespeare's is a thoroughly and pre-eminently aural theater. So, of course, is opera. An opera audience may applaud a pleasing stage picture when the curtain rises, but it will not stay happy long if the ensuing vocalism is inadequate. It has come to *listen,* as William Meredith charmingly explains in his poem "At the Opera." In its last stanza he describes our response to the prima donna's "repetitious furor" and "rallies to applause" (the emphasis is his):

> But no one minds her sawing
> The air and looking perfectly unreal,
> Or remembers what he's *seen.*

> In the foolhardy ordeal
> We are brought through by her being
> Every decibel a queen.

The consequences of being foremost an aural theater are significant and various. It is thus worth pausing here to explore how, over the centuries, the fundamental aurality of the two dramaturgies has been expressed, exploited, and (not infrequently) perverted or travestied. From this common foundation arise many of the remarkable similarities between Shakespearean and operatic style.

That Shakespeare's plays were intended more to be heard than seen has been observed by many—among them Ian McKellen, noted for his one-man program on Shakespearean acting: "I think that Shakespeare should on the whole be aural primarily rather than visual. If you don't get the language, then you've lost the heart of the matter." He adds, on another occasion, "We can take comfort from the fact that people who come to a theater are called an audience . . . *audio* . . . 'hear.' People who watch television are *viewers*." Several decades ago Granville-Barker confronted the astonishing fact that a twelve- or thirteen-year-old *boy* played Cleopatra at the Globe Theater. He concluded that this boy had no choice but to put all his eggs, so to speak, in an aural basket: "With the art of acting still dominantly the art of speech—to be able to listen an audience's chief need—[the boy] could afford to lose himself unreservedly . . . in the music of the verse, and let that speak." And the centrality of the actor's voice and musicianly delivery on the Elizabethan stage was even noted while Shakespeare was alive. His fellow playwright John Webster offered a thumbnail sketch of "An excellent Actor" in 1615 that includes this telling remark: "Sit in a full Theater, and you will thinke you see so many lines drawne from the circumference of so many *eares,* while the Actor is the Center. . . . He addes grace to the Poets labours: for what in the Poet is but ditty, in him is both ditty and *musicke*" (emphasis added).

Some excellent hints that vocalism took priority over gesture and "production values" are scattered throughout Shakespeare's canon. "I will *hear* that play," says Duke Theseus of "Pyramus and Thisby" in *A Midsummer Night's Dream*. "We'll *hear* a play tonight," says Hamlet of "The Mouse-trap." "Open your *ears,*" inveighs Rumor in the prologue to *2 Henry IV,* and the prologue of *Henry V* asks the audience, "Gently to *hear,* kindly to judge our play." *Hear* also figures in the prologues to *Henry VIII* and *The Two Noble Kinsmen,* while in the initial scenes of *The Taming of the Shrew,* characters speak of hearing a play three times. Only the oafish tinker Christopher Sly in the entire canon says of a play, "Well, we'll *see* it." It is rather fleet-witted Puck in *A Midsummer Night's Dream* who expresses the attitude of the typical Elizabethan theatergoer: "What, a play toward? I'll be an *auditor.*"

I am inclined to think the aural bias of an Elizabethan theatrical event is hinted in Caesar's remark about Cassius: "he loves no plays, / As thou dost, Antony; he *hears* no music." This privileging of the aural over the visual is certainly expressed in Volumnia's acerbic observation, in *Coriolanus,* about the Roman plebeians (the sort who, in London, would be the groundlings in a Globe Theatre audience): "the eyes of th' ignorant [are] / More learned than the ears." Hamlet, too, may be indicating to us the priority of vocalism on the stage of the day in his talk to the traveling players. For his very first concern is that they "Speak the speech, I pray you . . . trippingly on the tongue."

"Language shows the man," said Ben Jonson; "speak that I may see thee. It springs out of the most retired, and inmost parts of us. . . . No glass renders a man's form, or likeness, so true as his speech." If there is a single key to Shakespeare's stagecraft, it is in this recognition. And it is no wonder that language—utterance of the "inmost parts" of passion's slaves—dominated the Elizabethan stage: elaborate costumes made finesse of gesture or carriage difficult, and the enormous platform—a thousand square feet or so, thrusting out among the standees, barren except for a few portable props—was both dwarfing and illusion-busting. Also, daylight at the Globe ruined the sometime cover of imagined darkness (several plays, like *Macbeth* and *Othello,* take place almost wholly at night). Within these constrictions there was but one great means of salvation: charismatic projection. A student of Shakespeare's stagecraft, J. L. Styan, has concluded, "It is probable that Shakespeare saw his actors as playing to their audience most of the time. . . . The Elizabethan actor was at all times vulnerable, and compelled to communicate with the audience, provoking it and provoked by it." This approaches the ambience of the opera house, and Styan, in a discussion of "Shakespeare's aural craft," does indeed use such musicianly phrases as "variation of tempo," "staccato phrasing," "use of tone and rhythm," "crescendo," and "orchestration of speech." Almost inevitably, he is led to this conclusion: "At one extreme, Shakespeare's drama approximates to opera and its language to song."*

From the very beginning, performers were likely to be subjected to "musical" criticism. A contemporary of Shakespeare's, Thomas Dekker, quipped that in theatrical "Consorts many of the Instruments are for the most part out of tune," and he ridicules the typical miscreant thus: "let the Poet set the note of his Numbers, even to *Apollo*'s owne Lyre, the Player will have his owne Crochets, and sing false notes, in despite of all the rules of Musick." The era's greatest tragedian, Richard Burbage, was on the other hand praised for his musicality. Richard Flecknoe called his appearance on the stage "Beauty to th' Eye, and Musick to the Ear," and explained that there was "as much difference betwixt him and one of our common Actors, as between a Ballad-singer who onely mouths it, and an excellent singer, who knows all his Graces, and can artfully vary and modulate his Voice, even to know how much breath he is to give to every syllable."

The primal musicality of Shakespeare's verse was acknowledged in succeeding eras, too. In the preface to Purcell's *Fairy Queen* we are reminded, "he must be a very ignorant Player, who knows not there is a Musical Cadence in speaking; and that a Man may as well speak out of Tune, as sing out of Tune." In the following Augustan age, Colley Cibber—dean of Shakespearean adapters—set forth this musical idea: "The voice of a singer is not more strictly ty'd to time and tune, than that of an actor in theatrical elocution: the least syllable too long, or too lightly dwelt upon in a period, depreciates it to nothing." Some years later Cibber's son Theophilus praised Barton Booth by noting simply that "the Tones of his Voice were all musical." The Georgian period was dominated notably by Edmund

Kean, whose point-making emotional violence and attempts to naturalize the blank verse became notorious. (He perhaps of all eminent tragedians took Hamlet's famous advice least to heart.) In 1829, toward the end of Kean's reign, a critic observed forlornly, "Shakespeare has his music as well as Handel, but where upon the stage do we find an actor who does justice to his melody? Might not Shakespeare, so far as his cadences depend for their effect on actors, have written *Othello* and *Macbeth* in prose? . . . Who among [current actors] studies the principles that regulate the music of passion, as singers do the principles of their art?" More recently, Granville-Barker began his famous series of prefaces on the production of Shakespeare's plays with this simple admonition: "The text of a play is a score waiting performance."

Except in Shakespeare's flattest, least tunable prose passages—which do not occur all that often—the actor must bestir himself to recognize and exploit the elaborate composition of sounds and convey not only the bare sense but also the various pleasures of vocalization for its own sake. That aural pleasure is to be had everywhere, even when a ghastly character like Lady Macbeth, Richard III, or Iago is in the limelight, is both a Shakespearean and an operatic principle. Mozart expressed this pervasive aural pleasure-principle in a letter explaining why he gave the comically bloodthirsty Osmin in *Die Entführung aus dem Serail* an unusual key and tempo change in his first aria: "a man in such a towering rage exceeds all control, moderation and purpose; he does not know what he's doing—just so, the music mustn't know either. But . . . the passions, whether violent or not, must never be expressed disgustingly, and music, even in the most terrible context, must never insult the ear, but even there must give pleasure, that is, must always remain music." Iambic pentameter—even when it emerges as "Come, thick night, / And pall thee in the dunnest smoke of hell"—must, as Mozart urges, remain music and give aural delight. The opera repertory is full of examples of appalling evil musicalized gorgeously: for instance, the triumphal duet for that ghastly couple, Nerone and Poppea, at the end of Monterverdi's *Coronazione di Poppea,* most of Don Giovanni's role, or Count di Luna's "Il balen del suo sorriso" in *Il Trovatore*. This last villain is often asked to sing *dolce* ("sweetly"), a direction apt for many of the speeches of Iago and Richard III.

Getting at all the music in Shakespeare's text requires skill and attentiveness. The word that John Barton lands on in *Playing Shakespeare* by way of urging this challenge is *relish*. Quoting Hamlet, he says, " 'Words, words, words.' The Elizabethans loved them: they relished them and they played with them." Three of Barton's actors read a passage from *Love's Labour's Lost* that demonstrates this love of language, and he reiterates his theme: "Verbal relish . . . today we're a bit apt to fight shy of it. But until we love individual words we cannot love language. Later he stresses "a tendency in our acting tradition to run away from verbal relish, especially of vowels." (Incidentally, "relish" is the now-archaic musical term for the appoggiaturas or grace notes that ornament a melodic line: recall that Burbage was praised, singer-like, for knowing "all his Graces." Especially in the eighteenth and nineteenth centuries, relishes—also called shakes, mordants, and accacciaturas—were one of a composer's principal means of eking ever more nuance from his melodic line. One might even paraphrase Barton and say that until a singer loves individual notes, he or she will be unlikely to relish music.)

At the end of Barton's session, "Making the Words One's Own," he himself tackles a speech in heightened verse (the French King's list of knights in *Henry V,* 3.5.37–54), explaining, "I'll tend to pronounce every single sound within a word because I suspect that Elizabethan actors may have done that more than we do. So I'll overstress and over-relish the sounds." The several responses to Barton's reading are worth quoting here, for they approximate the kinds of praise one encounters for an operatic aria finely interpreted:

> *Lisa Harrow:* "There was this amazing thread that went right the way through from the beginning to the end. You never let it drop or let us flag for a moment."
>
> *Ben Kingsley:* "It never lets us off the hook when it's done that way. It swept us along."
>
> *Roger Rees:* "It's wonderful to hear something like that because so much of our literature and playwriting today seems to be obsessed with the lack of language. You know . . . the spaces and the pauses . . . the absence of text sometimes. To hear a bit of text so highly encrusted with all kinds of different shapes and movement in it and in the sounds is wonderful."

Making words in a promptbook or notes in a score resonate is a vocal responsibility; it requires a constant awareness and exploitation of lyric possibilities lying on the printed page. Actors who come to Shakespeare Cassius-like, hearing no music, are doomed to defeat at the crucial moments when lyric levitation is both possible and necessary. When this defeat occurs, an audience will be left earthbound and of Samuel Johnson's mind; for he accused Shakespeare of often clothing "trivial sentiments and vulgar ideas" in "sonorous epithets and swelling figures" . . . and of creating "declamations or set speeches [that] are commonly cold and weak." If an unmusical actor cannot "swell" sonorously, almost any Shakespeare will seem cold and weak indeed. His meaning alone is not enough. The more hostile critics of Shakespeare, too, almost always show evidence of a tin ear. Sounding for all the world like Emperor Joseph II, who complained of the "monstrous many notes" in Mozart's *Entführung,* was the infamous Thomas Rymer. In his *Short View of Tragedy* (1693) he fumed: "Many . . . of the Tragical Scenes in Shakespeare, cry'd up for the *Action,* might do yet better without words. Words are a sort of heavy baggage, that were better out of the way . . . especially in his *bombast Circumstance,* where the Words and Action are seldom akin." And he adds perhaps the most screwball pronouncement in the history of Shakespearean criticism: "In a play one should speak like a man of business." So much for verbal relish!

We can be thankful that very few Shakespearean characters speak like men or women of business. Not even his businessman of Venice speaks that way. They tend far more often to be creatures of verbal relish. This is most strikingly true of the earlier plays. Noting the "convention of word-spinning and thought-

spinning" in which most of *Romeo and Juliet* is cast, Granville-Barker flatly asserts
in his first sentence on the play, "[It] is a lyric tragedy, and this must be the key to
its interpreting." Samuel Johnson, not grasping this key, was repelled by the
highly wrought speeches ("miserable conceit") uttered by its "distressed" characters. Typically, he came to the exuberant little aria in which Romeo displays his
puppy-love for Rosaline—

> O brawling love! O loving hate!
> O any thing, of nothing first create!
> O heavy lightness, serious vanity,
> Misshapen chaos of well-seeming forms . . .   (1.1.176–179)

—and sourly commented in his note: of "all this toil of antitheses . . . neither
the sense nor the occasion is very evident." Far more appropriate to take
Granville-Barker's hint and enjoy the play as a musical entertainment. Of the
passage in which Juliet puns furiously on *I, Ay,* and *eye* (3.2.45–51) Granville-
Barker coolly urges: "Shut our minds to its present absurdity (but it is no more
absurd than any other bygone fashion), allow for the rhetorical method, and
consider the emotional effect of the word-music alone—what a vivid impression
of the girl's agonized mind it makes, this intoxicating confusion of words and
meanings." And he says of the "verbal embroideries" of the long "Gallop apace,
you fiery-footed steeds" speech (I think rightly) that they "owe their existence in
great part to the bravura skill of the boy actress who could compass such things
with credit." In the subsequent speech, with all its exclamation points, we seem to
be making the acquaintance of something very much like an operatic *cabaletta:*

> Beautiful tyrant! fiend angelical!
> Dove-feather'd raven! wolvish ravening lamb!
> Despised substance of divinest show!

About this speech, which runs for eight more lines in this superheated vein,
Granville-Barker suggested, "The boy-Juliet was here evidently expected to give a
display of virtuosity comparable to the singing of a scena in a mid-nineteenth-
century opera." I think Granville-Barker would not mind my adding that, as does
many a treasured opera, *Romeo and Juliet* contains a few dreary verse "arias" (for
example, Lady Capulet's lavishly extended simile comparing Paris to a book,
1.3.81–94) and the occasional overwrought ensemble that one would be perfectly
happy to see cut in performance (see 4.5.34–90, which I think is Shakespeare's
most awful scene).

All of which is not to say that the blank verse should be approached as thinly
disguised vocalise. The ideal must always be to convey both the sound and the
sense, trippingly on the tongue. Eric Bentley happened upon the rare ideal in a
fine Macbeth: "Mr. Keith has few equals in this country as a speaker of
Shakespearean verse; he does not yield to Maurice Evans in his eagerness to
render the music but he succeeds also in delivering the sense. Voice and carriage
take us back to the days of heroic acting." The interpretive point, rather, is to
know when sense and music exchange or share the lead as the action unwinds.

Shakespeare knew very well the limitations of pristine, immediately comprehensible, plain-style syntax and expression, just as he also knew, as Iago does, the enormous rhetorical powers of nonsense. Lodovico asks of Othello, "Are his wits safe?" and Iago replies with magnificent but impenetrable assurance:

> He's that he is; I may not breathe my censure
> What he might be. If what he might he is not,
> I would to heaven he were! (4.1.270–272)

As in the passage from *Titus* we have examined, Shakespeare knew when the very complex meaning of lines might not be fully grasped as they are declaimed, and yet—because of a careful concatenation of sounds and placement of words—a very powerful effect can still be achieved. The purely aural bedazzlement in Shakespeare's plays, as is well known, covers a multitude of loose plot-threads, incredible non sequiturs, and major and minor expediencies of all kinds.

The most brilliant display of Shakespearean blank verse is thought by many to be in *Antony and Cleopatra,* and one speech from this play has drawn special attention as an example of the author's aural alchemy. Sir John Gielgud has said, "[Y]ou have to know that a great speech like Cleopatra's speech on the death of Antony is written for the sound and not for the sense." Here it is:

> O, wither'd is the garland of the war,
> The soldier's pole is fall'n! Young boys and girls
> Are level now with men; the odds is gone,
> And there is nothing left remarkable
> Beneath the visiting moon. (4.15.64–68)

Granville-Barker on the speech: "This, in analysis, is little more than ecstatic nonsense; and it is meant to sound so. It has just enough meaning in it for us to feel as we hear it that it may be possible to have a little more . . . it gives us to perfection the reeling agony of Cleopatra's mind; therefore, in its dramatic setting, it ranks as supreme poetry." Shaw goes at this passage in a similar way, but, as might be expected, he drives home the parallels with musical performance more aggressively: "This is not good sense—not even good grammar. If you ask what does it all mean, the reply must be that it means just what its utterer feels. The chaos of its thought is a reflection of her mind, in which one can vaguely discern a wild illusion that all human distinction perishes with the gigantic distinction between Antony and the rest of the world. Now it is only in music, verbal or other, that the feeling which plunges thought into confusion can be artistically expressed. Any attempt to deliver such music prosaically would be as absurd as an attempt to speak an oratorio of Handel's, repetitions and all. The right way to declaim Shakespear is the sing-song way. Mere metric accuracy is nothing. There must be beauty of tone, expressive inflection, and infinite variety of *nuance*."* Shaw's views here, of course, are the perfect response to those who find opera librettos so hilariously lacking in "good sense."

Such vigorous analyses make clear that signification in Shakespeare (as in opera) is not all. They also suggest the value of approaching the plays on the page as one might an opera score; that is, with a constant awareness that the words were never meant to exist in black-on-white. The actor Alan Howard unwittingly makes something like this point as he discusses the ways one can fail to do justice to Shakespeare on stage: "Obviously you need to comprehend as well as you can what the lines *mean*. But I think that the other aspect of the actual *sounds*, the textures and the rhythms, invoke a word which perhaps we don't understand so well today. The word is 'apprehension' as opposed to 'comprehension.' Something we *sense*. I think that 'apprehension' to the Elizabethans was a very palpable thing. They were sensually highly aware of how rhythms, sound, and texture could combine with comprehension to bring about something which goes beyond just the sense." Getting beyond the sense (or nonsense) is what the reader of an opera libretto is often obliged to do, and Shakespeare's lines—even very memorable ones—require the same effort. Shaw points to Othello's tremendous speech, "Like to the Pontic Sea . . ." (3.3.453–462), and lays some Shavian flares about it: "The words do not convey ideas: they are streaming ensigns and tossing branches to make the tempest of passion visible . . . If Othello cannot turn his voice into a thunder and surge of passion, he will achieve nothing but a ludicrously misplaced bit of geography." Granville-Barker singles out Cleopatra's speech as servants and soldiers gather around Antony's corpse (4.15.82–88). It lies a bit flat of the page, but, interpreted "musically," it levitates in the end like Leonora's "Pace, pace, mio dio" or Aïda's "O patria mia": "Note how actual incoherence—kept within bounds by the strict rhythm of the verse—leads up to and trebles the nobility of a culminating phrase. . . . The compelled swiftness of the beginning, the change without check when she turns to the soldiers, the accordant discipline of the line which follows, so that the last two lines ["Let's do it after the high Roman fashion, / And make death proud to take us"] can come out clarion-clear; here . . . is dramatic music exactly scored."

Boito wrote of the "inherent, powerful musicality" in Shakespeare's *Othello*, and it is worth allowing Shaw to reiterate this fact of Shakespearean drama in an operatic context, before we pass on to specific "musicological" aspects of Shakespeare's poetry. Shaw thought only one thing worse on the stage than the Elizabethan "merry gentleman," and that was the Elizabethan "merry lady." So he wondered out loud, in a review of *Much Ado about Nothing*, why Beatrice and Benedick hold the stage so damnably well:

Before I answer that very simple question let me ask another. Why is it that Da Ponte's "dramma giocosa," entitled *Don Giovanni*, a loathsome story of a coarse, witless, worthless libertine, who kills an old man in a duel and is finally dragged down through a trap-door to hell by his twaddling ghost, is still, after more than a century, as "immortal" as *Much Ado*? Simply because Mozart clothed it with wonderful music, which turned the worthless words and thoughts of Da Ponte into a magical human drama of moods and transitions of feeling. That is what happened in a smaller way with *Much Ado*. Shakespear shews himself in it a commonplace librettist working on a stolen plot, but a great musician. No matter how poor, coarse, cheap, and obvious the thought may be, the mood is charming, and the music of the words expresses the mood. . . . Not until Shakespearean music is added . . . does the enchantment begin. Then you are in another world at once.

Shaw's talk of a poor, coarse, cheap libretto and stolen plot saved by poetic music should put one in mind of any number of librettos for estimable operas . . . most infamously that of *Il Trovatore*. Add the music, though, and we are in another world at once.

Another odd libretto is that of Verdi's *Un Ballo in Maschera*. Julian Budden says it is "often held up as a prime example of literary incompetence," but then he goes on to ask if this is evidence of Verdi's poor literary judgment as well: "Not at all. Verdi is writing music drama on terms which are strictly his own. What mattered to him above all was the power of the individual word to shape a situation, to act as a rocket firing the music across the footlights. He knew well enough that in any cantabile, as distinct from a declamation, a line of verse as such tends to become submerged, with only the important words standing out as landmarks. It was a matter of secondary importance that the sentences should make literal sense."* Budden's reasoning happily supports Shaw's and that of many others who have asserted that music often takes priority over mere syntax, √ coherence, and sense in Shakespeare. Play him any other way and the result is likely to be the equivalent of an evening of recitative in an opera house.

4

Sweet Smoke of Rhetoric

It is the hilariously feckless would-be courtier Don Adriano de Armado in *Love's Labour's Lost* who exclaims so admiringly over "the sweet smoke of rhetoric." Like everyone else in this play except the aptly named Constable Dull, he relishes words and rhetoric's power to make them coherent and eloquent. But because he is, like a good comic butt, too eager and self-regarding by half, Armado's rhetoric unintentionally provokes much laughter. The King of Navarre says he is "One who the music of his own vain tongue / Doth ravish like enchanting harmony." (We have all experienced opera singers of this ilk, and sometimes their self-involved performances are as risible as Armado's.) Still, Armado is an extrovert, and rhetoric is an extrovert's game. It languishes in the reticent and tongue-tied like Dull, just as it thrives in show-off spirits like the play's hero, Berowne. *Love's Labour's Lost* is populated mostly with the latter: this is why it ranks as perhaps the highest-calorie "feast of languages" in all Shakespeare. It is also, not coincidentally, a play he wrote when he was new to London and trying to make a name for himself. In no work that followed did Shakespeare play the rhetorical-poetical peacock as unabashedly.

Its characters striking the essential Shakespearean note of rhetorical display, *Love's Labour's Lost* unfolds in operatic fashion: four courting couples match wits in highly patterned scenes, the principals are apportioned their share of *tour de force* speeches, the background is filled in by several delightful comprimario figures, and throughout resounds the music of the "honey-tongued" poet-playwright.* Clothing images are prominent in this play and one might say that rhetoric—Shakespeare's "taffeta phrases, silken terms precise, / Three-pil'd [velvet] hyperboles"—clothes the play as elaborately as great crop-doublets, bombasted trunk-hose, and farthingales clothed aristocrats and players toward the end of Elizabeth's reign. Indeed, the first performers of the play must have reveled in showing off both their costumes and their pun-filled, elaborately alliterated, image-laden lines, not to mention over a thousand lines of rhymed couplets (the Shakespearean record).

Love's Labour's Lost is at once characteristically Shakespearean and operatic; the motto for both genres, and for the play's cast, could be borrowed from Oscar Wilde: "Nothing succeeds like excess." I pause over it here because it is so forthrightly and brilliantly rhetorical—and hence representative of a prominent element of Elizabethan dramaturgy. Rhetoric—the artificial composition of sounds and words—gives these characters and the actors who impersonate them their displays of eloquent virtuosity, just as an elaborate aria would give a coloratura soprano the opportunity to display her wares. *Aria di bravura* or *aria d'agilità* suit nicely as descriptions, for instance, of Berowne's seventy-five-line lover's-call-to-arms oration (4.3.285–362), the Princess of France's meditation on fame (4.1.21–35), and the Princess' and Rosaline's final speeches in which the loss of the men's amorous labors is announced. How were these arias delivered? One scholar has described the Elizabethan stage as thrusting forward "from its broad platform a sort of horn upon the auditorium." What he imagines occurred there, I suspect, errs only slightly on the side of caricature . . . and is reminiscent of occasional goings-on in the opera house: "Along this horn, or isthmus, a player who had some specially fine passage to declaim advanced and began . . . and, having delivered himself, pressed his hand to his heart again, bowed to the discriminating applause, and retired into the frame of the play. An Elizabethan audience loved these bravuras of conscious rhetoric, and in most of his plays Shakeapeare was careful to provide opportunities for them."

In our century there has been a considerable falling out of love with the ornate rhetorical style, as well as with the exhibitionistic aesthetic in which the style thrives. I have come to think of *Love's Labour's Lost* as Shakespeare's <u>finest</u> and—because he fully probes in it both the virtues and vices of rhetorical power—most <u>profound comedy</u>. Unsurprisingly, it is a very difficult play for a modern audience to enjoy. It is too rich for a time that is dominated by the 1,000-or-so-word vocabulary of television; it is too rich for those who are embarrassed or repelled by more rhetorical ingenuity than is displayed in the advertising tags and doggerel of Madison Avenue. Rhetoric produces a virtually alien language; its sweet smoke is likely to irritate most modern sensibilities.

Rhetoric has lost its hold on the stage, too. "The 'modern' theater . . . turned its back on the 'old-fashioned' theater," writes one old hand of the English stage, "and created a 'fourth wall' so it could ignore the audience, talk (mumble?) 'naturally' and end up making films and television. The art of rhetoric and declamation was lost." Actors try to approach Shakespeare's characters as if they were one of us and end up seeming oddly frigid. One scholar, discussing the acting of Marlowe and Shakespeare, observes that their plays "move predominantly on a rhetorical plane. Even *Othello,* the most intimate, most Ibsenian of Shakespeare's tragedies, is moving always toward the set speech, the moment of 'declaration.'" This same writer then looks about today and finds "our actors are shy of the large gesture, of the resonant speech." "Big" actors who can carry off the rhetoric seem for the moment to be a nearly vanished species, as several obituarists had occasion to remark when Olivier died in the summer of 1989.

In the article "The Old Acting and the New," written in 1895, Shaw reminisced about what he thought was the last English exponent of "the tradition of superhuman acting," Barry Sullivan. He "was a splendidly monstrous performer in his prime: there was hardly any part that was sufficiently heroic for him to be natural in it. . . . He was the very incarnation of the old individualistic, tyrannical conception of a great actor." Nowadays even bravura actors try to affect shyness. Derek Jacobi, for instance, described his approach to a very taxing star role he was preparing to open on Broadway: "The aim is the total elimination of self and especially of that awful thing, 'bravura' acting. You want people to leave the theater feeling they've seen [the character], not saying how good you are." Deluded fellow. Everyone—including this writer—left the theater fully cognizant, as the *New York Times* reviewer put it, of "a typically British bag of histrionic tricks" and full of admiration for the "restless bravery" of Jacobi's acting.*

In this century there has been much effort to tone down the garrulous rhetoric of Shakespeare and the bravura technique that must deliver it. Peter Hall, a leading director of Shakespeare and opera, notes one way: "By doing Shakespeare in a tiny room you actually sidestep the main problem we moderns have with Shakespeare—rhetoric. We don't like rhetoric, we mistrust it: our actors can't create it, and our audiences don't respond to it." When we cram the plays into the tiniest "room" of all—television—however, the victory is almost always Pyrrhic. The result is usually the same as with opera in that medium: a powerful, visceral resonance is impossible to "hear" . . . while we are left looking very closely at the physiognomic grotesquerie that is necessary to achieve this resonance in the actual performing space. Alternatively, actors will try to pronounce lines *pianissimo* that are clearly *fortissimo* in their rhetoric. This unfortunate but necessary accommodation dogged the indifferent, often quite tedious, forced march of the BBC through the entire canon several years ago. Apropos the *Julius Caesar* of 1953, Eric Bentley pinpointed the unsatisfying effect of Shakespeare on film in a similar way: "The actual filming of Shakespeare never fails to remind us of how utterly he belongs to the stage. A Cassius who walks through a real street talking loudly to himself (as in the film) can only seem demented. Even full voice-projection—by which the verse gains in dignity—seems absurd in a movie, like all effects in art when the necessity for them has been removed." Pertinently for us, Bentley takes a swipe at the Antony of this film, Marlon Brando, for his inability to grasp the rhetoric of the role: "While star actors with no technique can get along nicely on personality, Shakespeare demands more."

In other words: rhetoric lost, all lost. Consider, for example, Othello's climactic outburst at the end of the great scene with Iago at the center of the play:

> *Iago:* Patience, I say; your mind may change.
> *Othello:* Never, *Iago. Like to the Pontic Sea,*
> *Whose icy current and compulsive course*
> *Ne'er feels retiring ebb, but keeps due on*
> *To the Propontic and the Hellespont,*
> *Even so my bloody thoughts, with violent pace,*
> *Shall ne'er look back, ne'er ebb to humble love,*
> *Till that a capable and wide revenge*
> *Swallow them up. Now by yond marble heaven,*

> In the due reverence of a sacred vow
> I here engage my words.   (3.3.452–462)

The italicized lines—a fine "simile aria"—are absent from the quarto edition of the play, presumably because they were cut in performance. This avoided a small halt in the action toward the end of a very long scene and saved a few moments' time. Alexander Pope left these lines out of his edition, too, calling them "an unnatural excursion in this place." A rhetorical excursion the passage assuredly is: Othello goes the fancy way about saying that his intention is fixed. But unnatural! Deprive the play of a few more such unnatural excursions that merely obstruct the action—"Farewell the tranquil mind . . ." and "Had it pleased heaven . . ." for instance—and *Othello* would not be worth the bother of staging. Remove from an opera several such "unnatural excursions," otherwise known as arias, and the effect would be similarly preposterous.*

Othello thrusts (like most Shakespearean plays and, indeed, most pre-modern operas) toward set solos and moments of declaration. These eagerly anticipated moments take the form of rhetorical excursions in which, typically, more words and images are used than need be, consonant and vowel sounds are carefully composed, and conceits are fetched from afar. Small samples of the result of this rhetorical elaboration are Romeo's lines as he takes the poison—

> Come, bitter conduct, come, unsavory guide!
> Thou desperate pilot, now at once run on
> The dashing rocks thy sea-sick weary bark!

—or Macbeth on the subject of

>   innocent sleep,
> Sleep that knits up the ravell'd sleave of care,
> The death of each day's life, sore labor's bath,
> Balm of hurt minds, great nature's second course,
> Chief nourisher in life's feast.

This tendency of rhetorical invention to overflow the banks of expository necessity or efficiency and take on a life of its own is a foundation of Renaissance literature. The rhetorical term for such excursions reinforces the flooding image: *copia*, from the Latin for "plenty" or "abundance."

The *métier*, excitement, and pleasures of copious invention are thus of the Shakespearean essence. Several years ago Jacques Barzun published a manual titled *Simple and Direct: A Rhetoric for Writers*; but being simple and direct were honors an Elizabethan writer dreamed not of. Characters in Shakespeare's plays who waste no words are likely to turn up with names like Snout, Shallow, Slender, and Dull. The bias of the age was all the other way, as can be suggested by this stunning passage from John Donne's *Devotions*. Even abbreviated, it is a breathtaking rhetorical aria:

> My God, my God, Thou art a direct God, may I not say a literall God, a God that
> wouldest bee understood literally, and according to the plaine sense of all that thou

sayest? But Thou art also . . . a figurative, a metaphoricall God too: a God in whose words there is such a height of figures, such voyages, such peregrinations to fetch remote and precious metaphors, such extensions, such spreadings, such Curtaines of Allegories, such third Heavens of Hyperboles, so harmonious elocutions. . . . O, what words but thine can expresse the inexpressible texture, and composition of thy word?

Those who are more at ease with the literal and "plaine sense" than with such rhetorical "peregrinations" as this will be bound to find Shakespeare's verbal terrain daunting. They might find that the rather similar pleasures of opera do not appeal to them for the same reasons. Shaw suggested as much when he wrote to his prospective American biographer that he was sorry the biographer was not more familiar with eighteenth- and nineteenth-century music: "for my plays bear very plain marks of my musical education. My deliberate rhetoric, and my reversion to the Shakespearean feature of long set solos for my characters, are pure Italian opera." Shaw's comparison brings us to a crucial subject: the shared methods lying behind the rhetorically heightened, artificial displays—by actors in words, singers in words and music—that dominate Shakespearean and operatic dramaturgy.

"Friends, Romans, countrymen, lend me your ears!" According to classic oratorical theory, deliberate rhetoric should arrest attention, persuade, and then move the hearer appropriately to agreement or action. One might say simply that any deployment of extraordinarily artificial, elaborate verbal sounds is rhetorical. Such utterance is achieved by the so-called figures of speech, which are the building materials of rhetorical architecture. George Puttenham gives extensive attention to "figures and how they serve in exornation [i.e., decoration] of language" in the most elaborate Elizabethan treatise on the subject, *The Arte of English Poesie* (1589). He describes "figurative speech" in general as "a noveltie of language evidently (and yet not absurdly) estranged from the ordinarie habite and manner of our dayly talke." "Figure," he goes on to say, "is a certaine lively or good grace set upon wordes, speaches and sentences to some purpose and not in vaine, giving them ornament or efficacie by many manner of alterations in shape, in sound . . . [thus] tuning and tempring them by amplification, abridgment, opening, closing, enforcing, meekening or otherwise disposing them to the best purpose."

The methods listed by Puttenham, obviously, are not far afield from what a composer does as he sets his "lively or good grace" upon words with his quavers, dynamic markings, appoggiaturas, fermatas, and accents. Indeed, as one reads through Puttenham's *Arte*—which is in effect a treatise about Shakespeare's own poetic art—all the talk of his verbal "music" in the preceding chapters becomes less and less an airy conceit and more emphatically an accurate description of the way he wrote poetry. Puttenham asserts that "the chiefe grace of our vulgar [i.e., English-language] Poesie consisteth in Symphonie," and he speaks of "your Cadences by which your meter is made Symphonicall." Elsewhere he observes that "our speech is made melodious and harmonicall, not onely by strayned tunes, as those of *Musick,* but also by choice of smoothe words." Puttenham also praises "your figures rhetoricall" because they "containe certaine sweet and melodious manner of speech, in which respect they may, after a sort be said *auricular:* because the eare is . . . ravished with their current tune."

Auricular figures, according to Puttenham, work "by their divers soundes and audible tunes [an] alteration to the eare and not to the mind." Among these are many figures used commonly by composers: *aposiopesis* (an interruption or pause); *polysindeton* (using the same connective word or phrase several times in succession); *prolepsis* (setting forth an idea in summary, then more elaborately); or *irmus* ("speech drawn out at length . . . with imperfit sence till you come to the last word or verse, which concludes the whole premise with a perfit sence and full period"—the essence of Mozartean style). Repetition is also a common compositional practice, and Puttenham notes that the "figure that worketh by iteration or repetition . . . is counted a very brave figure with the Poets and rhetoricians." He describes several of these in charming fashion. *Anaphora* (repetition of a word at the beginning of two or more clauses) "is when we make one word begin and, as they are wont to say, lead the daunce to many verses." The figure of *anaphora* occurs in both the words and music of "Vissi d'arte, vissi d'amore." *Antistrophe* (repetition in inverse form) is called "the Counter turne." *Anadiplosis* is the repetition of a prominent phrase after intervening material and is named the "Redouble"; it is a commonplace of operatic word-setting. *Epizeuxis* is the immediate repetition of a phrase. This is also typical of operatic libretto language and music (much satire of opera makes fun of the binges of epizeuxis that occur at the end of arias and finales). Puttenham gives it the apt name of "Coocko-spell."

Poetic and musical rhetoric have very much in common; indeed, they are in many respects the same rhetoric. Gioseffo Zarlino, for instance, wrote in his musical treatise *Instituzioni harmoniche* (1558), "the cadence is of equal value in music with the period [full stop] in oration; thus it may truly be called a period of cantilena." "Rhetoric" for composers has always encompassed such concerns as phrase structure, chordal progression, rhythmic scansion, articulation, dynamics, and tempo: every one of these shaping elements has its prosodic equivalent in a passage of Shakespearean blank verse, or even in much Shakespearean prose. It cannot be surprising that dramatic poets and composers, seeking the same "melodious and harmonical" effects Puttenham was after, should share such rhetorical strategies. A leading student of early musical rhetoric, Leonard Ratner, has concluded, "To be persuasive, both linguistic and musical rhetoric had first to establish *coherence* and then promote *eloquence.*" As our brief excursion into *The Arte of English Poesie* suggests, poets and composers often achieved their coherence and eloquence by the same devices.

Their treatises, naturally, shared the same classical Greek nomenclature. Ratner writes, echoing Puttenham precisely, "The basic unit of melody for all eighteenth-century music was the *figure.* Melodic rhetoric [is] the distribution of melodic figures." He compiles from nearly a dozen musical rhetorics of the eighteenth century a sampling of terms that describe melodic relationships. With them, and Ratner's brief definitions, we return squarely to the realm of Puttenham's *Arte:*

abruptio (breaking off of a final note), *anadiplosis* (repetition of a figure after a
punctuation), *confirmatio* (reinforcement of an idea), *dubitatio* (broken or
unexpected turn), *epistrophe* (repetition of an idea), *gradatio* (sequence, climax),
periphrasis (use of many notes where one will do; the speeches of Romeo and
Macbeth quoted a few pages above are examples of verbal periphrasis), and so on.
Likewise, the arrangement of groups of notes, figures, measures, and phrases was
treated according to traditional modes of poetic scansion. Thus, a 1746 *Musicus
Compositorio Practicus* gives musical examples of poetic feet, along with indications of what kind of music was apt for a given foot: *iambus* ("rather cheerful"),
trochaeus ("naïve, not serious; mixed with iambic in minuets"), *dactylus* (one
heavy, then two light stresses; "serious as well as humorous"), *molossus* (three
heavy stresses; "serious, sorrowful"), and *amphybrachys* (light–heavy–light; "for
lively expression"), to name a few.

   This, then, is the highly complex and calculating *métier* of the verbal and the
musical rhetorician. Rhetoric, then as now, was the path to power and pleasure,
for, as Puttenham plainly put it, "Eloquence is of great force." Not only poets and
composers, but also preachers, lawyers, and politicians recognized this and
carefully cultivated their luxuriant figurative gardens. Forceful eloquence—
produced through cunning deployment of the figures of speech—abounded in
Shakespeare's time, and I would like to offer a small prose example of what I
mean, a sentence from Sir Walter Raleigh's *History of the World*. Far from prosaic,
it could easily be made to look like a poem by playing with the lineation. I draw
attention to it because, to my mind, it has the feel and trajectory of an operatic aria
text:

> O eloquent, just, and mighty Death! whom none could advise, thou hast persuaded;
> what none hath dared, thou hast done; and whom all the world hath flattered, thou
> only hast cast out of the world and despised; thou hast drawn together all the far-
> stretched greatness, all the pride, and ambition of man, and covered it over with these
> two narrow words, *Hic jacet!*

Raleigh rivets attention with the "O" (the figure of *ecphonesis,* which Puttenham
calls "the outcrie"), then with three examples of *epitheton* (choice of qualifier),
then the *exclamatio*. The first six words produce a miniature *exordium* (introduction). In the following development of the idea (*distributio* or *periphrasis*) there
are several examples of *anaphora* (the repeated *whom, thou*) and *anadiplosis* (the
three *all*'s). *Gradatio* is exemplified in the increasing length of the clauses (seven
words, seven words, seventeen words, twenty-seven words), and the first three
statements carefully balance internal subclauses, while the last one acts as a
peroratio (rhetorical counterpart to the musical coda). The entire sentence is
shrewdly aimed—like an aria that launches steadily toward a high C—at the
climax of the last two words. (Puttenham's definition of *irmus* quoted earlier fits
this sentence perfectly.)

   One might even say the sentence offers a syntactic allegory of what it is
expressing: all the vanity of the ornate, expansive rhetorical style is itself deflated
by the two "narrow," plain words ("Here lies!") at the end. This sentence, in sum,

offers us the essential methods of Shakespeare and most operatic composers up to and including Richard Strauss (and many after him): expansion of a simple idea, variety within the context of a unifying structure, carefully judged amplification and repetition, and special emphasis on the final phrase (Raleigh's last six words have the uncanny effect of a musical *rallentando*).

I have paused over Raleigh's sentence because it offers a modest first dip into the rhetorical complexities of Elizabethan style. As we move into Shakespeare, the shallows fall off soon and drastically. Since my concern has been with "arias" both verbal and musical, perhaps the logical next step is to turn briefly to Shakespeare's Sonnets. For their relation to the plays is exactly that of Mozart's many florid concert arias to his operas. In several respects, in fact, it is possible to think of the Sonnets as operatic effusions. We have already noticed that Shakespeare's dramas aim toward isolated moments of declaration, and many of the Sonnets— soliloquies all exactly fourteen lines long—find the speaker forced into an emotionally volatile situation and required to "perform" his way out of it. Like most memorable arias, many of the Sonnets are prodigal of passion; twelve of them begin with the vocative "O" that reminds us of the *Deh!* or *Ah!* or *Oh!* so favored by librettists. Because most of the Sonnets are in one way or another love poems, the situations of many are stereotypical and very familiar from the opera repertory: one can spot Fiordiligi's *"Come scoglio"* vein (Sonnet 107, and Sonnet 55's "Not marble, nor the gilded monuments . . ."); Manon Lescaut's "Sola, perduta, abbandonata" vein (Sonnets 97, 109); and Otello's "Dio mi potevi scagliar" vein (Sonnets 29, 66); as well as the familiar moments of self-doubt, melancholy, jealousy, and the like. As well, most operatic arias tend to be organized in suspenseful crescendo toward some kind of climax—usually a note or phrase above the staff. This is paralleled by Shakespeare's habit of creating suspense in the three quatrains, then resolving it in the final rhymed couplet. Finally, as with many an aria by some very great composers, certain of the Sonnets were obviously written when genius was fast asleep and are tedious, clichéd, and overwrought. Sonnet 24, the worst of the lot, is all three.

Let us look, then, at just one of Shakespeare's youthful concert arias (he wrote his sonnets at about the time of *Love's Labour's Lost*) and sample their customary and extraordinary rhetorical artifice. Sonnet 12, perhaps the most Mozartean of all in its complex, perfect balance, is one of the "breeding" sonnets in which the speaker seeks to persuade a handsome young man to marry and beget children:

> When I do count the clock that tells the time,
> And see the brave day sunk in hideous night;
> When I behold the violet past prime,
> And sable curls all silver'd o'er with white;   4
> When lofty trees I see barren of leaves,
> Which erst from heat did canopy the herd,
> And summer's green all girded up in sheaves,
> Borne on the bier with white and bristly beard: 8
> Then of thy beauty do I question make
> That thou among the wastes of time must go,
> Since sweets and beauties do themselves forsake

And die as fast as they see others grow,   12
And nothing 'gainst Time's scythe can make defense
Save breed, to brave him when he takes thee hence.

As with any operatic aria, several modes of organization are here being employed simultaneously. The first line sets up a regular duple meter with five aptly "clockwork" iambs; afterward this norm is varied contrapuntally for emphasis—for example, with the three strong stresses of "bráve dáy súnk," the trochee at the beginning of the ninth line where the sonnet's direction turns, and the two heavy stresses on the final crucial phrase, "Sáve bréed." This is one of the more metrically regular sonnets, and one imagines its tempo to be a deliberate *andante:* after all, the speaker wishes to appear soberly contemplative as he offers his advice. Then there is the manifest pattern of three quatrains in alternating rhyme and the aurally climactic couplet. The syntax here is organized suspensefully, with the three successive *when*'s leading to *then* in line 9 (Shakespeare used *when . . . then* in nine other sonnets, always effectively). As in Raleigh's sentence, there is a measured *gradatio* in clause lengths (two, two, four, then six lines). The argument itself amounts to a *tema con variazioni.* The young man is given six imagistic variations on "time flies." These lead to the application to the young man (line 10), a statement of the general theorem (line 13), and then the solution (line 14). Standing out from all these methods of organization, though, is the positively Mozartean dexterity of aural patterning in the poem. One writer has devoted three pages solely to Sonnet 12's phonetic effects; one can simply allude to some of them here: the unerring rhythmic positioning of consonantal and vowel relationships, the subtle effects of assonance and alliteration, and the classical aural balancing in nearly every line (note the symmetry of the *s-s-g/g-s-s* sequence in line 7).

One might well feel the Sonnets are more akin to Mozart's or Handel's world of splendidly excessive, acrobatic aural feats than to our own. The notions of form following function and less being more have perhaps caused us to lose our ear for such feats. Students, in my experience, generally find the Sonnets difficult and alien; they read them impatiently. One of these students chose to vent her frustrations in kind, with this brilliant parody of Sonnet 12:

When I do read the verse the Bard has writ,
And find his laurels tilt a bit askew;
When sugared words are irredeemed by wit,
Far worse than any taming of the shrew;
When self-indulgence has destroyed the sonnet,
And noble thoughts are ruined by conceit
(Some ostentatious bee inside his bonnet),
And rhythm merely acrobatic feat:
Then of his genius do I question make,
And wonder if this same hand held the quill
That penned the plays from which I pleasure take,
Or could these poems be of another Will.
　　Presumpt'ous one am I; the choice is clear:
　　The playwright keep but shed the sonneteer.

The *New Yorker*'s late poetry editor, Howard Moss, suggested how out of touch we are with the musicality of Shakespearean rhetoric in a similarly piercing fashion. He offered three "Modified Sonnets"—versions of Sonnets 8, 18, and 29—dedicated "to adapters, abridgers, digesters, and condensers everywhere." Shakespeare's Sonnet 18, another "breeding" sonnet, opens with these six lines:

> Shall I compare thee to a summer's day?
> Thou art more lovely and more temperate:
> Rough winds do shake the darling buds of May,
> And summer's lease hath all too short a date;
> Sometime too hot the eye of heaven shines,
> And often is his gold complexion dimm'd . . .

Here, *sans* rhetoric and verbal music, is the version Moss offers for our time:

> Who says you're like one of the dog days?
> You're nicer. And better.
> Even in May, the weather can be gray,
> And a summer sub-let doesn't last forever.
> Sometimes the sun's too hot:
> Sometimes it is not . . .

And Shakespeare's superbly assonant and lilting couplet—"So long as men can breathe or eyes can see, / So long lives this, and this gives life to thee"—becomes, "After you're dead and gone, / In this poem you'll live on!"

Though the plays may strike us as much less alien, the brutal fact is that the rhetoric and sound-patterning of the Sonnets courses as life's-blood through them. The Sonnets have been called "a workshop for the plays," and Shakespeare carried most of his habitual methods in that workshop into his dramas. He even managed to sneak a perfect sonnet into the action of one of them—for Romeo and Juliet when they first meet and kiss at the Capulet ball (1.5.93–106). The Sonnets' elaborate architecture can be seen reflected especially in the plays' longer, more challenging set speeches, and this is no doubt why John Barton devotes an entire chapter to "using the Sonnets" in *Playing Shakespeare*. He begins with this observation: "Sonnets can be excellent exercise pieces for actors. Most of the textual and verbal points that come up in working on the plays appear in the Sonnets in concentrated form. . . . They are like little self-contained scenes, fourteen lines long." A skilled speaker of the Sonnets will feel quite at home as Bolingbroke when he greets Richard II:

> See, see, King Richard doth himself appear,
> As doth the blushing discontented sun
> From out the fiery portal of the east,
> When he perceives the envious clouds are bent
> To dim his glory and to stain the track
> Of his bright passage to the occident.   (3.3.62–67)

And he will feel at home as Jachimo looking over the sleeping Imogen in *Cymbeline:*

> How bravely thou becom'st thy bed! fresh lily,
> And whiter than the sheets! That I might touch!
> But kiss, one kiss! Rubies unparagon'd,
> How dearly they do't! 'Tis her breathing that
> Perfumes the chamber thus. The flame o' th' taper
> Bows toward her, and would under-peep her lids,
> To see th' enclosed lights, now canopied
> Under these windows, white and azure lac'd
> With blue of heaven's own tinct.   (2.2.15–23)

Even the simpler passages where Shakespeare avoided the color purple, I think, reflect the effects of the sonneteer. Compare, for instance, the plain-style exhaustion of Sonnet 66 ("Tir'd with all these, for restful death I cry") with King Lear's

> And to deal plainly,
> I fear I am not in my perfect mind.
> Methinks I should know you, and know this man,
> Yet I am doubtful: for I am mainly ignorant
> What place this is, and all the skill I have
> Remembers not these garments; nor I know not
> Where I did lodge last night. Do not laugh at me . . .   (4.7.61–67)

Colley Cibber wrote in his *Apology* (1740) that opera is "not a Plant of our Native Growth, nor what our plainer Appetites are fond of." And the adapters, digesters, and condensers of the Restoration and the following Augustan period felt much the same about the Shakespeare of the 1623 Folio. They thus went dutifully about the task of rendering his "dazling . . . Disorder" (as Tate put it) neater for plainer appetites. My purpose in these pages has been to place Shakespeare firmly in the same aesthetic tradition as opera: his essential pleasures are not for those who, in Cassio's phrase, prefer to feed on a "nice and waterish diet." The "full liberty of feasting" proclaimed by Othello's herald is a phrase far more to the point of Shakespearean style. This style, like opera's, is one of spectacular excess, resounding utterance, and, as Puttenham readily admitted, estrangement from the ordinary "manner of our dayly talke."

It is rhetoric, generator of *copia,* that causes the Shakespearean and operatic horns of plenty, or cornucopias, to pour forth as they do. The most operatic and rhetorically figurative of all American poets, Walt Whitman, may have put his finger on the similarity I have been urging here in a boast from one of his several farewell poems, "So Long!" The boast is perfectly just, and its first adjective is especially apt: "I announce a life that shall be copious, vehement, spiritual, bold." It seems to me the great characters of the Shakespearean and operatic repertories announce this kind of life, too. After they have done so, the sweet smoke of rhetoric is usually left hanging thick in the air.

5

The Story of O

When Walt Whitman was in his early sixties he published some reminiscences under the title *Specimen Days* (1882). Here he observes that "certain actors and singers had a good deal to do with the . . . gestation of *Leaves of Grass*," and then calls to mind the theatrical pleasures of his New York City years. First he reels off the names of favored actors he saw at the old Park Theater: Ellen Tree, the younger Kean, Macready, Sheridan Knowles, and especially Fanny Kemble ("name to conjure up great mimic scenes withal—perhaps the greatest"). Then he comes to Shakespeare: "As a boy or young man I had seen (reading them carefully the day beforehand) quite all Shakspere's acting dramas, play'd wonderfully well. Even yet I cannot conceive anything finer than old Booth in *Richard Third,* or *Lear* (I don't know which was best) or Iago . . . or Tom Hamblin in *Macbeth* . . . or old Clarke, either as the ghost in *Hamlet* or as Prospero. . . ." Whitman writes that his "musical passion" followed his theatrical one. "I heard, these years, well render'd, all the Italian and other operas in vogue, *Sonnambula, The Puritans, Der Freischutz, Huguenots, Fille du Regiment, Faust, Etoile du Nord, Poliuto,* and others. Verdi's *Ernani, Rigoletto,* and *Trovatore,* with Donnizetti's [*sic*] *Lucia* or *Favorita* or *Lucrezia,* and Auber's *Massaniello* [*sic*], or Rossini's *William Tell* and *Gazza Ladra* were among my special enjoyments. I heard Alboni every time she sang in New York and vicinity—also Grisi, the tenor Mario, and the baritone Badiali, the finest in the world." He muses, too, on the performances of Maretzek's Havana company in the Battery's Castle Garden: "the fine band, the cool sea-breezes, the unsurpass'd vocalism—Steffanone, Bosio, Truffi, Marini in *Marino Faliero, Don Pasquale,* or *Favorita.* No better playing or singing ever in New York. It was here too I afterward heard Jenny Lind. (The Battery—its past associations—what tales those old trees and walks and sea-walls could tell!)"*

And what tales could they tell of the characteristic Whitman style as well. It would require a very close call, for instance, to decide whether Shakespeare or opera helped more to form the second stanza of his most famous poem, "When lilacs last in the dooryard bloom'd":

> O powerful western fallen star!
> O shades of night—O moody, tearful night!
> O great star disappear'd—O the black murk that hides the star!
> O cruel hands that hold me powerless—O helpless soul of me!
> O harsh surrounding cloud that will not free my soul.

The decision is much easier in a similarly mournful agony, as the "he-bird" ("the lone singer wonderful causing tears") reacts to the loss of his mate in "Out of the cradle endlessly rocking":

> O darkness! O in vain!
> O I am very sick and sorrowful . . .
> O troubled reflection in the sea!
> O throat! O throbbing heart!
> And I singing uselessly, uselessly all the night.

After the thrush's grand sixty-line *scena* reaches its climax—

> Loved! loved! loved! loved! loved!
> But my mate no more, no more with me!
> We two together no more.

—the poem's speaker tells of the "aria sinking . . . the notes of the bird continuous echoing" and of the "aria's meaning . . . swiftly depositing / The strange tears down the cheeks coursing."

 Whitman and these passages offer a perfect overture to what will be both the concern and the conceit of the following pages. My concern will be: the artistic expression of common emotions overwhelmingly felt, the primary material of Shakespeare and opera. (Teddy Roosevelt said that Whitman was the only nineteenth-century poet who "dared use anything that was strikingly and vividly typical of the humanity around him.") My preliminary conceit is that the story of Shakespearean and operatic experience is, in essence, the story of "O."

Or rather, the story of an auricular figure of our previous acquaintance: *ecphonesis,* from the Greek for "to cry out." Shakespeare made constant use of this favorite Whitman figure—2,434 times, to be exact—usually to give his characters some expressive leeway at moments of great emotional exaltation or trauma, elevation or depression of spirit. Salisbury says of the King in *King John* at a tense moment, "His passion is so ripe, it needs must break." The playwright carefully prepared for such moments, and a vocative *O* often signals that the breaking-point of passion has arrived. Thus, Paulina, deeply distressed over Leontes' jealous rage at her mistress in *The Winter's Tale,* cries, "O, cut my lace, lest my heart, cracking it, / Break too!" This launches her into a long, high-pitched recital of Leontes' vile acts, which climaxes with the revelation of Hermione's death.

Shakespeare reveled in such driven vocalism, and the *O* and *Ah* (the latter occurs 187 times) were among his favorite means of igniting "matter from the heart" when the bonds of politeness, restraint, or repression could no longer hold. "Ah, why should wrath be mute and fury dumb?" asks Aaron in *Titus Andronicus;* and indeed, they rarely are in Shakespeare's dramas.

A remarkable number of the most famous lines Shakespeare wrote begin with the exhaled syllable of overflowing emotion:

> O Romeo, Romeo, wherefore art thou Romeo?
> O no, it is an ever fixed mark . . .
> O Harry, thou hast robb'd me of my youth!
> O ceremony, show me but thy worth . . .
> O what a rogue and peasant slave am I!
> O, let me not be mad, not mad, sweet heaven!
> O brave new world that has such people in't!
> O, what a fall was there, my countrymen!
> O gull, O dolt, as ignorant as dirt!
> O world, thy slippery turns!

The *O* in Shakespeare is virutally an index of passionate extroversion, and so it will come as no surprise that it occurs far more often in his romantic *Romeo and Juliet* and volcanic *Othello* than in any of the other plays.

It has been happily said that Shakespeare "is always capable of observing the fineness of excess," and ecphonesis was the natural and supreme figure for introducing passion writ large. Puttenham describes it most pertinently for us in his *Arte:* "The figure of exclamation, I call him *the outcrie* because it utters our minde by all such words as do shew any extreme passion, whether it be by way of exclamation or crying out, admiration or wondering, imprecation or cursing, obtestation [solemn supplication] or taking God and the world to witness, or any such like as declare an impotent affection." Puttenham's nickname for ecphonesis and the several strenuous activities he mentions point directly to the world of opera. How many memorable moments in opera fall into the categories of taking God or the world to witness, cursing, supplicating, and declaring powerless affection! And how many of the repertory's great arias begin with a rhetorical outcry. They are everywhere:

> Deh! vieni alla finestra . . .
> Ach ich fühl's . . .
> Ah! quando rapita in estasi!
> Ah! bello a me ritorna . . .
> O don fatale!
> Ah! fuyez, douce image!
> O du, mein holder Abendstern . . .
> O paradiso!
> Ach! so fromm!
> Ah! leve-toi, soleil!
> Oh! mio babbino caro . . .

Almost paraphrasing Paulina's distress of heart, Aïda begs: "Love, fatal and terrible, break my heart, let me die!" Thus ends her great aria that begins "Ah! non fu in terra mai." Calaf's excitement when Turandot first appears before him is made fully clear: he is given eight O's in the first nine bars he sings. And when Wotan's anguish is most intense in *Die Walküre* he cries, "O heilige Schmach! O schmählicher Harm! ("O sacred shame! O shameful distress!"). At such moments as these, composer and librettist are insisting on the "fineness of excess." This excess is not merely emotive but also executive: opera audiences come in expectation of excessive—that is, physiologically highly abnormal—vocal utterance.

The dramaturgy of outcry, if we may call it that, focuses on the preparation for moments during which Puttenham's "extreme passion" is displayed. Before Hal turns over his new leaf, his father, the king, advises his other sons how to treat the heir apparent:

> being moody, give him time and scope,
> Till that his passions, like a whale on ground,
> Confound themselves with working.

Though the image is not the happiest, it does happen to capture Shakespeare's usual method with his characters, which was to find them "moody" and then give them "time and scope" in extended speeches or dialogue to exhaust their passion. *O* is rarely absent from such events. One might venture, for instance, that the entire action of *Othello* is thrust headlong toward the goal of that most difficult line beyond words in the last scene, which consists simply of the Moor's "Oh, oh!" (5.2.28; the quarto version raises the ante: "O, o, o"). A similar trajectory occurs in the career of Lady Macbeth, who utters three *O*'s at the end of one of her very last speeches.

O is the initial of passion's slaves. It is the escape valve for energy that can take myriad forms: perplexity (Viola's "O time, thou must untangle this, not I"), trauma (Cleopatra's "O cleave, my sides!"), regret (Lear's "O, I have ta'en too little care of this!"), wonderment (Ophelia's "O, what a noble mind is here o'erthrown"), or pathos (Hamlet's "O, I die, Horatio"). Shakespeare could also excite "the passion of loud laughter" with his O's. In *Love's Labour's Lost* we are left in no doubt that Berowne is suddenly on love's cloud nine, for Shakespeare gives him six successive one-line speeches that begin with *O* (3.1.143–157) just before he performs his "O, and I, forsooth, in love!" solo. Elsewhere in the play we learn what excites the passion of the pedantic schoolmaster Holofernes: "O, I smell false Latin." And perhaps there is some comment on dramatists who rely too much upon *ecphonesis* in Bottom's overripe first speech as Pyramus:

> O grim-look'd night! O night with hue so black!
> O night, which ever art when day is not!
> O night, O night! alack, alack, alack . . .

(Bottom seems to know about *periphrasis* and *epizeuxis,* too; but alas, his power of invention is nil.) Benjamin Britten, taking the cue of Bottom's subsequent apostrophe ("And thou, O wall, O sweet, O lovely wall"), provides in his *Dream* opera a parodic miniature aria for his baritone, marked *con espansione*.

O, Ah, Deh, Ach, and the like are also commonplace because composers and librettists also eagerly seek out moments for their "moody" principals that demand "time and scope" for elaborate displays of ecphonesis, otherwise known among musicologists as *melismatics* (defined by *Webster's Dictionary* as "the art of florid vocalization"). This similarity is reflected, unwittingly and in strikingly different terms, in an attempt by G. Wilson Knight to define the essence of Shakespeare's drama according to certain Nietzschean principles. I have emphasized Knight's most suggestive phrasing:

> We may say that the Dionysian, which is the dramatic essence, *jets up from below, moment by moment,* and that it is the Apollonian partner who arranges, as best he may, the Dionysian into rational coherence. A surface too lucid, logical, and Apollonian may even appear to be a disadvantage to drama, since the audience has to attend too sharply with its logical reason. Dramatic thought is made of a *succession of almost independent, transfixing moments, vertically motivated from the unseen.*

Most operas, too, can be described as offering a succession of independent, transfixing moments. The notion of vertical motivation from the "unseen" is also attractive, since this is one way of describing the effect of music upon the text of a libretto. The composer's crucial task is to see that the music "jets up from below" at appropriate and effective moments in the drama. If an *O* is justified by a situation, the time is very likely right. This is Shakespeare's method. Olivia in *Twelfth Night,* for instance, finds her Apollonian intellectual defenses crumbling as she falls in love with Cesario, and rises to one of the play's choice verse arias. It begins, "O, what a deal of scorn looks beautiful. . . ."

It should be added that *O* has for much of this century suffered an eclipse. We are, I suspect, more inclined to treat it as the gateway to garrulous bombast rather than to tragic sublimity. We have noted John Barton's observation of "a tendency in our acting tradition to run away from verbal relish, especially of vowels," and *O* is the most relishable vowel of all. In *O,* one might say, lies the essence of the Shakespearean challenge. The actor Anthony Quayle recalls a conversation on the subject of *O* he had with Olivier during a performance of *Titus Andronicus.* He said to Olivier, who was playing Titus: "You know what you hate about Titus, he's always saying 'oh, oh, oh, look at . . . [or] fancy them doing that to me, oh, oh, oh.' And how many ways are there of saying 'oh, oh'? It's very tough on your imagination, it's very tough on your resourcefulness of variations of all kinds, and therefore it's also a very great strain physically." According to Quayle, Olivier replied, "Othello is all of that and you have to be black as well." Today, however, the virtuosity of *O* is not often called for. I lamented in my pessimistic afterword to *Literature as Opera* the scarcity in contemporary literature of the qualities that are likely to elicit the outcry: lyric expansiveness, implicit or explicit largeness of scale, passionate force, and immediacy of impact. Cooler heads and a more understated rhetoric have prevailed. Fittingly, our century's most famous evocation of spiritual ennui should treat Shakespearean ecphonesis irreverently: "O O

O O that Shakespeherian Rag." A dessicated world such as Eliot's "The Waste-
land" describes will be unlikely to rise to the challenge of *O*.

If the penchant for "outcry" lies somewhere near the heart of Shakespearean
and operatic style, however, the challenge of *O* must be met. Successful interpreters will, as Quayle makes clear, be those who rise to the extravagant
demonstrations and declamations that *O!* and *Ah!* so often signal. This requires,
one might say, an imaginative recklessness, perhaps even a certain tastelessness.
George Santayana, a stuffily Apollonian philosopher, wrote that Shakespeare
"pays the price" for his "depth of passion and vividness of characterization [by a]
notable loss in taste, in sustained inspiration, in consecration, and in rationality."
There is some truth in the aspersion, and actors of Shakespeare (like singers in
opera) must pay a similar price. The Dionysian playwright, as Knight was eager to
suggest, urges his performers into situations where they will, if all goes well, lose
some rationality, suffer a "notable loss in taste," become slaves to passion. The
performers' challenge, in short, is to pluck heightened, transfixing apprehensions
of human experience from the jaws of "vulgarity," "exhibitionism," and "unreality."

This is not a task for the timid or prim. Rex Harrison once explained that his
inability to play Shakespeare was due to his reluctance to take chances; and the
director of Olivier's *King Lear* film has said, "A great actor has to expose himself,
and [Olivier] has never been afraid of revealing his nakedness." (Perhaps we have
a variation here on the "heroic nudity" extolled by Whitman.*) The same
distinction arises when James Agate describes the best Cleopatra of his experience: Janet Achurch "was equally overpowering whether what she did was
immensely right or immensely wrong. . . . There was majesty and there was
physical passion; there were looks which might have unpeopled a city and tones
which might have quelled provinces." Agate compared Achurch favorably with
Edith Evans because Evans "administers none of these shocks. She has not
enough passion and vulgarity for Cleopatra."

The belief that one must err, in Shakespeare, on the side of Dionysian vulgarity
and recklessness rather than the side of Apollonian reserve has been reiterated in
several amusing ways. Jorge Luis Borges, for instance, asserted that Samuel
Johnson "was a far more English writer than Shakespeare. Because if there's one
thing typical of Englishmen, it's their habit of understatement. Well, in the case of
Shakespeare, there are no understatements. On the contrary, he is piling on the
agonies." Borges then adds an afterthought pertinent to our subject: "Johnson,
Wordsworth, and Kipling also, I think they're far more typically English than
Shakespeare. I don't know why, but I always feel something Italian, something
Jewish about Shakespeare." Shaw was onto the same idea when he said the typical
Englishman "is musical, but he is not operatic." The notion that understatement
is un-Shakespearean is also reflected in Shaw's reaction to an actress who tried to
play Rosalind primly in *As You Like It:* "a dainty, pleading, narrow-lipped little
torrent of gabble will not do for Shakespear. . . . The effect is like a canary
trying to sing Handel." The folly of bringing a stiff upper lip to the major tragic
figures in the canon is even more apparent. "The part of Othello annihilates
pretentiousness and shows quietism the door," wrote Agate. "It goes against the

English grain, demanding *fougue* [fire, dash] from a race famous for phlegm; chopping into messes . . . is not what we call cricket." But perhaps no one has put this idea more bluntly that André Maurois, in the judgment he passed on the nineteenth-century actress, Ann-Françoise Mars: "Marivaux [author of delicate comedies], not Shakespeare, was her natural element. . . . Mlle. Mars lacked the ✳ bad taste which makes a good shrieker."

Bad taste. One is reminded of the hero's observation just before the performance of *Lucia* begins in *Where Angels Fear to Tread:* "There is something majestic about the bad taste of Italy. . . . It observes beauty, and chooses to pass it by. But it attains to beauty's confidence." "Bad taste," of course, is a phrase used about Shakespeare and opera by those who are unmoved or unmovable in their presence. Shaw might seem to be echoing Forster when he calls *Lucia* "a vulgar beast of an opera," but he hastens to add, "and yet what passion and what melody is there in every act of it; and what memories cluster for some of us about those melodies." On countless nose-thumbing occasions, Shaw called Shakespeare's plays vulgar beasts, but he usually fell back finally on praise of their passion and melody. Less pejorative phrases than "bad taste" might be "courageous willfulness" or "reckless self-assurance." Olivier, it would appear, was fully capable of the "bad taste" that Mlle. Mars lacked. The almost constant note in praise of him was his courage. He was called "the bravest actor of our time," and Ralph Richardson called him "very, very bold." Even a critic who disapproved of his Othello was obliged to grant, "There is a kind of bad acting of which only a great actor is capable." And here is Olivier on his own preparation of Othello: "I gave a full-blooded, all-out rendering at the very first rehearsal and from then on I tried out things extravagantly, way over the top. . . . I have done this most of my acting life, grasping the nettle early on and making a fool of myself." He adds, "If you are frightened of making a fool of yourself, you create huge barriers for yourself later on." Pluck lost, all lost, in other words. How else to create all the large Shakespearean roles that require Agate's *fougue*?

The Italian cognate for *fougue* is *foga* . . . "ardor" or "impetuosity." Composers, however, seem to prefer the nearly synonymous word *slancio,* which can be translated as "rush," "vehemence," "zest," or "enthusiasm." *Con slancio* is often found in opera scores, and it is to my mind one of the most exciting and quintessentially Shakespearean instructions that a composer can give to a singer and conductor. Passages marked *con slancio* require very good shriekers. There are, in fact, numerous admonitions in opera scores that could be put in the margins of memorable Shakespearean passages, admonitions that urge the performer to grasp the nettle with complete assurance: *con tutta forza di sentimento* (in *Trovatore*), *animando con molta passione (Traviata), dolcissimo con grande sentimento (Tosca)*, to give just a few examples. More common but in the same vein are *con energia, con anima, con violenza, con brio, con espansione, con molto calore, abbandonandosi, con entusiasmo, tutta forza, grandioso,* and so on.

These invitations to extravagance put the singer in the same position as the actor of a major Shakespearean role. Energetic risk-taking is the ideal in the opera house, too. In this respect one might call Maria Callas the counterpart of Olivier, and Gielgud's thumbnail summary of Olivier's gifts as an actor can serve quite aptly for her as well: "Attention to detail; complete assurance in his conception of character; athleticism; power; and originality." With singers, of course, the dangers lying in wait for the vocal instrument are greater, and thus a more poignant, even melancholy, dimension is added to the performance: there is an intimate relationship between the impetus to self-exposure within most great fictional stage characters and the willingness of the singer to expose his or her voice and reputation before an audience.

The horribly proper Mr. Casaubon in *Middlemarch* cares little for singing: "I never could look on it in the light of recreation to have my ears teased with measured noises."* This recommends him to Mr. Brooke as a husband for his niece Dorothea, who reacts very emotionally to music. He urges Casaubon, in fact, "to teach my niece to take things more quietly." We have seen that this is not the way of Shakespearean or operatic drama, which shows quietism and propriety the door. The urge is rather to take things very noisily indeed. This urge leads to the "big bow-wow strain," as Sir Walter Scott described his own style in contrast to Jane Austen's "exquisite touch." The principal characters in Shakespeare and opera, possessed of great passions in spite of their better judgment, are impressively willful creatures. They tend to display senselessness and sensuality rather than an Austenite sense and sensibility. Lincoln Kirstein has said that ballet is "about how to behave," and I am inclined to say that operatic and Shakespearean protagonists tell us about how not to behave . . . how to react with full emotion—propriety and "good taste" be damned. It is the Whitman style, too: "I must sing / With banner and pennant a-flapping."

And so Shakespeare and opera are loud. *Fougue* and *slancio* make them so. *O* and *Ah* can scarcely be projected by laryngeal half-measures or tight lips. They require lungs. "Almost the first thing a modern actor finds about playing Shakespeare," said the director Margaret Webster, "is that he hasn't enough breath." And breath control is of course the foundation of all vocal training. Great singing is simply impossible without very shrewd control of the column of air that carries the voice. That a similarly vigorous, demanding vocalism was at the heart of Elizabethan theater seems clear. As a leading student of this theater has summarized, "Implied emotion is not characteristic of the period. Today actors hint at unfathomed depths or suppressed drives which are ever on the verge of bursting forth. This was not the style of the Globe. Passions were immediately and directly expressed. A character revealed the full extent of his passions at once. . . . Audacity and vehemency were required."* Audacity of voice, especially. How else than by power of voice can an actor, as Hazlitt urged, give the role of Othello its "general swell and commotion" and make its passions "resemble the heaving of the sea in a storm"? Agate captures an effect familiar in the opera house when he insists that "the big speeches [in *King Lear*] should make you feel that not only the actor, but you, the spectator, are supported on a sea of sound. The part is largely a solo for the voice." Agate is even more "operatic" in his praise for a "really magnificent Shakespearean performance": "Cellier has presence, gesture, and, above all, voice; and every one of his ringing exits brings down the house in the old-fashioned way."

From thence to Leigh Hunt's manner of praising Giuditta Pasta as the heroine in Mayr's opera *Medea* is but a small step: "She never minces what she feels, or goes aside of it, or affects anything as superior to it, or has any doubt about it. . . . Madame Pasta's style is epic. She hits the great points and leaves you to feel the rest." Shaw's confessedly operatic dramaturgy led him to constantly reiterate the centrality of this kind of big bow-wow vocalism. He boasted of having brought "the forgotten heroic business and the exciting or impressive declamation" back into fashion, even though it all seemed to his Victorian leading actresses "to be unladylike barnstorming." He asserted that "there is much more of *Il Trovatore* and *Don Giovanni* in my style than of [William Congreve's] *The Mourning Bride* or [Richard Brinsley Sheridan's] *The School for Scandal*"; and on one occasion he even went so far as to suggest as a rule for casting that the four principals in any play should ideally be viewed as a soprano, alto, tenor, and bass. (*Major Barbara* and *Heartbreak House* can be effectively cast this way.) And on the subject of interpreting Hamlet and his own plays he advised John Barrymore: "There is no time for silences or pauses: the actor must play on the line and not between the lines, and must do nine-tenths of his acting with his voice." Spoken like a true opera composer.

In Shakespeare and opera, as in Shaw, vocalism must reign. Granville-Barker said of the most passionate of passion's slaves, Othello, that he is "built to an heroic scale of expression. . . . The gamut must be run, with no incongruous gap appearing, between the squabble over the handkerchief up to the highest pitch of imaginative emotion." Though not as drastically, most other major Shakespearean roles require that a similarly virtuosic gamut be run. Cleopatra tells us so of Antony, for instance, when she recalls that

> his voice was propertied
> As all the tuned spheres, and that to friends;
> But when he meant to quail and shake the orb,
> He was as rattling thunder.

Heroic scale of expression . . . running the vocal gamut from celestial harmonies to rattling thunder: these are the ways we usually think of the most challenging roles in the opera repertory—Norma, Peter Grimes, Violetta, Isolde, Tosca, Sachs, Falstaff, and many others.

There is a shared image here: not of the poor player strutting and fretting, but of a very rich player—descendant of the first Hamlet, Othello, and Lear, Richard Burbage, striding magnificently and suffering deeply in brilliant voice and at center stage. Then, presumably, the sound and fury signified something rather than nothing.

Perhaps this common dependence on the human voice can be emphasized one final way by returning *da capo*–fashion to Walt Whitman and a poem he titled "Vocalism." It begins,

> Vocalism, measure, concentration, determination,
> and the divine power to speak words;

> Are you full-lung'd and limber-lipp'd from long trial?
> from vigorous practice? from physique?

Not once in this poem's three dozen lines is acting or singing specifically mentioned. Yet, to my mind, these lines precisely express the rigors and rarity of great utterance on both the legitimate and the lyric stage:

> After complete faith, after clarifyings, elevations,
> and removing obstructions,
> After these and more, it is just possible there comes
> to a man, or woman, the divine power to speak words;
> Then toward that man or that woman swiftly hasten all—
> none refuse, all attend. . . .

In the same way, both of our theatrical genres await their moments of fulfillment—and the great performers who can deliver them. When they make their infrequent appearances, we "swiftly hasten":

> All waits for the right voices;
> Where is the practis'd and perfect organ? where is
> the develop'd soul?
> For I see every word utter'd thence has deeper,
> sweeter, new sounds, impossible on less terms.

When this occurs—and in my experience it does so very seldom—not a little *ecphonesis* will erupt from a grateful audience.

Dramaturgy

6

Passion Play

In February 1918, H. L. Mencken graced the pages of the *New York Evening Mail* with a knock-down assault on opera, written in his best pugilistic prose. Several of his jabs are below the belt, . . . and yet not much different from aspersions one hears about opera fairly regularly. He chides the "posse of fat madames throwing themselves about the stage" in *Die Walküre* and the coloratura soprano "who can gargle her way up to F-sharp *in alt*." The person "who actually delights in such spectacles," he avers, "is the sort of person who delights in gas-pipe furniture."* But one point he makes, however dyspeptically, contains a kernel of truth: "To write a successful opera a knowledge of harmony and counterpoint is not enough; one must also be a sort of Barnum. All the first-rate musicians who have triumphed in the opera house have been skillful mountebanks as well." Who would seriously deny that the Barnum elements of exploiting exaggerated humanity and shrewdly manipulating the audience's responses to it are essential in opera?

They are, of course, also essential in Shakespeare. Mencken asserts that the success of an opera performance is "in proportion as the display of notorious characters on the stage is copious," and we have already observed that the playwright often succeeds by copious means as well. Those who use the word *copious* pejoratively are bound to be uncomfortable at the opera or at a Shakespearean play: the *raison d'être* for both dramaturgies is the copious display of passion. Thus, Nicholas Rowe, the first editor of the plays, spoke in 1709 of "that Fire, Impetuosity, and even beautiful Extravagance which we admire in Shakespeare." This phrase could be applied to most of opera's eminent protagonists. . . . Debussy's Mélisande always excepted. Outcasts for whom such characters hold no imaginative appeal will, like Mencken, call them notorious; they will also call Barnumlike and mountebankish the manipulations by which their creators prepare the dramatic situations that make "beautiful Extravagance" possible. Mencken displayed his true colors as a slave of probity elsewhere, when he made a comparison very favorable to a composer who (we can be very thankful) never wrote an opera: "Brahms kept his belly-aches to himself. Tchaikowsky never failed to put them into his music." No wonder opera left Mencken cold: in opera, as in Shakespeare, belly-aches will out.

Hamlet gives us several hints that passionate extroversion—belly-aches—were the supreme element of Elizabethan drama. The traveling players arrive and he is soon egging them on, "Come give us a taste of your quality, come, a passionate speech." Alone, he speaks of the theater as offering a "dream of passion." He also implies that, in the kind of drama he has become devoted to, the crux of plotting is to find "the motive and the cue for passion." And in his subsequent admonitions before the performance of "The Mouse-trap," he describes perfectly what has been and always will be the supreme challenge of acting the Elizabethan repertory. This is to remain in artistic control of the huge emotions that the blank verse gives access to: "[I]n the very torrent, tempest, and, as I may say, whirlwind of your passion, you must acquire and beget a temperance that may give it smoothness." The typical Elizabethan play ends when all passion is spent . . . be it the "passion of loud laughter" in *A Midsummer Night's Dream* or the passion of "heartache" and the "thousand natural shocks" that flesh is heir to. The Athenian rustics get their dramaturgy right, at least in principle, when they see that Thisby's suicidal "passion ends the play," just as Shakespeare arranged for both Gloucester and Lear to die "['t]wixt two extremes of passion, joy and grief." Indeed, with Shakespeare's passion-riven protagonists there always hangs in the air at the end (spoken by an observer on stage or unspoken in the audience) Lodovico's awed question about the noble Moor: "Is this the nature / Whom passion could not shake?"

Toward the end of *The Tempest,* Ariel says that, were he human, the "men of sin" in Prospero's control would be shown mercy. Prospero responds by asserting his own humanity:

> shall not myself,
> One of their kind, that relish all as sharply
> Passion as they, be kindlier mov'd than thou art?

Relishing passion: this is the fundamental posture of Shakespeare's principals; just so is it the "butt / And very sea-mark" of his dramatic style. "It is mainly through the depiction of passions," Bernard Beckerman concludes about the fifteen plays written for the Globe Theater, "that Shakespeare individualizes his characters." The heart of this dramaturgy, Beckerman adds, is "overwhelming passion intensively portrayed." Thus, what the second-act Chorus tells us of Romeo and Juliet—that "passion lends them power"—could be the motto for all the playwright's other heroes, heroines, and villains. His audience, Granville-Barker imagines, was born to this manner: "How passionate the Elizabethans were! They were capable of intellectual passion, as Englishmen have hardly been since. And when poetry and rhetoric display it in the charged atmosphere of the theater, the effect—even the distant echo of it—is intense." Finding the motive and cue for passion and arranging for the plot-crises that might detonate poetical-rhetorical display, then, were surely in the forefront of Shakespeare's mind when he set about writing a new play.

Librettists and composers, eager for moments that elicit musical-rhetorical display, have traditionally approached their respective tasks in this frame of mind. The dramaturgies of Shakespeare and opera are tied to excessive passion with, in Iago's phrase, "cables of perdurable toughness." In both genres the manipulations that produce passionate outbursts are often as boldly and baldly calculating as Iago's. Consider, in this regard, an illuminating and very famous scene from *Julius Caesar*. It displays as brilliantly as one could wish the Shakespearean and operatic technique of arousing passion to spectacular effect. Indeed, I will venture that the successive "performances" of Brutus and Antony before the Roman plebeians after Caesar's assassination can be viewed as a parable of the stylistic paths to failure and to triumph on the Shakespearean and operatic stage.

First, consider Brutus. He is Shakespeare's most earnest and sober tragic hero, the only one in all the plays who is never given lines expressing impassioned rage. Early in the action he admits with his usual directness, "I am not gamesome; I do lack some part / Of that quick spirit that is in Antony," and his oration, unfortunately, is perfectly in keeping with such a personality. It is the speech of a man who plays no rhetorical games. "Romans, countrymen, and lovers, hear me for my cause, and be silent, that you may hear," he begins severely, and then he appeals coolly to reason: "Censure me in your wisdom, and awake your senses, that you may the better judge." Blood scarcely seems to flow as he arrives at his laconic, logical climax: "As Caesar lov'd me, I weep for him; as he was fortunate, I rejoice at it; as he was valiant, I honor him; but, as he was ambitious, I slew him." There are no poetical frills here, and Shakespeare mutes the oration's effect even more by casting it in prose rather than verse. Only as Brutus ends does he permit himself a flourish, and even this is wryly ironic: "as I slew my best lover for the good of Rome, I have the same dagger for myself, when it shall please my country to need my death." This elicits a unanimous "live, Brutus, live, live!" Brutus's speech, in sum, would seem suitable for musical realization as, at most, a lightly accompanied recitative.

Comes then Antony a few moments later, with Caesar's body, and the most notorious bit of upstaging in all theater (almost in the same league is Walther following Beckmesser in the prize-song competition of *Die Meistersinger*). This *tour de force* of persuasion, which runs 135 lines compared with Brutus's 35, is too well known to need rehearsing here. But it is worth pausing over as a prime example of the methods of the Elizabethan and operatic dramatist stalking high passions.

Shakespeare took his cue for Antony's oration from Plutarch: "Antonius . . . mingled his oration with lamentable wordes, and by amplifying of matters did greatly move their harts and affections unto pitie and compassion." "Amplifying" is but another way of saying that he elaborated his speech by the rhetorical methods and devices we explored in Chapter 4. And Antony does indeed deploy various auricular figures—figures of reiteration and *gradatio* (sequential augmentation), and well-chosen epithets—quite brilliantly . . . most famously in the slow, drippingly sarcastic metamorphosis of

> For Brutus is an honorable man . . .
> And sure he is an honorable man . . .
> Who (you all know) are honorable men . . .

into his climactic, "I wrong the honorable men / Whose daggers have stabb'd Caesar." As the temperature of his auditors slowly rises, Antony controls his rhetorical crescendo with a temporary access of emotion (and the telltale *ecphonesis*)—"O judgment! thou art fled to brutish beasts"—after which he lapses ostentatiously into silence. A plebeian observes, "his eyes are red as fire with weeping." Are these tears genuine? one suspiciously begins to wonder.

Also stagey is Antony's feigned reluctance to read Caesar's will; but the real masterstroke is his display of the blood-soaked mantle: "See what a rent the envious Casca made. . . ." With increasingly florid appeals to the imagination he rises quickly to what would be, if this were an operatic *scena,* an orchestral *tutti* introduction and a *cabaletta* (marked *con slancio e passione*):

> O, what a fall was there, my countrymen!
> Then I, and you, and all of us fell down,
> Whilst bloody treason flourish'd over us.
> O now you weep, and I perceive you feel
> The dint of pity.

The response from his audience is precisely as he intends: it reels in passionate empathy . . . and in *ecphonesis:*

> *1. Pleb.* O piteous spectacle!
> *2. Pleb.* O noble Caesar!
> *3. Pleb.* O woeful day!
> *4. Pleb.* O traitors, villains!
> *1. Pleb.* O most bloody sight!

After this the throng is in Antony's palm, and he plants the seed of action with almost arrogant irony: "let me not stir you up / To such a sudden flood of mutiny." Alone after the plebeians finally leave in high dudgeon, the master manipulator says with satisfaction, "Now let it work. Mischief, thou art afoot, / Take thou what course thou wilt!"

There is something peculiarly Shakespearean and operatic in the way Antony carries away his audience. Mencken was right: a touch of the mountebank, the wily manipulator of the motives, cues, and props of passion, is necessary. Cassius, early in the play, calls Antony a "shrewd contriver," and most memorable scenes in both our dramaturgies—however naturally they may seem to unfold in performance—are the product of shrewd, complex contrivance. When there is genius in this contrivance, we may even find ourselves fully engaged although we are semiconsciously aware that the playwright or composer is achieving his victories through legerdemain. In the Elizabethan theater Antony's power to make passion irresistible elicited the highest praise. What was said of the Beaumont and Fletcher first folio (1647) applies even more emphatically to

Shakespeare's Folio: "You may here find passions raised to that excellent pitch and by such insinuating degrees that you shall not chuse but consent, & go along with them." In worthy performances of the finest operas, one is similarly and happily deprived of choice.

Toward the end of his oration Antony confesses, "I have neither wit, nor words, nor worth, / Action, nor utterance, nor the power of speech / To stir men's blood; I only speak right on." This is daring and utterly coy, for he has just shown himself a virtuoso in each of these respects! It is Brutus who unwisely and ineffectively speaks "right on." All these means that help Antony to stir plebeian blood are shared by the composer, whose primary function—like Antony's in the Forum—is to stop the action and "amplify." It is not to the composer's purpose to speak "right on," but rather to raise his characters' passions to a high pitch and draw his audience along with them. Everyone will grant that the Forum scene of *Julius Caesar* is a typically Shakespearean success, but it is also a fitting image of success on the opera stage.

Passions in Elizabethan drama and opera, to summarize, are typically raised to an "excellent pitch" by means of rhetorical amplification. But not only is the focus *on* passions keen; these passions are themselves conceived as highly focused, discrete emanations of psychic energy. This is because passions or (to use a word not in Shakespeare's vocabulary) emotions were viewed according to the simplistic notions of human psychology that held sway during the Renaissance and through the eighteenth century. That is, they were conceived of as unitary, definite in character, and separable in the mind: joy, fear, grief, wrath, and so on. The main principle behind Elizabethan drama was that reason is the governing faculty in the mind, keeping the various passions in control. "We have reason to cool our raging motions," says Iago, and later Othello begins to anger over the night brawling that has disturbed Cyprus and invokes this same psychology with his usual poetic flare:

> Now by heaven,
> My blood begins my safer guides to rule,
> And passion, having my best judgment collied,
> Assays to lead the way.

Othello's wise self-analysis is ironic, though . . . a terrible premonition for us of how he will soon be allowing Iago to overwhelm his "safer guides." According to Elizabethan psychology, Beckerman observed in his study of Shakespeare's acting style, "an individual of extraordinary will could suppress [the] signs" of passion overwhelming reason, "but the vast majority of people was helpless to hide the play of passion within their souls." To this dramatically potent majority of human beings and to their visible play of passion Shakespeare's drama was deeply devoted, and it is largely through the successive bursting forth of a series of discrete and intense passions during an action that a Shakespearean character develops.

If the Stanislavskian or "Method" style of acting urges a seamlessly integrated, all-embracing consistency of effect on stage, then Elizabethan psychology urged a more disjointed, focused, and explosive manner. This, in turn, led to an artificial, self-conscious performance style that became more and more pronounced in the seventeenth and eighteenth centuries. Here is a summary of the style in the age of David Garrick (1717–1779): "The spectator's enjoyment came from viewing the 'passions' going through their paces, or being exquisitely delineated in a series of isolated actions and declamations, rather than from enjoying the development of a 'whole' character. . . . Set speeches were helpful and soliloquies seemed made for exhibiting virtuosity. The star performer could declaim both and rivet attention to his individual performance. Supreme art showed in doing it well." This, obviously, also characterizes the typical contour of an operatic performance. Indeed, our Garrick scholar adds this conclusion after the above sketch: "A parallel comes to mind in our present-day experience with grand opera." To make his point, he recalls the approval of an audience for the way Montserrat Caballé performed an aria in Verdi's *Vêpres siciliennes:* "approbation—not for the plot, not for the characterization, but for the quality of her performing the aria."

Garrick's performance practice is intriguingly echoed in the style of Kabuki theater, which sprang to life in Japan at precisely the time Shakespeare was flourishing in England. One student of Kabuki style writes, "a characteristic quality of all Kabuki acting is that it is not concerned with depicting subtle and fluid transition from one emotion to another; instead, the actor tends to present the character in a succession of unrelated, detached moments." The Elizabethan theater's artificial set speeches and soliloquies also come to mind as the climactic, nonrealistic "attitude" of Kabuki—called the *mie*—is described: "Kabuki's most significant moments are . . . not realized in movement but in the achievement of a static attitude." For example, the *mie* of the assassination of Caesar would occur at *"Et tu, Brute?"* The Kabuki audience "comes to see a succession of striking images" and "applauds at exactly the point it is most pleased; and these are, in general, passages which show brilliant technique and execution." One can easily begin thinking of the *mie* of impotent old age in Lear's "I am a very foolish fond old man"; the *mie* of jealous rage at Othello's "All my fond love thus do I blow to heaven"; the *mie* of renunciation at Prospero's "Ye elves of hills" speech; or the *mie* of humiliation in Wolsey's glorious "long farewell to all my greatness" soliloquy in *Henry VIII.*

The assumption that passions are almost palpable phenomena (typified, for example, by Gloucester stretched on a metaphorical rack between "two extremes, of joy and grief") encouraged Shakespeare and his contemporaries to create a dramaturgy of sudden, overwhelming responses to crisis, vivid set pieces momentarily arresting action, and the more-or-less autonomous scene that either is a set piece itself or sets one off like the foil for a jewel. This dramaturgy naturally produced very systematic theories of drama and acting, especially in the eighteenth century. Aaron Hill, a playwright and the librettist for Handel's first London opera, was pleased to reduce "the dramatic passions" to ten: joy, grief, fear, anger, pity, scorn, hatred, jealousy, wonder, and love. In his 1755 treatise, *Letters Concerning Taste,* J. G. Cooper described the effect of good theater on its audience almost exactly as Othello described the effect of passion on his judgment. The theater, Cooper wrote, stimulates a "good *Taste* [that] thrills through our whole Frame and seizes upon the Applause of the Heart, before the intellectual Power, Reason, can descend from the Throne of the Mind to ratify its Approbation." In less stately words, the theater makes us passion's slaves.

This mechanistic view of the passions produced similar results in music theory, especially in Germany with the *Affektenlehre,* or doctrine of the passions. To take but one English expression of this doctrine, Charles Avison postulated in his *Essay on Musical Expression* (1752) three main compositional modes and their respective sub-modes: the Grand (the Sublime, Joyous, Learned); the Beautiful (the Cheerful, Serene, Pastoral); and the Pathetic (the Devout, Plaintive, Sorrowful). And Avison sounds much like the writer who praised the Beaumont and Fletcher folio when he praises the power of music to "raise a Variety of Passions in the human Breast." "By the Musician's Art," he adds, "we are often carried into the Fury of a Battle, or a Tempest, we are by turns elated with Joy, or sunk in pleasing Sorrow, roused to Courage, or quelled by grateful Terrors, melted into Pity, Tenderness, and Love, or transported to the Regions of Bliss." This pigeonholing of the passions may seem ludicrous to us now. Still, one must acknowledge that a startling number of the most memorable Shakespearean set pieces are embraced by Hill's ten dramatic passions, and a large number of the opera repertory's most memorable arias and ensembles could find a fairly comfortable place under one of Avison's nine varieties of musical passion. We might also acknowledge that these old-fashioned ways of focusing emotional energy onstage reap the considerable rewards of impressiveness and intensity of impact, even while they give up something in sophistication and fidelity to so-called real emotional life.

One more important similarity in the Shakespearean and operatic play of passion requires mention: the passions in question are by and large from common human experience. Exotic though the personages and localities of so many of the plays and operas are, the focal emotions, ideals, and values that are celebrated or damned as action unfolds are virtually always accessible from general human experience. I wrote several years ago that Handel's strength as a composer "lay in the main line of human experience—filial devotion, jealousy, melancholy depression, dalliance, connubial bliss, courage," and so on. But the statement applies just as emphatically to Shakespeare (surely this was Verdi's point in exclaiming, "Ah Shaespeare, Shaespeare! Il gran maestro del cuore humano!") as it does to Mozart, Rossini, Donizetti, Wagner, Puccini, Strauss, and even Alban Berg, as well as to Verdi himself. Isaiah Berlin has written especially eloquently of Verdi as "the last master to paint with positive, clear, primary colors, to give direct expression to the eternal, major human emotions: love and hate, jealousy and fear, indignation and passion; grief, fury, mockery, cruelty, irony, fanaticism, faith, the passions that all men know." And he ends his encomium by explaining that the reason for Verdi's popularity lies in his expression of "permanent states of consciousness in the most direct terms, as Homer, Shakespeare, Ibsen, and Tolstoy have done."

Alessandro Manzoni, an idol of Verdi's and the honoree of his *Requiem,* wrote that "Shakespeare stands above all other artists because he is the most moral." This remark helps underscore the fact that the passions explored in Shakespeare's plays and most of the central operas in the repertory are generated by the commonplace moralities—and immoralities—of human behavior. Though Shakespeare was by no means a moral*ist*, a powerfully orthodox conception of the virtues and vices in human relations—of good and evil, in other words—underlies all his plays. Alfred Harbage has spoken of "the intensity and complexity of the moral content of Shakespeare's plays." He finds that "moral gleams play over the surface and under the surface of Shakespeare's words," and views his characters as "foci of a quickened moral interest . . . [who] are vouchsafed no ethical rest periods." Though Shakespeare's response to moral dilemmas may be complex, the values that govern them are from the mainstream. Malcolm, in *Macbeth,* neatly catalogues several of the most important ones when he speaks of

> The king-becoming graces,
> As justice, verity, temp'rance, stableness,
> Bounty, perseverance, mercy, lowliness [humility],
> Devotion, patience, courage, fortitude . . .

But of course these are demonstrated throughout the canon to be graces not merely in kings but in all human beings . . . just as their disgraceful perversion generates the full palette of human degradation and villainy in Shakespeare's *dramatis personae*.

Malcolm's humanist "graces" are central to countless Shakespearean and operatic actions; what makes them so effective time and again is simply that they are writ so large. Shaw made this point when he assessed Verdi's career shortly after his death. His genius, Shaw said, "like Victor Hugo's, was hyperbolical and grandiose: he expressed all the common passions with an impetuosity and intensity which produced an effect of sublimity." Shaw might have likened the method to Shakespeare's own, but the reference to Hugo is, I think, very germane. For Hugo was one of the first clarion voices of theatrical Romanticism, and his *Hernani* of 1830 served to announce a return to the robust, explosive dramatic style of Shakespeare—whose plays were being "rediscovered" all across the Continent in the 1820s and 1830s. Théophile Gautier described this ferment: "A movement analogous to that of the Renaissance was taking place. A sap of new life was running hotly; everything was germinating . . . intoxicating scents filled the air." Gautier called one of his contemporaries, Joseph Bouchardy, "the Shakespeare of the Boulevard" and praised him, in fact, in terms that perfectly fit Shakespeare's own plays, not to mention most nineteenth-century operas: "He loved to present the great and beautiful commonplaces which form the very basis of man's soul, and which bring an evil smile to the lips of the skeptic: paternal love, fidelity, chivalric loyalty, the defense of honor, and all the noble motives that can determine action."

Primal moral impetus—whether the great love or *agape* that overmasters self-interest in Desdemona, the forgiveness dispensed at last by Prospero, or the ruthless ambition of Macbeth—generates the primal passions of the typical Shakespearean and operatic protagonist. Take "honor," for example. The Folio, one might say, contains a huge, far-from-comic discourse on Falstaff's question

"What is in that word honor?" The word occurs upwards of a thousand times in Shakespeare. Many a character is of Brutus' mind: "I love / The name of honor more than I fear death." Some love it too much, like Hotspur . . . all too eager to "pluck up drowned honor by the locks." "Deed-achieving honor" is the motor in many plays, and some heroes are able, others unable, to show "the high sparks of honor." Unhappy ones find their "honor is at pawn." Honor is likewise a nearly constant theme in Verdi. "There is hardly a Verdi libretto in which honor, in some form, does not play a catalytic part," writes William Weaver. "Honor can be male—military, patriotic, or even racial (cf. *La Forza del Destino*)—or female and sexual (cf. *Stiffelio, Un Ballo in Maschera*). Most often it is linked with patriotism, a key Verdian motive as late as *Aida*, and often with love, with family affection." It is no small irony that Verdi crowned his career by setting Falstaff's cynical catechism on honor to music, for the preceding canon often and powerfully demonstrates that, *vecchio John* to the contrary, there was to Verdi's mind very much in that word "honor."

Many years after the premier of Hugo's *Hernani*, Gautier reminisced: "It would be a difficult task to describe the effect produced upon the audience by the striking, virile, vigorous verse, that had so strange a ring, and a swing that recalled at once both Corneille and Shakespeare." Gautier's epithets for the Shakespearean rhetorical style are very apt, but they could also describe the bold ways the playwright exploited the fundamental moral principles of Renaissance humanism. Something more than the grand style was rediscovered by the Romantics. With Wolfgang von Goethe and Friedrich von Schiller notably leading the way, the theater, and especially opera, was poised to leave behind the Voltairean smile of skeptical reason (and Voltaire's disgust with Shakespeare) for more extravagant utterance and the more salient, loftier ideals of liberal humanism. Shakespeare was a chief inspiration for those like Hugo, seeking to escape from the previous century's neoclassicism and its tyranny of rules, reticence, and rationalism. They wanted, as their heroes and heroines, slaves of passion. This renewed and now fully international heyday for Shakespeare came, as Auden has noted, just as opera was rising to hegemony among the performing arts: "The golden age of opera, from Mozart to Verdi, coincided with the golden age of liberal humanism, of unquestioning belief in freedom and progress." This nineteenth-century humanism, of course, derived in part from the humanism of Shakespeare's canon.

Not merely brilliance of poetic expression or cleverness of construction, then, can explain the potency of Shakespearean drama. It is also due in part to the intrinsic simplicity, nobility, and idealism of their concepts that give his characters their motives and cues for passion. There is, in other words, an elevation of spirit as well as of expressive power. This passionate idealization of the best and (in the villains) worst humanity is capable of surely goes far to explain Shakespeare's long continuance in the position of Top Bard. It may also explain this praise of Shakespeare from Walt Whitman, with its typical allusion to vocalism: "Only he holds the proud distinction . . . of being the loftiest of the singers life has yet given voice to."

This spirited grandiloquence is of the operatic essence, too, as Whitman also suggested when, in a poem called "The Dead Tenor," he honored an artist who gave him "Ferrando's heart, Manrico's passionate call, Ernani's, sweet Gennaro's . . . Freedom's and Love's unloos'd cantabile." Another Whitman poem, a short one, finely suggests how often the opera singer, like the Shakespearean protagonist, is called upon to celebrate the great and beautiful commonplaces. The poem leaves one thinking especially of works like *Le Nozze di Figaro, Fidelio, Don Carlo, Die Meistersinger, Wozzeck,* and *Peter Grimes:*

To a Certain Cantatrice

Here, take this gift,
I was reserving it for some hero, speaker, or general,
One who should serve the good old cause, the great idea, the progress
  and freedom of the race,
Some brave confronter of despots, some daring rebel;
But I see that what I was reserving belongs to you as much as to any.

7

Simple Savor

In one of the last sonnets addressed to the aristocratic young man, Shakespeare's speaker wryly tells of the "pitiful thrivers" he has seen lose everything by preferring "compound sweet" to "simple savor." This was the poet's culinary way of contrasting the sugary etiquette of courtly politics with simple, nourishing friendship. Earlier in the sequence this same speaker compared his own "true plain words" favorably with the "strained touches" of other poets' "rhetoric." We have paid much attention to the "compound sweet"—the ornate composition— of Shakespearean and operatic rhetoric, and we shall return to this rich style again. This is a good point, however, to take a short breather from excessive utterance. Too much of it, after all, can leave one thinking as Troilus does: "Words, words, mere words, no matter from the heart." Too many of them, and one begins to thirst for simple, plain words . . . a relaxing verbal *adagio* with no fancy harmonics or counterpoint whatsoever. One perhaps even begins to thirst for a little silence. Shakespeare well knew that, amid high-flown expressions full of sound and fury, simplicity could be a precious and useful commodity. Some of his most poignant moments of "simple savor"—when the palate is cleared at the feast of language—deserve a pause for contemplation here, for they remind us that in opera as well there are occasions when less is considerably more.

"Verdi, when he is simple," wrote Shaw, "is powerfully so," and this is also a Shakespearean hallmark. The drama of *King Lear* explodes into motion when, after the two fulsome rhetorical displays of Goneril and Regan, Cordelia refuses to heave her heart into her mouth and says only, "Nothing, my lord." Cordelia's whole role, one might add, is a triumph of simple savor. It is a mark of the playwright's power to be simple that she makes such a potent impression in only about a hundred lines (three percent of the play's total). The ultimate in simplicity of expression is mere repetition, and several times in the final scenes of *Lear* it is memorably used by Cordelia. Lear asks if she is his daughter, and she replies: "And so I am, I am." Lear allows she has good cause not to love him, and she says: "No cause, no cause." Lear's own repetitions in his last scene—notably the famous five "never"'s—reflect what a great fall he has suffered from the royal rhetorical heights of the abdication scene: "Howl, howl, howl! . . . No, no, no life! . . . Look on her! Look, her lips, / Look there, look there!" One is tempted to see a stylistic moral here: the King has become a "pitiful thriver" and lost all in this terrible world where speaking plainly what we feel, not what we ought to say, can lead to calamity.

Conversely, the high-flown style can make one very vulnerable. Plainness and reticence prove effective for Bolingbrook, who forces Richard II from his throne. Bolingbrook shrewdly allows Richard, self-imprisoned in a world of fancy rhetoric, to fall from the weight of his own ornate-style folly. Henry's son, Prince Hal, can also be devastatingly simple in expression: after Falstaff exhausts himself in the tavern with his "banish plump Jack, and banish all the world" oration, Hal responds with withering curtness, "I do, I will." The canon is rich in such juxtaposition of the ornate style's tendency to balloon and the plain style's power to deflate. When human illusions take rhetorical flight in Shakespeare, there is often a character nearby, like Viola in *Twelfth Night,* to observe, "They that dally nicely with words may quickly make them wanton."

Shakespearean plays are generally very heightened in expression, and this is perhaps why most of them end—with all tragic or comic passion spent—in a reversion to some form of plain speaking and style. These conclusions ease our return to verbal normality and the "real" world, thus avoiding a case of the aesthetic "bends" that might occur otherwise. Shakespeare's epilogues, like the sextet of *Don Giovanni* and the few lines spoken by Puccini's Gianni Schicchi, are the most obvious such devices; others are more subtle. The playwright very sensibly saw, for instance, that his audience would be full from the "feast of languages" in *Love's Labour's Lost,* and so he ends the play in his most rustic plain style—with songs by Spring ("When daisies pied, and violets blue") and Winter ("When icicles hang by the wall"). Similarly, *Twelfth Night*'s complex exploration of love's labyrinth ends with the exquisitely simple savor of the Clown's song, "When that I was and a little tiny boy." In *King Lear,* Shakespeare extracts his audience from the final pathos not only by Edgar's plain speaking but also by reverting at the last to the sing-song of four pairs of rhymed couplets.

The heightened expression of opera, too, has encouraged certain endings that triumph by their simple savor. The holy fool who laments the fate of Russia at the end of Mussorgsky's second version of *Boris Godunov* provides a touchingly plain coda to all the preceding musical and historical panoply. "The holy fool, alone at the end, literally clears the stage of falseness," writes a student of the opera. The mundane pantomime with the blackamoor in *Der Rosenkavalier* releases us from the pathos of the Marschallin's farewell to youth and love just as effectively as the epilogue of *The Tempest* releases us from the pathos of Prospero's farewell to his art and to Ariel. In tragic obverse is the plain savor of the nursery-tune hop-hop of the little boy who ushers us out of the oppressive tragedy of *Wozzeck.* In opera the plainest utterance is simply to speak, and Benjamin Britten, in his *Dream* opera, faithfully arranged for Puck to speak the epilogue over rhythmless sustained chords.

We cannot afford to be prejudiced either against "compound sweet" in

Shakespeare and opera (the rousing set speech or the *gran scena ed aria*) or against "simple savor" (the trenchant deadpan retort or the economical *arioso*). They are necessary parts of the feast; genius lies in serving them appropriately. It is a matter of learning the place for the chest-thumping expansiveness of, say, "Not marble nor the gilded monuments / Of princes shall outlive this pow'rful rhyme" (Sonnet 55), as well as a place for the plain, heart-rending admission, "My deepest sense, how hard true sorrow hits" (Sonnet 120). This was a dramatic skill that Shakespeare acquired only gradually. It took perhaps a dozen plays for him to hit his characteristic stride in *A Midsummer Night's Dream* and *The Merchant of Venice*. With the mature tragedies this skill is brilliantly displayed. The spectacular stylistic juxtapositions thus produced demand great skill in the actor—as when, for instance, Othello must erupt into the ornate fury of

> Blow me about in winds! roast me in sulphur!
> Wash me in steep-down gulfs of liquid fire!

only to subside into the utterly plain style of

> O Desdemon! dead, Desdemon! dead!
> O, O!*

Macbeth's role, to take another example, embraces the foot-long diction of "multitudinous seas incarnadine" as well as the eerily flat monosyllables of his later "She should have died hereafter; / There would have been a time for such a word." And perhaps nowhere in the canon are the two antipodal styles more potently deployed than in *Antony and Cleopatra*. The savor of Antony's first encounter with Cleopatra after the defeat at Actium is devastatingly simple:

> *Cleopatra:* Let me sit down. O Juno!
> *Antony:* No, no, no, no, no.
> *Eros:* See you here, sir!
> *Antony:* O fie, fie, fie! (3.11.28–31)

Temporarily ebullient later on, Antony returns to the superbly strained touches of rhetoric:

> O thou day o' th' world,
> Chain mine arm'd neck, leap thou, attire and all,
> Through proof of harness to my heart, and there
> Ride on the pants triumphing! (4.8.13–16)

Verdi, the most Shakespearean of all opera composers, also came gradually to his dexterity in juxtaposing ornate and plain modes of expression. Many have suggested, plausibly, that it was only with his tenth opera, *Macbeth,* that he came to view his creations as *opere ad intenzioni,* that is, operas formed according to dramatic rather than musical convention. With *Macbeth,* in other words, he began to cease thinking of operas as aural necklaces of cavatinas, duets, ensembles, and choruses strung one after another. In pursuit of these more pristinely theatrical intentions, Verdi became an often very urgent partisan of simple savor. He opened his *Macbeth* project with his librettist, Francesco Maria Piave, with the exhortation, "Brevity and sublimity"; but three weeks later he had to complain "how wordy you always are!!" and warn in capital letters, "ALWAYS BEAR IN MIND: USE FEW WORDS . . . FEW WORDS . . . FEW, FEW BUT SIGNIFICANT." The usually prominent tenor role became, in *Macbeth,* a lone aria because Verdi was properly convinced that the play offered only three significant roles: Macbeth, his wife, and (collectively) the witches. (Baritone Felice Varesi, the Macbeth, seemed very pleased to relate in a letter that "[t]he tenor has a very small part.") The short (but still un-Shakespearean) death scene for Macbeth also ignored operatic tradition: Verdi wrote to Varesi, explaining it would be "a very short death scene—but it won't be one of those usual death scenes, oversweet, etc. You'll well understand that Macbeth shouldn't die as Edgardo [in *Lucia*] and *his like* do." When Verdi revised *Macbeth* for Paris, he fought (vainly, it turned out) to avoid having the tenor's role beefed up by giving him part of Lady Macbeth's drinking song at the banquet in act 3. He also took the opportunity to remove the Garrick-inspired death scene of Macbeth entirely.

Verdi's next Shakespearean project was a *Re Lear*. Though he appears never to have begun composing, he worked earnestly on a libretto in the 1850s. Several draft versions of the libretto and many letters to and from his librettist, Antonio Somma, are fascinating because they show Verdi often urging simplifying solutions to musico-dramatic dilemmas.* After the banishment of Delia (Cordelia) in the first scene, for instance, Somma originally gave quatrains to each character on stage, obviously with a big, concerted ensemble in mind. Verdi reacted to them by saying he could only make "one of the usual strettos that provide a show of chords and instruments, which I don't care for at all." He suggested instead just two lines for each soloist, then Lear's brisk departure, followed by a *scena* for Delia. He carefully schematizes his idea in his letter, noting in the margin that he wants "not an ensemble . . . but a brief musical passage [*squarcio*] that will in no way interrupt the action."

Having very much his own opinion about Shakespeare's characters, Verdi was also quick to reject Somma's lines that seemed to derive from conventional operatic forms. For Edmund's "Thou, Nature, art my goddess" speech, Somma concocted several stock clichés of operatic villainy and poured them into the usual slow aria/fast aria mould. Verdi wrote back, demanding something with "a variety of colors: irony, contempt, wrath," and he added (making explicit his distaste for "compound sweet"): "Try to avoid an aria that is a decoration." I have already remarked on the simple savor of Cordelia's role, which Verdi recognized very well. Unfortunately, his librettist and the energetic nature of most nineteenth-century opera heroines urged upon him a more assertive Delia than Shakespeare created. Verdi bridled at those passages in the libretto, though, that positioned the soprano for pyrotechnics. In act 3, for instance, Somma dressed Delia—as leader of the French army—in military garb. Verdi never cared for the touch and wrote, "It strikes me that Delia's character, so sweet, so angelic, loses by making her a warrior and dressing her as an amazon. Like this, she takes on slightly too harsh and severe an aspect; we have too many willful and strong characters in this drama as it is." Verdi also pursued this faithfully Shakespearean opinion when it came time to think of a singer for the role. He rejected a soprano who had been a great success as the spirited heroines of *Ernani, I Lombardi,* and *Giovanna d'Arco,* and offered three new names: "all three have weak voices but great talent, spirit, and feeling for the stage." Verdi wanted simple savor, not a prima donna.

The *Re Lear* correspondence also shows Verdi looking for moments when the aural equivalents of simple savor might afford the audience some respite from the tremendous passions on stage. He worries about the panoply contemplated for act 2: "As it is, the whole act would be too full, that is to say, in too many places the eye would have no chance to rest." The result: "monotony." Verdi also saw potential danger in Lear's characterization in the first two acts. He feared that Somma had, though following Shakespeare, "almost always depicted him in bold colors," and he feared that Lear might "turn out monotonous in music." Two years later, Verdi was urging, "Try, try to mitigate in this second act the too consistently powerful phrases of Lear." He suggests that Lear might utter an "unpremeditated and confused memory of Delia" suitable for *mezza voce,* and points out another moment at which he might utter "broken, interrupted phrases" showing his loss of strength. The quest for simplicity also led to the eventual excision of the entire Gloucester subplot (the libretto is "100 percent improved by removing the two Gloucesters," Verdi wrote) and to the one large and unresolved impasse of the project: how to manage all the plot that unfolds late in the play when Lear is offstage for 500 lines. Somma made a last heroic effort to turn this into something playable, and Verdi responded with his usual candor in a way that perhaps sealed the opera's doom: "I am not certain whether the fourth act of *Re Lear* goes well in the version you recently sent me. What is certain, though, is that it will be impossible to make the audience swallow so much uninterrupted recitative, especially in a fourth act." Thinking, no doubt, of his tired audience, Verdi also felt the already simple savor of Somma's forty-six-word denouement after Delia dies could be improved upon: he produced a new one with only twenty-seven words!

Verdi's Shakespearean wisdom of sometimes seeking for powerful effects through stringent artistic economies is perhaps most emphatically expressed in his letters breaking in his librettist for *Aida,* the work that ushered in *Otello* and *Falstaff.* In the Nile scene, for example, Ghislanzoni composed lines for a lively *cabaletta* for Aïda in which she enthusiastically agrees to deceive Radames—this, after her father has harrowed her with the most terrible verbal abuse. Verdi responded, "In the duet between father and daughter [a *cabaletta*] seems especially out of place. Aïda is in such a state of fear and moral depression that she cannot and must not sing a *cabaletta.*" Verdi proposed instead that Aïda, mentally exhausted, only utter disjointed phrases: "I would abandon forms of stanzas and rhythm; I would not think of having them sing and would render the situation as it is, were it even in verses of recitative." Verdi won his point and produced a dirgelike *andante* during which Aïda sings in a very quiet and veiled voice ("*molto sottovoce e cupo*"). This pleased him: "Now Aïda, it seems to me, says what she ought to say and is truly in character."

Verdi bristled more succinctly when Ghislanzoni offered a trio at the end of the Nile scene for Amonasro, Aïda, and Radames, just after Radames discovers that Aïda has betrayed the plan of the Egyptian army: "Here you wanted to write a trio, but this is not the place to stop for singing." And I doubt that there is a more eloquent expression of what could be called the philosophy of simple savor than is found in a letter Verdi wrote to Giulio Ricordi while he was composing *Aida* in July 1870:

> I would not want the *theatrical word* to be forgotten. By theatrical words I mean those that carve out a situation or a character, words that always have a most powerful impact on the audience. I know well that sometimes it is difficult to give them a select and poetic form. But (pardon the blasphemy) both the poet and the composer must have the talent and the courage, when necessary, *not* to write poetry or music. . . . Horror! Horror!*

8

These Fierce Moments

In his poem "About Opera," William Meredith ponders his abiding fascination with this art form. After discounting "the tunes" he likes to hum and whistle and plots that "continue to bewilder," he concludes:

> An image of articulateness is what it is:
> Isn't this how we've always longed to talk?
> Words as they fall are monotone and bloodless
> But they yearn to take the risk these noises take.
>
> What dancing is to the slightly spastic way
> Most of us teeter through our bodily life
> Are these measured cries to the clumsy things we say,
> In the heart's duresses, on the heart's behalf.

Image of articulateness . . . a happy phrase for those moments when operatic characters—overwhelmed by wrath, madness, bliss, or dejection—rise to the superhuman expressiveness of, say, "Or sai chi l'onore," Lucia's mad scene, the *Liebestod,* or "Vissi d'arte." "In questi fieri momenti tu sol mi resti," says Gioconda of suicide in her great aria: "In these fierce moments you alone remain for me." For Gioconda and so many other characters who pass through "fieri momenti," something else invariably remains as well: the rhetorical and syntactic power to turn the heart's duresses into elaborate, eloquent, "measured cries." The primary image of articulateness in opera is that of the singer—the *solo*ist—alone on stage or in the presence of bystanders who leave him or her uninterrupted for a substantial time . . . during which self-will, slavery to passion, or "the great and beautiful commonplaces" are demonstrated in a highly focused, richly amplified fashion.

Richard Strauss urged this fundamental view of operatic dramaturgy when he wrote to Hofmannsthal concerning their last collaborative project: "Arabella must at all costs conclude the first act with a longish aria, soliloquy, or contemplation." Doubtless hoping to repeat the success of his Marschallin's poignant soliloquy on time's passing at the end of *Der Rosenkavalier*'s first act, Strauss explained: "a soliloquy [is] always particularly effective in an opera because it gives the audience the necessary rest after pure dialogue scenes, the composer an important opportunity of spreading himself lyrically, and the singer a chance to *sing* by herself for a little while." Strauss had urged this set piece for his heroine a few months earlier, with an even more blunt pronouncement: "Could a little more lyricism be fitted into *Arabella*? The *aria*, after all, is the soul of opera. . . . *Separate numbers with recitatives in between*. That's what opera was, is, and remains." Whether this was still entirely accurate—Strauss was writing in 1929—the assessment happily serves for most works from opera's golden age.

Strauss' rationale for Arabella's soliloquy approximates what must have coursed through Shakespeare's mind as he prepared his own many memorable solo arias: there are 144 soliloquies alone in the fifteen plays he wrote for the Globe Theatre, an average of almost ten per play. They offered respite from the forward thrust of often rather complicated plots; they gave him scope to "spread himself" and demonstrate his powers of invention; and they provided his Burbage, Heminge, Lowin, Condell, and lesser colleagues their chance to strut and fret at center stage a while . . . sometimes a very long while. Goethe offered a concise variation on Strauss' dictum when he said that "opera consists of significant situations in artificially arranged sequence," and this, too, is a way of looking at Shakespeare's dramaturgy. For he was ever on the lookout for the significant situation that could generate a grand, blank-verse aria. Thus, most fine performances of his plays are remembered as one remembers an opera—as a series of intense, focused, bravura events.

Thinking not of opera but of Greek and Shakespearean drama, G. Wilson Knight has argued that "dramatic thought is made of a succession of almost independent, transfixing moments," and he points by way of example to the ancient Greek convention of the *dithyramb,* the choric rhapsody whose "passion-impregnated song-music" eventually overwhelmed the spoken parts of Greek drama. The dithyramb, Knight suggests, prefigures the sublimely heightened moments in Shakespeare when passion-impregnated verse flows so copiously. The dithyramb, an ancient "image of articulateness," exerts its influence over opera as well. Functioning like an opera aria, the dithyramb of later Greek drama was employed to express ecstatic or irrational emotion more vividly and tumultuously; hence its prominence at Dionysian festivals. Comprimario confidants, nurses, parents, and bosom friends are forever urging patience, retreat, wariness, or sensibleness; principals just as regularly say them nay at extensive dithyrambic length and go their reckless, Dionysian ways.

Explicitly taking his cue from Shakespeare, Richard Wagner asserted in his essay "The Destiny of Opera" that "heightened Pathos" was the supreme object of drama and that "the technical means of securely fixing its delivery" was the supreme gift a dramatist must possess. "Fixing" the delivery of heightened pathos was, for Shakespeare, largely the task of the Big Speech. The timeless image of his actor is that of a figure striding the platform giving vent to the sound and fury of a good long speech. It is mainly the long speech that gives scope to judge whether the actor is a master (o'erstepping not the modesty of nature) or a journeyman in his trade (tearing his passion to tatters). Not so much the play as the speech was the thing, as one scene in *Hamlet* makes quite clear.

When the players arrive at Elsinore, Hamlet speaks about a play they recited for him once. Its title escapes his memory, but a major speech in it certainly doesn't: "One speech in't I chiefly lov'd, 'twas Aeneas' tale to Dido." The speech describes the attack on Troy and the murder of King Priam, and Hamlet proceeds to reel off thirteen lines of it perfectly, to Polonius' astonishment: "'Fore God, my lord, well spoken," says the old man, "with good accent and good discretion." One of the players takes over and launches into a full thirty lines more. Hamlet has told us the play was intended for the sophisticated ("caviar to the general"), so we are not surprised when the rather lowbrow Polonius interrupts and says, "It is too long." Hamlet explains to the player that the old fogey is "for a jig or tale of bawdry, or he sleeps," and then the Prince demands the recitation of another fourteen lines! With the speech now reaching nearly sixty lines and no end in sight, Hamlet finally promises to hear "the rest of this soon."

Clearly, the Big Speech of the actor impersonating Aeneas was intended by its nameless author to stick very prominently in the memory of his audience . . . just as Norma's "Casta diva," the Marschallin's monologue, or Isolde's *Liebestod* will (when superbly performed) linger in the memory of an operagoer. And so it did with Hamlet. It is a great and for us pertinent irony that the play in which Hamlet appears is itself "chiefly lov'd" for the big speeches that create its succession of independent, transfixing moments: not only the hero's famed soliloquies, but also the ghost's narrative *tour de force* (which stuck in Whitman's memory as performed by "old Clarke"), Polonius' gaseous advice to his son, the advice to the players, Ophelia's "O, what a noble mind is here o'erthrown" and mad scene, the King's anguished prayer, the Queen's narrative of Ophelia's drowning, and the graveyard philosophizing.

Though Shakespeare was almost always very solicitous of the lowbrows in the pit who doted on jigs, bawdry, and other matter suited to short attention-spans, the careful positioning of significant situations that would sustain impressive rhetorical explosions was foremost in his compositional strategy. This virtuosity in expression and, necessarily, in thespian execution is what brought an Elizabethan audience to the Globe: "Elizabethan drama was built upon vigor and beauty of speech. . . . We may suppose that at its best the mere speaking of the plays was a brilliant thing." The grander and more lavishly decorated the speech, the greater the invitation to such virtuoso display. The impetus toward these blank-verse arias (there are several great prose ones, too) is a principal shaping element of Shakespearean drama, and several consequences of such an impetus give this drama its operatic characteristics.

The most important consequence, perhaps, is the oft-remarked tendency of Shakespeare to tease the fullest possible effect from the *coups de théâtre* in individual scenes, even if this damages the integrity of the plot or action as a whole. L. L. Schücking long ago observed Shakespeare's penchant for "episodic intensification" and his "supreme interest in the individual scene."* More recently, James Hirsh, in a full study of the subject of Shakespeare's scenic architecture, has reiterated that the individual scene is "the most important common structural denominator found in all the plays." Hirsh's analytical categories, incidentally, are perfectly suitable for a study of opera: he speaks of solo scenes, duets, unitary group scenes, two-part scenes, and multipartite scenes. As in opera, too, the dramatic energies of a Shakespearean performance tend to be focused and released scene-by-scene. The Shakespearean character and plot develop not so much in a continuous, smoothly flowing cinematic fashion, but rather like a narrative series of carefully composed paintings or stop-time "frames." In their eagerness for spectacular and focused passion, Shakespeare and his contemporaries were happy to relinquish that famous hobgoblin of little minds, consistency. As a recent editor of *Much Ado* has summarized, "Elizabethans in general worried little about dramatic consistency: like Polonius, they would swallow almost any tragi-comical-historical-pastoral mixture."

And the crucial moment around which Shakespeare organizes the "perspective" of each scene is usually one that produces a crisis that challenges an expressive response. "Within the framework of an Elizabethan scene," writes Beckerman, "the most marked characteristic is the placement of emphasis not on the growth of action but on the character's response to crisis." The full intensity and impact of the typical Shakespearean scene, Beckerman also points out, is not realized in the events accomplished by the scene but rather "in the effects it produces." The supremely compelling effect, of course, was the big speech. Especially in his deployment of the soliloquy, Shakespeare showed himself a wily calculator of the prodigious speech in response to crisis. Says Granville-Barker: he "never pinned so many dramatic fortunes to a merely utilitarian device. Time and again he may be feeling his way through a scene for a grip on his audience, and it is the soliloquy ending it that will give him—and his actor—the stranglehold. When he wishes to quicken the pulse of the action, screw up its tension in a second or so, the soliloquy serves him well." This, of course, is the method of the librettist and composer. They are also poised for the moment of crisis when the character's emotion rises, when the artist's adrenalin begins to flow, and when—perhaps by an escalation from recitative to arioso to *cabaletta*—the stranglehold on the audience is tightened.

Impressive soliloquies or set-piece performances within the action are emphatically focusing events, and the Shakespearean play (like most operas) tends, no matter how complex the plot, to throw relatively simple relationships into very high relief. "The architectural superiority of Shakespeare can be seen," writes Beckerman, "in the way he raises his entire center of action to a markedly intensified level." This heightened center of the action (in *1 Henry IV* it is the positioning of Hal between cowardly Falstaff on the one hand and recklessly valiant Hotspur on the other; in *Rigoletto* it is the relation of father to only daughter) also typically leads to clear distinctions between the principals in the center of action and comprimarios who "assist" by providing convenient foils for the virtuosity of the principals. Aside from some plot-commencing narrative speeches, Shakespeare rarely favors comprimario figures with extended passages for the forestage. When he does, it is usually for very good reason. He looked, for instance, at the actions of both parts of *Henry IV* and recognized how overwhelmingly they were dominated by political and martial themes and the male voice. For the contrast of feminine pathos and domesticity, therefore, he sought out Hotspur's wife and gave her a charming scene in Part One showing the plight of a military wife, and a poignant speech in Part Two eulogizing her dead husband. The part is brief and yet requires a highly skilled actress. Similar "small star parts" (as Strauss called them) occur in opera—Macduff in Verdi's *Macbeth* and the Composer in Strauss' *Ariadne auf Naxos*—but also relatively infrequently.

The style of acting encouraged by the dramaturgy of the Big Speech will also be familiar to those who attend opera. "Altogether," summarizes Beckerman, "the evidence points to a type of acting which emphasized the individual performer, minimized his relationship to the other actors, and placed great emphasis upon the delivery of speeches."* This emphasis on *delivering* rather than *relating* dismantles the "fourth wall" separating actors from audience . . . a wall that did not exist for Shakespeare and that he would have found incomprehensible. This emphasis on delivery, rather, encourages intimacy between actor and audience. "It is probable," says J. L. Styan, "that Shakespeare saw his actors as playing to their audience most of the time"; and, if only for purely acoustic reasons, the opera singer, too, will gravitate to a position facing the audience rather than colleagues on stage. The soliloquizing player, alone on the 1,000-square-foot Globe stage, would also surely have moved to the forestage, which became in succeeding centuries the predominant site for the triumphs of Thomas Betterton, Garrick, Kean, Macready, Booth, and other star players. Mr. Puff makes this point in Sheridan's *The Critic* (1779): "Now, sir, your soliloquy—but speak more to the pit, if you please—the soliloquy always to the pit—that's the rule." It would appear that the rule is still in effect. For in *Playing Shakespeare,* John Barton ventures, "There are very few absolute rules with Shakespeare, but I personally believe that it's right ninety-nine times out of a hundred to share a soliloquy with the audience. I'm convinced it's a grave distortion of Shakespeare's intentions to do it to oneself."*

This emphasis on virtuoso delivery causes individual Shakespearean and operatic performances to be judged in similar ways. James Agate praised Gielgud's Lear—"he delivered the speech about 'looped and window'd raggedness' with extraordinary beauty and, in the last act, the 'Let copulation thrive' soliloquy with admirable point"—precisely as a music reviewer might comment on a soprano's traversal of "Sempre libera" and "Addio del passato" in *La Traviata*. More in the "bravissimo" vein of the opera house, perhaps, was the response to thespian fireworks in the eighteenth and nineteenth centuries. Here, according to the *Spirit of the Times* of 15 December 1862, is what happened after Edwin Booth rattled off the "curse of Rome" speech in Edward Bulwer-Lytton's *Richelieu:* "The men stood up and gave vent to cheer upon cheer, the ladies waved their handkerchiefs, and every man in the theatre helped to swell the violence of applause that greeted this unequaled gem of the player's art."

Such dramaturgies beckon artists down the primrose path to egomania, but the truth is that in both Shakespeare and opera the thoroughly unrealistic and undisguised "show" of delivery shrinks to inconsequence when the verse or notes are delivered with complete conviction and skill. A former music critic of the *New York Times,* Harold Schonberg, reminisced fondly about Sir Donald Wolfit and naturally was led to an operatic comparison: "When he played King Lear, he would brush aside most of the cast, advance to the footlights for his big moments like Corelli getting ready to sing *'Di quella pira,'* and pulverize the audience with the power of his voice, the beauty of his enunciation, and the innate musicality with which he phrased the great lines. I imagine that Burbage must have done much the same in London in the early 1600's."

Then as now, of course, underpowered or overly self-conscious delivery can fall horribly flat and destroy the fundamental theatrical illusion. In a 1640 play by Richard Brome, a character aptly named By-play is chided for addressing the audience instead of his "coactors in the scene" and for turning away after his speech is done, "not minding the reply." And one Quailpipe—like many a complacent singer with a gorgeous voice—is criticized because he "dreamt too long upon his sillables." In Shakespeare's time, too, apparently, angling for approval could be carried too far: "A Common Player" was described in 1615 in a way that also recalls to mind numerous operatic impersonations: "When he doth hold conference upon the stage, and should looke directly in his fellow's face, hee turnes about his voice into the assembly for applause-sake, like a Trumpeter in the fields, that shifts places to get an echo." It is but a small step from this to the droll thoughts that Shaw imagined running through the mind of a reigning tenor of his time: "Behold me in my elaborate frame! Here is excellent music written by men of talent to display my voice. Here are several worthy people behind me, whom you could see if I had not taken care to spread myself well across the footlights, engaged expressly to sustain my melody with judicious harmonies! But do not let them distract your attention from Victor Capoul!"

Shaw's plays, too, prominently feature the Big Speech. But since Shaw pretended to carry a very ostentatious chip on his shoulder where the Bard was concerned, he was pleased to thank opera instead for this aspect of his style: "Opera taught me to shape my plays into recitatives, arias, duets, trios, ensemble finales, and bravura pieces to display the technical accomplishments of the executants." And it is true that Shaw would annotate his actors' scripts with musical directions. Actor Robert Loraine's *Don Juan in Hell* script, his widow recalled, "twinkled with crotchets, crescendos and minims; with G clefs, F clefs, and pianissimos." He often gave long speeches a particular pitch on which to commence and suggested points for modulations, and on one occasion he urged an actor, "Begin at a low pitch and drag the time a little; then take the whole speech at a crescendo—*p* to *ff*." Such an approach would fit not only many a slow-tempo operatic aria, especially of the Puccini variety, but also many a Shakespearean speech.

Thinking of the typical Shakespearean soliloquy, Knight offers a quite different image to describe the same steady movement toward a climax: "The normal procedure is to start quietly, perhaps colloquially, and then, smoothly taking off like an aeroplane, to become poetically airborne, and finally return to earth." Shaw also aped the response-to-crisis method in staging his rhetorical "flights," as Bertolt Brecht noticed: "Shaw creates a play by inventing a series of complications which give his characters the chance to develop their opinions as fully as possible." This was the great Shavian difference: his characters express their *opinions* rather than their *passions* in the aftermath of crisis. The technique, however, is still thoroughly Shakespearean and operatic: creating complications that tease forth eloquence like a butterfly from its cocoon.

The means by which such complications are created were perhaps summarized as well as one could by the director Peter Brook, when he wrote that, in the Elizabethan period, "Drama was exposure, it was confrontation, it was contradiction." The huge rhetorical displays ignited by such events led Brook to conclude that Shakespeare's style "plunges deep and reaches high" and that his plays show a "roughness of texture and a conscious mingling of opposites." These brilliant, often quite sudden shifts in intensity and elaborateness of expression also led Brook to put Shakespeare's plays in his category of "Rough" theater. Indeed, he asserted that Elizabethan theater is "the greatest of Rough theaters" in history. Opera, too, is essentially "rough" theater, though Brook churlishly found another category for it: "Grand Opera, of course, is the Deadly Theater carried to absurdity." Opera is "Rough" in its conscious mingling of opposite means of expression: recitative and aria, solo obbligato and orchestral tutti, major and minor modes, *pianissimo* and *fortissimo, adagio* and *allegro molto,* unison and counterpoint, harmony and discord, and tessitura (placement of the vocal line) that "plunges deep and reaches high." And all of these mighty aural opposites are usually employed together to structure the action, as in Shakespeare, upon a series of isolated rhetorical displays—these also commonly arising from exposure, confrontation, and contradiction. In this respect, obviously, the aria or *scena* is the dramaturgical synonym for the soliloquy or set speech. It remains for us, then, to canvass the Shakespearean canon and the opera repertory in the following chapters to see just how extensively the purposes of the playwright's major set pieces and methods of deploying them are paralleled in the creative habits of composers and their librettists.

9

Soliloquy, Set Speech, Aria I

The heroines and heroes of our two dramaturgies are invariably urged to flirt, in Falstaff's phrase, "with the very extremest inch of possibility," and the set speech or aria is the single most striking means at a playwright's or composer's disposal for conveying central characters to these exciting frontiers of human experience. The centrality of the set piece is particularly demonstrated by the many Shakespearean and operatic protagonists who receive more than their fair share of them during the course of an action. These roles thus become a kind of decathlon in which the artist is confronted with a variety of challenges that must be met if the performance is to flourish. It is scarcely too much to say that the traversal of this series of set pieces is the performance's *raison d'être*. From the forty-one-line soliloquy beginning "This is the winter of our discontent," to that most famous of all exit lines ("A horse, a horse! my kingdom for a horse"), *Richard III* is dominated by the soliloquies, persuasions, and orations of Richard of Gloucester. As we have seen, this play was drastically altered for performance in the eighteenth and nineteenth centuries to make the virtuoso villain stand out even more spectacularly. Operatic protagonists regularly receive the same treatment. It is small surprise that, during the golden age of opera, *Richard III* was by far the most popular of Shakespeare's plays.

More subtly achieved but still in the manner of the thespian decathlon are the roles of Hamlet, King Lear, and Macbeth. And Othello: his impersonator must marshal his resources for the poised self-defense before the Venetian council in act 1; the ecstatic love scene of act 2; the anguish and then rage of the seduction by Iago in act 3; the alternating near-madness and Job-like suffering of act 4 ("Had it pleas'd Heaven / To try me with affliction"); then the last scene's panorama of eerie calm ("It is the cause . . ."), overwhelming remorse, and the final miserable attempt at self-exculpation. That Verdi's tenor is forced to traverse more or less the same emotional terrain is perhaps the most obvious reason why *Otello* is such a brilliant transformation of Shakespeare.

Starkly in the operatic style of serial "performances" in response to crises are

Richard II and *Henry V,* in which the dramatist offered, respectively, his exemplars of the impotent and potent monarch. As Richard—Shakespeare's most purely "coloratura" role—loses control of his kingdom and Bolingbroke begins to rise, his habit of deluding himself with the sound of his own very decorative rhetoric is displayed in what can only be called a series of exhibitionistic arias and *scenas.* He reacts to an early military defeat by philosophizing morbidly for thirty-two lines "of graves, or worms, of epitaphs." He confronts Henry at Flint Castle with some long-winded but highly "artistic" hysteria ("Down court! down king! / For night-owls shriek where mounting larks should sing!" etc.). At his deposition he asks for a prop, the hand-mirror, and then renders a highly theatrical moral on the theme, "A brittle glory shineth in his face." At play's end, Shakespeare gives Richard a dungeon scene that makes Florestan's in *Fidelio* pale in comparison. At sixty-six lines the longest speech in the play, this famous soliloquy comparing "This prison where I live unto the world" reminds one of the climactic vocal fireworks that conclude the careers of many heroes and heroines in opera.

A much happier tale is that of Prince Hal after he becomes Henry V. His play, in fact, is scarcely more than a pageant celebrating the various situations in which a superior monarch must perform brilliantly. Thus, his set pieces display him, first, shrewdly dominating courtly deliberations, then inspiring in battle ("Once more unto the breach, dear friends"), then commanding as a diplomat. Later, he is a wise teacher of his people as he walks disguised among his soldiers, an eloquent political philosopher in his fifty-four-line "O Ceremony, show me but thy worth" soliloquy, and—for good measure—a charming suitor in the bilingual love scene with Katherine, Princess of France.

Favoring one or occasionally two characters with several well-spaced set speeches, Shakespeare created plays that are typically star vehicles rather than ensemble pieces.* Just how high he catapulted his star above his assisting colleagues is suggested by the page in the first Folio that lists the "Principall Actors in all these Playes." Twenty-six names appear there. Aside from Shakespeare himself and the comedians Kempe and Armin, the *sole* name distinguished by contemporary reputation is that of Richard Burbage. Except in some of the more "democratic" comic actions, Shakespeare's habit was to focus his (and his audience's) attention upon the apparently great thespian athleticism of his star. As a result, many of his plays set a number of two-dimensional but extremely vivid and compelling lesser characters in constellation around a highly complex or simply charismatic central figure—Titus, Hamlet, Lear, Pericles, Timon, Coriolanus, and Prospero, to name a few. It is therefore easy to see the quintessential image of star quality in certain famous stage pictures from the plays: Falstaff hilariously dominating the "irregulars" of the Boar's Head tavern, Henry V dominating the procession at Westminster Abbey after his coronation, or Prospero presiding over the extended denouement of *The Tempest.*

The opera repertory quickly calls to mind numerous familiar examples of this same hub-and-spokes organization of the *dramatis personae,* which is usually achieved through a performance-dominating series of numbers for the principals. Verdi's *Falstaff* is brilliantly loyal to this Shakespearean method, even though the composer carefully miniaturized all of his baritone's moments in the vocal limelight. Because Aleksandr Pushkin consciously constructed his play *Boris Godunov* with Shakespeare's histories in mind, Modest Mussorgsky's opera based on it also comes to mind. There the Czar completely dominates a cast of twenty-one. Baroque and early nineteenth-century composers were particularly, often slavishly, tied to the necessities and advantages of tailoring their bravura arias to reigning singers. Handel and his contemporaries were just as scrupulous toward their principals as the adaptors and rewriters of Shakespeare were in setting forth the talents of David Garrick and other box-office stars. As opera progressed, though, star quality became more subsumed in the dramatic action. (One notes the same progress, incidentally, in Shakespeare's own career as he moved from *Richard III* to the mature tragedies.) But in spite of this, the domination of the stage by central, virtuosic characters has remained a nearly constant feature of the standard repertory. One thinks (to mention just a few of the more obvious roles) of Don Giovanni, Norma, Violetta, Carmen, Manon, Hoffmann, and Butterfly. And examples are scarcely lacking from later in this century: Berg's Wozzeck and Lulu, Britten's Peter Grimes and Gustav von Aschenbach, and Stravinsky's Tom Rakewell.

Two operatic epitomes of this Shakespearean tendency to allow heroically expressive protagonists to tower over lesser mortals stand out in my mind. One is comic: that most Shakespearean of *buffa* operas, Puccini's one-act *Gianni Schicchi*. The wily Florentine *babbino* dominates its cast of fifteen, and he even has a touching little spoken epilogue begging applause, *à la* Shakespeare. The tragic epitome is the first act of *Tristan und Isolde*. Doubtless it comes to mind because the most memorable single act of my operagoing experience was this one, as sung by Birgit Nilsson in San Francisco many years ago. This act is Shakespearean and operatic dramaturgy epitomized. It finds Isolde assisted by a young sailor, Tristan, Brangaene, and Kurwenal in the expression of a series of powerful, focused, but highly varied emotional responses to crisis. Specifically, they are: bitter dejection, then sarcastic irony, melancholy reverie, dignified calm, forceful recrimination, cool persuasiveness, and finally, amorous exaltation. The usual Shakespearean hallmarks are all present: the plunging deep and reaching high of the emotional range, the sudden transitions as one Kabuki-like *mie* or posture gives way to the next, the deployment of lesser characters (Tristan in *this* act is a mere comprimario) to allow Isolde and the audience to relax periodically from her thespian and laryngeal ordeal. Wagner's demands on interpretive and physical resources here are enormous. Consequently, the challenges facing a soprano in act I of *Tristan* can be viewed just as an actor would view certain of Shakespeare's greatest thespian challenges: "Each great Shakespearean role seems to have its defining set of acting problems and rewards, problems . . . of interpretations, of transition, of expressive coherence—the problem of finding a way to make sense of the innumerable, highly various, separate bits of execution that constitute the role."

Solving all these problems is frighteningly difficult. That is why there are sometimes just a few great Normas or Hamlets, Cleopatras or Tristans in a generation . . . and sometimes none. This is partly because, as Eric Bentley has observed, "the 'great' characters . . . have something enigmatic about them . . . their final effect of greatness is one of mystery." A more graspable reason for the difficulty of big roles is that it takes great energy to make their internal multifacetedness cohere both vertically in each high-profile set piece and horizontally through a full evening. One might even venture that the challenge presented by each individual set piece captures in miniature form the challenge of putting an entire role on the stage. This primary challenge of the set piece, described by the director John Barton in *Playing Shakespeare,* exists for operatic actors, too: "The problem is how to do justice to the rhetoric and yet to keep it human. Shakespeare often explores a single emotional moment and situation with text that at first blush seems *literary*. The speaker seems much more articulate than he could possibly be if he were really in the actual situation." Doing justice to the rhetoric of melodic phrases, *fioratura,* and high notes while preserving the essential humanity of the action seems a contradictory task—one requiring special genius. Ellen Terry felt certain passages in Shakespeare are so difficult to execute that they require in a performer a "state of grace," and one can think of many similarly fiendish passages in opera. Likewise, it could be said of many operatic solos what Margaret Webster said of Prospero's "revels" speech: "Here a beautiful voice will not do . . . clear analysis will not suffice." Something more—something not easily put into words—is necessary. The official production book for *Otello,* prepared by Boito and published by Ricordi, arrives at Jago's *Credo* and makes much the same point: "The actor who doesn't find in his own talent the way to express this passage of music and poetry will only manage to offer a mediocre interpretation if he allows himself to be guided by a scenic description [like this one]." The production book also depends on a state of artistic grace at the opera's tragic climax: "It is useless to say what expression in gesture and in voice must be found by the actor playing Otello in this ultimate catastrophe: if he cannot sense through his own genius and by intuition what the sublimity of the tragedy requires, he will surely succeed in offering nothing but a pallid interpretation."

Let us now look more closely at the kinds of Shakespearean set pieces and the purposes they fulfill in terms of characterization and plotting. Most of these genres of set piece find close equivalents in opera. As our examination of the passage from *Titus Andronicus* made clear, one of the most useful and common is the elaborate response to sudden traumatic threat. *The Winter's Tale,* for instance, flames to life at the moment the polite attentions between Leontes' bosom friend Polixenes and his wife Hermione spark jealous fears. Standing aside, Leontes displays the growing jealous "infection" of his mind: "I have *tremor cordis* on me; my heart dances, / But not for joy; not joy." The heart (and musical rhythm) dancing, but not for joy, is an operatic staple. One thinks of arias as disparate as the Queen of the Night's "hölle Rache" and the fury of Lucia's brother Enrico in his first-act entrance aria. Freely paraphrasing Leontes, Enrico sings, "fatal suspicion makes me freeze and shudder, my hair stand on end!" Aïda's aria after her lover is sent off to battle against her father's forces also ends with *tremor cordis:* "Amor fatal, tremendo amor, spezzami il cor" ("Fatal and powerful love, break my heart").

Hamlet's brutal "Get thee to a nunnery" outburst leaves Ophelia alone for a brief but touching exit soliloquy, "O, what a noble mind is here o'erthrown." Countless opera arias are reflected in her reverie of happy expectation suddenly destroyed, and the ululations of many a distraught soprano echo in one's memory when Ophelia speaks of herself as "of ladies most deject and wretched," or when she exclaims, with fitting *ecphonesis,* "O, woe is me / T'have seen what I have seen, see what I see!" One might mark her speech with the same *con grande sentimento* that accompanies Tosca's similarly prostrate "Vissi d'arte." Also responding to Ophelia's crisis and filled with *grande sentimento* are the Countess' two wistful arias in *Le Nozze di Figaro,* "Porgi amor" and "Dove sono."

There is space here for but a sampling from Shakespeare's memorable speeches responding to crisis, so many of which commence with the telltale rhetorical figure of outcry. Antony reacts to his defeat at Alexandria with "O sun, thy uprise shall I see no more, / Fortune and Antony part here"—a heaving combination of remorse, anger at Cleopatra, and the pain of humiliation. This soliloquy epitomizes the effect of countless arias sung by characters who, like Antony, are "Beguil'd . . . to the very heart of loss." The shattering crisis of the Ghost's appearance elicits Hamlet's solo, beginning

> O all you host of heaven! O earth! What else?
> And shall I couple hell? O fie, hold, hold, my heart . . .

These lines will be echoed at the end, like a musical reprise, with "now cracks a noble heart." In the Shakespearean and operatic dramaturgy of sudden crisis, the image of the heart bursting with emotion is inevitably a commonplace. Paulina, outraged at Leontes' treatment of her mistress Hermione, cries, "O, cut my lace, lest my heart, cracking it, / Break too." Many Shakespearean characters suffer such, coronary stress. *Lear,* as one might expect, offers an apotheosis of rended hearts: "goes thy heart with this?" . . . "my old heart is crack'd" . . . "my heart breaks at it" . . . "his flaw'd heart / (Alack too weak . . .)" and "Break, heart; I prithee, break!" Perhaps Lear's most powerful and difficult speech, which begins "O, reason not the need," responds volcanically to crisis, and ends: "this heart / Shall break into a hundred thousand flaws. . . ." In other words (Aïda's), "spezzami il cor" . . . a motto in Italian for several Shakespearean slaves to passion.

A second important kind of set speech might be called "characteristic." Shakespeare's plays are rich in characters who are mono-functional or whose psychological existence is simplified into a single emotional facet, mania, or blinkered view of the world. It was his habit to provide these figures prominent speeches showing their characteristic posture, usually at or near the beginning of their role. Their equivalent in opera is the entrance aria. A minute or so after *Othello* commences, Iago is given a twenty-four-line credo in which his character as a proud, cynical, brazen, and power-obsessed deceiver is fully displayed. For Caliban's second appearance in *The Tempest,* Shakespeare provides an aria that perfectly captures his vicious, antisocial mind: "All the infections that the sun sucks up / From bogs, fens, flats, on Prospero fall. . . ." Polonius' first major speech, running thirty-six lines, proves him terminally pedestrian, thus setting him up as a foil for Hamlet's acerbic wit. In *1 Henry IV* the playwright introduces his most wonderful hothead with a first speech of forty-one lines. Here Hotspur explains why he has refused to hand over prisoners to a royal envoy, "a certain lord, neat, and trimly dress'd . . . perfumed like a milliner." The speech thoroughly ingratiates Hotspur with us, but the intemperate pride it reveals will eventually seal his doom. Hal's famous first soliloquy in the same play—"I know you all, and will awhile uphold / The unyok'd humor of your idleness"—comes like an exit aria at the end of the same play's first tavern scene. This gives us our first hint that Hal is a virtuous manipulator of men who sincerely intends a "reformation glitt'ring o'er [his] fault."

Juliet's characteristic speech—indeed, its perfumed poetry and emphasis on haste characterize the whole of *Romeo and Juliet*—is her thirty-one-line "Gallop apace, you fiery-footed steeds." Notable here are the incantatory repetitions ("Come, civil night . . . Come, night, come, Romeo . . . Come, gentle night . . .") and its climactic outcry: "O, I have bought the mansion of my love, / But not possess'd it." One cannot help but think of Gilda's "Caro nome" in *Rigoletto* as Verdi's attempt to match the effect of Juliet's performance. But it is perhaps for his villains that Shakespeare indulged most forcefully in the soul-defining rhetorical effusion. He carefully positions one for Lady Macbeth in her first scene, just before she greets her husband for the first time. In these lines we see her typical recklessness. The formulaic repetitions, it might be noted, amount to a hideous perversion of Juliet's amorous invocation: "Come, you spirits . . . Come to my woman's breasts . . . Come, thick night, / And pall thee in the dunnest smoke of hell . . ." Verdi aptly turned these lines into the *allegro maestoso* of his Lady Macbeth's *cabaletta,* "Or tutti sorgete, ministri infernali." Surely the most vivid of Shakespeare's evil tirades, aside from Iago's soliloquies, is Edmund's "Thou, Nature, art my goddess" entrance soliloquy. The music that one can almost hear behind his brilliant final flourish—"I grow, I prosper: / Now, gods, stand up for bastards!"—makes one regret all the more Verdi's failure to bring his long-pondered *Re Lear* to the stage.

Arias revealing the essential nature of a character as the above speeches do are ubiquitous in opera. A small sampling shows their variety. Both of Despina's arias in *Così fan tutte* show her to be a kind of female Mercutio, dousing "true" love with exuberant satiric cynicism, while "Se vuol ballare," Figaro's first solo in the other great Mozart comedy, aptly introduces him as a comic Iago to the Count Almaviva's Othello. Don Giovanni's first solo, "Fin ch'han dal vino," perfectly captures the reckless, ostentatiously domineering core of his personality, just as "Or sai chi l'onore" embodies Donna Anna's usual wrath. In *Il Barbiere di Siviglia,* Rosina's dextrous "Una voce poco fa" very accurately warns us that she will run circles around pompous Bartolo (whose own characteristic aria is the perfectly blustering "A un dottor della mia sorte"). The jailor Rocco's aria praising gold in *Fidelio* speedily establishes his bourgeois mentality as a contrast to Leonore's nobler one; exactly the function of Polonius' advice to his son. "Questa o quella" at the opening of *Rigoletto* tells us all we need to know about the Duke's libidinous vanity and high spirits, while one might call the first-act *scena* for Violetta a perfect expression of the two emotions at war within her when *La Traviata* begins: depressed longing for true love in "Ah, fors' è lui!" and lingering addiction to the luxurious but shallow life of Parisian society in "Sempre libera." Norma's first scene, working in the same way, introduces her "official" position with "Casta diva" and her secret love for Pollione in the following *cabaletta,* "Ah! bello a me ritorna." Surely the most tremendous characteristic aria in the repertory is the heroine's obsessed, neurasthenic opening monologue in *Elektra,* with its climactic premonition of her final dance of death.

In a third common species of set speech, Shakespearean characters will respond to crisis not with an access of emotion but with a more formal declaration. Such speeches almost force the actor to plant his or her feet firmly as the character expounds some rigidity or certainty of mind. Most of these speeches tend to usher in significant decisions or changes in the direction of the plot. Though Macbeth says "Words to the heat of deeds too cold breath gives," Shakespeare often heaped words into set speeches as a suspenseful, kindling preliminary to important deeds. The very "dagger" speech in which Macbeth utters these words issues in the decision to commit regicide, and it is the one Macbeth soliloquy that Verdi and Piave made full use of in their opera. Shylock's "Hath not a Jew eyes?" speech in *The Merchant of Venice* eloquently declares his bitterness and firmly announces his un-Christian desire for revenge. After listening to his father lecture and browbeat him for well over a hundred lines, Prince Hal makes a superb declaration—in thirty lines of purple blank verse—of his intention to turn over a new leaf (*1 Henry IV*, 3.2.129–159). The King is won over completely. Thunder and lightning spectacularly usher in Ariel's thirty-line "You are three men of sin" proclamation in *The Tempest,* which pronounces 'ling'ring perdition" upon the shipwrecked usurpers. But we soon learn that it is not a true sentence, just one Prospero has instructed him to utter. The three real set speeches of formal declaration—and there are none more moving in the canon—will be uttered by Prospero himself in the last scene: the announcement to Ariel that he will take the "rarer action" of forgiveness; the abjuration of his magic in "Ye elves of hills . . ."; and the declaration of his pacific intentions as his former enemies emerge from their charmed sleep—over sixty lines in all.

Like Prospero's quasi-ritualistic renunciation of his powers, some stance-taking speeches rise, by virtue of highly figured rhetoric, to the level of oaths. Shakespeare sets up Troilus and Cressida for the ignominious collapse of their "true" love by having them "war" with each other in swearing fidelity. Thus, Troilus trumpets forth,

> True swains in love shall in the world to come
> Approve their truth by Troilus. When their rhymes,
> Full of protest, of oath and big compare,
> Wants similes, truth tir'd with iteration,
> As true as steel, as plantage to the moon,
> As sun to day, as turtle to her mate,
> As iron to adamant, as earth to th' centre,
> [Yet] after all comparisons of truth

(As truth's authentic author to be cited)
"As true as Troilus" shall crown up the verse . . .   (3.2.173–182)

Cressida responds with fourteen lines in kind! Several years earlier, in *A Midsummer Night's Dream*, Hermia indulged in the same verbal *fioratura* in a comic context:

> My good Lysander,
> I swear to thee, by Cupid's strongest bow,
> By his best arrow with the golden head,
> By the simplicity of Venus' doves,
> By that which knitteth souls and prospers loves,
> And by that fire which burn'd the Carthage queen
> When the false Troyan under sail was seen . . .   (1.1.169–174)

Britten, in his *Dream* opera, seized this obvious opportunity and turned it into a stately little duet for his Hermia and Lysander. Verdi and Boito, in a similar fashion, took Othello's "Now by yond marble heaven" and Iago's "Witness, you ever-burning lights above" oaths and produced the emphatic duet that brings down the second-act curtain of *Otello*.

Arias that brilliantly express decisive intentions are an operatic commonplace. Fiordiligi's "Come scoglio, immota resto" ("like a rock I stand firm") in *Così fan tutte* is an apt phrase for all arias in this category. The Pasha in *Die Entführung* threatens Konstanze with torture if she does not respond to his desires. "You do not frighten me," she replies, and then launches into one of the repertory's most pyrotechnical expressions of fearlessness, "Martern aller Arten." "Nothing, nothing, nothing, nothing," she sings, "will make me shudder." Equally grand but deriving from more threatening circumstances is the fearless determination of Leonore's "Abscheulicher!" aria in *Fidelio* (this is preceded by the villainous Pizarro's own ruthless aria *cum* declaration, "Ha! welch' ein Augenblick!"). In a boisterous *cabaletta* in *Nabucco*'s second act, Abigaille exults that she will "soon ascend the golden throne" and displace Nabucco's real daughter, Fenena. Leonora, in *Il Trovatore*, is given not one, but two arias—"Di tale amor" and "Tu vedrai che amore in terra"—announcing her loving devotion unto death, while Verdi's patented galloping rhythms accompany both Di Luna's declaration of his profane love for Leonora ("Per me ora fatale") and Manrico's rousing declaration of belligerence in act 3 ("Di quella pira"). In *La Gioconda* the spy Barnaba arrives more coolly at his fatal decision to denounce his enemy Enzo to the Venetian council in his monologue, "O monumento." Another famous declaration, Gioconda's "Suicidio," provides the opera's denouement.

Several of the most famous speeches in Shakespeare fall into a fourth category, a natural one for drama so redolent of the sweet smoke of rhetoric. These occur when a character is challenged to an extraordinary effort of *persuasion*. Miniature orations, these speeches usually bring the action to a dead stop. Mastery of forensic style was perhaps the single most important skill required for high place in the university, church, government, and professions of Shakespeare's time, and this mastery is likewise a prominent feature in Elizabethan drama. In the plays several such speeches are judicial or quasi-judicial, most notably Portia's plea for mercy in *The Merchant of Venice*. We are allowed a taste of an utterly self-possessed Othello when Shakespeare has him answer the hysterical charges made by Desdemona's father before the Venetian council. His forty-three-line defense is utterly charming and wins the Duke over completely: "I think this tale would win my daughter too." Hermione answers the indictment at her drumhead trial for adultery in *The Winter's Tale* with nearly a hundred lines of poised, well-reasoned, and passionate rebuttal. The divorce proceedings against Katherine in *Henry VIII* (2.4) amount to one of the grandest crowd scenes in Shakespeare's canon; it seems like something out of a Meyerbeer opera, with no fewer than twenty-two characters and supernumeraries named in an elaborate stage direction that also requires "Trumpets, sennet, and cornets." The Queen is called into court, and her grand speech of supplication is prepared for in a manner worthy of the Paris Opéra in its heyday: "The Queen . . . rises out of her chair, goes about the court, comes to the King, and kneels at his feet; then speaks."

The keen delight of Shakespeare's audience in persuasive virtuosity is suggested on many other occasions. There are the orations in *Julius Caesar,* the clarion summons to heroism before the battle of Agincourt in *Henry V,* as well as the two battle speeches in *Richard III.* Then there is the deep harmony of dying Gaunt's prophetic tongue in *Richard II.* His famous "This other Eden, demi-paradise" speech includes a sentence that rolls on endlessly like a Mozart concert aria: its one hundred forty-six-word subject is amplified by seventeen *this*'s before the predicate finally arrives. Volumnia, the thunderously civic-minded matriarch of *Coriolanus,* induces the play's catastrophe by pleading that her son seek peace between Rome and the Volsces. Though he senses it will eventually cost his life, Coriolanus is persuaded by her eighty-line oration. In a lighter vein, the four erstwhile academics of *Love's Labour's Lost* wish to break their oath of celibacy and court the visiting Frenchwomen. So they ask Berowne to "prove / Our loving lawful." He does so, seeking out all the right loopholes in a highly polished, lavishly ornamented fifty-four line *tour de force*. One of Shakespeare's most lugubrious speeches is the stilted effort of Ulysses to explain the failure of the Greek siege at Troy in *Troilus and Cressida;* it takes sixty-two marmoreal lines. But the prize for Most Tedious Big Speech in all Shakespeare goes, without doubt, to the Archbishop of Canterbury. His argument for the king's genealogical claim to the French throne, at the opening of *Henry V,* is also sixty-two lines long but infinitely more tedious.

Long-winded rhetoric is sometimes mustered on more private occasions and hence takes on an air of lecturing. Polonius' paternal lecture has been mentioned. When Romeo attempts suicide, Friar Lawrence extemporizes his fifty-line "Hold thy desperate hand! / Art thou a man?" speech. The Duke in *Measure for Measure,* disguised as a friar, comforts the soon-to-be-executed Claudio with thirty-six lines on the theme, "Be absolute for death," Shakespeare's finest contribution to the *contemptus mundi* tradition in poetry. Its stoic pessimism is quite out of character for the Duke, but this is far from the only instance of Shakespeare's risking neat consistency of character in his pursuit of an impressive rhetorical aria.

Far and away the longest lecture in the plays is the nearly hundred lines that Henry IV addresses to his seemingly profligate son Prince Hal on how "sun-like majesty" should "pluck allegiance from men's hearts."

In opera, too, the action often halts for impressive arias of persuasion. In comedy this often leads to long moments in the limelight for the *buffo* star: Leporello's "catalogue" aria intended to persuade Elvira that she has much company in her misery, for instance, or the blathery sales pitch of Donizetti's "dottore enciclopedico," Dulcamara, in praise of his "balsamico elisire." If the most compact of persuasions (and perfectly in scale for the one-act comedy of *Gianni Schicchi*) is Lauretta's "O mio babbino caro" addressed to her softhearted daddy, then surely the most extravagantly virtuosic is Zerbinetta's attempt, in *Ariadne auf Naxos,* to urge the "Grossmächtige Prinzessin" not to be so gloomy and moribund over unluckiness in love. "Deh! vieni alla finestra" gives us a sample of Giovanni's balcony-side persuasive manner, while notable persuasions from the French repertory are Dalila's "Mon cœur s'ouvre à ta voix" addressed to Samson, Micaëla's plea to José to return home, and Manon's "N'est-çe plus ma main," which lures Des Grieux from the seminary of St. Sulpice. There are several memorable arias of persuasion in Verdi: the elder Germont's "Di Provenza" and its following *cabaletta* in *Traviata* fail to persuade his son to return to Provence; Rigoletto's "Miei signori, perdono" pleading with Mantua's courtiers for his daughter falls on deaf ears, too. Perhaps the most noble attempt at moral suasion in all Verdi is Roderigo's "Quest' è la pace" in *Don Carlo,* where Roderigo urges King Philip to free Flanders. The public counterpart to this private interview occurs at the end of the Council scene in *Simon Boccanegra,* when the Doge—in a set piece based on two letters by Petrarch—urges peace between the Venetian republic's warring factions: "Plebe! Patrizi! Popolo dalla feroce storia!"

A fifth and last kind of set piece in the two dramaturgies shows most undisguisedly of all how they are tied to isolated, virtuosic "images of articulateness." This, as well as the significant and similar ways set pieces are deployed in Shakespearean and operatic plots, will be explored in the following chapter.

10

Soliloquy, Set Speech, Aria II

Perhaps the most fascinating and characteristic set piece of all is produced when the playwright or composer drops all pretense—or at least most of it—and allows a character to become a genuine "performer" within the action. In Shakespeare this occurs most obviously in plays within plays—Bottom and Flute, for instance, becoming the protagonists in "Pyramus and Thisby"—and whenever his action evokes the occasion for a song: the Clown's "Oh mistress mine!" in *Twelfth Night,* which embraces the play's central question ("What is love?"), and Ariel's "Full fathom five" (describing the apparent fate of Ferdinand's father) and "Where the bee sucks" (anticipating his imminent freedom) in *The Tempest.*

Other "performances" are integrated more subtly. Within moments of his first entrance, Mercutio is tossing off his fifty-line "Queen Mab" scherzo on the "vain fantasy" of dreams, which proves conclusively that he is the livest wire in Verona. Three of the men in *Love's Labour's Lost* recite their god-awful sonnets to their ladies in the spirit of several operatic "letter scenes." Both performances are superb when Falstaff and Hal impersonate the latter's father, King Henry, in the "play extempore" at the tavern. In *As You Like It,* Jaques gets some promising advance publicity as an extemporal satirist—

> *Duke Senior:*   But what said Jaques?
>   Did he not moralize this spectacle?
> *First Lord:* O yes, into a thousand similes.

—and, lo, a few scenes later he takes the hint in Duke Senior's reference to the world's "wide and universal theater" and produces his entertaining Seven Ages of Man set speech. Cleopatra's thespian talents are several times alluded to, and she approaches her death very theatrically indeed: "Give me my robe, put on my crown, I have / Immortal longings in me." She is naturally furious when, during her "long farewell," she is upstaged by her servant Iras' suddenly committing suicide before she does. Still, she recovers her wits for a wonderful cutting of the "knot intrinsicate / Of life." *Hamlet*'s graveyard scene is a comic performance on two levels: the lowbrow banter of the two "clowns" and the sophisticated, ironic *memento mori* meditation of the prince. In *King Lear,* the Fool's rhymed prophecy ("When priests are more in word than matter . . .") helps Shakespeare to ease us out of the first storm scene, while Cranmer's grandiose (and true) prophecy over the baby Elizabeth—future Queen of England—makes a perfect climax for all the pageantry of *Henry VIII.*

Many of Shakespeare's comic scenes—like those for Peter in *Romeo and Juliet,* Launcelot Gobbo in *The Merchant of Venice,* and Dogberry in *Much Ado*—are transparently solo turns for a skilled comedian. Interpolated as comic relief are several of Falstaff's speeches, which amount to stand-up routines: the "catechism" on honor, the theme-and-variations on Bardolph's red nose, the character assassination of Justice Shallow, and the praise of sherris sack. Doubtless, Falstaff and his death were kept offstage in *Henry V* because in this more solemn and serious play Shakespeare desired no comic or sentimental diversions of attention from Henry, the most perfect of all his stage monarchs.

A few splendid "duet" performances might also be mentioned: the perfect sonnet for Romeo and Juliet at the Capulet ball; the "merry war" in several duets for Beatrice and Benedick in *Much Ado;* and the lyrical exchange for the young lovers that opens the last scene of *The Merchant of Venice* (5.1.1–24). They try to out-"night" each other in a series of comparisons like these:

> *Lorenzo:*　　　　　　　　　　　In such a night
> Stood Dido with a willow in her hand
> Upon the wild sea-banks, and waft her love
> To come again to Carthage.
> *Jessica:*　　　　　　　　　　　In such a night
> Medea gathered the enchanted herbs
> That did renew old Aeson . . .

Peter Conrad says, with a sour grace, that the lovers here "cheapen mythology, dealing out its stale exempla like playing cards." But Berlioz, I venture, happily read this duet more feelingly, as we can tell from the gorgeous love duet for Dido and Aeneas in *Les Troyens* that takes this passage for its text.

This element of self-conscious performance in so many Shakespearean roles, which calls attention to the illusion of the whole theatrical event, is duplicated when composers allow their characters, whose natural "speech" is song, to become executant artists with in-set "numbers" to perform. These performances within the action are like the Shakespearean soliloquy in which the character steps forward toward the audience, and they are (also like soliloquies) a brazen "fourth-wall"–wrecking, but highly useful, way to create occasions for virtuosic display. They occur everywhere: in Lakmé's Bell Song, Olympia's Doll Song in *Les Contes d'Hoffmann,* Escamillo's Toreador Song, Marie's regimental song in *La Fille du régiment,* Méphistophélès' Song of the Golden Calf, Eboli's Moorish Song in *Don Carlo,* and the Tenor's aria in *Der Rosenkavalier,* to name a few.

These "indented" performances range from the utterly vestigial and gratuitous to the dramatically poignant. Rosalinde is asked, at the Orlovsky ball in *Die Fledermaus,* to prove she's Hungarian, which she does, resoundingly, with her Czardas; and in the last scene her maid Adele proves her talent for acting by offering an aria in which she becomes, in quick order, a girl from the sticks, a queen, and a lady from Paris. The Countess in *Figaro* commands a performance of Cherubino's "Voi che sapete," while near the opening of Puccini's *La Rondine* Magda takes up a new song that the poet Prunier can't remember the end of and performs it gorgeously ("Chi il bel sogno di Doretta"). It is the opera's best-known number. Massenet's *Manon* dominates the lavish Cour la Reine scene with her gavotte, while Vaarlam enlivens a seedy tavern in *Boris Godunov* with his rousing ballad of the czarist victory over the Tartars. In *Der fliegende Holländer,* the old nurse—satisfied with her comprimario lot—refuses to sing the ballad of the Dutchman, but Senta is happy to oblige. Among Wagner's other operas, both *Tannhäuser* and *Die Meistersinger* are plotted around vocal competitions that give notable opportunities for in-set performances.

There are several famous drinking songs (*brindisi* or *chants à boire*) in the repertory. One brightens the first moments of *La Traviata;* another ushers in the climax of *Cavalleria Rusticana*. Orsino's *brindisi* in the last act of Donizetti's *Lucrezia Borgia* provides an ironic prelude to that opera's climactic mass poisoning. The un-Shakespearean "Chanson Bachique" for the Prince is the most memorable number in Thomas' *Hamlet* opera. On the other hand, Verdi and Piave took Macbeth's line, "we'll drink a measure / The table round," and produced a fine *brindisi* for Lady Macbeth that frames the appearance of Banquo's ghost. Also common in opera is the *preghiera*—a performance for divine ears. One thinks of Norma's "Casta diva," the Queen's prayer in Donizetti's *Maria Stuarda,* Elisabeth's prayer in *Tannhäuser,* Elsa's in *Lohengrin,* and Leonora's "Madre, pietosa Vergine" in *Forza*. Verdi also planned a prayer, complete with anachronistic genuflection, for the soprano of his *Re Lear* opera, just after she learns that her father is still alive. Also quasi-performative are various other common operatic set pieces that can be briefly noted: recountings of dreams (Elsa's genuine one in *Lohengrin,* Jago's false one in *Otello*), or hallucinations (Delia was to get one of these in *Re Lear*'s last scene), accesses of insanity, letter scenes (notably in *Figaro, Onegin,* and *La Périchole*), prophecies (Erda's in *Das Rheingold*), and curses (Alberich's in the same work).

Having thus far isolated several prominent kinds of set speech and aria, it remains to look a little at how they are deployed within their respective actions. These highly focused, intensely expressive events carry heavy responsibilities. One can see this most obviously in the several plays and operas whose entire action seems launched in a very clear trajectory toward the target of a big, climactic utterance that may resolve suspense, release tension, complete a tragic fall or comic rise, or simply heighten the sense of an ending.

These "capstone" utterances, to change the image, often express the essence of the particular drama's movement and purpose. The title of *Love's Labour's Lost,* for instance, is a good hint that Shakespeare was carefully aiming all along at the moment, near play's end, when the blithe King of Navarre makes his final plea, "Grant us your loves," and the Princess responds with her heartfelt speech of rejection. Her premiere lady-in-waiting, Rosaline, seconds this speech with a fine one of her own to Berowne. Similarly, *The Taming of the Shrew* is aimed at the climactic moment when Kate—in the play's longest speech—displays her "new-built virtue and obedience." The first part of *Henry IV* moves steadily to Hal's triumph over Hotspur, with its series of three successive memorable speeches— one each for Hotspur, Hal, and then Falstaff. The sequel play moves just as inevitably toward the sobering twenty-five-line public rejection of Falstaff by the newly crowned king: "I know thee not, old man, fall to thy prayers . . ." The action of *Timon of Athens* is simply the huge crescendo of the hero's misanthropy; it clearly reaches *fortississimo* in a vitriolic forty-line soliloquy that ends, "grant, as Timon grows, his hate may grow / To the whole race of mankind, high and low!" The horrific suspense of Richard III's career is fully released near his play's end, when a mightmare vision of several of his victims induces a final soliloquy. Here at last, in thirty lines, his "coward conscience" finally begins to do its duty. One last example: at the opening of *The Tempest*'s last scene, Prospero announces, "Now does my project gather to a head." As so often before in his career, Shakespeare himself gathers his dramatic project to a head with a set speech . . . or, rather, as we have seen, a brilliant necklace of three grand speeches.

Likewise, many operatic projects gather to a head with extended arias or concerted set pieces that are clearly in view from the beginning of the action: Rossini's ebullient, happy-ending rondo, "Non più mesta," for his Cinderella; Lady Macbeth's sleepwalking; Elektra's last scene and *danse macabre*. Butterfly's "Tu, tu, tu, piccolo iddio" ushers in with devastating pathos the suicide that has been in prospect since act 1 . . . just as the heroine's final aria in *Manon Lescaut,* "Sola, perduta, abbandonata," predictably ends the fable of foolish self-indulgence leading to misery. The *Liebestod* similarly epitomizes the notion of true love bending not with the remover to remove, which is at the heart of *Tristan und Isolde,* while Walther's inevitable triumph as suitor and artist in his Prize Song is *Meistersinger*'s long-expected denouement. Mussorgsky's opera has the obvious goal of Boris's heartsick final monologue, with its many echoes of Shakespeare's ambitious rulers toward the end of their careers. Among the many set pieces that bring an entire action into final focus are the trio for aging Marschallin and youthful lovers at the end of *Der Rosenkavalier* and the exultant duet, "O namenlose Freude," for Leonore and Florestan in *Fidelio.* But the most stunning example in this kind must be Brünnhilde's *tour de force* before her immolation in *Götterdämmerung.* The dramatic arch of the entire *Ring* is spectacularly capped here: for her action signals the final triumph of selfless love over selfish love—the struggle between which is central to the tetralogy's action.

Another responsibility of the set speech or musical monologue is sometimes not so much to provide the climax to a linear progression, but simply to isolate the heart of the action in a powerful way. We have noted that "the architectonic superiority of Shakespeare can be seen in the way he raises his entire center of action to a markedly intensified level," and his typical recourse in doing so is often some kind of set speech. Many famous ones—and numerous musical numbers in opera—fall into this "center-raising" category. The thirty lines of weary, insomniac musings for the King in *2 Henry IV* that end, "Uneasy lies the head that wears the crown," for instance, exactly capture the play's general ambience of uneasiness, pessimism, and decay. The same ambience dominates *Don Carlo,* and Verdi explores it in a very similar monologue for King Philip in act 3 ("Ella giammai m'amo"). It, too, comes after a sleepless night. More familiar, perhaps, and in the same vein of regal melancholy is Boris Godunov's first monologue, "I have attained the highest power . . . but my tortured spirit knows no happiness." The most important speech in *King John* is the eloquently sarcastic soliloquy given to the Bastard, Faulconbridge (it contains the longest sentence in all Shakespeare, twenty-five lines). Here, in nearly forty lines (2.1.561–598), he creates a powerful picture of courtly corruption—the backdrop for the entire drama. His first line, "Mad world, mad kings, mad composition!" ought to remind us of the opening of Hans Sachs' monologue in *Meistersinger,* which is the weighty philosophical centerpiece of Wagner's massive, humane comedy: "Wahn! Wahn! Überall Wahn!" ("Madness! Madness! Everywhere madness!"). The famed "To be or not to be" soliloquy gives us the essential dilemma of *Hamlet:* reconciling clear, rational, moral thought with effective action. Similarly, King Marke's monologue in the second act of *Tristan* throws the lovers' moral quandary into excruciatingly high relief. Finally, one might point to Prospero's "Our revels now are ended," the speech with which he cuts short the marriage masque for Ferdinand and Miranda. Many themes and levels of meaning in *The Tempest* are focused in this speech, most notably the subject of time passing. *Sic transit gloria mundi,* Prospero is announcing, and that is precisely the theme of the Marschallin's monologue on her fading youth, which is the philosophical and emotional heart of *Rosenkavalier*'s drama.

A third dramaturgical purpose of a big speech or aria can be to achieve an emphatic entrance or exit. In common especially with baroque, classical, and early romantic operas, Shakespeare's plays—having the individual scene as the main structural unit—show great concern for clearly defined beginnings and endings. "Shakespearean drama," writes Beckerman, "depends a great deal upon the vigorous movement of actors coming on and off the stage. The actors themselves, rather than stage equipment, provide the impetus of the play's progression." Notwithstanding the habits of lavish producers like Franco Zeffirelli or Jean-Pierre Ponnelle, this is also true of operatic dramaturgy, and in both genres the set piece is a natural way to invigorate an entrance or exit. The ubiquity of exit and entrance arias in opera needs no demonstration, but one might point in Shakespeare to such famous lines as "Now is the winter of our discontent" and "If music be the food of love, play on" as ringing in two of Shakespeare's most famous entrance "arias." His exit speeches, usually soliloquies, are everywhere: several of Iago's solos are exit speeches, as are Hal's apostrophe to his father's crown ("O polished perturbation! golden care!"), Bottom's "I have had a most rare vision," and Falstaff's analysis of the word "honor."

The final exit of an important character can lead to especially grand displays. Shakespeare was well aware that the tongues of dying men can enforce an audience's attention, as Gaunt says in *Richard II,* "like deep harmony." Indeed, not once did Shakespeare deny a major or admirable character *in extremis* a final meal of blank verse. Wolsey's "long farewell to all my greatness" in *Henry VIII,* Romeo's fifty-line adieu before swallowing the poison ("Eyes, look your last!/Arms, take your last embrace! and lips, O . . ."), Othello's "Soft you; a word or two . . .", and Hotspur's "O Harry, thou hast robb'd me of my youth!" have their equivalents throughout the opera repertory. Singers, like their Shakespearean colleagues, must often be, in Whitman's phrase, "garrulous to the very last." One thinks of Edgardo in the final scene of *Lucia,* Elisabetta's "Tu che la vanità" at the end of *Don Carlo,* and Violetta in "Addio del passato." Composers also seem drawn to those final moments when, as Violetta says, "ogni speranza è morta."

A fourth and last important responsibility often fulfilled by the big speech or aria is the narration of events that occur before the action begins or offstage during the action. (These narrative set pieces might have been discussed as a variety of in-set performance.) Even more than the soliloquy, the narrative is an artificial, disillusionistic device. In Shakespeare its starkest form is the choric speech, which he used in *Troilus and Cressida, Romeo and Juliet, Henry IV,* and *Pericles.* In *2 Henry IV,* he harked back fearlessly to the morality-play style and trotted out Rumor, "painted full of tongues," and in his very operatic (in the sense of recklessly plotted) *Pericles* he required the services of the choric figure of Gower six times. Such choric speeches inevitably halt the action, for nothing "happens" while they are in progress. One need merely compare the beginnings of the plays just mentioned with the wonderfully kinetic openings in *A Midsummer Night's Dream, Antony and Cleopatra, King Lear,* and *Coriolanus* to sense the difference.

Crude as the narrative set piece is, it is a most efficient means of moving characters, plot furniture, and the plot itself: telling takes less time than showing. And because narratives both require and display rhetorical energies, they can become magnets of emotion in the speaker and attention in the audience . . . *tours de force* in their own right. Boito, knowing this, went back to *Othello's* first act, found its narrative centerpiece (the Moor's speech before the Council, describing the growth of Desdemona's affection), and quarried it for the text of the essentially narrative love duet at the end of his own first act. The most straightforward narratives in Shakespeare and opera are pure reportage. "The circumstances I'll tell you more at large," says the Third Messenger early in *1 Henry VI,* and he then runs on for over thirty lines describing the "general wrack and massacre" of the English forces in France. Many operatic narrations, like the long, three-part one for Ferrando at the beginning of *Il Trovatore,* are fashioned in this way, and they often begin with an *Ascolta!, Udite!,* or *Senti!* to a hastily convened chorus. In this vein is the Ghost's "List, list, O list!" which rivets Hamlet's attention for the hair-raising news of regicide at Elsinore. In *A Midsummer Night's Dream,* Shakespeare gives Titania a lovely description of the upsetting effect of her husband's "brawls" on their fairy realm, and this is followed shortly by Oberon's narration to Puck about the origin of the "love-in-idleness" juice he will use on his wife. It begins beautifully,

> Thou rememb'rest
> Since once I sat upon a promontory,
> And heard a mermaid on a dolphin's back
> Uttering such dulcet and harmonious breath . . .

Puck says he *does* remember, but Oberon happily continues on for our sake.

Shakespeare saved his most astonishing narrative for last. In *The Tempest* he dared to devote nearly twenty percent of his play's lines (1.2.1–375) to a purely expository rehearsal of past events. No doubt Prospero's "Thou attend'st not!" and "Dost thou hear?" and "Hear a little further" to Miranda are meant to keep his audience alert, too. There is, to my mind, but one outright disaster among Shakespeare's narrative speeches: Friar Lawrence's recapitulation at the end of *Romeo and Juliet*. The good man says, "I will be brief," but forty lines at play's end describing action we have already seen are far too many. Wagner, incidentally, courts similar disaster several times in the *Ring,* notably in the plot-rehearsal of the Norns in *Götterdämmerung* and, in what must be the longest narrative in all opera, Wotan's interview with Brünnhilde in the second act of *Die Walküre* ("Als junger Liebe . . ."). In uninspired performances or performances not in the audience's own language, these narratives can prove excruciatingly tedious.

Several marvelous Shakespearean narratives obviously exist because the playwright trusted his poetry more than he trusted the bare-staged "wooden O" to create spectacular effects. Prospero asks Ariel if he has "Perform'd to point the tempest that I bade thee," and Ariel's vivid narrative of the storm and capture of the mariners becomes a performance itself. Shakespeare knew he could not convey on stage the miraculous revelation of Hal's military prowess in *1 Henry IV* (live animals did not tread the Globe stage), so he prepared a glowing description for a comprimario, Sir Richard Vernon. Its effect is exhilarating:

> I saw young Harry with his beaver on,
> His cushes on his thighs, gallantly arm'd,
> Rise from the ground like feathered Mercury,
> And vaulted with such ease into his seat
> As if an angel dropp'd down from the clouds
> To turn and wind a fiery Pegasus,
> And witch the world with noble horsemanship.

Cleopatra witches the world in another way. Maecenas says that "She's a most triumphant lady, if report be square with her," and in moments Enobarbus is obliging us with the famous description that captures her mysterious charisma: "The barge she sat in, like a burnish'd throne, / Burnt in the water. . . ." It begins like many operatic narratives with an "I will tell you" that causes everyone on stage and in the audience to lend their ears.

The most memorable narratives in opera tend to be like these speeches. While they may convey important information, they also exist to make a heightened artistic effect in themselves . . . and often serve other important purposes as well. The price paid in loss of subtlety or dramatic momentum is by no means too high. There is, for example, a remarkable number of arias at the beginning of *Il*

Trovatore whose texts are purely narrative. Not surprisingly, Ferrando's, though the most elaborate and requiring the most performance time, usually makes the least effect. Woe to the *Trovatore* cast that cannot upstage him with Leonora's "Tacea la notte placida," Azucena's "Condotta ell'era in ceppi," and Manrico's "Mal reggendo all'aspro assalto"—all recounting past events. This is, of course, because each of these numbers requires more purely virtuosic singing and because each serves to build and shape individual character.

Giovanni tells Leporello to explain everything to Elvira ("Dille pur tutto") and, by God, he does in the "catalogue" aria. It's just a narrative, but also a *buffo* extravaganza and a dramatically useful summary of the Don's amorous marauding. Siegmund's long autobiography in *Die Walküre* tells us much factually, but we also learn fully of his emotional, vulnerable character. This long narrative also comes alive dramatically because of Sieglinde's growing sympathy as it unfolds (she seems as affected as Desdemona is by Othello's life story). "Die Frist ist um" in *Der fliegende Holländer* is pure narrative, but it is also a magnificent entrance aria—its heroic demands on the baritone finely conveying the Dutchman's will to domination. This same will is also manifest in Turandot's main aria "In questa reggia," its purpose being at first strictly narrative (telling the fate of Princess Louling). Elektra's monologue is also primarily a narrative of Agamemnon's murder, but—as with Isolde's long narrative in the first act of *Tristan* ("Wie lachend sie")—the heroine's musical character is developing between the lines. Lohengrin's "In fernem Land" and Tannhäuser's Pilgrim Narrative are interesting as set pieces that also succeed as climactic resolutions of their respective actions.

The overview offered in these last two chapters demonstrates, I think, the overwhelming prominence of set pieces in the two dramaturgies. This overview makes clear, too, the fundamental priority of powerful situations over plot. Before moving on to consider the larger structural features of Shakespearean and operatic plotting, it is worth making one further observation.

"It's not the plots," Meredith wrote in his poem about opera's attractions. True enough: looked at in the cold light (and silence) of dawn, most librettos divulge plots that offer plenty of incitement to mirth. There are loose ends, ridiculous suddennesses, roughshod expediencies, and sometimes not a little mystification. This is because nicety of plot construction is not a high priority in the two dramaturgies. Verdi, assessing what he thought was his best opera as of 1853, said *Rigoletto* had four crucial ingredients: "very powerful situations, variety, brio, and pathos." No mention is made of plot or dramatic architecture, though this is of course part of all opera-making. Plot, in short, was simply all the guy-wiring that would hold several "very powerful situations" in some kind of linear and cumulatively effective order. This all reminds one strikingly of Shakespeare's method of composition. Indeed, what William Weaver says in summing up Verdi's style could do service for Shakespeare with a simple change of "music" to "poetry": "It is not the plot that counts. . . . What matters is character. From the characters comes the drama and, indeed, the music."

11
The M-Word

Late in January 1890, Bernard Shaw attended a performance that left him, characteristically, fit to be tied: "Such an old-fashioned, shiftless, clumsily constructed, empty-headed turnip ghost of a cheap shocker as this *Tosca* should never have been let pass the stage door. . . . I do not know which are the more pitiable, the vapid comedy of intrigue, or the . . . sham torture, rape, murder, gallows, and military execution, set to dialogue that might have been improvized by strolling players in a booth." He then adds, aptly indulging in some *ecphonesis:* "Oh, but if it had but been an opera!" Almost exactly a decade later, on 14 January 1900, Puccini's opera based on Victorien Sardou's romantic melodrama received its Roman premiere. Shaw was by then no longer reviewing music and as far as I know never recorded his views about the opera. But we can safely guess he would have received it more kindly than the lamentable Joseph Kerman, who famously doffed it aside many years ago as "that shabby little shocker."

Who would deny the manifold pleasures of making light of opera as Shaw does, or as the Marx brothers did in *A Night at the Opera?* It is both pleasant and easy to indulge in this fun, not only for aficionados (most of whom do it well and often) but also neophytes. Harry Truman, for instance, wrote to Bess in 1911, "I have some cousins in Kansas City who affect intellect. They once persuaded me to go to a season of Grand Opera with them. It happened to include *Parsifal* and some others which I cannot spell. Well I haven't recovered from the siege of Grand Opera yet . . . when it comes to a lot of would-be actors and actresses running around over the stage and spouting song and hugging and killing each other promiscuously, why I had rather go to the Orpheum." Shaw makes almost the same point—about the overwrought incidents of the plot—from the initiate's perspective. Indeed, the droll suggestion that crude melodrama can be rendered palatable if sung rather than acted is, in part, accounted for by his habit of shadowboxing with whomever and whatever most profoundly evoked his affection. "I am an Irishman," he once wrote, "intensely proud of being Irish (quite unreasonably)." Stella Patrick Campbell, Shakespeare, and opera—to name a few of

Shaw's most famous loves—received barbs and busses in about equal measure.

Behind most opera satire, affectionate or not-so-affectionate, lies the cruel word *melodrama*. This is the umbrella epithet of choice to cover the cheap shock-effects that Shaw found in *La Tosca* and the promiscuous hugging and killing that Truman found in Grand Opera generally. Any serious discussion of operatic dramaturgy must sooner or later face the challenge of this loaded word, so often used to quicken disgust. W. H. Auden met the challenge coolly and flatly, simply by asserting, "A good libretto plot is a melodrama in both the strict and the conventional sense of the word." The strict sense he refers to is both etymological (Greek *melos* = "song" + *drama* = "deed or action on stage") and historical: Rousseau first used the word *mélodrame* in the 1770s to describe a stage-piece that combined soliloquy, pantomime, and orchestral accompaniment. In nineteenth-century France and Italy *mélodrame/melodramma* came to be more or less synonymous with *opéra/opera*: Rossini's *Tancredi, Semiramide,* and *La Donna del Lago* were called *melodrammas,* as were Verdi's *Rigoletto, Boccanegra,* and *Ballo*.

Unfortunately, the conventional and pejorative sense of the word has long since taken over. "My idea of heaven," wrote Ralph Waldo Emerson, "is that there is no melodrama in it at all; that it is wholly real." As usual, melodramatic = unreal. The dictionary defines melodrama somewhat prejudicially, too: "a dramatic presentation characterized by the heavy use of suspense, sensational episodes, romantic sentiment." And the word is frequently deployed by critics on the rampage. One writes of Jago "merely function[ing] as a melodramatic agent"; another speaks of Honoré de Balzac as "a vulgar melodramatist whose versions of life are cheap, overwrought, and hollow." In an 1875 book, *On Actors and the Art of Acting,* we are told that a "melodramatic actor is required to be impressive, to paint in broad, coarse outlines, to give relief to an exaggerated situation; he is not required to be poetic, subtle, true to human emotions." Charles Dickens spoofs this style with his Mr. Lenville, star of Mr. Crummles' troupe in *Nicholas Nickleby:*

> The tragedian [took] an upward look at Nicholas, beginning at his boots and ending at the crown of his head, and then a downward one, beginning at the crown of his head, and ending at his boots—which two looks, as everybody knows, expresses defiance on the stage.

> Mr. Lenville folded his arms, and treated Nicholas to that expression of face with which, in melodramatic performances, he was in the habit of regarding tyrannical kings.

And surely the most piercing and briefest review of the melodramatic style in performance is Pip's on Mr. Wopsle's Hamlet: "Massive and concrete." With a bit more delicate touch, the reviewer for the *New Yorker* calls Arthur Miller's *A View from the Bridge* "a neatly plotted melodrama" and goes on, with a faint air of deprecation, to observe, "The play is a thriller, and the superb cast squeezes the last drop of operatic intensity out of every scene."

The long stage history of melodrama makes it clear why aspersions are now so often cast against it. For throughout the nineteenth century—the century during which both opera and melodrama reigned supreme—melodrama steadily lost its inspiration, sophistication, and freshness as it catered to an ever-growing commercial audience. The melodrama of Schiller, Hugo, and (in prose fiction) the archetypal Balzac declined into the shallows of *East Lynne, The Drunkard, Nick of the Woods,* and *The Prisoner of Zenda.* Describing its "escapist" appeal, a student of nineteenth-century English melodrama sketches it as something like our soap operas today: "Essentially, melodrama is a dream world inhabited by dream people and dream justice, offering audiences the fulfillment and satisfaction found only in dreams." The same writer almost succeeds in putting its vacuity in a positive light: "Melodrama has a refreshing lack of pretension about it; there is no messing about with intellectuality." Of all opera composers, Puccini was the most shamelessly and successfully wedded to melodrama in its later, more decadent form, and this goes far to explain Virgil Thomson's conclusion that "Puccini's operas have probably the lowest intellectual content of any, though their plots are far from stupid."

Rescue was a favorite melodramatic device, and my purpose in the following pages will be to rescue melodrama, Florestan-like, from these largely ignominious and in some respects unjustly deserved hostile connotations. Such a rescue is necessary here because Shakespearean and operatic architecture is melodramatic in the best sense of the word. The crucial similarity can scarcely be appreciated, let alone explored, except by looking on the brighter, more efficacious, more potent side of melodrama and by encouraging a little reluctance to aim the epithet as a convenient cheap shot. Viewed more dispassionately and defined a little less tendentiously, melodrama is not entirely the rowdy black sheep of drama it has been made out to be.

Perhaps the easiest way to urge that melodrama cannot be all bad is simply to note how often admirers of Shakespeare, rendered fearless by his impregnable reputation, have boldly labeled him a melodramatist *tout à fait.* Kenneth Muir has ventured that "All his mature tragedies may be regarded as 'melodrama humanized,'" while A. C. Bradley bravely asked, "Is not *Hamlet,* if you choose to so regard it, the best melodrama in the world?" Shaw called *Julius Caesar* "the most splendidly written political melodrama we possess," and spoke of *Othello* and *As You Like It* as among "the half-dozen big popular melodramas which the Bard has sublimified by his tempests of grandiose verse." Thinking of Shakespeare in performance, Eric Bentley has written, "Time and again we witness the failure of the non-melodramatic rendering of Shakespeare." Olivier has called him "this tinkerer with melodrama," and Verdi made a rather similar point when he wrote to his *Rigoletto* librettist about the thoroughly melodramatic Hugo play they were adapting: "Oh, *Le Roi s'amuse* is the greatest subject and perhaps the greatest drama of modern times. Triboulet is a creation worthy of Shakespeare!!"

This defensive strategy, however, does not get to the main issues. Perhaps the foundation for an attack on "the vulgar prejudice against melodrama" and on the "narrow and philistine Naturalism" that so readily deplores the outrageous coincidences and exaggerations of melodrama has been best laid out by Eric

Bentley in *The Life of Drama*. It is no wonder to find him, a shrewd partisan, admiring *Richard III,* Shakespeare's most forthright hiss-the-villain melodrama, in glowing terms: "What we have here is a most skillful arrangement of scenes, a very bold imagining of events and confrontations." Putting such drama on stage, Bentley makes clear, requires a fearlessness of Brobdingnagian effects ("today the task of actors is to rediscover and recreate a lost grandeur"), and melodramatic methods make such grandeur possible. The notion of vulgar melodrama creating a "dream world" (discussed on p. 11) Bentley cleverly turns to advantage: "That we are all ham actors in our dreams means that melodramatic acting, with its large gestures and grimaces and its declamatory style of speech, is not an exaggeration of our dreams but a duplication of them. In that respect, melodrama is the Naturalism of the dream life." Follow Naturalism's impetus toward the "real"— toward total reportage—and, Bentley suggests, the farther one will get from the life of drama: "Naturalism is more sophisticated but Naturalism is not more natural. The dramatic sense is the melodramatic sense, as one can see from the play-acting of any child. Melodrama is not a special and marginal kind of drama, let alone an eccentric or decadent one; it is drama in its elemental form; it is the quintessence of drama." Bentley might have been thinking of Shakespeare when he urged, finally, that "tragedy does not discard the bizarre, macabre, and morbid elements of melodrama. It exploits them farther." Especially germane for us is Bentley's observation that Shaw's plays call "for the 'exaggerated,' sweeping movements of operatic (that is, melodramatic) performance."

With the battle lines so starkly drawn by Bentley, we need a cooler, more specific definition of this thing that is "the quintessence of drama." As it happens, there is a very good one from Shaw himself. Though he was a persistent critical terror to the debased melodrama of "Sardoodledum" and was even capable of summarizing Verdi's canon as being "mostly about the police intelligence melodramatized," Shaw came to be a devoted, discriminating admirer of melo-dramatic architecture and acting (the great idol of his youth was the Irish barnstormer, Barry Sullivan). What seems to have finally won him over was the experience of writing *Arms and the Man* (1894), in which he had very deliberately set out to discredit the conventions of melodrama. Instead, he was completely won over to them. By 1896 he was confiding to Ellen Terry, "A good melodrama is a more difficult thing to write than all this clever-clever comedy: one must go straight to the core of humanity to get it, and if it is good enough, why, there you have a Lear or Macbeth." And a few months later he described to Terry a new play—*The Devil's Disciple,* which is subtitled *A Melodrama*—in terms that make it sound almost exactly like the *Tosca* he had hooted at six years earlier: "this thing, with its heroic sacrifice, its impossible court-martial, its execution . . . its sobbings and speeches and declamations, may possibly be the most monstrous piece of farcical absurdity that ever made an audience shriek with laughter. And yet I honestly tried for dramatic effect."

Here, in any event, is Shaw's definition of melodrama. It was evoked by a review, "Two Bad Plays," for *The Saturday Review* of 20 April 1895 and opens our exploration on a happily unpejorative note:

a simple and sincere drama of action and feeling . . . depending on broad contrasts between types of youth and age, sympathy and selfishness, the masculine and feminine, the serious and the frivolous, the sublime and the ridiculous, and so on. The whole character of the piece must be allegorical, idealistic, full of generalizations and moral lessons; and it must represent conduct as producing swiftly and certainly on the individual the results which in actual life it only produces on the race in the course of many centuries.

Shaw's own dramaturgy is reflected in this recipe, and perhaps it hardly need be observed that this recipe would fit many an opera and Shakespearean play very comfortably. This may be why, when performances of Shaw, Shakespeare, and opera fail, they usually fail along the same lines. Dismal Shakespeare or opera usually leaves one complaining as Shaw did about a poor revival of *Pygmalion:* "I missed the big bones of my play, its fortissimos, its allegros, its precipitous moments, its contrasts, and all its big hits."

The essential likeness of Shakespearean and operatic architecture also resonates between the lines of a more recent attempt to define melodrama as a literary (and not exclusively theatrical) phenomenon: *The Melodramatic Imagination: Balzac, Henry James, and the Mode of Excess* by Peter Brooks. Indeed, a pleasantly startling number of Brooks' formulations not only prove to be illuminating variations on the main points of Shaw's thumbnail sketch but also reflect pertinently on our dramaturgies tied to the rhetorical exhibitionism of the set speech and aria. Here is a sampling of these formulations (note Brooks' consistent refusal to condescend in his choice of words):

[Melodrama is] a heightened and hyperbolic drama, making reference to pure and polar concepts of darkness and light, salvation and damnation . . . [with] characters at the point of intersection of primal ethical forces.

[The typical melodramatic scene] represents a victory over repression.

Melodramatic rhetoric . . . tends toward the inflated and the sententious. Its typical figures are hyperbole, antithesis, and oxymoron.

We come to expect and to await the moment at which characters will name the well-spring of their being.

The melodramatic utterance breaks through everything that constitutes the "unreality principle," all its censorships, accommodations, tonings-down.

. . . the emphatic articulation of simple truths and relationships . . . a dramaturgy of admiration and astonishment . . .

Brooks himself eventually draws the obvious conclusion: "melodrama finds one possible logical outcome in grand opera . . . where melody and harmony, as much as the words, are charged with conveying meaning."*

One other aspect of Brooks' study, incidentally, turns in opera's direction. This is its respectful focus on Balzac, whom we have earlier heard castigated as a "vulgar melodramatist." His novels more than his plays are a monumental achievement in melodrama; Bentley, in fact, asserts that "melodrama died with Balzac's generation." It is telling, I think, that Hofmannsthal, able man of both the legitimate and lyric stages and eloquent opponent of naturalism, held Balzac in the highest regard, calling him "the greatest, most substantially creative imagination since Shakespeare." It is also no coincidence that, when Brooks comes to describe the "radically ironic and anti-metaphorical" stance of melo-drama's critics, he is also describing how satirists of opera behave: "They set against the ambitions of melodrama an attitude of deconstructive and stoic materialism, and a language of deflationary suspicion."

Talk of melodrama's "mode of excess" and "victory over repression" brings Walt Whitman back to mind. As a few random Whitman lines suggest—

> O to drink the mystic deliria deeper than any other man!

> A man is a summons and a challenge,
> (It is vain to skulk . . .)

> Across the stage with pallor in her face, yet lurid passion,
> Stalks Norma brandishing the dagger in her hand . . .

—he was a poetic melodramatist of the most ecstatic and operatic kind. And it may be that a distinction he made in a little essay, "The Perfect Human Voice," will also aid in the defense of melodrama: "Emerson says *manners* form the representative apex and final charm and captivation of humanity: but he might as well have changed the typicality to voice . . . there is something in the quality and power of the right voice (*timbre* the schools call it) that touches the soul, the abysms." Voice, I would venture, is the essence of melodrama. The voice is the instrument for its charateristically full expressivity, emphatic articulation, and immediacy of response. The eruptive voice that refuses to go unheard is a principle element of melodrama, whose "intense, excessive representations of life," Brooks says, "strip the façade of manners to reveal essential conflicts at work—moments of symbolic confrontation."

This unwillingness to contain or mask emotion in a good-mannerly way is also a common trait of prominent Shakespearean and operatic characters. Emilia, in *Othello,* speaks for them all when she cries,

> No, I will speak as liberal as the north:
> Let heaven and men and devils, let them all,
> All, all, cry shame against me, yet I'll speak.

Here is an avatar of the melodramatic figure, refusing to be silent, at great personal risk (she is stabbed by Iago moments later). "Liberal" speaking is, in Whitman's phrase, the apex and final charm of the two dramaturgies. Hence the extraordinary demands on the "quality and power of the right voice" for their excessive utterance. No wonder Emerson, to manners bound, imagined a quiet Heaven with no melodrama in it at all. Whitman, writing elsewhere of Emerson's "singularly dandified theory of manners," comes to the conclusion that "to the real artists in humanity what are called bad manners are often the most picturesque and significant of all."* Shakespearean and operatic protagonists prove time and again that Whitman was right.

This polar antipathy between polite "manners" and liberal "voice" flared up in an amusing musical context on a Parisian boulevard in August 1829, when young Hector Berlioz ran into François Boieldieu, composer of over thirty operas, most notably *La Dame blanche*. Berlioz had submitted his third Prix de Rome cantata, on the subject of Cleopatra after the battle of Actium, and the jury, of which Boieldieu was a member, had just announced no first prize would be awarded that year. Berlioz records the conversation, he says, word for word in his *Memoirs:* "it was too remarkable for me to have forgotten it." Berlioz plays Whitman to Boieldieu's Emerson:

"My dear boy, what have you done? The prize was in your hands and you simply threw it away." [Boieldieu]

"I assure you, sir, I did my best." [Berlioz]

"That is exactly what we have against you. You should not have done your best. Your best is the enemy of the good. How can I be expected to approve of such things [striking rhythms, enharmonic progressions, etc.] when you know that what I like most is soothing music?"

"Sir, it's a little difficult to write soothing music for an Egyptian queen who has been bitten by a poisonous snake and is dying a painful death in an agony of remorse."

"Oh, I know you have an answer—you always do. But that proves nothing. It is always possible to be graceful."

"Gladiators could die gracefully, but not Cleopatra. She hadn't the knack—it was not her way. . . ."

The old man's parting shot was: "take this lesson to heart and be more sensible next year." As it turned out, Berlioz did win the next competition, but appearing graceful and soothing was not to be the Berlioz style or the style of the burgeoning Age of Melodrama. Berlioz's withering assessment of this conversation that had taken place so many years before makes clear how inappropriate the well-mannered style of Boieldieu was to the noisy, vigorous, vividly colored lyric stage of the mid-nineteenth century: "Yes, that was it, soothing music . . . even in the most violent situations; music that was not too dramatic, but lucid, rather colorless, safely predictable . . . a 'French gentleman's' art, dressed in tights and top boots, never carried away, always correct. . . ." Just a few months after this conversation occurred, Hernani's horn in Hugo's swashbuckling melodrama was heard for the first time, and Boieldieu's soothing neoclassical aesthetic vanished permanently from the scene.

It is hardly a wonder that Shakespeare's heroes and heroines should have proved such an inspiration at this time; for, as Berlioz says of Cleopatra, they do not have the knack of being graceful or quiet in moments of crisis. Getting carried away is of their essence. It would be nice to think that Berlioz' own deep affinity for them is partly explained by his own impulsive nature, which plunged deep and reached high. One contemporary wrote of him as a "true enthusiast, not at all talkative until the moment when his enthusiasm seizes him; then, he becomes full of eloquence."* Another described his "long brooding silences, with lowered eyes and a glance that seemed to plumb unimaginable depths—then a sudden dazzling recovery of spirits, a stream of brilliant, amusing or touching remarks, bursts of Homeric laughter." These assessments, in a curious way, might serve to describe the volatile dramaturgy of Shakespeare and opera, both tied to moments when passion erupts and eloquence flows dazzlingly forth. (Isaiah) 53:7—"He was oppressed, and he was afflicted, yet he opened not his mouth"—could not possibly be a motto for their greatest characters.

Perhaps this brief effort to rescue melodrama a little from its detractors should be prolonged by one more strategy. It is the same strategy Sir Philip Sidney used in his *Defence of Poesy* (1595), when he rebutted the "imputation laid to the poor poets" that poetry is the "nurse of abuse" because it is often put to vile uses by some of them. To this Sidney sensibly retorted, "But, what! Shall the abuse of a thing make the right use odious? . . . Do we not see skill of physic, the best rampire [defense] to our oft-assaulted bodies, being abused, teach poison, the most violent destroyer? Doth not knowledge of law, whose end is to even and right all things, being abused, grow the crooked fosterer of horrible injuries?" Melodrama, too, deserves to be judged by its finest, most honestly earned effects, not its more debased and risible manifestations. This means distinguishing between, say, *Othello/Otello* and *La Gioconda.* Ponchielli's opera (libretto by "Tobia Gorrio"—that is, Arrigo Boito), as a matter of fact, elicited a Shavian outburst on the "right use" of artistic means by geniuses and their "abuse" by second-raters. *La Gioconda,* Shaw wrote, is an "instance of the mischief which great men bring into the world when small men begin to worship them. Shakespear set all the dramatic talent in England wasting itself for centuries on bombast and blank verse; Michael Angelo plunged Italian painting into an abyss of nakedness and foreshortening. Handel plagued our serious music with a horrible murrain of oratorio . . . and Verdi is tempting many a born quadrille composer of the South to wrestle ineffectually and ridiculously with Shakespear and Victor Hugo."

Granting melodrama its best and worst scenarios might, to take another example, mean contrasting Shakespeare's *King Lear* or Verdi's *Rigoletto* with a typical libretto by the hugely successful Eugène Scribe. Wagner had a noisy good time satirizing Scribe's "bombastic and extravagantly odd" stage pieces with their long series of effects without causes. Of the mute heroine in Scribe's drama for Auber's *La Muette de Portici,* for instance, he wrote acidly, "This dumb girl was the dumb-struck Muse of Drama, who wandered broken-hearted between singing, raging throngs, and, tired of life, made away at last with herself and her hopeless sorrow in the artificial fury of a stage volcano!" But the true marriage made in the hell of crude melodrama was between Scribe and Meyerbeer. For them, Wagner wrote, the ideal libretto was "a monstrous, piebald, historico-romantic, diabolico-religious, fanatico-libidinous, sacro-frivolous, mysterio-criminal, autolyco-sentimental dramatic hotch-potch."

Hofmannsthal also had occasion to articulate the contrast between Shakespeare and Scribe—between high and crude melodrama—when Strauss half-jokingly wrote him about their *Arabella*-in-progress: "I should like to inject you with a little Scribe and Sardou, at the risk of your berating me as an old-fashioned trashmonger!" At the end of his letter Strauss added, "Don't be angry with me for wantonly goading you." Hofmannsthal responded, in a rather more serious vein, with an excellent way of distinguishing between the "right use" and "abuse" of melodrama: "I am by no means annoyed by your letters and suggestions, *my dear Dr. Strauss;* they are on the contrary of real service to me, and it is after all perfectly true that from time to time I can do very well with an injection of Scribe and Sardou. I have, on the other hand, the gift—which these two never possessed—of kindling in the characters a breath of life. Sardou's plays have been summed up as 'life through movement'—and against this the naturalists have pitted their motto 'movement through life.' The true aim of course is to combine the two, as Shakespeare did." In other words, the powerful effects and virtuosic utterances—the "movement"—of *high* melodrama will derive credibly from the imagined "life" on stage; whereas in *crude* melodrama "life" is a mere derivative of or pretext for the "movement" . . . mere stage busyness. Thus, Wagner in "The Destiny of Opera," a rather windy encomium to Shakespeare, called works of this latter sort *Effektstücke*. He mordantly defined an "effect" as "a stunning of the spectator's senses, to be documented by the outburst of applause."

Sidney's defense should apply, of course, to the *performance* of melodrama as well: judge it by its most expert, not its clumsiest, executants. John Webster, in fact, closed his 1615 sketch of "An excellent Actor" by echoing Sidney's argument: "I value a worthy Actor by the corruption of some few of the quality, as I would do gold in the ore; I should not mind the dross, but the purity of the metal." In any theater—like that of Shakespeare and opera—essentially focused on an executant's "soloistic" displays, the skill-less or self-indulging artist will almost inevitably be drawn toward wretched excess. Then, as now, he is often the party responsible when melodrama collapses into ludicrousness. Hamlet apparently saw in Wittenberg his share of players who "strutted and bellowed" and imitated humanity "abominably." Shakespeare must have, too, for he makes fun of the eyeball-roller in *Richard III* when Buckingham boasts,

> Tut, I can counterfeit the deep tragedian,
> Speak and look back, and pry on every side,
> Tremble and start at wagging of a straw;
> Intending deep suspicion, ghastly looks
> Are at my service, like enforced smiles . . .

Criticism of Shakespeare in performance through the centuries is rich in satirical deprecation of the crudely exaggerated interpretation. It is another reminder of the likeness of our two dramaturgies that this satire often glances provocatively at bad acting in the opera house. One thinks especially of several Verdi and Wagner baritones of past experience, for instance, at reading of James Quin, who flourished between 1715 and 1750 (and was ridiculed by Pope in *The Dunciad*): "he rolled out his heroics with an air of dignified indifference." One is almost tempted to name some operatic names now before the public after reading this about Garrick from the poisonous pen of his jealous rival, Charles Macklin: "His whole action, when he made love in Tragedy or Comedy . . . consisted in squeezing his hat, thumping his breast, strutting up and down the Stage, and pawing the Characters that he acted with." And how many a climactic moment in opera has one seen unfold as it did, according to Charles Churchill's *Rosciad,* when Spranger Barry held the stage?

> Some dozen lines before the ghost is there,
> Behold him for the solemn scene prepare:
> See how he frames his eyes, poises each limb,
> Puts the whole body into proper trim—
> From whence we learn, with no great stretch of art,
> Five lines hence comes a ghost, and, ha!, a start.

Byron likewise despised the strutters and fretters of the next generation and said of his play, *Marino Faliero,* "It is . . . nothing *melo*dramatic—no surprises, no starts, nor trap-doors, nor opportunities 'for tossing heads and kicking their heels.'" Walt Whitman, repeating the litany yet a generation later, complained of tragedians who "enact passion" by employing "all kinds of unnatural jerks, swings, screwings of the nerves of the face, rolling of the eyes, and so on."*

The caricature of melodrama is all too often the consequence of such caricatured acting. Though *melodrama* was not in Shakespeare's vocabulary, he recognized the dangers of melodramatic style. He knew that drama relying mainly on thespian virtuosity could lead to immodest o'erstepping of the "modesty of nature." This is surely why his trenchant connoisseur of acting warns the players to keep their wits about them: "do not saw the air too much with your hand, thus, but use all gently, for in the very torrent, tempest, and, as I may say, whirlwind of your passion, you must acquire and beget a temperance that may give it smoothness." The note of compromise that Hofmannsthal discerned in Shakespeare is present here in the call for "temperance"—with its implication of not only sobriety but also moderation and artistic control. Webster echoed this call for temperance when he wrote of the excellent actor as presenting nature "neither on Stilts nor on Crutches, and for his voice, tis not lower than the prompter's, nor louder than the Foile and Target [the sound of stage battle]."

Hamlet's plea for temperance is explicitly intended to assure that "The Mousetrap" will rise to an imitation of "humanity" and not degenerate into an *Effektstück.* It is a clear call for high rather than crude melodrama, and it has echoed often in the following centuries. In most cases, the admonition is germane to operatic performance. For instance, William Macready (to whom Dickens dedicated *Nicholas Nickleby* as a "token of admiration and regard") must have followed Hamlet's advice exactly when he assumed the title role in Byron's *Werner.* For he confided in his journal after a performance: "[A]cted very well. Preserved an erect deportment in the midst of passion, and let the mind act." Shaw perhaps made Hamlet's point most succinctly when he advised Laurence Irving, as Othello, to "shout cautiously." And much more recently the actor

Michael Pennington, during the episode "Passion and Coolness: A Question of Balance" in the *Playing Shakespeare* series, had occasion to rephrase the central dilemma of Hamlet's plea. His remarks occurred, appropriately, in a discussion of the role of Hamlet itself; change "language" to "music" and the dilemma becomes the central one for the opera singer: "The actor's need [is] to control the flood-tide of emotion and to discipline it mentally and technically. The means is the language through which, and only through which, that emotion can be fed. That is the essential eye of the needle . . . the language in the play is terrifically wrought and elaborate. And it is possible through an excess of feeling to distort that language instead of working through it." Distort the language or the notes with undisciplined emotion, and, to use Hofmannsthal's terms, thespian "move-ment" will overwhelm the "life" of the drama.

This failing afflicts Shakespearean and operatic performances all too often, simply because the large majority of artists lack the ability, as Pennington puts it, to thread the essential eye of the needle. But almost every generation affords a handful of superlative performers—the Callases, Oliviers, Fischer-Dieskaus, Giel-guds, Söderströms, Chaliapins, Duses, Vickerses, and so on—who make us see clearly the difference between dross and the purity of true interpretive metal.

Acting or singing of especially pure metal can sometimes, by a sort of alchemical magic, even make the cruder forms of melodrama glister as if they were golden. "The possibility of good acting in a bad play is familiar," writes Bentley; "less familiar is the fact that acting can actually change the genre of a piece, turning cheap melodrama, for example, into high melodrama." This is less familiar, of course, because the challenge is so much greater. Shaw suggested as much when he offered his recipe for the optimum performance of Meyerbeer's coarsely melodramatic works. Meyerbeer, he wrote, "requires extreme mechanical preci-sion of execution and unflagging energy and attention. Everybody must be smart, polished, alert; there must be no hitch, no delay, no fluking or trusting to *laisser-aller*." Because of their intrinsic greatness and dramatic interest, Shakespeare's more estimable plays can often succeed very well on considerably less than what Shaw here specifies. But his cruder efforts demand much more of the kind of concentration that Shaw demands for Meyerbeer. Successful stagings of them (one thinks of *Troilus and Cressida, Pericles, All's Well, Timon of Athens*) are rare and all the more to be admired. In 1955, for example, Olivier apparently raised the gory melodrama of *Titus Andronicus* to something of a very high order: "We knew that Sir Laurence could explode," wrote one admiring reviewer; "now we know that he can suffer."

Almost every operagoer can recall performances like Olivier's *Titus* where, through some lucky alchemy, a work of dubious artistic merit was transformed by unflaggingly energetic and alert performers. This seems to have happened at the Metropolitan Opera on the evening of 21 December 1945. Virgil Thomson reported:

> There is no denying that Ponchielli's *La Gioconda* makes a good show when they really sing it. It is tommyrot from beginning to end, skillfully varied, exciting tommyrot. But since hokum is chiefly what makes a good show anyhow, and since the musical

part of this particular hokum was written for tiptop show-off, the piece is capable of producing shivers no end when the musical execution is a bang-up one. And that is exactly what Emil Cooper, who conducted, and a brilliant cast of singers gave us last night. . . . [The final applause] was the most prodigious theatrical climax I have ever witnessed in that house. It wasn't about anything, because Victor Hugo's play and Arrigo Boito's libretto and Amilcare Ponchielli's music are not about anything, unless one counts the provoking of applause sufficient motivation for an opera. But the moment was thrilling and the applause whole-hearted; and we were all, I think, grateful for an experience so rare at the Metropolitan. Besides, hokum properly performed has a purity about it that is refreshing. It makes one feel good, like a shower bath, leaves a clean taste in the mouth the way a good murder story does. From that point of view Ponchielli's opera is one of the best.

Performances like this one—the cast that evening was Zinka Milanov, Risë Stevens, Margaret Harshaw, Richard Tucker, Leonard Warren, and Ezio Pinza— should make us pause and consider before we leap to condemn melodrama, even in its cruder "shabby little shocker" manifestations.

12

The Nature of Melodrama

The first memorable critic of melodrama was Macbeth, when he compared life itself with

> a poor player
> That struts and frets his hour upon the stage,
> And then is heard no more. It is a tale
> Told by an idiot, full of sound and fury,
> Signifying nothing. (5.4.24–28)

Macbeth covers all the usual complaints: the noise and mayhem vastly in excess of their cause, the idiotic storyline, the excessive gestures (strutting, not striding) to diminished effect (fretting, not suffering). "Poor" opera can often strike us in the same way. Emerson attended a performance in Naples in 1833 and wrote in his journal afterward in Macbeth's deprecating vein: "I could not help pitying the performers in their fillets & shields & togas, & saw their strained & unsuccessful exertions & thought on their long toilette & personal mortification at making such a figure. There they are—the same poor Johns & Antonios they were this morning, for all their gilt & pasteboard."

But hold . . . consider the performance and the critic. Macbeth is speaking, as Hamlet does, only of the *poor* actor, and what play is immune from the miserable cheapening of a "robustious periwig-pated fellow" who seeks to "tear a passion to tatters, to very rags"? Consider, too, the wisdom of applauding a critic so terminally jaded by life and the theater. Macbeth's harsh reduction of the theatrical event comes *after* he has ceased to be a slave of passion. He has arrived at a flat, dessicated nihilism: "I have liv'd long enough: my way of life / Is fall'n into the sear, the yellow leaf." It is, I think, not entirely coincidental that Shakespeare put his most wonderful expression of cynical negation in the terms of a theatrical "bad review." This is perhaps the slyest and most poignant of several Shakespearean slaps at those skeptically inert, unhelpfully unimaginative members of his audience who would sit on their hands and resist being moved or delighted.

Anyone who is tempted to paint the tragedy or high melodrama of Shakespeare in crude terms is thus bound to sound, to his or her discredit, a little like Macbeth. Thomas Rymer certainly did, in his notorious lambasting of *Othello* in 1693. He called the play "a Bloody Farce, without salt or savour" and drew only these two "morals" from its action: "A caution to all Maidens of Quality how, without their Parents' consent, they run away with Blackamoors" and "a warning to all good Wives that they look well to their Linnen." Shaw, in one of his on-the-rampage fits, sounded a similar note, referring to *Othello* as containing "its noble savage, its villain, its funny man, its carefully assorted pathetic and heavy feminine interest, its smothering and suicide, its police-court morality and commonplace thought."

But *Othello,* though full of sound and fury, is not a tale told by a theatrical idiot. It signifies something rather than nothing; though often a strutting Othello (like James Earl Jones several years ago on Broadway) can obscure this fact. No amount of Rymerite or Shavian bluster will convince us otherwise, even though the bluster may contain undeniable grains of truth. So it is very pleasant to read Dryden's magisterial put-down of Rymer: "Almost all the Faults which he has discover'd are truly there; yet who will read Mr. Rymer, or not read Shakespeare? For my part I reverence Mr. Rymer's Learning, but I detest his Ill-Nature and his Arrogance." And it is also pleasant to add hastily on Shaw's behalf that in the same passage just quoted he calls *Othello* a "splendid melodrama." Elsewhere he praises the play to the skies.

How do the seemingly unpromising materials of melodrama issue in such splendor? Clearly, Macbeth—beyond caring about life or death—is the least likely of Shakespeare's characters to tell us. His enervated "to-morrow, and to-morrow, and to-morrow" suggests how distant he is from melodrama's impulsive energies. His famous speech is really more appropriate for *Waiting for Godot* than it is for the other great plays of the canon. Much more in the spirit of Shakespeare's melodramatic style is another very famous imaging forth of human life: the "To be or not to be" soliloquy. For the playwright's most passionately engaged and volatile hero, life does not creep in a "petty pace" but rather is a "calamity." The impassioned sound and fury of many a Shakespearean action echo in Hamlet's arousing diction: "outrageous fortunes . . . sea of troubles . . . heart-ache and the thousand natural shocks / That flesh is heir to." And several Shake-spearean protagonists must, in Hamlet's phrasing,

> bear the whips and scorns of time,
> Th' oppressor's wrongs, the proud man's contumely,
> The pains of despis'd love, the law's delay,
> The insolence of office, and the spurns
> That patient merit of th' unworthy takes.

There is nothing petty about the pace of such theatrical life as this; it is the pace of melodrama. Thus, Robert Heilman in *Tragedy and Melodrama* seems to be shadowing Hamlet's soliloquy when he summarizes, "Melodrama typically of-fers . . . the exaltation of victory, indignation at wrongdoing [Lear in a nut-shell], the pitiableness of victims [Desdemona], the frustrations of the indeterminate outcome [Hamlet], the warming anticipation of courage [Prince Hal's career], the despair of defeat, the shock of disaster, the sadness of death." We have seen in earlier pages how such huge events as these elicit poetic and musical effusions. It remains, then, to explore some of the elements of melodramatic style that make possible this potent focusing of expressive energies.

Perhaps the most obvious and important melodramatic hallmark of our two dramaturgies is the radically simplified context of their complex rhetorical set pieces. Simplification—the suppression of multiplicity of effect for unity of effect—is a fundamental melodramatic mode, whether simplification of person-ality, of morality, of human relationship, or of psychology. Both Shaw and Brooks emphasized this in their discovery of "broad contrasts between types" and "pure and polar concepts" in melodrama. Bentley says flatly that "melodrama presents the struggle between right and wrong," and Shakespeare's plays are in this respect often thoroughly melodramatic. After all, only a few generations separate them from the time of the morality plays, in which Good Deeds encountered Riot, or Sensual Appetite did battle with Humility. Hal's friendship with Falstaff is, at its core, the long and ultimately virtuous duel of Youth with Vanity. To take another example, Humility (Octavia) hasn't a prayer against Sensual Appetite (Cleopatra) in the struggle over their Everyman (Antony). And it is easy to view the skeletal plot of *Othello* as Everyman's failure to choose correctly between the "love" of Good (Desdemona) and Evil (Iago). Such rigid, clear moral distinctions are a Shakespearean and operatic commonplace, though of course there are many characters who embody or relate to them ambiguously. The lines from *Hamlet* just quoted, for instance, imply several very melodramatic antitheses: oppressor vs. oppressed, proud vs. humble, lover vs. despiser, and violent spurner vs. patient sufferer. The great advantage of factoring human experience into such "primal ethical forces," as Brooks calls them, is that the exciting moments when they burst forth or collide with each other can be more easily translated into focused, immediate, compelling utterance.

The exhibition, by heroic executants, of primal ethical or emotional forces is the heart of most Shakespearean and operatic actions. The "king-becoming graces" enumerated by Malcolm are balanced by the tyrant-becoming graces of Macbeth; Richard's otiose vanity contrasts with Bolingbroke's efficient *realpolitik* in *Rich-ard II*. Hamlet's acute sense of right and wrong is effectively set against the amoral complacency of almost everyone else at Elsinore. Hotspur and Falstaff are the perfect Aristotelian excess and defect of honor in *1 Henry IV,* while Edmund and Edgar (like Goneril, Regan, and Cordelia) in *King Lear* embody the worst and best in filial devotion. Perhaps the most spectacular antithesis of all came last: Ariel and Caliban in *The Tempest*—the "airy spirit" as agent of human wisdom, the "salvage and deformed slave" a distillation of human ignorance. Cosmic antitheses of this sort are prominent everywhere in opera, too. One thinks (to mention but a few) of Venus and Elisabeth in *Tannhäuser,* Turandot's *eros* and Liù's *agape,* Pizarro's ruthlessness and Don Fernando's magnanimity in *Fidelio,* Sachs' lyric wisdom and Beckmesser's *buffo* rule-mongering in *Meistersinger,* Sarastro's booming humanism and the Queen of the Night's ear-piercing vindictiveness in *Die Zauberflöte,* or Posa's plangent liberalism and the Grand Inquisitor's abysmal, *basso profondo* cynicism in *Don Carlo.*

Melodrama, being founded on such striking but common moral antitheses, leads naturally to simplified modes of characterization. "Melodramatic good and evil," writes Brooks, "are highly personalized: they are assigned to, they inhabit persons who indeed have no psychological complexity but who are strongly characterized." Detractors of melodrama and opera often call their characters flat, two-dimensional; sometimes even one-dimensional (an impossibility, since nothing can exist in only one dimension). Bentley, with some sarcasm, took the offensive against those who prefer Real Human Beings to allegedly flat characters. "An alternative expression" for the RHB, he writes, "is the Believable Human Being, and the opposite of this is a Type. A Type is a Bad Thing." Bentley ends by challenging, "is there not a place for the flat character?"

There certainly is in Shakespeare and opera. Many of the most memorable characters—to say nothing of most of the great comprimario roles—are, as radical simplifications of normal human volition and expression, very flat indeed: Giovanni, Lady Macbeth, Carmen, Amneris, Desdemona, Falstaff, Wozzeck, and countless others. If these characters are flat, then flatness is clearly not such a Bad Thing after all. We can be very grateful that Verdi was not put off in the least by the fact that Desdemona is not an RHB but a Type. "A female who allows herself to be mistreated, slapped, and even strangled," wrote Verdi, "appears to be a stupid little thing! But Desdemona is not a woman, she's a type! She's the type of goodness, of resignation, of sacrifice! They are creatures born for others, unconscious of their own ego! Beings that only partially exist and that Shakespeare has poeticized and deified." Or, as Shaw said of Desdemona's dreadful opposite, Lady Macbeth, "there never was no such person!"

Heilman calls flatness by a more unusual but perhaps more useful name when he comes to describe the heart of a typical melodramatic character: "monopathy." Seeking to distinguish tragedy from melodrama, Heilman suggests that "the identifying mark of the tragic character is dividedness . . . his nature is dual or multifold," and he is always tending toward "a world of self-awareness and contemplation." Melodrama, on the other hand, postulates in its characters a "wholeness or oneness of being." Heilman's term for this is monopathy, or "the singleness of feeling that gives one the sense of wholeness." And so he speaks of the "monopathic pleasures of melodrama."

Monopathic characters might, then, be called not flat but "centric" . . . in the sense of being completely integrated within a very specific compass. Their stage existence is limited to feeling emotions, expressing their desires, or viewing the world in one way. This dominant trait or worldview fills out their character completely and is their dramaturgical reason for existence. Such is the case with Canio's jealous paranoia, Coriolanus' reckless refusal to do in Rome as Romans do, Rosina's coquettishness and Bartolo's domineering bluster in *Il Barbiere,* Lady Macbeth's and Iago's lust for power (political for her, social for him), the outrage of wronged Donna Anna, and so on. Often, through poetic or musical methods of intensification, monopathy grows into mono*mania:* the feckless

macho fury of Osmin in *Die Entführung,* the comic and tragic zeal of Strauss'
Zerbinetta and Ariadne, Leontes' jealous neurosis in *The Winter's Tale,* the
remorse of conscience in Prince Hal's father and Boris, the unstoppable pedantry
of Holofernes in *Love's Labour's Lost* and Beckmesser in *Die Meistersinger,* and of
course Elektra's terrible fixation.

Writing exclusively about operatic characters, W. H. Auden comes very near to
paraphrasing this notion of monopathy. Like the monopathic character of melodrama, Auden's typical operatic figure must shy away from dividedness of
contemplation: "Opera . . . cannot present character in the novelist's sense of
the word, namely, people who are potentially good *and* bad, active *and* passive,
for music is immediate actuality and neither potentiality nor passivity can live in
its presence." By way of example, Auden says he finds Rossini's Figaro more
satisfying than Mozart's because Da Ponte's barber "is too interesting to be
completely translatable into music." Thus, "co-present with the Figaro who is
singing one is conscious of a Figaro who is not singing but thinking to himself.
[Rossini's] barber of Seville, on the other hand . . . goes into song exactly, with
nothing [left] over." Even if one is inclined to balk at this particular judgment,
one can accept Auden's general point that opera's monopathic characters bring
their own rewards: "The quality common to all great operatic roles is that each of
them is a passionate and willful state of being. . . . In recompence for this lack of
psychological complexity, however, music can do what words cannot, present the
immediate and simultaneous relation of these states [of being] to each other."*
The focused and forthright expressions of the monopathic character are also
reflected in Auden's assertions that "music is immediate, not reflective; whatever
is sung is the case" and that "music cannot exist in an atmosphere of uncertainty;
song cannot walk, it can only jump."

In a curious and highly suggestive way, Auden's observations have some
relevance for Shakespeare's characters, who are usually most compelling when
they are jumping in heightened blank verse or prose rather than merely walking in
naturalistic language. Their memorable moments at center stage often find them
expressing their "passionate and willful states of being" in undoubtful, energetic
declamation. Shakespearean rhetoric, like music, can scarcely thrive in an atmosphere of uncertainty. Even Hamlet, the most famously indecisive of all tragic
heroes, spends much of his time uttering magnificent declarations of certainty:
certainty that he is behaving horribly ("O, what a rogue and peasant slave am I!"),
certainty that fear of death is a great thing ("To be, or not to be"), certainty about
proper acting style, certainty that his mother is a ghastly woman, certainty of what
he ought to do ("How all occasions do inform against me"), and certainty in the
graveyard that all greatness passes to dust. Shakespeare's characters, like many in
opera, are most in their element when declaring themselves with dead-certain and
extravagant abandon. Howard Moss captured this melodramatic essence in his
witty poem, "Elizabethan Tragedy: A Footnote":

> . . . tears or throwing things or being pushed downstairs,
> Have value in the long run. Caution has its place.
> Premeditation, though, I think when face to face

With *sturm und drang* can never win the race.
Although the Prince is on the angels' side,
What got him there is wholesale homicide.

The "storm and stress" movement among German Romantics of the late
eighteenth-century was heavily influenced by Shakespeare's plays, and this movement in turn was a principal influence on nineteenth-century librettos and
melodramas. Moss's reference to *sturm und drang*, then, offers a convenient
transition to the second aspect of melodrama that is ubiquitous in Shakespeare
and opera: the eagerness for bold, powerful effects that will elicit emotional storm
and stress. This eagerness to reach for sublime effects—*sublime* derives from the
Latin for "uplifted"—is so commonplace in our two dramaturgies that we need
not linger long on the point. One quick entry, in fact, might simply be to notice
how an ancient treatise, *Longinus on the Sublime*, written nearly two millennia
ago, captures the central goal of uplifting the core of an action to a level of high
intensity. Longinus praises sublimity for creating "noble exaltation," "extravagance of grandeur," and "emotional intensification." Much of our previous
discussion of set speeches and arias is ratified, in a sense, by Longinus' observation
that nothing contributes to sublimity "as much as noble passion in the right place;
it breathes the frenzied spirit of its inspiration upon the words." Longinus
describes two main styles of sublime expression: Cicero's, which is grandiose and
oceanic, and Demosthenes', which strikes more like lightning. Clearly, the
Demosthenic style is that of drama tied to powerful local effects: "he is always
forceful, rapid, powerful . . . more suited to moments of intense and violent
passion when the audience must be altogether swept off its feet."

Shakespeare's devotion to such a dramaturgy needs no demonstration. *Richard
III*, so aptly reflected in all the above phrasing, was his first big success, and it
parades a series of one bold effect after another. As we shall see in the following
three chapters, virtually all the lesser and greater plays that followed can be viewed
as shrewd serial arrangements of multiple *coups de théâtre*. His final plays, *The
Tempest* and *Henry VIII*, are especially brilliant displays of this method. The
theory behind it, one might say, was expressed by Shaw when an actress
complained that an exit speech he had written for her was "terribly stagey" and he
replied, in the words of a true Shakespearean melodramatist, "You can't be too
stagey on the stage."

The impetus toward impressive local effects is also central to libretto plotting.
Joseph Kerman said of Claudio Monteverdi's *Orfeo*, which was first performed
just about the time Shakespeare was writing *Antony and Cleopatra:* "One virtue
of the libretto is that it gives Orpheus many occasions to react violently in
moments of crisis: the news of Eurydice's death, the departure of Hope, the
rejection of Charon, the return of Eurydice, and her second death." This virtue—
and it is the same virtue Shakespeare's scripts offered his protagonists—marks
most librettos in the following several centuries. Verdi, for instance, constantly
hounded his librettists for big effects. Contemplating *Trovatore*, he said (sounding a bit like Shaw on staginess) how he hoped his librettist would operate: "the
more novelty, the freer the forms he presents me the better I shall do . . . the

bolder he is the happier he will make me." Antonio Garcìa Gutiérrez' *El Trovador* appealed to him, he said, because it offered "fine theatrical effects and above all something original and out of the ordinary," and when Salvatore Cammarano complained about a coup that pleased the composer, Verdi defended it as a moment for sublime impression: "The scene where Leonora takes the veil must certainly be left in; it's far too original for me to give it up; rather, we must make as much of it as possible and get all the effect we can." Verdi's view prevailed . . . and the scene that resulted is one of the opera's great moments.

Inevitably, this exploitation of bold effects often succeeds at the expense of credible plotting and neatly tied plot-threads. This price Shakespeare was obviously often content to pay, simply because he knew a powerful effect well managed would, in performance, annihilate niggling objections. The most famous example of his cavalier attitude toward mere credibility occurs in *Othello*. Shakespeare, knowing that a scheme like Iago's must work quickly or not at all, arranges for the terrible events on Cyprus to unfold in thirty-three hours ("short time"). In order to allow time for the alleged adultery to take place and for the order recalling Othello to Venice to reach Cyprus, however, Shakespeare was forced to scatter several lines through the action that give the impression of a "long time" elapsing. This legerdemain, outrageous to credibility, is impressive, not least because it went unnoticed for over two hundred years. An editor of the play has justly said that this double time scheme "throws light on Shakespeare's astonishing skill and judgment as a practical craftsman. He knew to a fraction of an inch how far he could go in playing a trick on his audience, and the measure of his success is precisely the unawareness of the audience in the theater that any trick is being played."* He knew the bold effects and cumulative momentum of *Othello* would make chronological discrepancies vanish into thin air.

One often feels this same *force majeure* at work in opera performances. Clearly, Verdi was depending on this force in *Trovatore*, which, as William Weaver says, "tends to sweep us away. The experience of this powerful music is so immediate, so compelling that we have no time to worry about the presence or absence of nuances or subtleties." Julian Budden also finds occasion to observe how minor a defect illogic of plot can be when great music engulfs it as he compares the 1857 *Boccanegra* with the much finer version of 1881: "the fact that no writer has ever drawn attention to the flaws in the revision, though the first libretto with its more logical plot and clearer motivation was a constant target of abuse, is surely proof of how unimportant such flaws are in the context of the opera as a whole."

When the presentation of riveting effects is a primary order of business, detail and integrity of characterization can also suffer. Shakespeare could not—perhaps would not—have complained at Samuel Johnson's conclusion that *Macbeth* contains "no nice discrimination of character." Crudeness there assuredly is in the rather expeditious hardening of Macbeth's metal and the melting of his wife's— especially in her sudden appearance, as if lobotomized, in the sleepwalking scene. It is a brilliant but shameless coup. Utterly unconcerned that some fools might stop to wonder over such a drastic change in character or some other will-o'-the-wisp riddle like how many children had Lady Macbeth ("I have given suck," says she), Shakespeare was after much greater theatrical returns. He would surely have judged himself successful by the praise that Johnson lavishes on *Macbeth* in the same breath as he deplores the characterization: "This play is deservedly celebrated for the propriety of its fictions, and solemnity, grandeur and variety of action." Johnson's conclusion that "passions are directed to their true end" in *Macbeth* would have especially pleased him.

The melodramatic calculus that urges boldness at the expense of complexity of character Granville-Barker also confronted head-on when he came to Iago's defense: "Points of view will remain, for which a line drawn between Iago and the villain of melodrama is so fine as to be invisible. But melodrama is not necessarily false to life; it may only unduly simplify it. And Shakespeare's problem was to retain the melodramatic simplicity with the strength which belongs to it, and give this an inward verity too." This same "inward verity" is what distinguishes first-rate from second-rate characters in opera. Hofmannsthal, in fact, repeated this general idea of theatrical strength-through-simplicity when he warned Strauss not to encumber their *Arabella* scenario with too many details: "otherwise we shall have a room all cluttered up! Each motif must have space enough to unfold itself in terms of character, that is Shakespeare's wonderful secret."

A third characteristic follows inevitably from melodrama's urge toward simplification and eagerness for bold effects: the tendency toward immediate and exhaustive vocalism. This is a natural consequence of richly rhetorical stage writing, which exists in part to display executive virtuosity. The typical melodramatic scene, according to Brooks, will present "a climactic moment at which the characters are able to confront one another with full expressivity, to fix in large gestures the meaning of their relations and existence." Life in melodrama, he continues, tends "toward ever more concentrated and totally expressive gestures and statements." What will make these moments of victory over repression all the more effective, of course, is the suddenness or unexpectedness of their onset, and this is a common event in Shakespeare and opera. The "spontaneous and uninhibited way of seeing things" that Bentley finds at the heart of the melodramatic style is the great highway to the spectacular speeches and arias or ensembles of the two dramaturgies. Auden, as we have seen, several times identifies music's immediacy of effect as both the great limitation and the glory of opera, and Shakespeare's rhetoric, too, gains much of its power by sudden and full eruption. Bernard Beckerman's conclusion about Elizabethan theater is worth quoting again in this regard: "Implied emotion is not characteristic of the period. . . . Passions were immediately and directly presented. A character revealed the full extent of his passion at once." Beckerman goes on, by way of summary, to say that within the conventional framework of the Globe Theatre style of storytelling and scene-building "there operated a spontaneous, lyrical, and intensely emotionalized reality." A happier phrasing of the primary method of opera could hardly be wished.

Highly rhetorical drama that grants emotions complete expressibility carries with it some obvious risks. One is that characters eager and adept at externalizing their passions may begin to be perceived as not having much interior. Their heart will often seem to be on the sleeve rather than in its more normal and "realistic" position. Brooks states this debit flatly: "It is delusive to seek . . . the 'psychol-

ogy of melodrama,' because melodrama exteriorizes conflict and psychic structure." The only response to this fair aspersion must be that the genius by which
complete expression is achieved must succeed by *force majeure*. We have seen
Shaw boast of his ability to create characters notable for their "freedom from
inhibitions," and this ability, he said, set him "or Shakespear" apart from the mere
"representationist or realist." On another occasion he explained that he learned
this power from Handel: "It was from Handel that I learned that style consists in
force of assertion. If you can say a thing with one stroke unanswerably you have
style. . . . Handel had this power. You may despise what you like, but you
cannot contradict Handel."* This irresistible eloquence, this ability to make the
"unreal" persuasive by sheer force of rhetoric, is akin to the sublimity of
Longinus, which "appears suddenly [and] like a thunderbolt carries all before it."
This sudden sublimity lies at the core of the Shakespearean and operatic aesthetics. Perhaps this is why we tend to enjoy their heroes and heroines in performance much as Auden enjoyed Shaw's: "His plays are a joy to watch
. . . because the energy shown by any of his characters is so wildly in excess of
what their situation practically requires that if it were to be devoted to anything
'worthwhile,' they would wreck the world in five minutes."

Budden captures this dilemma of externalized emotions sapping inner complexity in an illuminating contrast of the styles of Verdi and that least Shakespearean of the great opera composers, Wagner. "Even at its most extrovert,"
writes Budden, "there is an 'inward' quality about Wagner's mature music. . . . His characters are concerned with thought rather than action. Tristan,
Wotan, Sachs, Gurnemanz—even Siegfried and Brünnhilde in their later moments are above all ruminative creatures." In contrast, "Verdi's characters are less
introspective. Initially they are concerned merely to vent a primitive, undifferentiated emotion whose force harmonic elaboration would only diminish." Verdi's
typical concern, Budden adds, is to depict "the release of emotion; Wagner prefers
to dwell and brood on the emotion itself." Verdi's characters, in other words, tend
to believe, in Auden's phrase, that "what is sung is the case." In this they are
profoundly allied to Shakespeare's characters, who make their mark when
expressing their passionate slavery with forthright immediacy. Thus, Budden's
conclusion about the essential Verdian style has some application to the Shakespearean style as well: Verdi, he writes, was not "concerned to present dramatic
experience on more than one level. Strongly felt emotion must come across the
footlights unencumbered by premonition or memory."

Another risk accompanying "freedom from inhibition" has more to do with
performance and will be discussed in Chapter 18. But it ought to be noted briefly
here: the risk that characters whose stage life is a series of huge, exhaustive
eruptions of emotion will demand extraordinary technique of the performer.
Martin Meisel has observed that Shaw "asked for a new mode of heroic acting
[and] sustained verbal heroics," and Shaw himself warned that the "precipitous
moments" in his plays "require a special technique of acting, and, in particular,
great virtuosity in sudden transition of mood." This same virtuosity is a *sine qua
non* for the many great Shakespearean roles that either require a chameleonic
assumption of multiple roles or require the striking of diverse, isolated emotional

postures during a performance. One thinks, in this regard, especially of Lear, Iago, Hamlet, Prince Hal, Cleopatra, Richard III, Edgar, and Prospero. Many operatic roles, as we have seen, require a similarly kaleidoscopic vocalism and thespian resourcefulness. And many a heroic operatic role makes demands that urge an artist to the limits, as, say, Othello does. "There are too many climaxes, far too many," said Olivier of the role, its "fits and raging . . . all demanding you to scream to your utmost."

Shaw says that *Othello* is "pure melodrama," and yet he quickly adds that it "remains magnificent by the volume of its passion." Containing this huge volume, as Olivier's complaint shows, is a daunting task. This brings us to a last characteristic of melodrama, which follows as a corollary to the urge toward full expressivity: the impetus toward excess and exaggeration. Brooks, we have seen, simply defines melodrama as the "mode of excess," while Heilman writes that melodrama tends to "cater to longings that can easily get out of control." The volume of passion, as it were, is always threatening to overflow the role or the play itself. When it does, the effect is usually ludicrous, bombastic, or sentimental. As with our discussion of rhetoric's heightening powers, we are again entering the arena of emotional hyperbole. Heilman, in fact, calls hyperbole (the word for exaggeration or "loud lying") the "melodrama of rhetoric," and he nominates Marlowe's *Tamburlaine the Great* as the quintessential Elizabethan melodrama because it is "dramatized hyperbole." The sources of expression in this play, Heilman suggests, are "unity of energy, drive, the violent pressure from within . . . [the] exuberance of physical strength and psychic zest." This seems to me the *métier* of many Shakespearean and operatic characters as well; the most memorable ones invariably suffer violent emotional pressures from within and show tremendous psychic zest. Heilman also finds a constant "pressure for escalation" in *Tamburlaine*'s melodrama, and this pressure we have hitherto noticed is always at work in our two dramaturgies. Indeed, Heilman finally comes to the comparison himself: in *Tamburlaine* "a speech is an aria; the atmosphere is operatic; every king is a vocal virtuoso."

As is the typical Shakespearean play or opera, melodrama is founded upon an aesthetic of escalation. Melodrama is "constantly reaching toward the 'too much'" (Brooks) and always arranging for characters to be "swept off their feet by placing them in situations which are too tragic or fantastic for 'words'" (Auden). Tested by such trials, these characters become themselves hyperboles of experience . . . bigger than life. Sibyl Thorndike thought Oedipus was Olivier's finest role because, though he tried for realism, "something in himself made it larger and bigger." The most successful performers share this power of enlarging characters beyond life-size. They must have a little fearlessness of hyperbole, as Olivier did with Othello: "I tried out things extravagantly, way over the top. . . ." These performers must have a little of the attitude that Shaw urged upon Stella Beech in playing the role of Raina in *Arms and the Man:* "You do not play the part with sufficient enjoyment: you should wallow in it, and pile passages up to the skies. You are only 6-1/2 feet high when you say them: you ought to be 16."

Of all the invitations to hyperbole in Shakespeare and opera, one perhaps stands out above the rest and deserves mention by way of conclusion. This is the

hyperbole of idealism . . . idealism that makes characters unwilling to compromise their monopathic visions of happiness and of the so-called real world. It is
often idealism that sends characters into the upper atmosphere of blank verse or
high notes. And heroic idealism is often given scale by the life-size sensibleness of
an attending comprimario. The Nurse tells Juliet that, with Romeo banished, she
should snap up Paris ("Romeo's a dishclout to him"), and Juliet asks, incredu-
lously, "Speak'st thou from the heart?" The great characters, of course, do: it is
Juliet's speaking from the heart that produces some of her play's most levitating
poetry. Suzuki offers Butterfly some similarly sensible, down-to-earth advice.
Butterfly replies, "You are lacking in faith! Listen" . . . and then launches into
"Un bel dì." Some of the most riveting conflicts in Shakespeare or opera are
between such idealists with access to higher planes of existence and such sensible
realists, whose roles tend to creep in a petty rhetorical or musical pace.

This powerful antagonism or antithesis between the heroic-scale voice of
idealism and the human-scale voice of realism or "normality" Hofmannsthal had
occasion to describe to Strauss as they planned *Ariadne auf Naxos*. The distinction he makes gives us many of the great roles in both repertories:

> [*Ariadne*] is about one of the straightforward and stupendous problems of life:
> fidelity; whether to hold fast to that which is lost, or to get over it, to transform
> oneself, to sacrifice the integrity of the soul and yet in this transmutation to preserve
> one's essence, to remain a human being and not to sink to the level of the beast, which
> is without recollection. It is the fundamental theme of *Elektra,* the voice of Elektra
> opposed to the voice of Chrysothemis, the heroic voice against the human.

It is, of course, the heroic voice that takes center stage in the two dramaturgies.
More recently, the composer John Eaton ventured an even more categorial
version of this antithesis in an essay on how to choose plots for opera: "Part of my
quest has been to restore heroism and nobility . . . to opera. That banal,
ordinary persons should sing as the major characters is absurd and misses part of
the mission of opera: to infuse into a society, through the powerful means of
music and poetic vision, high values and purpose."*

It is the hyperbolic thrust of melodrama to seek out moments of heroic human
consciousness (whether noble or ignoble): moments that, through rhetorical
supercharging, assume an almost hallucinatory, larger-than-life magnitude.
Eaton expresses his eagerness for such moments when he ends his essay with the
demand for "subjects at a sufficient remove from ourselves and resplendent with
dignity." The child who defined opera as "a song of bigly size" was saying the
same thing. Melodrama, seen in its best light, is an art of epiphany rather than
lucidity; it depends on strong feeling rather than strong intelligence or strong
powers of ratiocination. It is not (to recall Kirstein's aperçu about ballet) about
how to behave, but rather about how to feel ideally and articulate completely,
propriety and safety be damned.

The experience of Shakespearean or operatic melodrama can seem incredible
and downright pitiable when we detach ourselves, Macbeth-like, from the
transactions on stage or when the performance is mediocre. But when the
melodrama is melodrama of genius and the performers attack it with full

conviction, the effect should be levitating. It would appear that this happened, against all of Emerson's expectations, when he attended that opera in Naples in 1833. For the journal entry I quoted at the outset continues:

> But the moment the prima donna utters one tone or makes a gesture of natural passion, it puts life into the dead scene. I pity [the singers] no more. It is not a ghost of departed things, not an old Greece & Rome but a Greece & Rome of this moment. It is living merit which takes ground with all other merit of whatever kind—with beauty, nobility, genius, & power. O trust to Nature, whosoever thou art, even though a strutting tragedy-prince. Trust your simple self & you shall stand before geniune princes.

Melodramatic hyperbole is a most efficient means of producing this kind of exaltation. And, happily for the present argument, not only the *ecphonesis* but also the admonition of Emerson's "O trust to Nature" is Shakespearean: "hold as 'twere the mirror up to nature."

DRYDEN" Beware the fury of a patient man"
quoted by Arthur Ashe

13

Kaleidoscope;
or, Ringing the Changes

"To use two hours of public time," the director Peter Brook has said, "is a fine art." Shakespeare, who spent much effort accommodating the desires and limitations of his audience, recognized this. Indeed, it is tempting to imagine that he wanted to offer his thoughts on the fine art of play-making when he came to his last one, *The Tempest*. For Prospero's success is not only moral, paternal, and political, but also dramaturgical. All theatrical producers and directors should like to be able to say, as their work hastens to its denouement, what Prospero says at the beginning of the final scene:

> Now does my project gather to a head:
> My charms crack not; my spirits obey; and Time
> Goes upright with his carriage.

In the preceding chapters we have encountered in isolation, rather like a handful of balloons, the rhetorical elaboration of passion, personality, volition, and belief in the form of set speeches and arias. And we have discussed at some length the theory and nature rather than the practicalities of melodrama. We must now begin to observe how these balloons are anchored in the reality of the theater. We must consider how they are kept in one dramatically useful location . . . how, that is, the expressive virtuosities of characters are embraced within a "project" that gathers to an effective head and also unfolds with an "upright" timeliness. It is therefore necessary to explore how Shakespeare and the opera composer give their rhetorical extravaganzas (to borrow a phrase form Duke Theseus in *Midsummer Night's Dream*) a "local habitation and a name."

The composer quite literally inhabits the "aery nothing" that Theseus speaks of: the sound waves upon which his aural "forms of things unknown" travel. This gives him extraordinary mobility and freedom: the power of instant transition, worthy of a Puck or an Ariel, is at his command. Shakespeare used these airwaves, too, but another "aery nothing" that similarly liberated him was the empty platform on which his actions transpired. At the Globe it was 1,000 square feet or so. Brook, a brilliant exploiter of the acting-place-as-airy-nothing, has called the absence of scenery one of the "greatest freedoms" of Shakespearean drama and has noted how the neutral, open platform "enabled the dramatist effortlessly to whip the spectator through an unlimited succession of illusions."* Well . . . numerous, but not unlimited, for an audience's power of endurance is finite. As any theatrical performance moves into its third hour, the contest between engagement and weariness becomes very keen. Even art of genius does not always survive much beyond this threshold of boredom . . . or the threshold of pain induced in the *gluteus maximus* by too-long sitting or in the legs by too-long standing. I hope to go to my grave without having experienced a completely uncut *Hamlet* or *Richard III*. Doubtless this would leave one feeling like Tchaikowsky, who exited from *Götterdämmerung* "as though I had been let out of prison." And he added, "The *Nibelungen* may actually be a magnificent work, but it is certain that there never was anything so endless and wearisomely spun out." And even so devoted a Mozartean as Richard Strauss could assert with, to judge from my experience, some reason: "No unbiased listener can avoid a considerable sense of fatigue during the third hour of a *Don Giovanni, Figaro,* or *Così fan tutte.*"

Herein lies the art of using "public time": recognizing, in purely practical terms, how strictly limited it is and how counterproductive it can be to overstay one's welcome in the theater. Creators for the stage must listen carefully to time's winged chariot. Prospero certainly does: to Ariel he warns at the outset, "The time 'twixt six and now / Must by us both be spent most preciously" (Elizabethan plays began at 2:00 P.M.). He shows constant care that no planned effect comes tardy off. The actor who plays Prospero says in the epilogue that his real "project" was "to please" the audience, and the entire action of the preceding play has (among other things) demonstrated how theatrical magicians achieve this goal: split-second timing, alternative exercizing and relaxing of attention, and the aesthetically harmonious gathering of a variety of effects into, as Ferdinand says of the masque, "a most majestic vision."

The most crucial word when one comes to assess Shakespeare's dramatic architecture—the word that suggested the title of this chapter—is *variety*. This word, which lies behind Brook's notion of Shakespeare whipping his spectator through a succession of illusions, figures in many attempts to summarize the playwright's method. Colley Cibber in 1740 wrote of the "high and fresh variety" of his plays. Granville-Barker more elaborately observed that "clarity of statement, a sense of proportion, of the value of contrast . . . these, it will be found, are the dominant qualities of Shakespeare's stagecraft." More recently, Jean Howard has described the variety of this stagecraft in pertinently musical terms in a study called *Shakespeare's Art of Orchestration*. She writes of "aural shifts and contrasts," "shifts in tempo and tone," "changes of key," "aural bridges," and "crescendo–decrescendo" patterns. She notes, too, the "marked aural dialectic between clamor and calm, cacophony and harmony" in *Othello*, a dialectic that Verdi was able to realize so brilliantly in the music for his opera.

Shaw wittily drew attention to the rich variety of expression latent in Shakespearean lines when he said how much he envied all the directions that composers can put into their "scripts": "I turn over a score by Richard Strauss and envy him his bar divisions, his assurance that his trombone passages will not be played on the triangle. . . . What would we not give for a copy of *Lear* marked by Shakespeare 'somewhat broader,' 'always quieter and quieter,' 'amiably,' or, less translatably, *mit grosser Schwung und Begeisterung, mit Steigerung,* much less Meyerbeer's *con esplosione,* or Verdi's *fffff* or *pppppp.*" Such contrasts as suggested by *fffff* and *pppppp* are not unknown in Shakespeare. "Contrast, ringing the changes," says John Barton; "Shakespeare does it over and over." Contrasts of character, emotion, situation are everywhere in the plays, but Barton is—as a director—especially preoccupied with assuring that the variety inherent in profoundly contrasting language is fully projected. He notes the "violent switches between naturalistic and heightened language" in *Othello* and the poignant contrast at the end of *Antony and Cleopatra* between the prosaic banter of the country clown about his asps and the heroine's "I have / Immortal longings in me."

The distinctive character of Shakespearean drama derives in large part from an incessantly contrastive method. Barton believes that the most local form of this method—antithesis—is fundamental to the playwright's style and must be played to the hilt. Taking as his cues Hamlet's exemplary antitheses—"To be, or not to be," and "suit the action to the word, the word to the action"—Barton says: "If I were to offer one single bit of advice to an actor new to Shakespeare's text, I suspect that [it] would be, 'Look for the *antitheses* and play them' . . . Shakespeare was deeply imbued with the sense of it. He *thought* antithetically." Countless memorable effects, both large and small, derive from some kind of antithesis: Prospero's irresistible "art" versus the impotency of the "men of sin"; the "real-world" ducal court and the "fantasy" Forest of Arden in *As You Like It;* heroic-scale Lear and human-scale Gloucester, and so on. It would seem hard to go wrong following Barton's advice and playing for the antitheses—which is to say, for maximum variety—whether creating an entire role or speaking a single speech.

This is perhaps why I have always thought a short passage from *Othello,* composed of stupendous antitheses, stands out as a perfect embodiment of the Shakespearean style, as well as its challenges to performers. In Othello's responses to Iago we have Shakespeare doing what he often does: ringing the changes up to and almost beyond the credible, thus challenging the actor to turn an exaggerated contrast into (in Peggy Ashcroft's happy phrase) "a little bit of super-reality." Each of Othello's responses here to Iago's coaching has a tremendous contrast within it between volcanic rage and utter tenderness. One is tempted to add something like Verdi's *fffff* and *pppppp* where they clearly belong:

> Othello: [*fff*] She shall not live. No, my heart is turn'd to stone; I strike
>   it, and it hurts my hand. [*ppp*] O, the world hath not a sweeter
>   creature! she might lie by an emperor's side and command
>   him tasks.
> Iago: Nay, that's not your way.

Othello: [*fff*] Hang her, I do but say what she is. [*ppp*] So delicate with her
          needle! an admirable musician! O, she will sing the savageness out of
          a bear. Of so high and plenteous wit and invention!
   Iago: She's the worse for all this.
Othello: [*fffff*] O, a thousand, a thousand times. [*ppp*] And then of so
          gentle a condition!
   Iago: Ay, too gentle.
Othello: [*fff*] Nay, that's certain. [*pp*] But yet the pity of it, Iago!
          O Iago, the pity of it, Iago!   (4.1.182–196)

Verdi was probably thinking less of local effects like this than of variety in
character types and scenic spectacle when he made, in a letter to his prospective
librettist for *Un Ballo in Maschera,* one of his several deep bows of homage to
Shakespeare:

> I find that the opera of our day sins through an overwhelming monotony; in fact I
> myself would refuse nowadays to set subjects like *Nabucco, Foscari,* etc. They present
> some very interesting theatrical moments, but without variety. Everything is on one
> string—a lofty one if you like but always the same. To make myself clearer: Tasso's
> poetry may be better, but I prefer Ariosto's a thousand times. For the same reason I
> prefer Shakespeare to all other dramatists.

Preferring Ariosto's witty, kaleidoscopic *Orlando Furioso* to Torquato Tasso's
solemn, lugubrious *Gerusalemme Liberata,* Verdi makes it clear why he was such
a partisan of Shakespeare's contrastive dramaturgy. For Verdi as well, variety was
the great weapon in the battle for the attention of "the *great tyrant,*" as he once
called the public. His letters are filled with asides and admonitions that show he
accepted this tyranny as a necessary fact of life in the commercial theater. When a
librettist wrote too much recitative for a final act, for instance, he let off some
steam: "What is certain, though, is that it will be impossible to make the audience
swallow so much uninterrupted recitative, especially in a fourth act. These
exigencies have nothing to do with composition; I could set a newspaper to
music, or a letter etc., etc. But the public will put up with everything in the theater
but one thing—boredom." One can almost imagine Shakespeare—who like
Verdi made a handsome fortune with projects aimed at attracting a lot of paying
customers—thinking along these same practical lines. Serve up too much caviar to
the general public, after all, and the public will begin to mutter, as does the tinker
Christopher Sly in *The Taming of the Shrew* when a play begins to weary him,
"Would 'twere done!"

Verdi's insistent call for variety (his three other main desiderata were powerful
situations, pathos, and *brio*) has echoed throughout operatic history. Rare is the
successful libretto that fails to consider the attention span of Verdi's "great
tyrant." It is necessary to pace and vary the stage action in a way that avoids the
potential sources of boredom: too great a consistency of sound-mass, stage
picture, ensemble, orchestration, scene length, and the like. The history of lyric
writing is rich in recognitions of this simple necessity. The very first great operatic

composer, Monteverdi, wrote in 1638, "I am aware that contrasts are what stir our souls, and that such stirring is the aim of all good music," and Berlioz reiterated this view two centuries later: "Variety wisely ordered is the soul of music." Though Wagner despised the way early nineteenth-century operas achieved variety by "moments of lyrical display and mostly violent arrest of the action," several of his earlier works are thinly disguised "number" operas; some early editions describe these numbers with the usual *gran scena, cavatina, duettino,* and so on. And his later operas—in spite of the masking effect of through-composition—are masterpieces of calculated pacing and variety of effect. Wagner's derogatory term, *Effektstück,* applies to his own canon, too. Hofmannsthal, a generation later, thanked variety for the success of *Der Rosenkavalier:* "What makes the charm of a comedy like *Rosenkavalier* . . . that special brio which, for seventeen years now, has never and nowhere failed, is the concentration of colorful contrasts and incidents in each separate act." In sum, it is hard to imagine an effective evening of opera without a variety of aural and visual effects. Amalgamating these—by recourse to such means as exposition, "small" and "large" scenes, transitions from "real" to accelerated to "stop" time and back, climax and false climax, comic relief, divertissement, and denouement—was always the librettist's main task.

After making his plea for variety and invoking Shakespeare, Verdi urged upon Antonio Somma the possibility of a *King Lear* opera. In this play—the most spectacularly contrastive of Shakespeare's tragedies—Verdi says he found the variety that his "crazy brain would like." A few days later Somma replied, agreeing to the idea and expressing the hope that "with your most exquisite tact you will be able to show me how the form and succession of musical numbers ought to clothe the drama." My present purpose is to suggest that the fundamental operatic notion of a "form and succession" of set pieces bears a striking affinity to the architecture of Shakespeare's plays. Indeed, as the following analysis of the thirty-five plays in the canon and thirty-five operas from the (mostly) standard repertory will show, it is possible to say that the Shakespearean and operatic modes of dramatic architecture are virtually identical. So nearly identical are they that one almost begins to wonder whether similarities in construction rather than Shakespeare's sublime poetry and observations of human nature might explain the long-standing appeal of his plays to composers.

Let us consider, then, the main contours of a typical Shakespearean performance. First is the rather mysterious matter of the running time of plays, which, for the purpose of making other helpful calculations, we must try to estimate. There is a remarkable and perhaps also suspicious unanimity of opinion in Elizabethan references to running time. The prologue for *Romeo and Juliet* speaks of "the two hours' traffic of our stage," while the prologue for *Henry VIII* promises spectators they "may see away their shilling / Richly in two short hours." The audience for *The Two Noble Kinsmen* is promised it will "hear / Scenes . . . Worth two hours' travail," and Ben Jonson's prologue says *The Alchemist* (a play slightly longer than the Shakespearean average) will unfold in "two short houres." Jonson's *Bartholomew Fair,* is, at nearly 4,000 lines, longer than Shakespeare's *Götterdämmerung, Hamlet;* its induction refers to "the space of two houres and a halfe, and somewhat more." Can the assertion that a Shakespearean play filled only two hours possibly be accurate, even if we keep in mind that it would have been performed without intermission?

There are several reasons to doubt. "Two hours," for one thing, sounds like a convenient rounding off. If this timing was accurate for *Romeo and Juliet,* it could scarcely have applied to *Henry VIII,* which is over 400 lines longer and features several time-consuming scenes of pageantry. For another, it might make good psychological sense to err on the short side when addressing those who would be standing in the pit throughout the performance. Perhaps it was also best to err on the short side at a time when London's city fathers and commercial establishment viewed playgoing as a reprehensible waste of daylight hours. The most forceful reason to think two hours a decidedly minimal approximation, however, is that the speed of recitation such a running time would entail simply boggles belief.

It has been estimated that an Elizabethan play averaged about 2,500 lines.* A period of two hours, then, yields an average delivery of twenty-one lines per minute, which seems breathtaking enough. Shakespeare's thirty-five plays in fact average over 2,850 lines, which would require a recitation of twenty-four lines per minute to bring a performance in at two hours. This is incredible. Take, say, the first twenty-four lines of Hamlet's "O, what a rogue and peasant slave am I!"—a speech that should go at a pretty fast clip—and speak it in a minute. Sustaining such a pace is rather like listening to a 33⅓ record at 45 rpm . . . or like trying to perform Chopin's "Minute Waltz" as written in the prescribed time. Even if we grant that a faster rate of speech was acoustically possible at the Globe, where it is estimated that no spectator was more than seventy-five feet from the stage (the audience of 1,500 to 2,000 must have been tight-packed), it still seems necessary to imagine a slightly less frantic pace. A slight move in this direction can be achieved by rounding off the average play length to 2,800, in order to account for passages included in the Folio that were probably cut in performance. I also think it is within the realm of plausibility to assume, for our purposes, a running time of a bit over 2½ hours, or 155 minutes.* This yields the much less breakneck rate of eighteen lines per minute—though I think a modern audience, even in the 400-seat theater that Ian McKellen has specified as the ideal modern Shakespearean venue, would find this rate disconcertingly driven. Let us assume for the sake of argument, therefore, an average of eighteen lines per minute.

Possessed of estimates for the average play length, running time, and rate of recitation, we can now calculate approximately the full range of running times for all Shakespeare's plays. This range extends from the stupefying 218 minutes without intermission that *Hamlet* would have required if (as seems unlikely) it had ever been presented as the Folio prints it, through the three other three-hour-plus juggernauts (*Richard III, Henry V,* and *Coriolanus*), and down to the much more compact *Macbeth* (117 minutes), *A Midsummer Night's Dream* (120), and the rather tedious but very brief *Comedy of Errors* (99). More important, we can now also begin to see much more clearly how, in Somma's phrase, Shakespeare clothed his dramas in the "form and succession" of a series of individual scenes . . . which scenes, we have noticed hitherto, are the dominant unit of construction in his dramaturgy. With an estimated running time for each play and some notion of an average rate of recitation to hand, we can begin to get some sense, too, of the range of Shakespearean scene lengths and of the average running time for scenes in individual plays and in the entire canon. This exploration points to some startling similarities with operatic dramaturgy.

There are 677 substantial scenes in the thirty-five plays. (I am leaving out of account about three dozen very minor scenes that are fifteen lines or less.) The number of such scenes in a play varies remarkably: from 36 in *Antony and Cleopatra,* 28 in *Henry V,* and 27 in *Coriolanus,* to eight in *The Tempest* and *Love's Labour's Lost* and nine in *A Midsummer Night's Dream.* Twenty-two of the plays fall within the range of 15 to 25 scenes, and the average for all the plays is 19. Thus, we can conclude that the typical Shakespearean play is about 2,800 lines long; it is arranged into 19 scenes that run on average a little under 150 lines and last a little over eight minutes each.* Actually, the average scene-length in the individual plays varies widely too: from *Love's Labour's Lost* (19 minutes per scene), *The Tempest* (14^{1}/$_{4}$) and *Othello* (14), to *Antony and Cleopatra* (4^{3}/$_{4}$) and *Macbeth* (4^{1}/$_{2}$). It should be noted, too, that what constitutes a Shakespearean scene is often open to debate, especially in the longer scenes in the canon. Shakespeare very seldom wrote scenes with clear through-action that would last more than twenty minutes. Only fifteen of them (two percent) are over 450 lines long, and in most cases Shakespeare protected them from tedium by sectioning them internally. The second-longest of all his scenes, in *The Winter's Tale,* is 842 lines long, but a total of thirteen entrances and exits, three songs, and two dances make it impossible to feel its massive effect in performance. And the champion scene of 931 lines, at the end of *Love's Labour's Lost,* could be described as beginning with a quartet (for the four women), an aria (Boyet's warning), a nonet, then the play-within-a-play, the *deus ex machina* arrival of news of the death of the Princess' father, two more duets, and two play-ending songs.

   With these data, the terrain of a Shakespearean performance comes into sharper focus: the action is apportioned artificially into several scenic modules; these modules are arranged in a series, usually according to some mode of contrast; and the headlong "succession of illusions" Brook refers to is achieved by an over-whelming preference for scenes that make their effect and are superseded within six to ten minutes or so. This is a dramaturgy of elaborate but still highly compact rhetorical fits and then sudden starts produced by the emphatic entering and exiting of characters. Also making this terrain more variegated are sudden shifts and contrasts of place (most spectacularly in the Rome, Alexandria, and battle sites of *Antony and Cleopatra*), of social milieu (the tavern, corridors of power, battlefields of *Henry IV,*) or stage forces (from the soliloquizing Wolsey to a ten-part coronation procession in *Henry VIII*). I will examine in the next chapter another means Shakespeare uses to effect striking changes in his action: the scrupulously calculated movements between various modes of expression like prose, heightened prose, blank verse, and sometimes rhymed iambic pentameter.

   This theatrical terrain, in short, is full of variety, and it is perhaps rendered even more rugged by the very rapidity with which Shakespeare plunges his spectator through it. Brook speaks of the "roughness of texture" in Shakespeare's plays, and he is speaking in part of this kaleidoscopic mobility that a bare stage makes possible. The price paid for this mobility, this highly compressed "roughness," of course, is the heavy burden laid upon the dramatist's poetry to give the action and the characters the local "habitation" and "name" of stage reality. As in opera, the crucial transaction—the creation of the illusion—must be an aural one: both dramaturgies depend on superlative vocalism to bring the "forms of things unknown" into existence on stage. And both require imaginative reception on the other side. Shakespeare's Rosaline says, "A jest's prosperity lies in the ear / Of him that hears it," and that is where the prosperity of Shakespeare and opera essentially lies: in the ear. Thus, the challenge of Shakespeare's poetic "numbers" and the opera composer's "notes" are, in effect, both addressed by the exhortation of the Prologue in *Henry V*:

> . . . let us, ciphers to this great accompt,
> On your imaginary forces work.
> . . . 'tis your thoughts that now must deck our kings,
> Carry them here and there, jumping o'er times,
> Turning th' accomplishment of many years
> Into an hour-glass.

Let us ask now how opera composers and librettists have been inclined to use the two-and-a-half or three hour-glasses at their disposal. What we discover from a survey of thirty-five works, most of them from the core of the standard repertory, is that these men also tend to envision a dramatic action as unfolding, on average, in a dozen-and-a-half arbitrarily focused, discrete situations or scenes. As in Shakespeare, the *raison d'être* for the individual scene is often a set piece in heightened rhetoric for one person or more. And the running time for these operas, while showing the full Shakespearean range (*Tosca*'s 118 minutes and *Bohème*'s 108 nearly match *The Tempest*'s 115 and *Dream*'s 120, while *Lohengrin*, at 215 minutes, matches *Hamlet*), average out to a figure somewhat under the Shakespearean norm: 136 minutes, or $2^{1}/_{2}$ hours. This nearly accords with Hofmannsthal's assertion to Strauss that "an over-all duration of some two-and-a-half to two-and-three-quarter hours, plus entr'acte, is just right" for an opera.

The works in my sampling can mostly be described as war-horses, and most are from the nineteenth century. I have, however, included a few works at random from the repertory's second tier; among these, a few operas based on Shakespeare. They are: *Aida*, Britten's *A Midsummer Night's Dream*, *Il Barbiere di Siviglia*, *La Bohème*, *Butterfly*, *Carmen*, *Così fan tutte*, *Don Giovanni*, *Elektra*, *Falstaff*, *Faust*, *Die Fledermaus*, *Der fliegende Holländer*, *La Gioconda*, *Les Contes d'Hoffmann*, *Lohengrin*, *Lucia di Lammermoor*, *Die Lustigen Weiber von Windsor*, *Macbeth*, *Manon*, *Manon Lescaut*, *Norma*, *Le Nozze di Figaro*, *Eugene Onegin*, Rossini's and Verdi's *Otellos*, *Rigoletto*, *Roméo et Juliette*, *Samson et Dalila*, *Tannhäuser*, *Tosca*, *La Traviata*, *Il Trovatore*, *Turandot*, and *Die Zauberflöte*.

The running times for these operas are easily obtained from recordings. In several cases, Somma's "form and succession of musical numbers" is easily discovered, too, since it was a common nineteenth-century habit to number the constituent solos, ensembles, and choruses in the printed score. The Wagnerian revolution made this less attractive as the century wore on, but it is still usually possible to discern in very long dramatic structures the clear bracketing of integral scenic units. For some of the later operas in my sample, like *Elektra* or *Tosca,* I made my own estimate of the number of such integral scenes. If we assume—as seems reasonable—that the operatic "number" (which includes any dialogue or recitative pertaining to it) is the appropriate dramaturgical equivalent of the Shakespearean scene, then the average number of them in the thirty-five operas (18) is astonishingly close to the Shakespearean norm (19).* The number of musically and dramatically focused events, as with Shakespearean scenes, varies widely: from 12 in *Turandot* and 11 in *Elektra,* to 26 in *Carmen,* 28 in *Figaro,* and 31 in *Così fan tutte.* But the average length in minutes for the numbers in the thirty-five operas is, if anything, less wide-ranging than the avearage length of Shakespearean scenes. No opera in the sample approaches the 4½ -minute average of *Antony and Cleopatra;* seven of the plays have scenes averaging over ten minutes, but none of the operas does. Twenty-eight of the operas have average number-lengths falling between six and nine minutes, and the overall average is 7¼ minutes (the average Shakespearean scene, it will be recalled, is a bit over eight minutes).

The rhetorical expansions in Shakespeare and opera, of course, fall within a very wide range. Hamlet's "O, what a rogue and peasant slave am I!" launches over a fifty-five-line trajectory, while Ophelia's "O, what a noble mind is here o'erthrown" a few moments later lasts just ten lines. Barbarina, in *Figaro,* sings about her lost pin for a minute, the Countess about her lost love for several. Individual scenic units, we have seen, can also vary considerably in length. Still, the above calculations show that the Shakespearaean and operatic kaleidoscopes tend to move at very nearly the same dramaturgical speed: the vast majority of their discrete scenic events suppose an optimum attention span of between six and ten minutes. An evening of Shakespeare or opera typically consists of somewhere near fifteen or twenty riveting several-minute passages situated in headlong proximity to each other by expeditious, recessive (and hence aurally relaxing) intersitial maneuvering of characters and plot. From every Shakespearean play one could pluck the equivalent of the "highlights" disc that is often derived from a full-length opera recording. In these dramaturgies, in other words, the pleasures of "vertical" rhetorical expansion are mined to the fullest . . . at the expense of such "horizontal" values as the "normal" ticking of the clock in real time, the seamless development of the plot, or consistent, organic growth of character.

This necklacing of prominent moments is obvious enough from the table-of-contents page of any score. The first act of *Così fan tutte* progressess, in part, from duet, to chorus, to quintet-and-chorus, to trio, to recitative-and-aria, to aria, to sextet, and so on. *La Gioconda's* first act consists of: an introductory chorus; scena and trio; recitative, chorus, and romance; scena and duet; scena, recitative, and monologue; and a final chorus, *furlana,* and prayer. Shakespeare's plays are perfectly amenable to such terminology and analysis. Indeed, similar table-of-contents pages for them might be a helpful study aid. There is, for instance, something of *La Gioconda's* symmetry in the first act of *Romeo and Juliet,* which runs thirty-five to forty minutes: ensemble and chorus (fight in the Veronese street, climactic entrance of Prince Escalus "with his Train"), duet (Capulet and Paris), duettino (Romeo and Benvolio), trio (Lady Capulet, Juliet, Nurse), duet (Romeo and Mercutio), and finale (Capulet ball). The Shakespearean kaleidoscope moves at its accustomed pace in the brilliant and also symmetrical first act of *Othello*. Its eight discrete parts also unfold in about thirty-five minutes: duet (more an aria, really, with Iago given 72 lines, Roderigo eight), trio (Iago, Roderigo, Brabantio), duettino (Iago, Othello), chorus (Cassio and officers, then Brabantio and attendants), council scene and aria (Othello's self-defense), duet (Roderigo and Iago), and a final exit aria (Iago's soliloquy).

In the last several pages my concern has been to achieve a general sense of the typical constraints of performance-time, the usual lengths of scenes and numbers, and to establish the rather brisk pacing in the two dramaturgies. This has been worth drawing attention to at first, because all the other nontemporal sources of variety are heightened in their impact by the kinetic urgency—the relentless *con slancio*—that will drive any superlative performance. Performances of Shakespeare and opera, like a satisfying roller-coaster ride, must attain tremendous momentum at a series of crucial points in the action. When they don't, they can fall tediously, miserably flat. These exciting, vertiginous effects are won, in great part, by careful timing, and that is why the best performances usually display the work of directors or conductors listening attentively for the sound of time's winged chariot. The speaker in Shakespeare's Sonnet 76 frets that his verse is losing steam and drifting "far from variation or quick change." Now that we have seen how crucial timing is to the achievement of the kaleidoscopic effects in Shakespeare and opera, we can turn in the next chapter to the modes of variation and quick change themselves.

14

The Pyramid Game

*B*ore, *boring,* and *boredom* are not in Shakespeare's 29,066-word vocabulary, but he surely knew the thing itself from theatrical experience. His satire on the "tedious brief" play concocted by Bottom and his colleagues suggests that there were some tedious *long* ones on the Elizabethan stage. Ensconced in his Stratford retirement, he might even have been willing to admit that a few of those boring plays were of his own devising. In general, though, his canon shows an increasingly dextrous care to avoid tedium. Like any sensible man of the theater, he recognized that a tedious play would have the same repellent effect as the tedious man described by Ben Jonson: "a man [one] would leap a steeple from, gallop down any steep hill to avoid." And so he relied on several modes of variation in preparing his plays for the stage.

Many of these modes of variation have been time-honored in all theater: the presentation of strongly contrasting stage pictures and stage forces, as well as multiple plots. Several other modes of radical contrast in expression, however, are uniquely available to the composer. He can produce sudden changes of sound-mass along the continuum from solo obbligato to full orchestral tutti; he can shift suddenly in tempo along the continuum from *largo* to *prestissimo;* and his decision as to which instruments sound together can produce a vast range of orchestral timbres. More subtly but often very affectingly, he can shift in mode from major to minor or shift to a different key, often establishing key "identities" for particular characters or themes. He may also achieve variety by careful juxtaposition of musical rhythms. Verdi, for instance, cast much of the music of Desdemona in *Otello* and the elder Germont in *Traviata* in common, or duple, time ($\frac{2}{4}$, $\frac{4}{4}$), giving them a very solid, four-square cast; the music for Jago and Violetta is typically in triple meter ($\frac{3}{4}$, $\frac{6}{8}$), which gives their characters more energy and delicacy of nuance. The composer also works with the advantage of several starkly defined vocal categories, which allow brilliant aural contrasts as soloists sing alone, in ensembles, and against massed choral backgrounds.

None of this is available to dramatists of the spoken stage, though they and theater critics delight in borrowing musical metaphors to describe how their own variety is achieved. Shaw particularly loved to convey his methods this way. "Stick to my plays long enough," he admonished, "and you will get used to their changes of key and mode. I learnt my flexibility and catholicity from Beethoven; but it is to be learnt from Shakespear to a certain extent. My education has really been more a musical than a literary one as far as dramatic art is concerned. Nobody nursed on letters alone will ever get the true Mozartean joyousness into comedy." One Shavian boast in particular points to a way—surely the most prominent and important shared way—that Shakespearean and operatic variety is achieved: "My deliberate rhetoric, and my reversion to the Shakespearean feature of long set solos for my characters, are pure Italian opera." We have already seen Shaw asserting that opera taught him to shape his plays into arias, duets, ensembles and "bravura pieces" . . . and demanding from his actors "great virtuosity in sudden transitions." This is in great part due to his reversion to the Shakespearean technique of deliberately shifting between several distinctive "planes" of verbal expression.

Such a shift occurs in Shaw's *The Man of Destiny,* when Napoleon is about to launch into a bravura speech on the English. Shaw manages it with the same forthright artifice as Donizetti setting up Lucia's first big aria, "Regnava nel silenzio," a rapturous confidence to her companion, or as Verdi setting up "Tacea la notte placida," which Leonora sings to her companion in *Il Trovatore.* Shaw's stage direction tells us Napoleon's "audience composes herself to listen" and he, "secure of his audience . . . at once nerves himself for a performance. He considers a little before he begins, so as to fix her attention by a moment of suspense. . . ." Shakespeare's canon is rich in rhetorical gear-shifting as stagey as this. After speaking to Polonius and one of the traveling actors in prose, Hamlet is left on stage and says to himself, "Now I am alone." Then he bursts into the high-flying pentameter of "O, what a rogue and peasant slave am I!" with its rhymed-couplet flourish as a coda: "the play's the thing, / Wherein I'll catch the conscience of the King." In *1 Henry IV,* for another instance, after the noble blank verse of Hotspur's death speech and Hal's eulogy over his body, Falstaff rises from his feigned death to speak in his typically bumptious prose: "Embowell'd! if thou embowel me to-day, I'll give you leave to powder me and eat me too to-morrow. 'Sblood. . . ." And what could be more stagey (but also superbly startling) than Shakespeare's having Iago, after a long prose conversation with Cassio, shift suddenly to blank verse and cheekily ask the audience, "And what's he then that says I play the villain?"

The distinctive planes of rhetorical expression on which Shakespeare depends bear a general resemblance to the planes of musical expression at the composer's disposal. They might be arranged, in ascending order of eloquence and thrilling laryngeal exertion, as a pyramid:

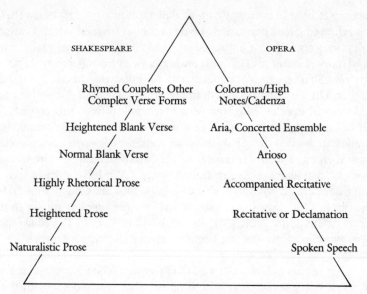

SHAKESPEARE   OPERA

Rhymed Couplets, Other   Coloratura/High
Complex Verse Forms   Notes/Cadenza

Heightened Blank Verse   Aria, Concerted Ensemble

Normal Blank Verse   Arioso

Highly Rhetorical Prose   Accompanied Recitative

Heightened Prose   Recitative or Declamation

Naturalistic Prose   Spoken Speech

The movement, during the course of a dramatic action, up and down this edifice of expressivity gives Shakespearean and operatic theater their alternately suspenseful and vertiginous likeness to a roller-coaster ride. The "concentration of colorful contrasts and incidents" that so pleased Hofmannsthal in *Der Rosen-kavalier* (and that occurs in most operas) is achieved in large part by judicious movement up and down this pyramid. Only a few visits to the opera house will make this technique perfectly familiar, so it needs no further demonstration here. Shakespeare's deliberate shifts between the planes of rhetoric are not so immediately apparent, however, and deserve more attention. A brief canvass of the canon will be pertinent, I think, not only because these highly artificial and sudden shifts largely account for the aural variety of Shakespeare's plays, but also because the occasions for such shifts (and the dramaturgical purposes behind them) are often strikingly reflected in the habits of composers and librettists.

Shakespeare's shifts back and forth from verse, to prose, to the occasional rhymed couplets are usually motivated by several intentions, but one that seems almost always to be present is simply the desire to avoid the aural tedium of remaining in one mode too long. Though there are several plays with almost no prose at all (*Richard II, 1 Henry VI, 3 Henry VI, King John*) or very little (two percent in *Richard III,* seven percent in *Macbeth* and *Julius Caesar,* nine percent in *Antony and Cleopatra*), prose does account for about thirty-eight percent of all Shakespeare's lines. And prose passages are often clearly intended to have the effect of opera's recitative, "sherbet" aria (during which one could safely leave one's box for refreshments), or comic-relief scene; namely, to allow the ear and the mind to relax their concentration.

In several of the plays the tension achieved by the antithesis between prose and poetry seems to be a primary principle of composition. For example, in *Love's Labour's Lost,* the four super-sophisticated aristocratic couples seem completely isolated in their fantasy world of leisure; Shakespeare achieves this isolation by having them speak mostly in blank verse and rhymed couplets. Their scenes alternate with scenes featuring the play's lower-class characters and clowns (Armado, Holofernes, Nathaniel, Moth, Costard, and Dull), whom Shakespeare confines to prose. The play thus presents an agreeable alternation of ballooning, ornate-style "arias" and deflating, plain-style "recitative." Similarly, the prosaic, fleshly world of the Boar's Head Tavern in *1 Henry IV* allows Shakespeare brilliant contrasts with the high-toned blank verse that resounds in the royal corridors of power. Hal, who moves in both worlds easily, of course can perform ably in both styles as occasion requires. The play in which he becomes England's great warrior king, *Henry V*, contains some of the most stagily exaggerated of the playwright's alternations in style. Henry's ringing "Once more unto the breach, dear friends" oration is immediately followed by a prose vignette from the very lowest ranks, featuring Bardolph, Nym, and Pistol. It begins with a prosaic parody of Henry's line: "On, on, on, on, on! To the breach, to the breach!" Earlier, Henry squelches a palace rebellion in golden blank verse; the scene ends with the flourish of a trumpet and a rhymed couplet: "Cheerly to sea! The signs of war advance! / No king of England, if not king of France!" Seconds later we are listening to the Hostess' homely but touching prose description of the death of Falstaff.

Shakespeare was obviously convinced that the restorative powers of prose were vital to drama centered upon fits of heightened expression. This is perhaps best shown in the many occasions when he went far out of his own and his plot's way to provide prosaic rest stops in the blank-verse journey. The *non sequitur* of the Porter's comic scene in *Macbeth* (2.3)—the first prose in the play aside from Macbeth's letter to his wife (Shakespeare's letters, unlike some of those in opera, are *never* cast in heightened style)—comes from so far out of the blue that some have suggested it is not geniune. But the deftness of its placement, just after the electrifying post-regicide "duet" for the Macbeths, is typical of Shakespeare. The slapstick prose of the Gobbo scenes in *The Merchant of Venice* appears to have been injected as relief from the Venetians' blank verse and Shylock's bitterly passionate prose outbursts. In *All's Well*, Shakespeare invented several characters not in his source who speak in prose: Parolles, Lavatch, and Lafew. This allowed him several occasions to drop relaxingly from rhetorical heights. Toward the end of *Cymbeline,* just after the very ornate dream-vision of Posthumus in which Jupiter appears (Jupiter's speech is in alternating rhymes, a Shakespearean rarity), the dramatist hales in sixty prose lines of gallows humor spoken by two jailers. This is utterly gratuitous, since Posthumus is freed even as they speak, but Shakespeare clearly felt the need for a prosaic interlude between the vision and the play's immediately following denouement, a rich 485-line blank-verse feast.

One more example of Shakespeare's devotion to passages of aural relaxation, however extraneous they seem to be, must suffice: that peculiar passage in *Lear* (2.2.1–67) where Kent hurls his long-winded, comic Insult Direct at Goneril's steward, Oswald. In scathing prose he calls him, among many other things, "a lily-liver'd, action-taking, whoreson, glass-gazing, superserviceable, finical rogue." It is an indication of the play's grim world that Shakespeare had to fetch so far for comic and prosaic relief, but he clearly desired a little variety amid the blank-verse scenes of act 2 that lead to Lear's expulsion into the storm.

Composers are always rising to arias and concerted ensembles to create effects
of formality and ritual. One thinks of Sarastro's arias and their accompanying
choruses in *Die Zauberflöte*, *Aida*'s Triumphal Scene, or Elisabeth's aria and the
grand procession in *Tannhäuser*. Shakespeare, likewise, typically ascends to
purple rhetorical levels when he wishes to achieve a sense of ornate stasis or high
moral, military, or judicial tone. His Halls-of-Power style can be seen at its most
verbose and marmoreal in the council of the Greeks Agamemnon, Nestor, and
Ulysses in *Troilus and Cressida* (1.3) on their conduct of the Trojan war. Similarly
laden with elaborate verse but more urgent is the *1 Henry IV* council-of-war scene
that begins with the King intoning,

> So shaken as we are, so wan with care,
> Find we a time for frighted peace to pant
> And breathe short-winded accents of new broils
> To be commenc'd in stronds afar remote.

The Venetian council-chamber scene in act 1 of *Othello* is cast in formal blank
verse; only when Roderigo and Iago are left alone after adjournment does
Shakespeare move into prose. The drumhead trial of Hermione by her jealous
husband in *The Winter's Tale* and the adjudication of Shylock's pound of flesh are
cast in the dramatist's stateliest blank verse—as are the *Turandot*-like riddle scenes
that Portia has with Morocco, Arragon, and Bassanio in *The Merchant of Venice*.

This penchant for large-scale scenes of pomp and rhetorical circumstance in
Shakespeare and opera is familiar enough. More interesting for our purposes are
the sudden shifts from prose to poetry that occur when Shakespeare wishes to
throw a crucial theme or "moral" into high relief, or when he wishes to raise a
character's emotional temperature with a challenge posed by the action. Sometimes the sudden access of poetry simply signals high (or very low) morality. The
deployment of blank verse at moments of accelerating passion, dedication to
action, or sheer high spirits is, of course, very near the composer's deployment of
arioso passages and more formal arias. A few examples of the Shakespearean
technique will show the kinds of similarity I have in mind.

A grandly scaled instance occurs when Henry V, in disguise, talks with three of
his lowly troops about proper military conduct (4.1). This discussion is cast in 150
lines of unadorned, closely reasoned prose "recitative." This turns out to be
preparation for a startling stylistic levitation. For when the soldiers leave, Henry
shifts instantly into some of the most resonant verse Shakespeare ever wrote: a
grand solo "aria" of fifty-five lines on "idol Ceremony" and "What infinite heart's
ease / Must kings neglect that private men enjoy!" It is no wonder all stops are
pulled out here, for this speech—showing a king free of all moral and intellectual
disabilities on the English throne at last—is really a climax not only to *Henry V*
but to the three preceding plays in this tetralogy as well. This same recitative-and-
aria technique inaugurates Hal's long ascent to glory. For in *1 Henry IV,* after
nearly two hundred lines of prosaic banter and idle jesting in the first tavern scene
with Falstaff, Hal is left alone for his promise, in an image-rich verse soliloquy, to
turn over a new leaf:

> . . . like bright metal on a sullen ground,
> My reformation, glitt'ring o'er my fault,
> Shall show more goodly and attract more eyes
> Than that which hath no foil to set it off . . .   (1.2.242–245)

Most other instances in which heightened blank verse is made to "show more goodly and attract more eyes" by juxaposition with prose are more subtle, but they are still very effective, and often in ways that are distinctly operatic. A notable example of this technique occurs in the "willow song" scene of *Othello*. During almost ninety lines of dialogue Emilia displays her common sense and cynical view of human nature entirely in prose, while Desdemona keeps steadfastly to the high road of iambic-pentameter idealism. The latter wonders if there be women who do commit adultery and concludes, "I do not think there is any such woman." To which Emilia unmetrically replies, "Yes, a dozen, and as many to the vantage, as would store the world they played for." Then something striking occurs. Emilia drops her comprimaria status and leaps into blank verse herself for an eighteen-line "aria" passionately condemning, in modern feminist terms, the double standard to which men hold women. This ends with an emphatic couplet *cum* coda: "Then let them use us well: else let them know, / The ills we do, their ills instruct us so." Like an operatic aria, this speech stops the action in its tracks; indeed, it must have been cut in some performances, since it does not appear in the acting version printed in 1622. Still, it is a splendid, exciting speech and typifies Shakespeare's operatic habit of giving a character lavish freedom of utterance when passion moves him or her . . . even where, as here, consistency of characterization or forward momentum seem to be sacrificed in the process.

A few other examples of purposeful escalation into poetry, chosen at random, can be briefly noted. In the middle of *Timon of Athens,* Timon's fortunes wane and his servants must ask those who have enjoyed his generosity to return the favor. Shakespeare arranges several rejections, spoken in eerily emotionless prose, to be followed immediately by passionate outbursts of blank-verse outrage on Timon's behalf. One servant exclaims, "Has friendship such a faint and milky heart / It turns in less than two nights? O you gods! / I feel my master's passion." The first scene of *Coriolanus* offers a more gradual escalation. It opens with a group of mutinous Roman plebes shouting in prose. The patrician Menenius soon appears to quell emotion with some leather-lunged blank verse. In several exchanges the plebes hold fast to prose, Menenius to verse. But as his famous comparison of Roman society to a human body gains momentum and a higher political tone is reached, blank verse begins to be spoken on both sides for the remainder of the scene. Soon the "Citizens steal away," says a stage direction. The sleepwalking scene in *Macbeth* is cast—as all genuine madness is in Shakespeare—entirely in prose. But at the end the Doctor is allowed to slip back into blank verse to emphasize the moral of the scene and to give his diagnosis eloquent weight:

> Unnatural deeds
> Do breed unnatural troubles; infected minds

> To their deaf pillows will discharge their secrets.
> More needs she the divine than the physician.   (5.1.71–74)

To end on a lighter note, consider rascally Parolles in *All's Well*. Prose is his usual *métier*, but for the moment when he finally acknowledges the moral of his own career and promises a new leaf, Shakespeare gives him a blank-verse aria at center stage, with several rhymed couplets to boot:

> Simply the thing I am
> Shall make me live. Who knows himself a braggart,
> Let him fear this; for it will come to pass
> That every braggart shall be found an ass . . .   (4.3.333–336)

In addition to providing aural variety and spotlighting important themes or plot crises, Shakespeare's sudden movements upward or downward on the rhetorical pyramid are often operatic in another important respect. In opera, the greatest vocal excitement almost always comes from characters who inhabit an "unreal" emotional world they have created by virtue of special powers of imagination, idealism of spirit, or egomania. Their sensibilities are unconfined by common sense, and so their actions almost never sink to the commonplace. The mere realm of possibility is for them a little claustrophobic; their resounding voices are constantly testing its borders. It is this other world of exhilarating *artifice* that Giovanni, Faust, Jago, Senta, Leonore, Cavaradossi, and so many other operatic figures inhabit. Characters with no access to this other and "higher" realm of experience become the decided second fiddles and peripheral comprimarios, who stay pretty much near the bottom of the pyramid: the Sacristans and Spolettas in *Tosca*, Suzukis and Goros in *Butterfly*, and countless nurses or ladies-in-waiting. The great operatic characters are, in sum, like the lunatics, lovers, and poets described by Theseus. They have

> Such shaping fantasies, that apprehend
> More than cool reason ever comprehends.

Heightened verse is often the Shakespearean gateway to more intense, more passionately idealized, and sometimes more dangerous, risk-laden "apprehensions." Blank verse is time and again the expressive mode for the liberation of the most profound emotions and desires; this liberation produces the plays' thrilling poetic arias and ensembles. "O, how full of briers is this working-day-world!" says Rosalind to Celia in her uncle's ducal palace, where prose seems the usual style. It is only with their liberating escape to Arden Forest that emotional excitement and blank verse spring to life. "Now go in we content / To liberty," says Celia as they set out. In Arden the many splendid poetic set pieces and several actual songs of *As You Like It* become possible. Similarly, the lovers of *A Midsummer Night's Dream* take not only their supercharged romantic energies but also Shakespeare's most graceful lyric style with them into the Athenian forest. The "hardhanded men" who rehearse there bring only their very literal minds . . . and prose. Even when enchanted, Bottom remains rooted in the "real" world: his speeches remain prosaic, while Titania flares with the sweet smoke of rhetoric: "The moon, methinks, looks with a watery eye, / And when she weeps, weeps every little flower. . . ."

The distinction between the mundane, prosaic "real" world and the exciting pulse of the poetic "other" world is often strikingly underscored at the beginning of a Shakespearean action. *The Winter's Tale* begins with a short conversation in prose between two lords at Leontes' court; this establishes a mood of normality, calm, and plain speaking: "I speak as my understanding instructs me," says one, "and as mine honesty puts it to utterance." But Shakespeare rarely sojourns long in the working-day world. After forty-five lines of normality to give some sense of scale to the exciting abnormality to follow, the scene changes. Polixenes introduces a rich blank-verse style—

> Nine changes of the wat'ry star hath been
> The shepherd's note since we have left our throne
> Without a burthen . . .

—and Shakespeare introduces the "other" world of liberated passions. Within five minutes Leontes' jealous rage ignites in rugged verse—

> Too hot, too hot!
> To mingle friendship far is mingling bloods.
> I have *tremor cordis* on me; my heart dances,
> But not for joy; not joy.   (1.2.108–111)

—and Shakespeare has his project well in hand.

A far more spectacular juxtaposition occurs in *The Tempest*. Shakespeare took pains to make the first scene a true-to-life evocation of the mariner's working-day world, and it is cast almost entirely in prose. The contrast with Miranda's first lines of poetry, after the storm is suddenly silenced—

> If by your art, my dearest father, you have
> Put the wild waters in this roar, allay them.

—makes a powerful effect. In *Troilus and Cressida,* for a last example, the "real" and "ideal" worlds are in constant battle. This war, not the Trojan one, is the true subject of the play, and Shakespeare sets up the tension between them in his first scene. Troilus, like Romeo in love at first with Rosaline, is full of romantic idealism and love-sonnet comparisons (and blank verse), while Pandarus parades his earth-treading cynicism in plain prose:

> *Troilus:* But I am weaker than a woman's tear,
>   Tamer than sleep, fonder than ignorance,
>   Less valiant than the virgin in the night,
>   And skilless as unpractic'd infancy.
> *Pandarus:* Well, I have told you enough of this. For my part, I'll not

> meddle nor make no farther. He that will have a cake out of
> the wheat must tarry the grinding.   (1.1.9–15)

Many other touches underscore Shakespeare's habit—shared with composers —of identifying uncommon potency of mind, emotional energy, (ig)nobility of purpose, or sheer charisma, with the ability to rise to heightened utterance. To characters with none of the above the dramatist could be amusingly ruthless: obviously in the metrical horrors of "Pyramus and Thisby," but also in the erstwhile courtier Armado and the learned poetry critic Holofernes of *Love's Labour's Lost* (neither of whom comes within hailing distance of an iambic pentameter line). Blank verse and earth-treading minds sometimes never meet at all . . . as is the case with "deformed and scurrilous" Thersites in *Troilus,* the humorless Welshman Fluellen in *Henry V,* or officious Constable Dogberry in *Much Ado.*

On several occasions Shakespeare carefully manipulated prose and poetry to pointed effect. Consider the venal Poet in *Timon of Athens*. In his last scene he speaks entirely in prose, except when performing unctuously for his former meal-ticket, Timon, whom he now thinks is wealthy again. In *Henry V,* Shakespeare shamelessly takes sides with the English by casting a French court scene (3.7) in arrantly mindless prose banter. And he has Cassio collapse into prose right after his first sip of Cypriot wine. When a sheriff suddenly interrupts the prose hijinks of the big *1 Henry IV* tavern scene, Hal takes charge and puts him off Falstaff's trail in a brief conversation that rises to the full princely dignity of blank verse. But perhaps the charisma of Shakespearean blank verse is most delightfully captured in *Twelfth Night,* when Viola disguises herself as Cesario and sues on the Duke's behalf for Olivia's hand with a "poetical" speech. To underline how affected and insincere it is, the playwright casts it in prose. Only when Viola, in exasperation, begins to speak from the heart does he allow her to shift (at 1.5.251) into truly eloquent verse. This, of course, has the effect of enamoring Olivia with "Cesario" instead. Such is the magic of Shakespearean poetry itself, which, in a phrase of Olivia's, works its effects as music does, with "an invisible and subtle stealth."

The reader may already be objecting that Shakespeare often did what a composer cannot: invest a prose speech (the equivalent of, say, a lengthy recitative) with such thrust and eloquence that it, too, achieves the effect of an aria. He could, in other words, remain near the prosaic bottom of the pyramid and yet produce riveting, bravura rhetoric. This ambidextrous eloquence, not available to the composer, was possible simply because he could deploy all the figures and flowers of rhetoric in his prose, too. He was not tied to the pulse of meter, as the composer is generally tied to the pulse of rhythm for his grand effects. And so there are many prose *tours de force* in the canon that could only be done full justice in opera as arioso or aria. One thinks, for example, of Edmund's potent soliloquy attacking astrology in *Lear* ("This is the excellent foppery of the world"), Shylock's "Hath not a Jew eyes," or the virtuoso speeches for Hal and Falstaff as they perform the "play extempore" in the tavern. Shakespeare makes Lear's great prose speech, "Is man no more than this," the centerpiece of his entire action, and one cannot imagine a *Lear* opera failing to make a riveting musical moment out of it.

Several of these impressive prose set pieces have, in fact, been translated handsomely into aria form. Bottom's "I have had a most rare vision" was turned by Britten into a lovely miniature aria, complete with a falsetto high F climax for his baritone. And the "merry war" betwixt Beatrice and Benedick, conducted by Shakespeare entirely in prose, Berlioz transformed into splendidly impetuous lyric duets in *Béatrice et Bénédict.* Lady Macbeth is the only major tragic figure to whom the playwright denied poetry at the end, yet Verdi turned her fractured, eloquent prose to haunting lyric effect in his opera's sleepwalking scene. He also took off in lyric flight from the prose soliloquies of Falstaff in praise of sack and dispraise of honor. In fact, *Falstaff*'s one full-fledged aria in the early Verdi style, Ford's jealousy monologue ("È sogno? o realtà?"), Boito derived from several prose soliloquies in *Merry Wives* (2.2, 3.5).

While we must grant this disparity between Shakespearean and operatic deployment of prose and recitative, which upsets the symmetry of my pyramid, a similarity far more important deserves emphasis. As I have already suggested, prose is for Shakespeare primarily a vehicle for dismantling illusion, fantasy, idealism, or pretension. His prose is often satirical and reductive. It usually impels us, as it were, toward ground level. Prose is often the playwright's means for laying the human spirit low; verse almost as constantly gives it flight (good or evil flight, be it said). The essential motion of his prose, then, is downward; that of his verse, upward.

All the prose speeches mentioned above bear this out, and several other instances of the dichotomy leap to mind. Enobarbus, for one, speaks in prose when he is satirizing Cleopatra's manipulative histrionics (1.2.120 ff), but when he tries to convey her charisma in the famous description of her barge he rises to blank verse that is itself charismatic (2.2.190 ff). In *1 Henry IV*, Hotspur celebrates the romance of war gloriously in buoyant verse, while a few moments later Falstaff describes in a long prose speech (4.2) the carnage of his beggarly, unfit troops (all the able-bodied have bribed him to avoid service). In *The Merchant of Venice* is a fine example of the author's prose-or-poetry discriminations. Portia mercilessly satirizes her prospective suitors in prose (1.2), but when she falls for Bassanio and he has just chosen the right casket, her exhilaration pours forth in rhymed verse:

> O love, be moderate, allay thy ecstasy,
> In measure rain thy joy, scant this excess!
> I feel too much thy blessing; make it less,
> For fear I surfeit. (3.2.111–114)

The similarity of Shakespeare and opera I am leading up to, of course, is their fundamental preoccupation with, indeed, bias toward, the *upward* mobility of heightened verse and elaborate cantilena, respectively. Falstaff's entire igno-minious career in prose over two whole plays symbolizes this bias, for this career is very specifically aimed toward the climactic moment when Hal, just crowned king, rejects Falstaff in magisterial iambic pentameter on the steps of Westminster Abbey. The Duke in *Measure for Measure* says, "Spirits are not finely touch'd / But to fine issues," and one can generalize that Shakespeare achieved his fine issues by moving up the pyramid . . . touching his finest spirits with poetic utterance. Lear is not abandoned on the heath speaking near-gibberish prose, but is returned to a more profound humanity—and to blank verse—for his last, devastating scenes. Similarly, Othello is not left sputtering the impotent, enraged prose of the fourth act, but is brought back, "finely touch'd" by poetry, for his tragic end. And, while Bottom's prosaic discourse on his dream is charming, the last and most ravishing word on dreams and the *Dream* play itself comes in sixty gossamer lines of poetry for Puck and Oberon at the end.

Recitative and declamation are the handmaidens of spectacular vocal displays that occur when principle characters in opera are finely touched. Prose, I would urge, frequently bears the same relationship to poetry in Shakespeare. The "real" world to which prose gives the readiest entry was of secondary interest to him. The sole play he ever wrote that is set literally in the "real" world (a nearby London suburb), *Merry Wives of Windsor* is, perhaps inevitably, by far the most prosaic of all his plays. An enormous eighty-eight percent of its lines are prose. Many feel it is one of his least inspired efforts, and it is tempting to imagine that he might never have bothered with it, had not Queen Elizabeth (so the legend goes) compelled him to bring Falstaff back from the dead and parade him as a comic, aging lover.

Rather than mire himself in prose, Shakespeare gravitated toward the mighty poetic line; except in a few of the comedies, prose assumes a decidedly serving status. Most of the vicious, scurrilous, or simply lamebrained characters in his canon are confined entirely to prose: Cloten in *Cymbeline,* Stephano and Trinculo in *The Tempest,* Ajax and Thersites in *Troilus,* Autolycus in *Winter's Tale,* to name a few. They are foils who "set more goodly off" the more important characters who speak in verse. Though, like many a comprimario in opera, these coarse-minded fellows may make striking, colorful impressions, their spiritual and expressive limitations are very marked. It would be as unimaginable for one of them to utter a beautiful iambic pentameter line as it would be for Puccini's Sacristan or Spoletta to utter a musical phrase like Cavaradossi's "Recondita armonia" or for Mozart's evil servant Monostatos in *Die Zauberflöte* to launch into one of Prince Tamino's arching melodic lines.

The precedence of poetry over prose is also obvious in several characters who move freely up and down the pyramid, speaking prose or poetry as need be. With them, too, Shakespeare usually aims for the poignant or climactic moments to rise into verse. Among the major roles, one thinks of such charismatic and charming (or, alternatively, cunning and manipulative) characters as Brutus, Edmund, Portia, the Duke in *Measure for Measure,* Iago, Shylock, Hamlet, and —most brilliantly of all—Prince Hal, who makes a point of boasting about his gift for being able to speak every man's language.* There are some exceptions: Rosalind's prose in *As You Like It* is more ravishing than her poetry, and one might say the same of Mercutio. But it is with comic characters—and plays—that Shakespeare was likely to give prose freer reign.

In sum, the more one attends to Shakespeare's manipulations of prose and poetry, the more strikingly apparent becomes his affinity for the composer's method of creating highly focused, emphatic effects by the sudden onset of rhythmic, rhetorically heightened vocalism. And these bursts of verbal music are usually placed with a surgeon's finesse. Hamlet's graveyard *scena*—all in prose—leaps to a rousing close with this rhyming cadenza:

> Imperious Caesar, dead and turn'd to clay,
> Might stop a hole to keep the wind away.
> O that the earth which kept the world in awe
> Should patch a wall t' expel the winter's flaw! (5.1.213–216)

Similarly, amid all the prose of *Merry Wives,* Mistress Page is spurred to this little exit aria in neon couplets that announces the play's moral:

> We'll leave a proof, by that which we will do,
> Wives may be merry, and yet honest too:
> We do not act that often jest and laugh;
> 'Tis old, but true: still swine eats all the draff. (4.2.104–107)

A bit later Shakespeare helps her convey the mysterious story of Herne the hunter and tell of the "urchins, ouphes, and fairies" that will gather around the oak, by shifting into iambic pentameter. In *Much Ado,* after Beatrice has overheard in the "pleached bower" that Benedick loves her madly, she steps forth alone for almost her only verse in the play. It is clear from the aural surprise of her lilting lines—rhymed, very unusually for Shakespeare, *abab cdcd ee*—that she, too, is experiencing a lover's "high." One scene in *Love's Labour's Lost* posed an unusual problem: Shakespeare wished to give Boyet a little *aria d'agilità* at one point, but the scene was already cast in very artificial rhymed couplets. His resourceful solution was to shift into poetical overdrive, with several lines of galloping dactylic hendecasyllables:

> Methought all his senses were lock'd in his eye,
> As jewels in crystal for some prince to buy,
> Who tend'ring their own worth from where they were glass'd,
> Did point you to buy them, along as you pass'd [etc.]. (2.1.242–245)

Occasionally, the aural collisions of poetry's "shaping fantasy" with the "cool reason" of prose occurs on a grand scale. Shakespeare's shrewd effect-consciousness in managing these collisions, perhaps more than any other aspect of his playbuilding, gives his drama its operatic style. Often one can scarcely refrain from "scoring" his aural transitions and juxtapositions as one hears them. One imagines that those formal, judicious couplets with which the Prince condemns Romeo to banishment might be set to a slow dirge, while Juliet's spree of lovesick fantasy just following ("Gallop apace, you fiery-footed steeds") begs a sudden shift to *allegro assai.* Marina, the chaste heroine of *Pericles,* finds herself in a brothel at one point. Though surrounded by vice and prose, she speaks "holy words" that never stray from iambic pentameter: we ought to hear her voice arching soprano-like above the surrounding chorus. In *The Winter's Tale,* Shakespeare carefully keeps the "notable passion of wonder" at the reunion of Leontes with his daughter in the form of prose "recitative" spoken by three anonymous gentlemen, well knowing this would contrast splendidly with the real and the poetical music of the climactic moment when the "statue" of Hermione comes to life:

> Music! awake her! strike!
> 'Tis time; descend; be stone no more; approach;
> Strike all that look upon with marvel. . . .   (5.3.98–100)

The prosaic scene thus provided an oasis of aural calm, from which emerges the uplifting ensemble finale.

One last example of Shakespeare's indulgence in the pyramid game should come from the play that, as Shaw pointed out, was written "in the style of Italian opera." The scene I have in mind shows him ringing the stylistic changes, and it begins as Desdemona and Iago nervously pass time on Cyprus, awaiting Othello's bark during the "desperate tempest." She wittily asks him for a little Petrarchan-style praise of her womanhood, and after his "Muse labors" a bit he launches into a few dozen rhymed couplets. Desdemona playfully tries to subvert his conceits, and she slips, as usual for a Shakespearean character in a satirical mood, into prose . . . and prose is very rare for her. Following these aural surprises, Iago steps aside and watches Cassio take her hand. In this speech, prose becomes—as it often does in the plays—a vehicle for displaying a vicious, reductive sensibility: "With as little web as this," says Iago, "will I ensnare as great a fly as Cassio. Ay, smile upon her, do; I will gyve thee in thine own courtship."

Shakespeare might have allowed Iago to speak these thoughts in verse, as he did in all his other solo speeches. But I think he wished to hug the stylistic ground, as it were, so as to heighten the contrast of what is to follow immediately: the entrance of Othello and his attendants in triumph. Appropriately, the Moor and Desdemona share the line that returns the action to iambic pentameter: "O my fair warrior!" / "My dear Othello!" In the next eighteen radiant lines (2.1.184 ff) the dramatist, with superb economy, raises the Moor to the blissful height from which he will soon fall:

> O my soul's joy!
> If after every tempest come such calms,
> May the winds blow till they have waken'd death!

All the poetical stops are pulled out here, and the ecstatic interlude reaches a climax with the kiss:

> I cannot speak enough of this content,
> It stops me here; it is too much of joy.
> And this, and this, the greatest discords be
> That e'er our hearts shall make!

Shakespeare gently relaxes from this high emotional pitch in the next few lines, as Othello gives orders to the garrison and leads his wife off. Iago immediately drops into prose with Roderigo, thus setting the other bracket of prose in place as a foil for this exhilarating love duet.

It is important to add that this love duet—unlike the one in Verdi's opera—is actually interrupted by an aside from Iago. With the voices of the two lovers stylistically in seventh heaven, though, Iago clings to the recitative of prose. He menaces, *sotto voce:* "O, you are well tun'd now! But I'll set down the pegs that make this music, as honest as I am."* Iago's image provides an apt conclusion to this chapter. For in the preceding pages we have explored how Shakespeare—a master manipulator too—was himself very shrewd in setting up or down the pegs that made his verbal music. We have explored how, in other words, he gave such effective variety in his plays to the "music of men's lives."

15

Operatic Bard, Bardic Opera

Thus far we have focused on the various methods and materials of the two dramaturgies by ranging at large through both repertories. It would seem appropriate, therefore, to look, by way of summary, at a single play and opera in some detail with an eye to their fundamental similarities. What more provocative way to do so than to nominate the most thoroughly operatic of Shakespearean plays and the most Shakespearean of all operas?

"A musician only," wrote Shaw in 1889, "has the right to criticize works like Shakespear's earlier histories and tragedies. The two *Richards, King John,* and the last act of *Romeo and Juliet* depend wholly on the beauty of their music." Taking a deep breath, Shaw continued about these plays in a way that brought him finally to a discussion of *Richard III:* "There is no deep significance, no great subtlety and variety in their numbers [i.e., prosody]; but for splendor of sound, magic of romantic illusion, majesty of emphasis, ardor, elation, reverberation of haunting echoes, and every poetic quality that can waken the heart-stir and the imaginative fire of early manhood, they stand above all recorded music. These things cannot be spectated . . . they must be heard. It is not enough to see *Richard III;* you should be able to *whistle* it." This vintage Shavian gusto was elicited by a performance of the play with incidental music by one Edward German. Shaw did not care for the musicianship of the actor who played the crookback (he "gave him up as earless" when "I am determinèd to be a villain" was pronounced "deter-min'd"), and German's music sank beneath contempt: "whenever Richard enters you hear the bassoons going: pum-pum-pum, pum, pum Paw!"

Though my approach will not follow Shaw's emphasis on the line-by-line delivery of Shakespeare's word-music, I will take his cue and choose *Richard III* as the most operatic of the plays. I make this choice knowing full well that its supremacy as the most popular acting-play, which it enjoyed virtually from Shakespeare's day to the late nineteenth century, is now somewhat tat-tered . . . and knowing full well, too, that this play (like all the histories) has not appealed to composers. Only two operas have been based on it, compared with twenty-four on *Romeo and Juliet* or thirty-three (and counting) on *The Tempest*. In spite of this, I am convinced *Richard III* displays, very starkly, the most impressive variety of "operatic" features. And, being a work of Shakespeare's apprenticeship, the play displays these features more vividly than do his later, more mature works. Even more important, *Richard III* is the most brilliant epitome of his early, more melodramatic style. Though Robert Heilman grants that the play's "fresh and variegated" characterization raises the play above "trite" melodrama, he convincingly asserts that Richard is a melodramatic villain and that the play's series of "victims to be pitied" and "monstrous villains to be loathed" make melodrama its dominant mode. Such a mode is what made it possible for Shaw to praise the early Shakespearean style for its "magic of romantic illusion, majesty of emphasis, ardor." We have seen that such a mode is also operatic, and it can be small coincidence that late eighteenth- and nineteenth-century opera, when ideally executed, ought to evoke precisely Shaw's terms of praise.

Which opera, then, might one single out as the most "Shakespearean"? Can there be the slightest hesitation in turning to Verdi? Though, to make this a sporting game it should not be one of the operas actually based on Shakespeare. Which of Verdi's other works suggest themselves? *La Forza del Destino* comes quickly to mind. Verdi found *Don Alvaro,* the Angel de Saavedra play that was its source, to be "powerful, singular, and truly vast . . . something quite out of the ordinary"; and this neatly describes many a Shakespearean action—especially that of *Richard III,* his second-longest play, with its singular villain and truly vast cast of fifty-two speaking parts. William Weaver has, in fact, nominated *La Forza del Destino* as Verdi's "most Shakespearean opera," saying it is a work "of great variety, vast scope, juxtaposing comic and tragic, employing a number of unusual characters, all sharply defined . . . perhaps Verdi's boldest attempt to portray an entire, complex, contradictory world." Julian Budden seconds the motion when he finds in *Forza's* "portrayal of humanity on a vast canvas ranging from the highest to the lowest in the land . . . the unmistakable sense of a Shakespearean chronicle play." Others to whom I have posed this question have plausibly opted for *Don Carlo*—based on a dramatic poem by Schiller, who was in turn influenced by Shakespeare—or *Un Ballo in Maschera,* the opera Verdi wrote instead of pursuing his *Re Lear* project. Verdi himself, as we have seen, adds *Rigoletto* to the competition as well: "Oh, *Le Roi s'amuse* is . . . perhaps the greatest drama of modern times. Triboulet is a character worthy of Shakespeare!!" After much thought, however, I have settled in favor of *Il Trovatore*.

Given all the *Night at the Opera*-style fun that has been had at the expense of *Il Trovatore's* "extravagant and bizarre plot" (Budden), my choice may seem perverse. After all, Heilman's thumbnail sketch of the "standard" (i.e., hostile) definition of melodrama happens to fit the opera perfectly: "conventional or straight-forward conflict, decked out in the various excitement of threats, surprises, risks, rival lovers, disguises, and physical combat, all against a background of ideas and emotions widely accepted at the time." But in fact there are numerous reasons to urge this comparison, which may seem at first blush farfetched. Some of these reasons are minor but rather suggestive. *Richard III* unfolds in twenty-three scenes, *Il Trovatore* in twenty-three discrete musical numbers. Both actions are set in the 1400s, amid caparisoned horses and heavy armor (Leonora, we learn, first met Manrico as the stock mysterious Black Knight at a joust). Both actions are also set against the grim backdrop of civil war, as well as medieval superstition and brutality. Witchcraft, in fact, sets both plots in motion: in the play a "wizard" has warned King Edward that his issue will be displaced by someone with the initial "G" (Richard is the Duke of Gloucester), while Azucena's mother casts the evil eye upon (as it turns out) baby Manrico. Ghostly hallucination also plays a part in each drama. More important, both actions are, at bottom, part of the Renaissance revenge-tragedy tradition. The dying words of Azucena's mother— "Mi vendica!"—obsess the gypsy, whose last words in the opera are, inevitably, "Sei vendicata, O madre!" As the action of *Richard III* progresses, the voices crying for revenge become a resounding chorus. Old Queen Margaret is especially long-winded in her vengeful vituperation; her "I am hungry for revenge, / And now I cloy me with beholding it" would have made an excellent alternative curtain-line for Azucena. But perhaps the most important reason to urge the present comparison is simply that, however crudely and arbitrarily conceived this play and this opera may be in comparison with other works in the canons, they sharply and clearly demonstrate the characteristic theatrical tactics of their creators. No one would seriously urge them as the supreme masterpieces by Shakespeare or Verdi, and yet one might make a case for them as the most splendidly *typical* ones. This, indeed, might explain why *Richard III* was so very popular through the centuries with actors and audiences and why, too, *Il Trovatore* was the most loved of Verdi's operas in his own day.*

Other bases for comparison will emerge in due course. For the moment, let us consider *Richard III* separately and attempt to throw into relief some of its operatic qualities. One can be dispatched quickly, simply because it is so ubiquitous: Shakespeare's extraordinary care in the disposition of the "figures auricular"—assonance and alliteration. A positively Mozartean attentiveness to aural balance and antithesis pervades very many of the play's lines and gives the work its "poetical" flavor. (The play is one of six that Shakespeare wrote entirely or almost entirely in verse.) This elegant patterning of sounds is apparent in Richard's opening soliloquy, when he says that now

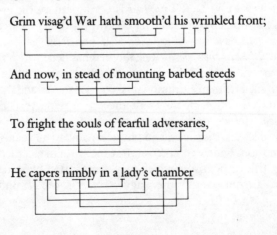

Grim visag'd War hath smooth'd his wrinkled front;

And now, in stead of mounting barbed steeds

To fright the souls of fearful adversaries,

He capers nimbly in a lady's chamber

To the lascivious pleasing of a lute.

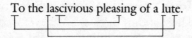

Such attentiveness to minute aural effects is surely what led Shaw to insist this play must be heard rather than spectated.

We have also seen both verbal and musical rhetoric depending on figures of repetition, extension, or augmentation, and this is very common in *Richard III*: syntax tends to replicate itself ("I had an Edward, till a Richard kill'd him; / I had a husband, till a Richard kill'd him"); epithets tend to multiply ("By underhand, corrupted, foul unjustice"; "The wretched, bloody, and usurping boar"); speakers consistently elaborate with verbal ornamentation upon their simple meaning—the verbal equivalent of appoggiaturas, shakes, and "relishes" on a simple musical line. This all, of course, reflects the habits of the lyric stage, where expansion is of the essence, both horizontal along the staff and vertical from bottom to top. Indeed, in the play's most famous line (it occurs not once but twice) one can see at work the common compositional technique of repeating a phrase once, then again, and finally with a climactic extension: "A horse, a horse! my kingdom for a horse!"

A playwright so attentive to the interplay of sounds will also be eager to rivet attention by virtuosic verbal display, and perhaps the most impressive feature of *Richard III*'s operatic profile lies in the numerous times characters are given center stage for elaborate vocalism. The Duchess of York speaks not only for herself but for Shakespeare's dramaturgy when she admonishes,

> . . . be not tongue-tied; go with me,
> And in the breath of bitter words let's smother
> My damned son that thy two sweet sons smother'd.
> The trumpet sounds, be copious in exclaims. (4.4.132–135)

Copious in exclaims . . . the phrase nicely embraces several arias in *Il Trovatore*. This is the posture that Shakespeare is usually aiming his characters toward. "Be not tongue-tied" is certainly the motto of his principals, and the primary mode of copious exclamation is the long, uninterrupted speech performed for others present on stage or for the audience in soliloquy. Such speeches fall into familiar categories, and *Richard III* offers a very full assortment of these genre pieces that perform quasi-ritual or quasi-performative functions.

Consider Richard's solos. The first, which opens the action, performs several functions brilliantly. It provides an informative entry into the historical circumstances, establishes the villain as an ostracized satirist-figure, and offers a "credo" rather after the fashion of Boito's for Jago. Richard boasts of loving to "descant on mine own deformity," and indeed this vicious "song of myself"–style speech performs the function of an operatic entrance aria (Shaw does call it a *scena*), in which we are vouchsafed the fundamental moral posture of a character. One can imagine an arresting front-of-curtain prologue—like those in *Pagliacci* and

Lulu—deriving from this speech. It is neatly balanced, incidentally, by Richard's solo at the end of this first scene, which seals the fate of his brother Clarence. In a third soliloquy he exults after the ghastly courtship of Anne, who knows Richard has murdered her father-in-law and husband. It begins, "Was ever woman in this humor woo'd?" Shakespeare uses it to drive home Richard's antipathy to goodness, bitter consciousness of his deformity, and tremendous egomania ("I am crept in favor with myself"). This is the stuff of many an operatic exit aria, expecially in the flourishing gesture of its final couplet: "Shine out, fair sun, till I have bought a glass, / That I may see my shadow as I pass."

"Arias" occur in *Richard III* with such regularity and such premeditated care for their theatrical effect that one can almost experience them as one experiences—to think of another hunchback—the arias in *Rigoletto*. Some of these are obviously "sherbet" arias that could be (often are) cut without great loss. The acting versions of the play, represented by the quarto editions, in fact show that certain speeches sometimes were cut. Other big speeches, however, are tremendous and pivotal. Some of these outbursts of vocalism, like certain of one's least-favorite opera arias, do stretch patience. Shakespeare gives Lady Anne thirty-two lines in which to curse Richard and "obsequiously lament" over the coffin of her father-in-law, Henry VI. The speech is stiff and bombastic, but its *raison d'être* becomes clear only after it ends and (in the play's first bit of arrant stage-managing that boggles credibility) Richard suddenly enters. In about two hundred breathtaking lines he manages to melt Anne's resistance: from the height of outrage in those thirty-two lines—worthy of Mozart's Donna Anna—Anne falls under Richard's spell and becomes, like Zerlina, putty in the courting villain's hands.* A far grander aria is produced by old Queen Margaret's "quick curses" of Richard, which roll on for nearly fifty lines (1.3.187 ff). One is very grateful that Richard's second attempt to cut her off succeeds. More cogent and amenable to musical transformation is the Duchess of York's "grievous curse" of a mere nine lines toward the end of the play. One can almost hear an imperious vocal gesture--a high C, say--and an emphatic harmonic resolution to accompany her exit couplet: "Bloody thou art, bloody will be thy end; / Shame serves thy life and doth thy death attend" (4.4.195–196).

Richard III also boasts a number of speeches that prefigure venerable operatic traditions. The speech in which Clarence, imprisoned in the Tower of London, recounts a dream to his keeper, combines the dungeon scene of the rescue-opera tradition with the commonplace "dream narrative" *scena*. Arias narrating past events are common in opera, and Shakespeare gives one of twenty-two lines (a soliloquy) to the murderer Tyrrel. The dramatist knew well that the murder of the two young princes in the Tower was the most sensational and famous of Richard's crimes, and Tyrrel's narrative was his way of scoring theatrical points with it. Lord Hastings' speech of exit from the play—and the world—combines elements of a scaffold performance and a "prophecy" aria. As often in the play, the couplet gives a satisfying sense of closure of the kind composers achieve with a few tutti chords:

> O bloody Richard! Miserable England!
> I prophesy the fearfull'st time to thee

That ever wretched age hath look'd upon.
Come, lead me to the block; bear him my head.
They smile at me who shortly shall be dead.

A more elaborate scaffold performance is naturally created for the more important figure of Buckingham later in the action (5.1.3–29).

Arias in which a character swears an oath are reflected in Richard's rhetorically gaudy and of course false promise to Queen Elizabeth that he will love her daughter with "[i]mmaculate devotion, holy thoughts" (4.4.397–411). The operatic *preghiera* is also part of the play's action: Clarence utters a short prayer before falling asleep in his cell, and a noble prayer puts God on the side of the emerging hero Richmond in act 5. Richmond also indulges in the "Ritorna vincitor!" mode for his oration to his forces before the victory at Bosworth (5.3.315–338). Though not as sabre-rattling as "Di quella pira," its admonition "Draw your willing swords" is the same as Manrico's "all' armi!" Richard gets his own battle speech, too, which is characteristically full of meanness of spirit and appeals to the worst human instincts of his soldiers.

Another terrible human instinct—for revenge—practically consumes both Azucena and Count di Luna, and the vendetta mode is spectacularly well represented in the play, notably when Margaret returns once again late in the action for a massive sixty-line tirade. "Earth gapes, hell burns, fiends roar, saints pray," she says, for the time when she "may live and say 'The dog is dead.'" There is even a rather near approach to an operatic "mad scene" here: Queen Elizabeth enters "with her hair about her ears" and erupts, "Ah! who shall hinder me to wail and weep, / To chide my fortune, and torment myself?" She comes to announce the sudden death (amazingly, it is a natural one) of her husband, Edward IV, and promises in hysterical grief "[t]o make an act of tragic violence." As with many a musical mad scene, though, her violence turns out to be highly poetical, as in the conclusion of this little *scena*:

> I am not barren to bring forth complaints.
> All springs reduce their currents to mine eyes,
> That I, being govern'd by the watery moon,
> May send forth plenteous tears to drown the world!   (2.2.67–70)

No opera or play can function with arias and set speeches alone, and there are many ensemble scenes in *Richard III* that break the monotony of individual vocalism. The courtship of Richard and Anne, for instance, dominates the beginning of the play, much as do the love duets in the first acts of *La Bohème* and *Butterfly*. This scene Shakespeare carefully modulates by moves back and forth between stichomythia (rapid-fire one-line exchanges) and longer speeches for each actor. Crucial moments are registered in such stage directions as "Spits at him," "He lays his breast open, she offers at it with his sword," and "She falls the sword." This duet would seem especially ripe for music, which can be such an effective means for conveying the differing intentions of characters interacting on stage. Music would also be well able to capture the huge decrescendo of Anne's wrath. A duet perhaps even more curious balances this one toward the play's end, when Richard cajoles Queen Elizabeth for the hand of her daughter (4.4.197 ff). This duet is impossibly long (the quarto acting version cuts nearly twenty-five percent of it), but one can still easily imagine in musical form its modulations from stichomythia to expostulation, then to dialogue, extensive persuasion, back to stichomythia, and finally to an oath. This kind of scene Verdi would surely have conceived and organized as a *gran scena concertato a due*.

Larger ensembles in the play are structured in highly operatic fashion as well. The ecstasy of grief experienced by Clarence's mother, sister-in-law (just widowed), and two children, for instance, rises to a perfectly formulaic and choric climax:

> *Elizabeth:* What stay had I but Edward? and he's gone
> *Children:* What stay had we but Clarence? and he's gone.
> *Duchess:* What stays had I but they? and they are gone.
> *Elizabeth:* Was never widow had so dear a loss.
> *Children:* Were never orphans had so dear a loss.
> *Duchess:* Was never mother had so dear a loss.   (2.2.74–79)

An impressive trio for three female voices might be made from the first scene of act 4, when Elizabeth, the Duchess, and Anne begin to harden in their resistance to Richard. It begins with lines from Elizabeth—

> Ah, cut my lace asunder,
> That my pent heart may have some scope to beat,
> Or else I swoon with this dead-killing news!

—of just the kind that are guaranteed to set an opera orchestra going at full tilt and send singers into high dudgeon . . . and unison high notes. This hundred-line scene comes to a surprising end as the lavish coda of the Duchess's big-gestured "farewell" aria is suddenly upstaged by Elizabeth's poignant and (given events to come) ironic apostrophe to the Tower of London that it "use my babies well": the stuff of operatic solos.

Certain of these larger ensembles produce other kinds of operatic effects. In the first scene of act 2, for instance, Richard suddenly informs the eight other figures on stage that the "gentle duke" of Clarence is dead. The Folio stage direction is "They all start," and in the next several lines reactions are registered all around. This is the kind of theatrical thunderbolt that Verdi and other bel canto composers looked for to indulge what Budden calls the "*largo* of dramatic shock," during which action halts and the moment is expanded by various forms of amplified musical rhetoric. "They all start" would be a perfect stage direction to introduce several famous concerted ensembles in the repertory. One thinks of the moment Edgardo flings open the door at the wedding in *Lucia di Lammermoor,* which sets off the famous sextet; the moment Alfredo tosses his gambling winnings at Violetta, which ignites the big finale of the third part of *La Traviata;* or the moment Otello wrenches Desdemona to the ground, setting in motion *Otello*'s only big stop-time ensemble. Another highly elaborate scene of response to "dead-killing news" that seems operatic is the mass lamentation after the murder of the two princes is discovered (4.4). One might note that—as Shakespeare does several times in the play—the stage is split, with Margaret for a time speaking in asides and the other women together. This, of course, is a common occurrence on the opera stage.

Also easily imaginable in an opera house is the scene in which, after his coronation, Richard ascends the throne and transacts his first royal business: "The trumpets sound a sennet. Enter Richard in pomp, crowned" (4.2). This would not only allow the obligatory big tableau scene, but the ensuing split-stage action—which includes private conversations with Buckingham and the assassin Tyrrel and some soliloquizing asides—would also lend itself to the usual methods of musical differentiation of simultaneous stage action. If set to music, the remaining large ensemble scenes would be highly reminiscent of Verdi's *Macbeth*. One such scene is the appearance of no fewer than eleven ghosts of Richard's victims in an elaborate dream vision (5.3; compare the pantomime vision in Verdi's third act). Comparable to the final alarums and skirmishes near Macbeth's castle, of course, are those at Bosworth Field.

Finally, the ways Shakespeare deployed his set pieces and ensembles within larger scenic structures deserve some notice. For, again, they bear strong resemblance to how an operatic action usually unfolds. This can be most efficiently demonstrated by "translating" a few scenes from *Richard III* into the typical structural units of opera. The almost classical symmetry of the second scene emerges, for instance, when we view it as consisting of an entrance *aria d'agilità* for the prima donna (Anne's furious lament, 32 lines), followed by a multi-movement *gran duetto d'amore* (Anne and Richard, 198 lines), and then a solo *aria di sortita* for Richard (36 lines). Clarence's dungeon scene unfolds in almost exactly the same number of lines, running about twenty minutes, and might be rendered operatic as follows:

Clarence mentions dream; Keeper asks, "What was your dream, my lord? I pray you tell me." (8 lines)	----------Recitative
Dream narration. (54 lines)	----------Aria (*andante*)
Clarence and Keeper speak, latter departs. (5 lines)	----------Recitative
Clarence prays, falls asleep. (8 lines)	----------Arioso (*adagio*)
Lieutenant of Tower enters, admits two murderers. (22 lines)	----------Recitative
Murderers debate with each other, in prose. (52 lines)	----------Duet (*allegretto giocoso*)
Clarence awakens, pleads for life, is finally stabbed and dragged away. (over 100 verse lines)	----------Trio (*crescendo poco a poco*)
Two murderers express remorse, soon leave. (12 lines)	----------Duettino (*sotto voce*)

The kaleidoscope turns much more swiftly in the fifty-seven lines (perhaps four minutes) of act 4, scene 3. Here Tyrrel has his emotional twenty-two line aria describing the smothered princes; there is a drop to recitative as Richard enters and dispassionately questions Tyrrel about the deed (13 lines); left alone for eight lines, Richard rises to a more heated arioso as he plans to win the hand of his dead brother's daughter; and finally Ratcliffe enters with threatening military news that sends Richard into a vigorous nine-line *cabaletta* in the "Di quella pira" vein. It ends not with one but with two emphatic couplets:

> Then fiery expedition be my wing,
> Jove's Mercury, and herald for a king!
> Go muster men. My counsel is my shield;
> We must be brief when traitors brave the field.

Act 4, scene 4 is comprised of several massive ensembles. But it must be admitted that this scene, the play's longest at over 500 lines, plays very lugubriously (like certain stultifying acts in Wagner) and need not detain us here. Perhaps Shakespeare was already preoccupied with the brilliant series of "bel canto" numbers that he was planning for the climax of his action. I refer in particular to the series of theatrical coups beginning with Richmond's noble *preghiera* (5.3.108). It is followed immediately by the several guilt-bearing ghosts who appear to Richard, then the soliloquy in which, at long last, he begins to lose his nerve and feel the bite of "coward conscience," then the two battle orations, and the final onstage demise of "the bloody dog."

A few other more general remarks about *Richard III* can perhaps be made before we turn to *Il Trovatore*. One has to do with the plot, which is centered around the virtuosic villainy of a single character, and which—unlike other Shakespearean plays—is not complicated by any subplots. But it does exemplify the playwright's habit of setting a number of two-dimensional but extremely vivid secondary figures around a highly complex or charismatic central figure . . . like Hamlet, Henry V, Lear, Coriolanus, or Prospero. Brief reflection on the opera repertory produces countless examples of this same hub-and-spoke organization of the *dramatis personae*. Typical and pertinent for several thematic reasons is the Tsar in *Boris Godunov,* since Pushkin consciously based his play on the style of Shakespeare's histories.

Also, a good case might be made for Richard as a quintessentially operatic protagonist. What is he, if not a perfect combination of Don Giovanni and Rigoletto? Like most great operatic characters, he is quite simply riveting. A recent editor of the play has observed that Richard's "attractiveness lies chiefly in his ability to make us admire him, even while our better natures know perfectly well that what he is doing is monstrous." Most operatic characters are also, like Richard, hermetically sealed from the world around them, whether by superior intelligence, extraordinary beauty, mania, goodness, villainy, or idealism of spirit. Their conflict with this outer world is figured forth in opera through the isolation of the vocal line from the orchestral mass. This isolation flourishes precisely because of the impressive detachment of the protagonist from the world in which he or she exists. Richard's detachment is impressive indeed; he is the complete ironist and aptly boasts, "Thus, like the formal Vice, Iniquity, / I moralize two meanings in one word." He is, in other words, a superb equivocator—a figure of two voices: his own voice, which we hear in asides and soliloquies, and the chameleonic role-player's voice he deploys in public.

Music is a splendid means for giving life to such figures who exist both in their own self-created world and in the great world . . . figures who speak in several voices, depending on the dramatic situation. One thinks of some of the other memorable inhabitants of these two "worlds": Giovanni, Norma, Lady Macbeth, Violetta, Rigoletto, Wotan, and Peter Grimes. Verdi boasted of the power of music to give artistic life to this profound equivocation when he wrote to Antonio Ghislanzoni, "A melody, based on words that seem to be spoken like a lawyer, will seem strange. But beneath these lawyer's words there is the heart of a desperate woman burning with love. Music is splendidly able to succeed in depicting this state of mind and, in a certain way, in saying two things at once."

Il Trovatore, as it happens, does not offer such a cynosure of attention as Richard. Though Verdi did on one occasion assert that Azucena was "*the* principal role" (and when a *Trovatore* performance is stolen, the culprit is usually the mezzo), the four principals in fact share approximately equal executive burdens.* But in several other important respects Verdi's opera is eminently Shakespearean. Perhaps we should begin with the most general: the very similar *tinta* of each work. That is, the primal coloring or ethos conveyed by each work as a whole. It is a particular manifestation of Shakespeare's and Verdi's genius as men of the theater that they were able to give each of their finest creations a distinctive, salient individuality of character, or *tinta*. Hofmannsthal was, I believe, unwittingly speaking of this *tinta* when he urged Strauss that "what is vital is to find the right atmosphere for the whole thing, a certain general atmosphere in which the whole piece lives."* Budden's final word in his chapter on *Trovatore* is that it is "a coherent masterpiece," and this phrase applies as well to the imagined world of *Richard III*. Both works cohere ultimately because the *tinta* of each is so effectively and completely impregnated in the text and music. The *tinta* is a very dark one, too. Both dramatists were to achieve some of their most powerful effects with the *tinta nera,* the color of darkness: Shakespeare in *Macbeth, Othello,* and the "cheerless, dark, and deadly" *Lear;* Verdi in *Boccanegra* and *Don Carlo.* Thus, *Richard III* and *Trovatore* might be considered their creators' first great successes at looking through the glass darkly.

"Bad is the world," says a scrivener in one of Shakespeare's choric comprimario arias, and *Richard III* does present a world of brutality and treachery where the only potent ideals, if such they may be called, are Machiavellian. Elizabeth captures the *tinta* of this world when she cries, "Welcome destruction, blood, and massacre! / I see (as in a map) the end of all." The Duchess of York responds with her own passionate précis: "O preposterous / And frantic outrage." Some might wittily use this phrase to make fun of the incredibilities of plotting in both the play and Verdi's opera, but I think it can be turned to advantage, too. For it captures the extremely cynical view of human existence that the two works share. The entire action of *Richard III* offers a premonition of Albany's frightening image of cannibalism in *Lear:* "It will come, / Humanity must perforce prey upon itself, / Like monsters of the deep." The pace of outrage is, if anything, even more frantic and inhumanely "preposterous" in *Il Trovatore,* which has well been called "the darkest, perhaps the most pessimistic" of Verdi's works. And another writer has ventured that this opera reflects the composer's own "furiously pessimistic view of the world". . . his conclusion that "there is no sense in human striving or human suffering, for life means nothing." We might be tempted to this conclusion by Verdi himself, who wrote after *Il Trovatore*'s premiere, "People say it is too sad, that there are too many deaths in it. But death is all there to life. What else is there?" Though this outlook may remind us more of Macbeth's nihilistic comment on life's "petty pace," it also points to the world of *Richard III,* a world, as Jan Kott puts it, "full of murder, violence, brute force and cruelty." It can be added, too, that there is no comic relief whatsoever in *Il Trovatore* to lighten the *tinta*—and that what precious little there is in *Richard III* is pure gallows.

This talk of the oppressive *tinta* in the two dramas must bring us hastily to the obvious irony that their effect on us is far from oppressive. Rather, they leave us *im*pressed and exhilarated. The tenor of their actions may be grim, but the effect of the artistic vehicle (richly patterned, lavishly artificial modes of utterance) is ravishing, pleasurable, and—at the most brilliant moments—even ecstasy-producing. E. M. W. Tillyard was perhaps aiming at this point when he observed that a fine performance of *Richard III* can achieve "the solemnity we associate . . . with the Dionysia [festivals] of Athens and the Wagner festival at Bayreuth." A superb performance will send us away with a deep feeling that a momentous episode has taken place, and that is ideally how one should leave *Il Trovatore* or any of the repertory's great tragic actions.

Such a heightened effect, we have seen, is a product of the constant urge in this rhetorical, quasi-ritualized drama toward eruptions of heroic passion. This goes far to explain why *Richard III* and *Il Trovatore* show such small interest in realistic character development or credible motivation—and why their plots are so transparently expeditious as they move as speedily as possible from one set piece to the next. Shaw once tried to cheer up an actor who had just had a leg amputated, with what could have seemed ridiculously buoyant solicitude: "my pieces are not leg pieces." Insofar as *Richard III* and *Il Trovatore* make their big effects when characters plant themselves firmly and vocalize at emphatic length, they are also far from being "leg pieces." *Il Trovatore,* indeed, is notorious as the repertory's ultimate stand-and-sing opera.

This fundamentally rhetorical style will also explain why both works are bound to be ludicrously served by any attempt to perform them with the petite gestures of naturalistic acting or the illusionism of realistic sets. For the ritualizing effect of rhetoric is to constantly remove us from the "real" world into that of what we might simply call surreal expressiveness. Shaw amusingly made this point about *Il Trovatore*'s unreality when he said Di Luna must never *ever* be allowed to sit down during a performance, because, as an incarnation of malignity, he cannot possibly assume such an ordinary posture. An editor has recently made rather the same point in his advice on staging *Richard III*: "The element of ritual is crucial to the work, and should not be eschewed in favor of naturalistic action, especially in the battle scene, where balletic rather than realistic action suits better the sacrificial nature of the elimination of Richard from the world."

The ritual formality of *Richard III,* we have seen, derives in part from the almost constant tendency of characters to fall into "performance." Shakespeare carefully arranged situations and dilemmas for his characters from which they could exit only by way of performing oaths, curses, prophecies, prayers, narratives, orations, or self-vaunting soliloquies. This is the very method of *Il Trovatore.* Its principals are cast—or cast themselves—in intensely focused roles, and most of their important moments in the spotlight take on the features of a performance-within-the-action. Several arias, we have already observed, narrate past events to interested bystanders. Manrico begins his career in the opera, fittingly, as a troubador serenading his love, and he ends it serenading his mother with a lullaby in their prison cell. She in fact dreams of the performances he will give with his lute back in the mountains of Biscay. Leonora's *cabaletta* in act 1, ("Di tale amor") is a spirited lover's performance intended to prove to Inez that she cannot "forget" the troubador. Azucena's "Stride la vampa" is a *canzone* crafted and performed to ease the arduous smithy-work of act 2. Di Luna's *cabaletta* a little later ("Per me ora fatale"), in which he steels himself for the abduction of Leonora from the convent near Castellor, has a text with much of the flavor of Richard's satanic solo speeches. Like Scarpia's diabolical asides during the Te Deum in *Tosca,* Di Luna's boastful performance is for our benefit, not for those on stage: "The joy awaiting me is not a mortal one," he sings; "in vain a rival God opposes my love!"

The last two acts bring other kinds of performances. The opening of act 3, for Ferrando and Di Luna's soldiers, ought to remind us of the battle orations at the end of *Richard III:* its text runs, "Let the war-trumpet sound and call us to arms, to battle, to the assault!" Manrico's "Ah sì, ben mio" is an eloquent lover's oath elicited from a very fearful Leonora. But as soon as its lyric, idyllic effect is achieved, Verdi and Cammarano hurry us on with some sudden military news from Ruiz. Within seconds, Manrico leaps into a bellicose posture for the warrior's oath of "Di quella pira." Finally, both Leonora's big act 4 aria and its *cabaletta* are "performances": "D'amor sull'ali rosee" is her turn to serenade Manrico, imprisoned in a tower cell at Aliaferia, while in "Tu vedrai che amore in terra" she gives us (Manrico shows no sign of hearing it) a vivid performance of affirmation in which she swears fidelity unto death. In all these salient, high-relief events of the play and opera we have perhaps the true "realism" of the two dramaturgies, which is simply the dedication to moments when *both* the imagined character and the executive artist inhabiting it on stage can—and must— "perform."

Scriveners and murderers, kings and queens get their moments for rhetorical display in *Richard III.* Often one can easily picture the actor moving, opera-style, closer to the audience (especially in asides and soliloquies) whenever "exclaims" become copious and the poetic temperature rises. Verdi's splendid series of arias and ensembles work in just the same way. Which is why, had Shakespeare taken to the habit of adding interpretive directions to his scripts, he could not have done better than translate several of Verdi's typical admonitions to his singers. When Leonora sings of a deliriously loving heart ("Il cor s'inebriò") she must do so *brillante,* not bad advice for Elizabeth's scene that begins with the request that her "pent heart" be given "some scope to beat." A wise actor playing Richard will give the final flourish of his several buoyantly vicious solo arias *con entusiasmo,* just as Verdi expects Leonora to end "Tacea la notte placida." He would also do well, in his two honey-dripping courtship scenes, to try for the *cantabile con espressione* of Manrico's "Ah sì, ben mio."

Concentrated expressivity and rhetorical expansion give Shakespearean drama much of its power, and so the very common Verdian directions in *Il Trovatore* and elsewhere are especially apropos: *con espansione, con forza, tutta forza,* and so on. Leonora describes, *con espansione,* the first sounds of the troubador's lute and her "angelic" joy at first seeing him; so, too, does Di Luna at the climax of his love-aria ("Il balen"). But perhaps the most evocatively Shakespearean of Verdi's directions comes after he has riveted attention on stage and in the audience by some arrant stage-management timed to a split second. I am thinking of the moment Manrico suddenly appears to interrupt Di Luna's abduction of Leonora. Everyone on stage emits an "Ah!" Then comes the standard pregnant pause and an *andante* of dramatic shock, which builds into the act 3 finale. It begins with Leonora singing of her amazement and rising joy *con tutta forza di sentimento,* then, as the melodic line becomes airborne, *con espansione e slancio.* Actors who approach Shakespeare—especially the more rhetorical plays of which *Richard III* is the supreme example—must, it seems to me, do so in the vigorous and forceful spirit of these two Verdian directions. In other words, *slancio* lost, all lost.

By the time Verdi came to compose *Il Trovatore* he was beginning to feel restless with the "number opera" style, which we have seen bears such similarity to Shakespeare's play-building habits. He even wrote to Cammarano, sounding rather like a proto-Wagnerian, "If in opera there were neither cavatinas, duets, trios, choruses, finales, and so on, and the whole work consisted, say, of a single number, I should find that all the more reasonable and proper." Cammarano, however, did not oblige with anything revolutionary, and Verdi fell back into thoroughly conventional fashions. (Though he did rule out a "mad scene" for Azucena as old hat.) One must note especially the prominence in *Trovatore* of the large rhetorical structure of the *scena,* which, we have seen, usually consisted of a recitative, *cavatina* (slow aria), bridge section, and a final rousing *cabaletta* (fast aria). There are four of them, each presenting a great executive challenge and each cast in this highly artificial form:

LEONORA (Act 1)	DI LUNA (Act 2)
"Tacea la notte placida"	"Il balen dul suo sorriso"
Andante (♩. = 50*)	*Largo* (♩ = 50)
↓	↓
Bridge: *Allegro assai molto* (♩ = 80)	Bridge: *Allegro assai mosso* (♩ = 80)
↓	↓
"Di tale amor"	"Per me ora fatale"
Allegro giusto (♩ = 100)	. . . *un poco meno* (no metronome mark)

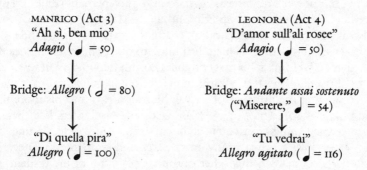

MANRICO (Act 3)	LEONORA (Act 4)
"Ah sì, ben mio"	"D'amor sull'ali rosee"
Adagio (♩ = 50)	*Adagio* (♩ = 50)
↓	↓
Bridge: *Allegro* (♩ = 80)	Bridge: *Andante assai sostenuto*
	("Miserere," ♩ = 54)
↓	↓
"Di quella pira"	"Tu vedrai"
Allegro (♩ = 100)	*Allegro agitato* (♩ = 116)

*Metronome numbers from Verdi's manuscript

In spite of these action-halting events, however, Verdi was still successful in approximating the very swift succession of scenes in a Shakespearean action: the twenty-three numbers average about six and a half minutes each, while the play's scenes run eight and three fourths minutes each. Clearly, Shakespeare depended on a similar hurtling momentum for *Richard III* (Samuel Johnson referred to "the continual hurry of the action"). The sole props he required were themselves time-savers: the two tents for the armies at Bosworth, which allowed rapid scene-shifting for the battle. As usual, the playwright well knew this relentless pace would help annihilate whatever puzzlement and scruples his incredible plotting and characterization might leave in their wake. Achieving a similar momentum became increasingly important to Verdi, and this momentum, or *slancio,* is absolutely vital to *Il Trovatore* in performance. ("If it allowed you to think for a moment," Shaw said of *Il Trovatore,* "it would crumble into absurdity like the garden of Klingsor" in *Parsifal.*) Hoping to get the opera off to a fast start, Verdi urged Cammarano (not, in the end, successfully) to avoid the usual opening chorus and to run the first two acts together: "it would be a good thing because these separate numbers so isolated by scene-changes seem to be designed for the concert hall rather than the stage." Though the new production of *Il Trovatore* at the Metropolitan Opera in 1987 was much reviled by the critics, Verdi would have been very pleased with two aspects of it: there were almost no stage properties to impede the action, and the rapidly moved scenery made it possible to take only one intermission all evening.

The two villainous soul-brothers deserve a few words. As with most Romantic opera villains, Di Luna's evil nature needs little explanation, but is rather a *fait accompli*—one of the monopathic givens that Heilman identified with the melodramatic style. Ferrando says that the abduction from the convent will be a daring venture ("ardita opra"), and Di Luna's reply is nearly as brusque and forthright as Richard's "I am determined to be a villain": "Daring, yes, and such as furious love and outraged pride [irritato orgoglio] demanded of me." Shakespeare, too, is generally far more interested in the theatrical effects than in the causes of evil nature in his characters.* Richard speaks hazily of "another secret close intent," but we never learn much more than that he possesses an overwhelm-

ing desire to exert power over others. "Irritato orgoglio" perhaps sums him up as
well as need be. The phrase sums up Edmund, Lady Macbeth, and Iago, too. Di
Luna is also like Richard in his stupendous changes in expressive style and mood.
Verdi never gave a more luscious melodic phrase to one of his tenor heroes than
the one he gives to Di Luna at his first entrance beneath Leonora's window, when
he sings, "Ah! the amorous flame burns in every fibre of my being!" Yet within
seconds, recognizing the presence of a rival, he transforms into his more usual
fulminating style—"Io fremo!" . . . "Oh gelosia!"—and then launches *ag-
itatissimo* into what counts as macho brutality in the bel canto style, singing "The
pangs of jealous love scorned ignite in me tremendous flames!" There is a
Richardian flavor in many of his other outbursts ("Victim of my disdain, you
must die at my hands!" and so on) and, of course, in the climactic way all the two
villains' *ardite opre* finally recoil upon them in the end. Though Di Luna, in sum,
may not display nearly as many facets as Richard, he certainly descends from the
same theatrical species.

   In certain other large respects and tiny details, *Il Trovatore* calls Shakespeare to
mind, and a few can be noticed here by way of conclusion. It should first be
observed that for all the sensational violence entailed in the action, *Richard III* is,
as an editor has rightly said, "an unusually chaste play." Only the stabbing of
Clarence and Richard's death occur onstage. Shakespeare's interest, typically for
him, is rather to bring the emotional and verbal *responses* to violence and crisis to
center stage. *Trovatore* works this way, too: for all its *tinta* of brutality, every act
of violence takes place offstage or between the scenes. One might also remark on
the Shakespearean touches in Cammarano's much-maligned libretto. One is the
very careful and extensive reiteration of certain imagistic themes. Especially
effective in conveying the story's romantic ardor is Cammarano's frequent use of
actual and figurative sparks, flames, and fires. Another passage reminds one of
Shakespeare's several entries in the ages-old poetic tradition of the *aubade,* or
description of the coming of dawn. Here, for example, is Friar Lawrence's entry,
from *Romeo and Juliet:*

>   The grey-ey'd morn smiles on the frowning night
>   Check'ring the eastern clouds with streaks of light,
>   And fleckled darkness like a drunkard reels
>   From forth day's path and Titan's fiery wheels.

Well, the gypsy men who open the beloved "Anvil Chorus," it turns out, are
poetical, too, and give the Friar some good competition with an elaborate simile
of their own:

> See! the huge vault of heaven puts off night's dark mantle, like a widow who sheds at
> last the weeds in which she wrapped herself.

Di Luna's desire that Manrico "die a hundred deaths amid a thousand agonizing
spasms" might well remind us of Othello's rage at Cassio: "O that the slave had
forty thousand lives! / One is too poor, too weak for my revenge." And Azucena's
narration of her filicide ends with a haunting *pianissimo* coda—one of Verdi's

masterstrokes—for this line: "Sul capo mio le chiome sento drizzarsi ancor!" ("I still feel my hair start upright on my head!"). What a cliché! one might be inclined to hoot. And yet . . . Shakespeare was not above indulging in it himself on no fewer than a half dozen memorable occasions.* In *Richard III,* for instance, Buckingham reacts to Queen Margaret's vitriol: "My hair doth stand on end to hear her curses."

When all has been said, though, I think one last similarity is perhaps the most important of all. Shaw once very sensibly warned that "the one thing that is not permitted to the champion of a lost cause is too small a style of fighting." The overwhelming interest of *Richard III* and *Il Trovatore* lies in watching characters who are fighting lost causes. It is their glory that they fight in a consistently big style—epitomized in Manrico's "Heaven and earth cannot hold me back [from rescuing Leonora]!" or Richard's last-ditch "fair Saint George, / Inspire us with the spleen of fiery dragons!" The natural *métier* of Shakespeare's and Verdi's characters is the big, heavyweight style of vocalism . . . vocalism that is not afraid, when necessary, to ignore the Marquis of Queensberry rules of "natural" or "realistic" expression. And this is surely why Shaw's phrase for the characters in *Trovatore*—"unbounded and heart-satisfying"—is also perfectly suitable for those in *Richard III* and many another Shakespearean play.

Performance

E flat minor

16

Heroic Utterance;
or, Be Not Too Tame

Thus far we have explored a variety of artistic means by which the unbounded and heart-satisfying characters of Shakespeare and opera are created, as well as the highly artificial fictional environment that is required to sustain their few hours' traffic of the stage. It is now necessary to consider how such characters are given life in performance. What, in other words, are the specific challenges attending the impersonation of these creatures? How do these challenges of performance affect the complex relationship between the actor or singer and the imagined personality he or she is re-creating? How do they affect the even more mercurial relationship between the character/artist and the audience? And to what extent are the executive challenges of the typical Shakespearean and operatic protagonists comparable?

In 1890, Shaw attended a performance of our opera of the moment, and this evoked from him as eloquent a general answer to the first question—what is the essential challenge of performance?—as one could wish for by way of opening the subject: "Now a subscription night *Trovatore* is a dreadful thing. This is not an explosion of Wagnerian prejudice. I know my *Trovatore* thoroughly, from the first drum-roll to the final chord of E-flat minor, and can assert that it is a heroic work, capable of producing a tremendous effect if heroically performed. But anything short of this means vulgarity, triviality, tediousness, and fail-ure. . . . The artists must have immense vital energy, and a good deal of [the great Italian actor] Salvini's impetuosity, without wrecking themselves or becom-ing ridiculous." Eleven years later, in a summation occasioned by Verdi's death, Shaw returned to *Il Trovatore* and reiterated his view that success in this drama simply cannot be had without tremendous performative vigor: "If you ask, What is it all about? the answer must be that it is mostly about the police intelligence melodramatized. In the same way, if you check your excitement at the conclusion of the wedding scene in *Il Trovatore* to ask what, after all, 'Di quella pira' is, the answer must be that it is only a common bolero tune, just as 'Stride la vampa' is only a common waltz tune. Indeed, if you know these tunes only through the barrel organ, you will need no telling. But in the theatre, *if the singers have the requisite power and spirit,* one does not ask these questions" (italics mine).

"Immense vital energy," "impetuosity," "power and spirit": all these phrases demand brimming executive force in Verdi's operas. And there are hints throughout the plays that such force is also the supreme *sine qua non* for superlative Shakespeare. Hamlet asks what the professional actor might do if he were given "the motive and cue for passion / That I have," and the hyperbole of the answer he supplies precisely expresses the Shakespearean thrust toward larger-than-life passion:

> . . . He would drown the stage with tears,
> And cleave the general ear with horrid speech,
> Make mad the guilty, and appall the free,
> Confound the ignorant, and amaze indeed
> The very faculties of eyes and ears.

Such is the *métier* of Shakespearean and operatic heroes and heroines: amazing us with (to borrow Richard II's phrase) "sky-aspiring and ambitious thoughts" and the astounding rhetoric necessary to convey them. Rising to such passionate motives and traumatic cues is no artistic task for the "dull and muddy-mettled rascal" that Hamlet accuses himself of being. Rather, it is for the actor who has the nerve to ride the tempests of grandiose verse, the physical ability (as Shaw demanded of any Othello) to turn the voice into a "thunder and surge of passion," and sufficient expressive range (as Granville-Barker insisted for the same role) to run the "gamut . . . up to the highest pitch of imaginative emotion."

Several aspects of Shakespeare's drama conspire to make its major roles, especially the tragic ones, so extremely challenging. Most obvious is the kind of central protagonist that the Elizabethan period first brought to light. Michael Goldman, in *The Actor's Freedom,* remarks that the "sudden achievement from about 1588 on would seem to be the creation of heroes . . . and the tremendous [poetic] line of unprecedented size and power." He also finds that "the figure of the man (or woman) thrown on his own resources, free or compelled to invent ceremonies, to improvise a role for himself, runs through all the drama of the period." The character Othello must do precisely this, but the actor himself is also thrown upon his own resources and is forced to invent effective thespian "ceremonies" and improvise his own responses to the role's executive crises. The actor unable to make heroic use of this extraordinary freedom will, as Angus says of Macbeth at the end of his career, begin to "feel his title / Hang loose about him, like a giant's robe / Upon a dwarfish thief."

The actor most accustomed to wearing the giant's robes on the Globe stage was Richard Burbage, and it is thus hardly a coincidence that the role of Othello— which demands executive courage well beyond the Shakespearean norm—should have been singled out in a Burbage eulogy as his "chiefest part, / Wherein beyond the rest, he moved the heart." The artistic heroism imbedded in Shakespeare's most imposing roles is wittily suggested by Olivier's fantasy about how the role of Othello came to be written: "Shakespeare and Burbage went out on a binge one night, each trying to outdrink the other, the yards of ale slipping down their throats and thickening their tongues. Just before one or the other slid under the alehouse table, Burbage looked at Shakespeare and said, 'I can play anything you write—anything at all.' And Shakespeare said, 'Right, I'll fix you, boy,' and wrote Othello for him."

Shakespeare's mellifluous and mighty lines, then, challenge the performer in various clarion ways to largeness of scale. But how were they delivered? As to how Burbage and his colleagues wore their resplendent rhetorical robes on the Globe stage we have no certain knowledge. Not a single substantial contemporary description of Elizabethan acting style exists. What meager surviving hints there are—like Hamlet's advice—can be marshalled to support very different conclusions. (Shaw noticed the irony of Shakespeare's making Hamlet "hold forth" about naturalistic acting "quite Wagnerianly.") We might well conclude that a hero who feels the urge to "drown the stage with tears" and "cleave the general ear with horrid speech" must mean something rather special when he talks about holding the mirror up to nature! Shakespeare himself, by the way, did mean something special by "mirror" here: not the toilet article that might seem to give the cue for naturalistic acting, but rather an exemplary or perfect pattern (as when Henry V is called "the mirror of all Christian kings" or Buckingham the "mirror of all courtesy").* This meaning perhaps gives the cue for a more operatic, formalistic acting style. Indeed, in spite of the absence of hard evidence, I think there are hints enough to venture an answer to the plain question posed fifty years ago by the eminent Shakespearean Alfred Harbage: "did the companies of Burbage and Alleyn act like [Stanislavsky's] Moscow Art Theater players or like the performers in Italian opera?"

Finding in Shakespeare the implied demand for actors able to symbolize rather than portray, recite rather than converse, and represent rather than impersonate, Harbage opted in the end for Italian opera. That is, Harbage imagined on the Globe Theatre stage a formal, audience-conscious style, as opposed to an illusionistic style that presumes the presence of a "fourth wall" rather than an audience. He thought the apparent adequacy of boys in female roles had to be explained "simply by formal acting," and he could imagine no other style as being suitable to the gifts of a Burbage, whom he assumed to be a tremendous reciter of iambic pentameter. "Formal acting," wrote Harbage, "would have released him from the fancied necessity of making the lines sound like natural speech (which they are *not*)." While a few voices have dissented, subsequent imaginers of Shakespearean acting have also generally tended to arrive at "operatic" conclusions. They suggest that a grand-gestured extroversion, rather than a close-to-the-vest, Method-ical introversion, was the norm. Bernard Beckerman, we have already seen, pictured "ritualistic acting" that expressed "overwhelming passions" with "audacity and vehemency." More recently, David Bevington has summarized in *Action is Eloquence: Shakespeare's Language of Gesture* that "Elizabethan acting was at once ceremonious in a conventional framework, romantic in dramatizing high passions, and epic in its scope and elevation of tone." Even more evocatively, Bevington suggests that acting in Shakespeare's time was "passionate, socially mimetic, virtuoso, imaginative, individual, embracing a kind of heroic naturalism." The paradoxical phrase "heroic naturalism" may seem a hedge, but it is perhaps a happy way of noting how intertwined in Shakespeare's scripts are both natural and heroic speech patterns. The same heroic naturalism is present in the opera singer's frequent shifting from declamation or recitation to virtuoso vocalism.

Even those like Marvin Rosenberg who attack the notion of a "non-personal, stereotyped acting in Shakespeare's time," however, agree on one highly artificial and formalistic (i.e., operatic) aspect of Shakespeare's scripts: their combination of several distinct modes of expression. Rosenberg could be speaking of any challenging operatic role, for instance, when he concludes, "The hard parts for Elizabethan actors—as for today's actors—were those that were complex, that ranged widely in emotional experience, that demanded sharp changes of tempo and technique." S. L. Bethell reaches a similar conclusion in an essay titled "Shakespeare's Actors": "I would suggest that the acting, like the drama itself, was in a mixture of styles. Shakespearean drama has long passages of high poetry and also long passages of colloquial prose; there is also a formal prose, demanding a third type of delivery." Moving self-consciously and with the audience's conscious complicity between these styles with skillful ease—as an opera singer will move between recitative, arioso, and aria—was the essence of Elizabethan acting. Bethell makes this point and draws the obvious comparison: "The Elizabethan actor must have appealed frankly to the audience from time to time, inviting, as it were, their active co-operation. . . . [T]he nearest modern parallel would be opera."

Other hints abound, it seems to me, that suggest that both Shakespearean and operatic execution ideally depend on healthy amounts of adventuring nerve and what Kipling called "juice de spree." One can guess from bully Bottom's "lofty" thundering in "Hercules' vein" or from Hamlet's disgust at certain "robustious" players who "spleet the ears of the groundlings" that purplish poetry was always tempting actors to go "over the top" and err on the side of overmuchness . . . just as William Meredith's soprano is urged to prove herself "every decibel a queen." In Shakespeare's plays and in opera, the performer is regularly tempted into heroic exertions. It is a very dangerous business: when a state of grace unites the lines or notes and the deliverer, the effect can be wonderful, even sublime; but abysmal possibilities of achieving merely ridiculous bombast and bathos yawn everywhere. When Shakespeare and opera fail, they usually do so in this wise, and it is small surprise that descriptions of failed "classical" acting can so easily recall mediocre opera performances to mind. A 1615 complaint about the "wantonizing Stage-gestures" and "Stentor-throated bellowings" of certain actors exactly captures how many an operatic impersonation unfolds, while Hazlitt's more clinical criticism of Eliza O'Neill's Belvidera in *Venice Preserved*—"she is compelled to make up for the want of genuine feeling by the external vehemence of her manner"—describes a common fault of operatic stage deportment. Even the vision of Garrick "thumping his breast . . . and pawing the Characters that he acted with" is guaranteed to remind every regular operagoer of several periwig-pated singers of dismal acquaintance. Nor can I help thinking of two or three very famous divas of recent years when I read Kenneth Tynan's summary of an Edith

Evans appearance as the Countess in *All's Well:* "tranquilized benevolence cascading from a great height, like royalty opening a bazaar."

One can also discern the invitation to heroic scale in the kinds of artists who fail or succeed in rising to the mighty pentameter or vocal line. We have already seen James Agate warning that Shakespeare "shows quietism the door," and just a generation ago Peter Brook urged a return to the "ambition and pretension" implicit in Shakespearean dramaturgy. John Gielgud may have underscored this point when he admitted his lack of success in modern plays and then suggested, "if one could find something that was powerful and heroic . . . I think it would be better for me." This impetus toward heroic scale is also suggested by Sybil Thorndike's remark (quoted earlier in a different context) on Olivier's Oedipus: "Larry went for pure realism in the part . . . but something in himself made it larger and bigger." No actor can take comfortable hold of the Shakespearean or operatic stage without a genius for enlargement, and without a certain fearlessness of the vocal and gestural hyperbole he is inevitably drawn to by the poetic or sung text. Shaw especially reviled the reticent, "ladylike" or "gentlemanly" style of Victorian acting, and thus professed stupefaction when the actress Ellen O'Malley—"a woman with a first-rate punch who has been wasted patting kittens when she should have been playing Lady Macbeth"—was criticized for playing a role in *Heartbreak House* to the Shavian hilt. By way of insisting upon thespian juice de spree, Shaw, as usual, rings in opera: "The part of Ellie is written with obvious operatic solos in it; Ellen quite understood that game . . . and delivered them accordingly. And immediately the old outcry begins against recitations, against sing-song, against talking as if in a trance, against pontifications, and all the rest of it. . . . How anyone can conceive those particular lines being spoken conversationally is beyond my comprehension."*

Tragic characters in the two dramaturgies—whose utterances often pour forth (to borrow from a Whitman title) in "pent-up aching rivers"—require not only emotional bravura but also heroic physical resources. Because the poetry and music are so much larger than life-size, the pressure on performers is toward the limits of physical capability. "Emotion has to be bigger in Shakespeare," says John Barton, and producing it, as the actress Lisa Harrow puts it, "requires a huge amount of energy . . . a fantastic drive." It is hardly surprising that images of mountain-climbing present themselves when the challenge of Shakespeare is being urged. "Mountaineers who attempt Everest are assessed not by that which they achieve but by the amount by which they fail. So the actor attempting to scale Mount Lear," wrote James Agate; and Peter Brook said of the same play, "it is like a mountain whose summit has never been reached," where the way up is strewn with the shattered bodies of earlier climbers: "Sir Laurence Olivier here, Charles Laughton there . . . it's frightening." The soprano Ester Mazzoleni made the same point more bluntly about the *Lear* of her repertory, *Norma,* when she referred simply to "this terrifying score."

The note of physical danger in such comparisons is apt, for what makes the expense of thespian energy all the more nerve-racking is the delicate but crucial task of articulating or channeling enormous expressive power in a way that rises to the dramatic occasion but does not also prove physically ruinous . . . and lead to a fall from the mountain. Olivier said Othello's fits and rages demand that "you to scream to your utmost," but Shaw pointedly warned that an "actor who should habitually attempt to shout his way through Othello, Macbeth, or Lear, would by the ruin of his voice and the wreck of his moral and physical health, remove himself expeditiously from the stage. . . . You cannot play Shakespear and live, unless you know how to speak rationally." It can only be by virtue of the fact that Olivier *did* know the secret of screaming to his utmost *rationally* that he survived so long and happily on the Shakespearean stage. One thinks naturally of long-lived singers like Beniamino Gigli, Joan Sutherland, or Alfredo Kraus, who learned the same secret. The splendid dramatic soprano Anita Cerquetti, who vanished after a few spectacular years, made exactly Shaw's point when she observed, "I am a highly emotional person, and that is perhaps why I was able to stir audiences when I sang. One pays for everything in life, and had I been more rational and less impulsive, I would not have sung the way I did."

Shaw, in fact, made his remarks on the "non-destructive method" of Shakespearean acting in the course of a discussion of the apparent decadence, in the 1880s, of the art of singing in opera, where the temptations and risks of laryngeal heroism are an important part of the "drama" of a live performance. "Risk management" in operatic performances is a fascinating and in some respects melancholy matter. Shaw was very hard on Verdi in this regard. While he considered Handel and Wagner "beyond all comparison the most skilled and considerate writers of dramatic vocal music," he felt that Verdi's style, with its cruelly high *tessitura,* was inclined to "smash voices." The Handelian or Wagnerian singer, he amusingly observed, "thrives on his vocal exercises and lasts so long that one sometimes wishes that he would sing *Il Trovatore* once and die." In his obituary for Verdi, Shaw also observed that he used his orchestra "unscrupulously to emphasize his immoderate demands for overcharged and superhuman passion, tempting executants to unnatural and dangerous assumptions and exertions."*

While these aspersions may have special force for Verdi, I would venture that immoderate and tempting demands for risky exertion are characteristic of most great operatic literature (as they are, to a lesser extent, of most great Shakespearean roles). Operatic roles that do not insist on an element of physical daring are, almost by definition, comprimario roles. Scaling foothills, after all, is small cause for exultation. One might also venture that the sole path to greatness on the Shakespearean and operatic stages lies in the acceptance of immoderate demands—along with their harrowing physical consequences. No career has captured the irony and poignancy of such heroism better than that of Maria Callas. "The anxiety, the sadness, and also the willfulness of her greatest portrayals were at the same time her own anxiety, sadness, and willfulness," wrote Paul Robinson in a memorial shortly after her death in 1977: "Had she limited her repertoire to one of the recognizable vocal categories, and had she sung more prudently, she would probably still be on the stage today, as are many of her famous contemporaries. But she would also have become a much lesser singer: merely great, and not the epochal figure she in fact became."

One might also locate the heroism of performers in the sheer sound and fury they must produce. Elizabethan drama, Granville-Barker observed, "was built upon vigor and beauty of speech," and opera is just as firmly founded upon vigorously and beautifully sung speech: the essence in both cases lies in vocal power, power deriving from huge volumes of breath and maximally effective breath-control. A visitor at the Milanese premiere of *Macbeth* was struck by the "Herculean lung-power" Verdi demanded of his two protagonists, but by our mundane conversational standards, almost all big operatic roles require such heroic lungpower. Both genres demand of executants a genius for *extension*— whether of vowels, of emotional climaxes, or of the normal vocal range and dynamics (on one page of his *Trovatore* MS Verdi wrote "sempre ppppppp-pianissimo"). And all such extensions require much breath. Without it, the "almost outrageous outpouring of energy" in operatic singing and Shakespeare's "acclamations hyperbolical" (an apt phrase from *Coriolanus*) would be impossible to create.

What is more, the large vocal effects of the two genres put the actor very much at the mercy of the venue in which he must operate. Burbage was very fortunate to have performed his heroics in a theater where, we guess, the most distant auditor was about seventy-five feet away. And the great tragedian of the Restoration was also apparently able to cultivate an un-robustious style—"Betterton kept his Passion under"—because the theaters remained relatively small: the first Drury Lane Theater (1662) seated about 700, the second one (1674) about 2,000, which is probably a bit more than what the Globe accommodated. Ian McKellen, we have noted, has suggested 400 as an ideal audience size, which, given the more spacious seating and the requirements of modern fire laws, perhaps approximates the auditorium space in which Burbage and Betterton worked. Move the actor or the opera singer drastically in either direction from this happy medium, however, and the results can be traumatic or risible. Bring the spectator within a few feet or inches of the actor, as happens with the television or film camera, and the big gestures intrinsic to the poetry or the music will exaggerate the necessary physical contortions and show how claustrophobic Shakespearean plays and operas really are. Or—equally ludicrous—the performers will be encouraged to deliver their high-flown utterances more naturalistically, rather as one might play a Lizst Hungarian dance or a Chopin polonaise with one's foot never leaving the soft-pedal. The heroic exertions of Shakespearean and operatic protagonists demand a large measure of performatory *Lebensraum,* which the eye of a camera only grudgingly affords.

But the greater danger today is movement in the other direction. Post-Betterton theater history is one of ever larger auditoriums and reliance on more grandiose, self-conscious acting styles. The third Drury Lane Theatre (1794) seated 3,600, while the 1808 Covent Garden building held 2,800 and later (as the Royal Italian Opera House) reopened with a capacity of 4,000. Under these intimidating circumstances the charismatic star who could carry the evening by sheer thespian heroics became ever more vital, and he or she moved ever farther toward the forestage or apron to indulge in a privileged relationship with the audience. By Garrick's time—he flourished from 1742 to 1776—the forestage was the predominant acting area. Here the great stars of the eighteenth and nineteenth centuries faced a challenge recently described by the soprano Kiri Te Kanawa, when she explained the feeling of performing in a venue like the Metropolitan Opera, which seats 3,800: "We seem so far away, with the audience out in this incredible black hole. You feel as if you are not quite getting to all those faces . . . that a large gesture is better than no gesture." Claire Bloom attended a Te Kanawa performance in *Otello* at the Metropolitan and was naturally struck by the "very old-fashioned," big-gestured acting style.

The comparison works the other way, too. We have been left several trenchant and witty descriptions of how the big stars of the Augustan and Romantic Shakespearean stage conquered their vast spaces, and one notices often how these are reminiscent of acting styles on today's big opera stages. The urge toward marmoreal grandeur perhaps began with Barton Booth (1681–1733), about whom Theophilus Cibber wrote, "Mr. Booth's Attitudes were all picturesque. He had a good Taste for Statuary and Painting." But it reached its climax in the school of Charles Kemble, who flourished from 1794–1832. William Hazlitt wickedly satirized him as the "very still-life and statuary of the stage; a perfect figure of a man; a petrifaction of a sentiment, that heaves no sigh, and sheds no tear; an icicle upon the bust of Tragedy." The tone of cool "dignity" affected by many singers is also reflected in the amusingly polite phrasing of *The Spectator,* in 1829, about an actress of the Kemble school: "Of her attitudes generally we must however say that they border a little too closely on distortion."

At the other end of the spectrum was the school of Edmund Kean (who flourished from 1814–1833), patron saint of all the operatic scene-chewers of one's acquaintance. Expressive violence of emotion, detailed attention to pantomime and byplay, and incessant making of points were the hallmarks of his style—which much influenced the barnstormers later in the century. In 1814 Byron wrote to a friend, "If you are fond of the drama you must see [*nota bene,* not "hear"] Kean— he is the triumph of mind over matter for he has nothing but countenance & expression." One might find in the Kean style the type of the singer whose thespian exertions distort the music, for at the height of the Kean fashion one critical voice lamented, "where upon the stage do we find the actor who does justice to [Shakespeare's] melody . . . ?" W. C. Macready, who followed after Kean, has been credited not only with restoring some of Shakespeare's original tragic conceptions to the stage, but also with combining the declamatory beauty of the Kemble school with the intensity of the Kean School. Perhaps in the style of Macready—whose combination of pathos and melodrama so appealed to Dickens—we may have something like the proper *via media* for operatic impersonation.

Be that as it may, the crucial point here is that larger performing spaces exacerbate the already heroic demands upon Shakespearean and operatic actors. Ian McKellen has observed that "it wasn't easy for 19th-century actors who were working in large theaters, or in America today where Shakespearean acting is different from ours, mainly, I think, because [American] theaters are much larger. This leads to a grander, more generalized, open style of acting."* And John Barton also reminds us of the all too common distortion of voices, and of operas themselves, induced by performance in a cavernous site, when he muses, "We are so often controlled by the buildings we perform in. When a theater is big it takes us over and transforms us and becomes our lord rather than our servant."

Large theaters, which dwarf the performer and make the struggle for larger-than-life stature all the more daunting, emphasize another heroic aspect of performance in the two dramaturgies; namely, the isolation of principle performers. Consider Cassius' famous description of Caesar:

> . . . he doth bestride the narrow world
> Like a Colossus, and we petty men
> Walk under his huge legs, and peep about . . .

Most of Shakespeare's plays and most operas offer us colossal protagonists, beneath whom men and women of petty comprimario powers tend to peep about. The finest moments for these protagonists, even when they are by no means alone on stage, are fundamentally soloistic and are often directed outward beyond "the narrow world" of the acting space itself. Inevitably, an element of executive stress and artistic vindication is intertwined in the emotional joys and traumas of these fictional characters. The exaltation of human isolation was a hallmark of Elizabethan drama, credited first to Marlowe; but Shakespeare spun virtuoso variations on the theme. "Exalting the solitary dignity of the soul," Granville-Barker observed, is "a recurring note in all Shakespearean tragedy." And Michael Goldman describes Shakespeare's tragic figures as passing "from isolation to isolation," a phrase that not only describes their imagined lives but also the technical challenge to the actors inhabiting them: the challenge of rising to expressive heroics at a series of moments when they are isolated and thrust audienceward from the comprimario ensemble. Shakespeare's and opera's dramaturgy, we have seen, is predicated on this privileged bond between virtuoso performer and audience; larger theaters, which require even more virtuosic projection and urge the actor to the forestage and away from the imagined world, serve but to heighten the self-consciousness of the dramatic event. It would be folly to deny or to try to evade the executive "isolationism" intrinsic to both dramaturgies, but one can still be aware of the damage caused when their colossal soloists, presented with huge theaters, are provoked to levels of vociferation their creators could scarcely have imagined as they prepared their works for more intimate stages.

A final and very general kind of heroism that Shakespearean and operatic performances will ideally entail deserves mention, though it is almost impossible to talk about. This has to do with "genius," charisma, or the ineffable saliency of individual artistic personality. The potential for sublime effects is discernible everywhere in the great plays and operas, but only rarely are executants vouchsafed access to these sublimities. Describing how it happens when they gain such access is a hopeless task, for it is quite as difficult to anatomize the sublimities of Shakespeare, Mozart, Verdi, or Wagner as it is to explain how gifted artists actually bring them to life on stage.

Ellen Terry spoke simply of those passages that require a "state of grace." James

Agate offered his own Indescribability Theory when he said that he was willing to call Donald Wolfit an "immensely talented" but not a great actor because "he does nothing which we cannot explain." Verdi, too, had a favorite but vague phrase for the kind of performer he most admired. Envisioning the role of Amneris in *Aida,* he wrote to Ricordi: "Keep in mind that this Amneris is driven by the devil, has a powerful voice, is very emotional, and very, very dramatic." A few months later, he offered a more general credo and used the phrase again: "So-called *polished singing* matters little to me. I like to have the roles sung as I wish; but I can't supply the voice, or the soul, or that certain I-don't-know-what . . . which is commonly called *being driven by the devil [aver il diavolo adosso]*."* Perhaps it is enough to say that the great creators make diabolical executive demands, and the great *re*-creators—those driven by the devil—will be the ones with sufficient nerve and energy to fulfill them.

In these two dramaturgies of "elevated and compressed style" (August Wilhelm Schlegel's summary of Shakespeare) and of human experience with, as Hazlitt said, "the dull part left out," we find the stimuli to enormous executive effort described in myriad ways. On the Shakespearean side we hear of the essence of Elizabethan heroes resting in "largeness of imaginative vision and expressive power." We hear, too, of the impersonator of an heroic role exercising a "dangerous freedom—with aggression, escape, exposure, self-assertion beyond normal limits." Borges talks of "piling on the agonies," while Erich Auerbach in *Mimesis* refers to a "dynamic throbbing of elemental forces" in the plays. Wagner writes of the playwright's supreme objective of "a heightened Pathos" and the necessity of possessing the "technical means of securely fixing its delivery." On the operatic side, the formulations have a distinctly similar ring. The supreme ability required in an opera composer, according to Wagner—that of being able to delineate "profound and energetic passion" in "large and forceful strokes"— excellently states the requirement for superlative impersonators as well. The composer John Eaton made it clear what kinds of characters are necessary to contain such energetic passions when he insisted they possess "superhuman" qualities. And Auden extends Eaton's assertion to the matter of the stage environment when he explains what is suitable to the colossal operatic personality and impersonator: "Only one thing is essential, namely, that everything be a little over life size, that the stage be a space in which only the grand entrance and the grand gesture is appropriate."

The big Shakespearean and operatic roles, in sum, require the extraordinary executive tricks of "strong imagination." It is a great task: giving these superhuman "shapes" a local habitation and a name, making the "antic fables" in which they appear seem credible for the time being, and allowing the audience to "apprehend / More than cool reason ever comprehends." So daunting is the task at these rarefied heights that some have fallen back on tropes of impossibility. The pianist Artur Schnabel once said that "great music is music that is better than it can be played," which is something of a variation on Charles Lamb's notorious assertion that Shakespeare's greatest plays are of imagination so compact that they cannot possibly be realized on stage in anything near their true glory. Much as it would have appalled the playwright to hear such nonsense spoken, the notion is rather tempting . . . given how rarely the thespian heights are scaled. Everyone, I assume, has certain favorite plays and operas—or even particular scenes, speeches, or arias—that one half expects (with sadness but not great surprise) to live a whole lifetime without experiencing ideally executed in a live performance.

One might simply fall back and say of the challenge of live performance that there must be, as with Desdemona's handkerchief, "magic in the web of it." One might also note, by way of a negative approach, how mercilessly the central roles in the two dramaturgies expose merely modest talent to disaster. One might well think that simply mastering the projection of iambic pentameter or a melodic line is burdensome enough in itself. The tenor Giuseppe Fancelli—about whom Verdi once wrote, "Beautiful voice, but a blockhead"—apparently felt this way. After an especially frustrating rehearsal, Verdi seized Fancelli by the neck and shouted, "When will anything get into your head? . . . Never!" He stormed out and Fancelli is reported to have said to the stunned bystanders: "A great Maestro . . . Yes sir! . . . I agree. But he wants the impossible, the impossible! You bet! He wants people to read his music as he wrote it and to sing on pitch and in tempo, even to pronounce the words! How can you do all that at the same time? I couldn't, even if I were the Almighty God himself!"* But even this threshold technical *savoir-faire* (and note that Hamlet's demands on the players exactly match those Verdi made on Signor Fancelli) finally is not enough. *Im*modest talent is also necessary. The dictionary defines *expletive* as "a syllable or phrase inserted to fill a vacancy," and Thomas Wilks used it in this sense when he wittily described a bad actor as "a mere expletive on the Stage." The expletive actor is the bane of Shakespearean and operatic performances; we earnestly wish him deleted. And it is the rare Shakespearean play or opera that can survive satisfactorily on stage when one or two of its principal roles are only partially inhabited. *Hamlet* becomes *Hamlette, The Tempest* will turn teapot size in the event. And *Norma* can turn into *Mrs. Normer, Lucia* into *Lucy Did Sham Amour, Ernani* into *Herr Nanny*—to recall the titles of a few operatic travesties that were actually performed in nineteenth-century New York City.

Hazlitt, too, insisted on specially salient thespian energy in his excellent description of the merely "respectable actor." He is "one who seldom gratifies, and who seldom offends us; who never disappoints us, because we do not expect anything from him . . . who never risks a failure because he never makes an effort; who keeps on the safe side of custom and decorum, without attempting improper liberties with his art; and who has not genius or spirit enough to do either well or ill."* This sketch neatly describes the expletive opera singer or conductor. When several such expletives foregather, the consequence is a run-of-the-mill or hack performance. Sometimes a single such performer, manifestly overparted, can sink an entire evening. Hazlitt, by the way, goes on to say that the "respectable" actor may do very nicely in respectable plays (today he would have banished him to television soap operas) and in "indifferent characters." But "whenever he plays Shakespeare, we must be excused if we take unequal revenge for the martyrdom which our feelings suffer." Hazlitt's revenge on this particular occasion was for a mediocre impersonator of Prospero, but we have all experienced the similar martyrdom occasioned by the merely respectable singers who venture their Norma, Carmen, Violetta, or Tosca.

Prescriptive rather than proscriptive ways of expressing the ideal in heroic performances will provide more cheerful food for thought, though they are scarcely more specific. One way might be to suggest that actors themselves must possess the "Fire, Impetuosity, and even beautiful Extravagance" that Shakespeare's first editor, Rowe, found in the plays. This seems remarkably close to the qualities that Verdi demanded "before all else" in performances of his operas: "fire, spirit, muscle, and enthusiasm." Shaw was happy to boast that he revolutionized Victorian theater by returning to such elements: "I went back to the forgotten heroic stage business and exciting or impressive declamation I had learnt from oldtimers like Ristori, Salvini, and Barry Sullivan. . . . [This] would have seemed like the merest routine to Kemble and Mrs. Siddons, but to the Victorian leading ladies this seemed to be unladylike barnstorming." The fell business of punching out Lady Macbeth in one telling blow after another rather than patting kittens requires a reservoir of energy. In practical terms it also requires a special bond between playwright and actor that Shaw found notably in Shakespeare's plays (and which he sought after in his own): the "bond of stage method, of rhetoric, of insistence on exceptionally concentrated personal force and skill of execution, of hammering the play in by ceaseless point-making." This assumes the extraordinary confidence and force of character likewise required of opera singers: as the great German director Walter Felsenstein observed, "a strong personality can help tremendously to close the gap between human reality and the artificiality of singing."

Strength of personality perhaps also lies behind the question posed by Hazlitt after attending a performance by Edmund Kean as Richard II: "It may be asked, then, why all great actors choose characters from Shakespeare to come out in; and again, why these become their favorite parts?" The answer Hazlitt supplies also explains the attraction of Violetta and Tosca for sopranos, Hoffmann and Manrico for tenors, Giovanni and Wotan for baritones, Boris and Méphistophélès for basses: "First, it is not that they are able to exhibit their author, but that he enables them to show themselves off. . . . [The actor] perceives how much he has to do, the inequalities he has to deal with, and he exerts himself accordingly; he puts himself at full speed, and lays all his resources under contribution; he attempts more, and makes a greater number of brilliant failures; he does all he can and bad is often the best." As Hazlitt suggests, it may be an occupational hazard of Shakespeare and opera that, attracting and rewarding artists of enormous willpower as they do, these stage genres sometimes suffer when this willpower takes a performance in a direction (we may think) the dramatist never intended. When an actor of sufficient stage presence does this, however, the effect can still be extremely exciting. And it is a peculiarity of the most powerful artistic personalities among actors and singers that their occasional depredations or aberrant impersonations are often far more riveting than those by performers who, as Hazlitt put it, never take improper liberties with their material. Agate confessed as much when he wrote, "I personally had always more pleasure in seeing [Sir Henry] Irving fail than any other six actors succeed."

Agate put the English actress Janet Achurch (1864–1916) in the same category. He remembered her as the "best Cleopatra I ever saw . . . Mr. Shaw put it on record that everything that Janet Achurch did as Cleopatra was mistaken, but the fact remains that her performance was unforgettable." The Shaw remark that Agate recalls was no doubt a bit of typical Shavian grandstanding, for he was also on record as saying that Achurch was her generation's "only tragic actress of genius." If we turn to an 1897 review of her Cleopatra, we find Shaw at his most enchanted and enchanting. I will quote from it at length not only because he makes much of Achurch's musicianship, but simply because most memorable operatic performances are the product of such comprehensive and irresistible executive charisma as this actress displayed as the Egyptian queen. Such performances tend to leave one just as stunned and discombobulated as Shaw was:

> Miss Janet Achurch, now playing Cleopatra in Manchester, has a magnificent voice, and is as full of ideas as to vocal effects as to everything else on the stage. The march of the verse and the strenuousness of the rhetoric stimulate her great artistic susceptibility powerfully. . . . Of the hardihood of ear with which she carries out her original and often audacious conceptions of Shakespearean music I am too utterly unnerved to give any adequate description. The lacerating discord of her wailings is in my tormented ears as I write, reconciling me to the grave. It is as if she had been excited by the Hallelujah Chorus to dance on the keyboard of a great organ with all the stops pulled out. I cannot—I dare not—dwell on it. I admit that when she is using the rich middle of her voice in a quite normal way, the effect leaves nothing to be desired; but the moment she raises the pitch to carry out some deeply planned vocal masterstroke, or is driven by Shakespear himself to attempt a purely musical execution of a passage for which no other sort of execution is possible, then—well then, hold on tightly to the elbows of your stall, and bear it like a man. And when the feat is accompanied, as it sometimes is, by bold experiments in facial expression which all the passions of Cleopatra, complicated by seventy-times-sevenfold demoniacal possession, could but faintly account for, the eye has to share the anguish of the ear instead of consoling it with Miss Achurch's beauty. I have only seen the performance once; and I would not unsee it again if I could; but none the less I am a broken man after it.

Strenuousness, hardihood of ear, audacious, stops pulled out, bold, demoniacal ("il diavolo adosso" indeed!) . . . such is the primary lexicon of high praise for Shakespearean and operatic performances that live in the memory.*

As befits the mystery of thespian heroism, attempts to capture its rare manifestations tend not toward coolly reasoned analysis but rather toward such catapulting flights of eloquence as Shaw's. Thus, for instance, Hazlitt's reminiscence of his youthful encounters with Sarah Siddons: "I was stunned and torpid after seeing her in any of her great parts. I was uneasy, and hardly myself, but I felt (more than ever) that human life was something very far from being indifferent, and I seemed to have got a key to unlock the springs of joy and sorrow in the human heart. . . . Death and Life played their pageant before me. The gates were unbarred, the folding doors of fancy were thrown open, and I saw all that mankind had been, or that I myself could conceive, pass in sudden and gorgeous review before me." Or consider Arthur Symons' sketch of the effect of Sarah Bernhardt at her best: "It was as if the whole nervous force of the audience were sucked out of it and flung back, intensified, upon itself . . . this very artificial acting seemed the mere instinctive, irresistible expression of a temperament; it mesmerized one, awakening the senses and sending the intelligence to sleep." Those who remember great operatic performances tend to speak in rather the same radiant fashion.

Though in prose not quite so empurpled, John Coveney takes Hazlitt's approach as he recalls Lotte Lehmann's Met debut in 1934: "The event remains as vivid now as it was then, not because I have a good memory but because I can never forget it. That night I realized opera could be a stunning experience, that singing could rank with the noblest of human achievements." Another Lehmann admirer wrote later of her Marschallin that it was "spun from within like the spider's web, and the beholder had nothing to do but to marvel and weep." Lucrezia Bori thought of Claudia Muzio as "a case apart: you simply cannot classify her, for in the end you had been so emotionally destroyed by her performance, you did not even know anymore what kind of instrument she had." Callas has left behind descriptions of her overwhelming stage presence too well known and numerous to need mention here. And some operagoers are fortunate enough to be able to recall performances in which several singers conspired in heroic tandem. John Ruskin, for one example, recalled nearly fifty years after the event the experience of hearing *I Puritani* in Paris with Grisi, Lablache, Rubini, and Tamburini: "I then heard, and it will always be a rare, and only once or twice in a century possible, thing to hear, four great musicians, all rightly to be called of genius, singing together, with sincere desire to assist each other, not eclipse; and to exhibit, not only their own power of singing, but the beauty of the music they sang."*

Finally, though, it is the more specific and local moments of artistic epiphany that Shakespearean and operatic performances can leave etched indelibly in the mind . . . glorious memorial baggage that we shall take to the grave. Nothing, I suppose, better symbolizes the ineffable effect than the High Note spectacularly executed. Auden, in fact, rebutted the notion that opera is impossible to write in our troubled, pessimistic, unheroic age by reference to the daring singer and the High Note. The impossibility of writing opera, says Auden, "would follow only if we should cease to believe in free will and personality altogether. Every high C accurately struck demolishes the theory that we are irresponsible puppets of fate or chance." I cannot resist drawing attention to three remarkable descriptions of the soul-invading High Notes as a way of departing from the subject of heroic utterance, not only because several great High Notes continue to ring in my memory, but because they are the ultimate test for the "brave confronter" that Whitman thought every admirable vocalist must be. Great lines in famous speeches are the High Notes of Shakespeare's plays. Though they do not require the extreme laryngeal exertions of a singer undertaking Norma, Butterfly, or Brünnhilde, I hope the state-of-grace task of "attacking" these lines and, as it were, hitting them dead on pitch is sufficiently comparable to allow the following, strictly operatic coda to carry a Shakespearean point as well.

The first high note occurs in hilarious circumstances, James McCourt's novel *Mawrdew Czgowchwz,* a satire on the opera world written *con amore* and *con abbandono.* The novel follows the meteoric career of the title character, whose last name is pronounced "gorgeous" and, it will be noted, whose initials are the same as those of a great prima donna of recent memory. Several of the funniest pages on opera I have ever read describe a Czgowchwz performance as Violetta at the Metropolitan Opera. The note in question comes at the end of act 1:

> The "Sempre libera" began; it built. The voice grew; the sides of it fell off, the bottom opened (like the portals of doom), and Czgowchwz soared in flames to B-naturals full-voice. There were involuntary screams, shock upon shock, fresh denials from every tier. . . . The final measures were upon her; the optional E-flat hung fire. She rose higher and wider by turns. The voice seared, shooting out of the whirling smoke of her consumptive waltz. "Il mio pensier . . . il mio pensier . . . ah . . . ah . . . ah!" For an instant there was no sound; then something unheard since the creation—a Czgowchwz fortissimo A-natural above high C the color of the core of the sun. Mawrdew Czgowchwz ripped the camellia from her hair, which then cascaded over her face like a flaming veil, threw the Baccarat bell goblet against the wall, and collapsed. The gold drapes fell.
>
> She appeared for one solo call during which she knelt amid screams for one half hour while many in the audience were removed to the sidewalks and fell about the pavements. Mawrdew Czgowchwz wept.
>
> The performance redid history.

Outrageous perhaps—A-natural above high C, for heaven's sake!—and yet . . . there is something about the passage that is decidedly from real life, albeit very rare real life: the huge gestures, the huge outpouring of energy, the suspenseful gathering for the climax, the shock of hearing something one has heard dozens of times before but *never like this.* In sum, the ecstasy of it all.

The second high note I have in mind is one that Mary Garden actually heard issue from the throat of Nellie Melba. Garden's description of this note, which Mimi sings at the end of the first act of *La Bohème,* captures the gooseflesh excitement that heroic utterance can create. It captures, too, the reality behind McCourt's high-camp spree:

> The way Melba sang that high C was the strangest and weirdest thing I have ever experienced in my life. The note came floating over the auditorium of Covent Garden: it left Melba's throat, it left Melba's body, it left everything, and came over like a star and passed us in our box, and went out into the infinite. I have never heard anything like it in my life, nor from any other singer, ever. It just rolled over the hall of Covent Garden. My God, how beautiful it was!
>
> Since then I always wait for that note when I hear the first act of *Bohème,* and they reach and reach for it, and then they scream it, and it's underneath and it's false, and it rolls down the stairs, and it never comes out from behind the door, never. That note of Melba's was just like a ball of light. It wasn't attached to anything at all—it was *out* of everything.

The most famous high note in all literature produces a similar moment of aural ecstasy. It comes in an aria well off the beaten path—Lionel's "M'appari" from the third act of Flotow's *Martha*—and the performance does not even take place in an opera house, but in Dublin's Ormond Bar. Simon Dedalus sails it out in the "Sirens" chapter of *Ulysses*. The siren attraction of heroic and beautiful human utterance is caught by Joyce—great devotée of the tenor John McCormack and an erstwhile tenor himself—in the inimitable paragraph that spins out the pleasure of the aria's climactic high B-flat:

> It soared, a bird, it held its flight, a swift pure cry, soar silver orb it leaped serene, speeding, sustained, to come, don't spin it out too long breath he breath long life, soaring high, high resplendent, afflame, crowned, high in the effulgence symbolistic, high, of the ethereal bosom, high, of the high vast irradiation everywhere all soaring all around about the all, the endlessnessnessness . . .

17

Star Quality; or, Risky Business

Laurence Olivier ventured in his book *On Acting* that "film is the director's medium, television the writer's, but the theater is the actor's." One might go a step further and suggest that Shakespearean and operatic theater are the domain of the *star* actor. Byron attended a performance of *Coriolanus* by Kemble and reported that "he was glorious & exerted himself wonderfully." This is precisely what we hope for from every executant of a major role: wonderful, stellar *exertion*. Nearly every *dramatis personae* in the Top Bard's canon and in the standard opera repertory breaks down clearly into a hierarchy of minor comprimario, major comprimario, and principal or stellar roles—with the demand for stamina and bravura exertion greatly increasing at each level.

Some have fervently wished that this bravura stardom were not part of the equation. The composer Roger Sessions warned in a 1938 essay that "if there is to be any general revitalization of opera, the star system will have to be eschewed and an ideal of ensemble substituted for that of highly paid and spectacular voices." Though we always and fervently hope that divas and divos will subdue their egomania and respect the collegial ensemble during a performance, Sessions' operatic New Deal was folly: opera without stars is as inconceivable as a contending baseball or football team with no star players. Ensemble, or team-work, is an admirable thing, but one must still have bravura artists for bravura tasks. No amount of well-integrated ensemble work will save *Il Trovatore* or nearly any other major opera if spectacular voices are not also on duty.

And bravura will out, too. Derek Jacobi, undeniably a bravura actor, announced that his aim in taking on the huge central role of a play titled *Breaking the Code* was "the total elimination of self and especially of the awful thing, 'bravura' acting." Jacobi's intention proved as deluded as Sessions': reviewers were unanimous in admiring the "feat" of his performance. Whether or not Jacobi's intention was appropriate for *Breaking the Code,* one can venture that it would be either preposterous or disastrous to approach the great Shakespearean roles by seeking to disguise the bravura nature of the task at hand, or even by seeking to make the artistic "self" disappear completely.

One hopes that the remark about "awful" bravura acting by Jacobi—who turned down the opportunity to play the bravura roles of Macbeth and Richard II to do *Breaking the Code*—was merely polite self-deprecation. Having succeeded so often by bravura, it seems rather late for him to turn ingrate. At any rate, shying coyly away from invitations to bravura declamation rather than grasping them *con slancio,* as he seems to urge, would go against the Shakespearean grain. Nothing impressed an Elizabethan audience—whether in the theater or in real-life law courts, government chambers, or pulpits—more than the impressive delivery of eloquent speech, and the plays can be seen as brilliant vehicles for delivering such goods. They were manifestly written to show off the abilities of actors; after all, a familiar colleague of these actors, a sometime actor himself, wrote them. When we consider how intimately Shakespeare was involved in the life of his company, in which he was a shareholder, it can come as no shock to be told that his "stagecraft concentrates, and inevitably, upon opportunity for the actor" or that Shakespeare's art "consists in making the exercise of . . . technique and talent possible—and valuable."

This, of course, is the composer's art too. In both cases the specific art proves "valuable" not only because it enhances the meaning of the fictional event unfolding on stage, but because it may give a favored artist the opportunity to exert himself or herself wonderfully. One may not wish to go as far as Michael Goldman. In *Acting and Action in Shakespearean Tragedy* he suggests that "an actor's performance, *qua* performance, may itself constitute an important part of the action of a play." But one can more easily grant that the performance, *qua* performance, will constitute a significant part of *the theatrical event.* Coleridge asserted in one of his lectures on Shakespeare in 1811–1812 that "those who went to the theatre in our day, when any of our poet's works were represented, went to see Mr. Kemble in *Macbeth,* or Mrs. Siddons' Isabella," and can we doubt that part of the draw in Shakespeare's own day was expectation of glorious comic exertions from Kempe or Armin or tragic exertions by Alleyn or Burbage?

Evidence that the star system was well in place in Shakespeare's time comes from several sources. Michael Hattaway, in *Elizabethan Popular Theater,* imagines that "a great player like Alleyn could assume not simply that his audience had come to see Tamburlaine but to see Alleyn playing Tamburlaine. The 'star system' was part of the play." Alleyn, in fact, was the first notable exemplar of Goldman's observation that acting offered "one of the lightning careers of a speculative and chaotic age": he retired a wealthy man in 1603 at the age of thirty-seven and, after twenty-three years passed, left a fortune large enough to endow Dulwich College. Shakespeare's comfortable circumstances at his retirement are well known: good actors—but star actors in particular—made his success possible. The playwright was surely acknowledging this fact when he made bequests for the purchase of mourning rings to the three surviving main colleagues of his theatrical salad days—Heminge, Condell, and Burbage. One might remark, too, that it was Burbage, not Shakespeare, whose death elicited elegies in gratitude for pleasures given at the Globe and Blackfriars.

Finally, though, the shape of the plays themselves makes their "star" equation most abundantly clear. Consider, for example, how Shakespeare distributed the lines to his actors in *Love's Labour's Lost:*

Star	Principal	Comprimario	Walk-on
Berowne, 619 lines (23 percent of total)	King of Navarre, 314	Dumaine, 85	Jaquenetta, 12
	Princess of France, 283	Longaville, 71	Forester, 6
	Armado, 261	Maria, 46	Marcade, 4
	Boyet, 230	Nathaniel, 44	
	Costard, 205	Katherine, 44	
	Holofernes, 185	Dull, 32	
	Rosaline, 178		
	Moth, 165		

T. W. Baldwin speculates, reasonably I think, that Burbage played the juicy, pyrotechnical role of Berowne. For *Love's Labour's Lost* is often dated about 1594–1595, which is just when Shakespeare's company, the Lord Chamberlain's Men, first emerged and when the playwright and Burbage are first known to have been present in the same company. Perhaps Berowne was the first role he tailored to the actor's extraordinary talents. If so, this organization of the cast around a star turn was an historic foretaste of the great Burbage roles that were to follow over the next two decades, up to and including what is nearly the last of them—Prospero. In *The Tempest* Shakespeare gives about 630 lines (thirty percent of the total) to his star. In all, Shakespeare wrote seventeen other single-star vehicles; five more are essentially two-star affairs.*

That a man of Burbage's talent fell into collaboration with a colleague, fellow shareholder, and dramatist who was to become the greatest in the language is without doubt the single most important and ramifying accident in English— perhaps even world—theatrical history.* For Burbage is responsible for the "self-evident" fact observed by Schlegel in 1808 that "most of [Shakespeare's] principal characters require a great player" . . . as well as for the corollary expressed much later by Harbage: "From the time of Burbage in the early seventeenth century until that of Irving in the late nineteenth, every leading English and American actor was a Shakespearean actor." But for the presence of the thespian realms of film and television, the corollary might still hold today. Because Burbage bestrode the Elizabethan stage like a Colossus and because Shakespeare was able to create roles that fully displayed his thespian abilities, posterity was vouchsafed a range of theatrical Himalayas. On their slopes the talents of each generation's high-aspiring actors have been tested. One might even say that Burbage (with Edward Alleyn notably paving the way) was responsible for the essential features—if not for some of the more wildly egomaniacal abuses—of the star system of the following centuries. Joseph Donohue, for example, describes the English theater of the Romantic era as being "emphatically not a playwright's theater but an actor's . . . the successful playwright was one with the knack of tailoring his piece to the abilities and tastes of players . . . the play was chiefly a wagon for a star." But, taking the cue of Olivier's opening observation, we can safely assume this also describes how plays were prepared backstage at the Globe Theatre.

We have been left no reason to think that Burbage was other than a spectacular star actor. In a play of 1614, when Burbage's career was nearing its end, Ben Jonson included this exchange:

> *Cokes:* . . . which is your *Burbage* now?
> *Lanterne:* What meane you by that, Sir?
> *Cokes:* Your best *Actor.*

Surely Shakespeare would not have dared the blatant irony of giving Hamlet's browbeating critique of "villainous" acting to any lesser man than his most exquisite holder of the mirror up to nature. And we may guess from his bequest that they parted on good terms and that Burbage was fondly remembered at Stratford. The mourning-ring bequest, after all, put Burbage on the same footing as those two devoted, painstaking friends who produced the 1623 Folio and thus saved nineteen of the plays from oblivion.

But O, what appalling thespian sins have been committed by actors who have followed in the tradition initiated by Burbage! He has much to answer for as the inspiration for so many invitations to thespian megalomania. Shakespeare seems to have been alive to the dangers presented by juicy roles that require full exertion. One might guess as much from Hamlet's disgust at actors who strut and bellow to make their effects and at clowns whose "most pitiful ambition" has led them to enlarge their speeches extemporaneously. One might also guess so from that most *noblesse oblige*–ridden, tirelessly self-promoting player, Bottom. But it is very hard to imagine that Shakespeare's experience with Burbage and some of his lesser stars could have prepared him to imagine some of the monstrosities imposed upon his plays by star-actor egotism in the next few centuries.

Though Betterton, Burbage's first notable successor, cultivated a reputation for extreme devotion to Shakespeare's memory, he was happy, during the Restoration, to perform in versions of the plays spectacularly mangled and prettified by D'Avenant (Lady Macbeth's "Screw your courage to the sticking place" became "Bring your courage to the fatal place") and, later, in adaptations of his own devising. But it was in the age of Cibber, the first half of the eighteenth century, that most pitiful ambition among star actors became truly rampant. Beginning with the Restoration, scenery was becoming more and more elaborate and requiring more time to move. Interstitial musical numbers were also becoming more prominent (in *Macbeth* the witches' scenes grew into a full-fledged masque). This necessitated wholesale cutting of the plays. Invariably, a raft of minor characters disappeared, and several scenes not featuring the protagonist were likely to vanish. Rather than catalogue all the horrors here, I may point simply to Cibber's adaption of *Richard III* as the supreme example of the eagerness in the eighteenth and nineteenth centuries to enhance the starring role or roles at the expense of Shakespeare's original language and dramatic conception. Published in 1700, it shrinks the original 3,600 lines to about 2,050—of which half are Shakespearean and the rest either modified or wholly concocted by Cibber. No scenes with Richard are cut, but five without him vanish, and he gains seven new soliloquies. His share of the lines rises, in the process, from thirty-one percent to forty percent of the total. As a recent editor of the play concludes, Cibber's Richard is "an egomaniac's delight." As has been noted hitherto, this star-struck version caused *Richard III* to remain the most popular of Shakespeare's plays for about two hundred years. Indeed, the 1955 Olivier film was "to a considerable extent influenced by Cibber's adaptation," and a dramaturg's program note indicates that it was consulted in preparing a 1988 revival of the play at the Guthrie Theater in Minneapolis.

We shall never know whether Burbage was in truth an egomaniac or whether he peremptorily demanded colossal roles from his company's in-house playmaker. Nor shall we ever know (to imagine another scenario) whether Shakespeare, whose commercial acumen has never been questioned, made his plays hostage to hugely demanding roles in cynical exploitation of one of the hottest commodities on the London theatrical scene: Burbage, the "delightful Proteus" of the day. But one thing is clear: by concentrating the executive burdens of performance so intensely in one major role or two, he was indulging in a very risky business. Risky, because executive self-indulgence—the "noble art of self-pretence," in Joyce's phrase—becomes the very essence of the theatrical event. Opera composers, of course, run much the same risk. Hofmannsthal and Strauss agreed that the "fundamental truths" of ideal opera were "(a) mimetically convincing situations and (b) good parts," and once one begins thinking in terms of "good parts," as Shakespeare manifestly did, one must tie one's fortunes to the courageously self-indulgent artist who will be drawn to the challenges and risks of the forestage and the limelight (a piece of stage technology Shakespeare would have used effectively, if it had been available to him).

Shaw, following as usual his Shakespearean and operatic habits, thought very consciously of good parts and courted precisely the same all-or-nothing dangers of finding the right actors to assume them. His view of *Pygmalion* thus expresses pretty nearly the attitude of any sensible impresario planning to mount a Shakespearean tragedy or a major nineteenth-century opera: "*Pygmalion* is essentially a star play: unless you have an actress of extraordinary qualifications and popularity, failure is certain." Composing his plays as he did, Shaw also inevitably caused the odds against superlative performances to soar—another parallel with the history of Shakespeare and opera in performance. Eric Bentley's observation upon the need for star quality in Shaw applies emphatically to our two dramaturgies as well: "One actor in a dozen will have the tone and rhythm without which Shavian comedy does not properly exist. The memory of a Shavian playgoer is studded—not, unhappily, with great productions—but with individuals who transcended a general competence or incompetence. . . . Mere competence is not enough for Shaw."

The most obvious risk of hitching one's theatrical wagon to a star is that the star will run away with it in an unexpected or untoward direction. We have at least one clue that Shakespeare was alive to the risks that a star actor and a star-struck audience could subvert the whole performance. In *Richard II*, York describes Richard's failure to hold the royal stage against his more charismatic rival, Bolingbroke:

> As in a theatre the eyes of men,
> After a well-graced actor leaves the stage,
> Are idly bent on him that enters next,
> Thinking his prattle to be tedious,
> Even so, or with much more contempt, men's eyes
> Did scowl on gentle Richard. (5.2.23–28)

An actor's "well-graced" self-indulgence, in the best possible sense of this admittedly dubious term, might be defined as the thoroughly confident, generous, complete deployment of all one's intellectual and physical thespian energies. But, human nature being what it is, self-indulgence can easily devolve into self-aggrandizement—and the dismal spectacle of Hamlet or Tosca becoming the performer, rather than vice versa. Because the temptations to this sort of collapse are so enormous in Shakespeare and opera, it happens fairly often. The annals of criticism and satire are rife with descriptions of odious self-indulgence. We have Bottom, of course; we have one By-Play, in Brome's *The Antipodes* (1640), who is ridiculed for courting the audience, ignoring "his coactors," and "not minding the reply" to his speeches; and we have Jonson fuming in one of his plays about players who "over-act prodigiously in beaten satin and, having got the trick on't, will be monstrous still, in despite of Counsel."

In real life, the temptations to self-indulgence have constantly been courted even by the greatest exponents. Indeed, it may be their most common occupational hazard. Jealous dominion over center stage must have caused Garrick never to "speak warmly in the commendation of any actor, living or dead." Emil Devrient, the eminent nineteenth-century German actor (whose Hamlet pleased London), has been described as "the virtuoso supreme. His unmistakable figure and bearing ensured that audience interest focused upon him as an actor, not on the characters he played." And Shaw had several occasions as a reviewer of plays and operas to produce wicked sketches of pseudo-impersonations by divas and divos. In 1890 he wrote of Adelina Patti, for instance:

> The operatic stage is improving, like other things. But it is still possible for a prima donna to bounce on the stage and throw her voice at the heads of the audience with an insolent insistence on her position as a public favorite, and hardly a ghost of a reference to the character she is supposed to impersonate. . . . Madame Patti's offences against artistic propriety are mighty ones and millions. She seldom even pretends to play any other part than that of Adelina, the spoiled child with the adorable voice; and I believe she would be rather hurt than otherwise if you for a moment lost sight of Patti in your preoccupation with Zerlina, or Aïda.

Shaw found virtually the same phenomenon among star actors on the legitimate stage. Sarah Bernhardt, who incidentally succeeded in several star roles that were eventually to find their way into the opera repertory (Dumas' lady of the camellias, Fedora, Francesca da Rimini, La Gioconda, Adrienne Lecouvreur, Juliet, Santuzza), elicited a highly unflattering comparison with Eleonora Duse in which Shaw concluded that Bernhardt was possessed of the "art of making you admire her, pity her, champion her, weep with her, laugh at her jokes, follow her fortunes breathlessly, and applaud her wildly when the curtain falls. She does not enter into the leading character: she substitutes herself for it." Even Janet Achurch, whom Shaw had once admired so extravagantly, appears to have succumbed to the hazards of self-indulgence. Seven years after he was shaken to the core by her Cleopatra, he wrote in a letter, "Janet has now lost all power of doing anything but her own particular *Io son io* [I am myself]."

And yet: *Io son* Burbage . . . or *Io son* Storace (Mozart's first Susanna), or *Io son* Pasta (Bellini's first Norma), or *Io son* Tamagno (the first Otello) is a part of the equation of performance. The ignominious side of self-indulgence is all too familiar and needs no further notice here, but perhaps its more positive, exciting, charismatic side does deserve some special emphasis, for without it there can be no truly heart-satisfying performance of Shakespeare or opera.

One way to achieve this emphasis might be to suggest that the great role demands the nerve, the daring of the *alazon*. This Greek word means "boaster" or "impostor" and was used to define a class of characters in ancient drama. The typical *alazon* is someone who pretends to be something more than he is; this is in contrast to the *eiron,* who typically is self-deprecating, unobtrusive, and given to issuing counsels of prudence. The *alazon*'s expressive style is broad-brush, hyperbolic, and tends toward the grandiose; the *eiron*'s is more subtle, oblique, and ambiguous (our word *irony* is related to him). The bias in Shakespeare's plays and in most operas is shamelessly in favor of the *alazon*. In these dramatic worlds, charismatic *alazons* (Norma, Lear, Siegfried, Antony, Butterfly, Lady Macbeth, Carmen) who create some kind of magnificent imposture or delusive role for themselves and prove extravagantly "abnormal" are often brilliantly contrasted with lesser *eiron* figures (Adalgisa, Kent, Mime, Enobarbus, Suzuki, Banquo, Micaela) who urge prudence and "normality."* Only Hamlet among Shakespeare's protagonists has a strong flavor of the *eiron;* the rest—from Berowne, the two Richards, Prince Hal, and Falstaff, down through Iago and Prospero—are thoroughgoing *alazons* who are adept at role-playing (i.e., imposture) and at convincing others that they are nobler, angrier, greater, or more sincere than they really are. This is perhaps why Verdi never seems to have been tempted to compose a *Hamlet* opera . . . why, indeed, this Shakespearean hero is the most difficult to imagine successfully inhabiting an opera. The grand emotional, rhetorical, and physical gestures of the *alazon* provide Shakespeare's and the opera composer's more-customary material.

I mention the *alazon* character because the star performer must, at heart, be one, too. The *alazon*'s eager but risky display of the noble art of self-pretence reflects on star quality, for the star is nothing if not a magnificent impostor who parades as a character grander and more heroically articulate than the common sort. And the essence of success in a major acting or singing role is the creation of the illusion that the executant is *not* an impostor, but rather can convincingly establish artistic "rights" to the heroic identity. Failure occurs when telltale signs of imposture begin to emerge during a performance. When, in other words, we are left feeling as Costard does about the country curate's impersonation of Alexander the Great in *Love's Labour's Lost:* "He is a marvelous good neighbor, faith, and a very good bowler; but for Alisander—alas, you see how 'tis—a little o'erparted." Sweet-natured Nathaniel was doomed, of course, to being too tame. Good-neighborliness is not the path to glory in stellar roles.

What is needed, rather, are the hugely self-indulgent instincts that Walt Whitman—American literature's supreme poet-*alazon*—urged in a poem called "To a Pupil." He first asks the pertinent rhetorical question (in one of your longer poetic lines): "Do you not see how it would serve to have such a body and a soul that when you enter the crowd an atmosphere of desire and command enters with you, and every one is impress'd with your Personality?" Giovanni, Norma, Tosca, Richard III, Cleopatra, or Coriolanus, if they are to succeed, must have this overmastering effect upon us. So, too, must the artist who assumes their identity. To achieve this effect, Whitman urges his pupil: "Go, dear friend . . . commence to-day to inure yourself to pluck, reality, self-esteem, definiteness, elevatedness, / Rest not till you rivet and publish yourself of your own Personality."

Some such process as this—today it would be called "psyching up"—is an element of star quality. It is hard to imagine a theatrical diva or divo venturing on a heroic role unable to take Whitman's admonition to heart. Cibber, I think, saw the "publishing" of Personality at work in the famous operatic rivalry of his day: "Caesar and Pompey made not a warmer Division in the Roman Republick than . . . Faustina and Cuzzoni blew up in our Commonwealth . . . by their implacable Pretensions to Superiority." And while Cibber saw the obvious temptations to "fantastical Pride and Caprice" in their self-images, he could still allow that the "Greatness of Soul" with which they pretended to superiority was their "unalterable Virtue." Ljuba Welitsch, a most charismatic performer during her rather brief career, also strikes the Whitman note when she observes that "Young singers . . . don't have time to build their own personality [today]. A performer has to come onstage and *be there,* and tell your audience, 'I am here! You have to look at me, you have to follow me all the time, and you have to be with me!'" Derek Jacobi, a leading Shakespearean of his generation, has been praised as precisely the kind of performer Welitsch demands: "Jacobi is always in the giving vein. Theatergoers need never doubt that he will be *there* for them."

Welitsch's admonition almost exactly fits the "Self-Expressive Mode" of acting as defined by Bert States in his study of the "phenomenology of theater." In this mode, "the actor seems to be performing on his own behalf. He says, in effect, 'See what I can do.'" States finds the Self-Expressive Mode dominant mainly in the classical repertory—Faust, Falstaff, Lear, Hamlet, Medea, Cyrano—but he might well have included some operatic roles as well. For operatic roles also, as States observes of the Self-Expressive Mode, "seem to charge the event with the electricity of competition between actor and character." (This competition is the subject of the following chapter.)

Entering into this competition by exposing and publishing one's own artistic personality in a live performance is fraught with several dangers: danger that one's own personality will be perceived as violating the character's; that discrepancies between the two personalities will emerge and become distracting; that the audience will resent the demand to "watch me all the time" if it is not made to do so by artistic *force majeure;* and, finally, the danger that the exertion of personality will in one way or another strike a false note and collapse into a repellent display of pride. Such is the price of "vaulting ambition," which, as Macbeth remarks, sometimes "o'erleaps itself" and fails of its purpose. To assume a virtuoso role, in other words, is always to court outrage and outrageousness. Just as great comic or tragic characters are always testing, forcing, sometimes violating the standards of normal, good-neighborly behavior (courting the outrage of their fellow men, the gods, or fate), so is the star actor or singer urged to test and probe the limits of normal human utterance, the limits of his or her own physical resources, and the limits of the audience's credulity. Michael Goldman puts this notion the other way round when he writes, "As acting itself seems provocative, risky, magnetic, so the heroes of drama appeal to us by going to extremes." Obituarists, in the summer of 1989, repeatedly paraphrased this assertion in assessing the genius of Olivier. The *London Times,* announcing the death of this actor of "spellbinding power," observed: "His playing disdained veiled nuances in favor of unambiguous clarity, with every decision fearlessly exposed and pushed to the limit. 'Danger' is an overworked word when applied to actors; but the sense of risk was integral to his work. . . . The sense of risk was no illusion. When Olivier did fail . . . he failed on a scale that obliterated the surrounding play."

Major Shakespearean and operatic characters—and their finest impersonators—are most at home on the risky frontiers of propriety and plausibility, where soul-invading eloquence can shade into bombast, prodigious individualism into affectation, or self-confidence into complacency with quite appalling ease. Josh Billings, that wise if spectacularly unlettered philosopher of the last century, had something to say about this fascinating frontier: "It seems tew be necessary that there should be sumthing outrageous in evrything, tew show us whare propriety ends and impropriety begins. This iz melancholly in the case of the rooster affair, for we hav the shangi rooster, the gratest outrage, in my opinyun, ever committed in the annals ov poultry." It is a happy coincidence that Billings makes his point by referring to an animal famous for its extreme vocal exertions. But the crucial point to draw from him is that there is "sumthing outrageous" in the stellar roles of our two dramaturgies—not only in the behavior of the fictional characters themselves, but also in the outrageous executive demands that are typically made in order to bring them into complete, satisfying existence on stage. And perhaps "sumthing outrageous" is but another suitable phrase—like Verdi's "il diavolo adosso"—for star quality.

There was certainly something outrageous in Walt Whitman, and it had mainly to do with the exuberance of his self-indulgence and his fearlessness of garrulity or vulgarity (for which he was often lampooned). This species of fearlessness is necessary to the Shakespearean actor and the opera singer, whose respective dramaturgies are premised upon hyperbole (lingua franca of all *alazons*), rhetorical elaboration, and excess. In a note to himself, Whitman once wrote, "Make the poems of emotions, as they pass or stay, the poems of freedom, and the expose of personality." This "expose of personality" is usually significant in the major characters in our two dramaturgies. This exposure presents a challenge to the artistic spirit: an executant's Whitmanesque "song of myself" must somehow resonate silently in performance behind the "deep harmony" of the role's poetry or music itself.

"Courage is needed to be romantic," Stendhal wrote in an essay on Racine and Shakespeare, "for *risks must be taken.*"* Thus far, I have paused over the more general and admittedly more impalpable risks—spiritual risks, one might say—that face the star performer. But there are also risks of a more specific and graspable kind that require courage, too, and these deserve some attention here.

Beyond the exposure of personality, most obviously, is the exposure *to* enormous technical challenges and exposure *in* them as one performs. Each great Shakespearean and operatic role has its unique array of executive problems and rewards, but all of them have something in common with a steeplechase course. Prowess, stamina, and a sure, full command of technique are essential as one traverses it in the course of a performance. In both dramaturgies the performer is compelled to be aware of and shrewdly analyze the hidden and not-so-hidden dangers in the role, as well as thoroughly conscious of his or her own physical resources. The director Peter Hall speaks eloquently of what the great role thus imposes upon the actor: "Part of the epic scale of the great dramas is the way they stretch the individual heroic actor to his limit. If you have played the first two-thirds of Hamlet or Macbeth or Lear correctly, the tragic intensity of the last third is that much heightened because of the physical marathon you've already gone through, which leaves you calm and clear-headed. Each of the tragedies is a microcosm of man's life, full of effort and then exhaustion. *So is the actor's performance.* It is a metaphor as potent for the life of the individual as the metaphor of the Globe Theatre is for the world itself."* Olivier, we have seen, captured the sort of heroic challenge Hall speaks of when he said, "there are too many climaxes, far too many" for Othello and when he complained of the role's "fits and raging . . . all demanding you to scream to your utmost."

This stretching to (and it sometimes seems beyond) the limits is even more strikingly apparent in opera, as one might gather from several sopranos famous for their impersonation of Floria Tosca. They express their fears of Puccini's demands in *Tosca*'s second act rather as an actor might shiver at the difficulty of Othello's volcanic outbursts in acts 3 and 4 or the thespian terrors of Lear's scenes in the storm. Renata Tebaldi observes, "If you want to do something real on the stage, you have to forget your voice sometimes, even if you know you are doing something that is not good for your throat—which can be the case in the second act of *Tosca.* But when you are sure of your technique, you can sing what you want." Birgit Nilsson makes the counterpoint: "Sometimes you must separate the voice from the emotions. You have to protect the throat." And she adds, "More than one singer has killed her voice in the second act of *Tosca.*" Dorothy Kirsten is more succinct: "One has to have every kind of technique one can muster." Sounding Olivier's note, Grace Bumbry is more analytical: "It is such an incredible part, because of all those outbursts, which are really quite exposed, you know, in the second act. You have an A, B, high C, and not just one of them—they come constantly. Then you've got that very delicate '*Vissi d'arte*' to sing. How do you scale that big dramatic voice down to this pianissimo?" Bumbry's conclusion is, one might venture, spoken for all who attempt star roles: "You've got to have your wits about you."

When one thinks of stellar roles comprised of a series of set speeches or musical numbers, it is natural to think of obstacle-course images . . . or of a vigilant riverboat captain keeping his wits about him in a dangerous stretch of the Mississippi. Also apropos of these dramaturgies of escalation, though, are visions of death-defying risks courted at great heights: the false step taken at the edge of a yawning crevasse or airborne gymnastics launched from a flying trapeze. Luigi Pirandello, assessing Italian theater, has written that "it is always possible to detect a Shakespearean influence whenever there is an elevation of tone in any of our theater works." This urge toward elevation—Whitman, remember, called Shakespeare "the loftiest of the singers life has yet given voice to"—and the vertiginous exposure of the artist's resources serve to make the sense of imminent danger a part of the heroic performance. This sense of danger is heightened in opera, with its more extreme laryngeal exertions. It can be merely the danger of failing to give such-and-such a speech or aria its due, or the more general danger described by Wesley Balk in his analysis of acting on the musical stage: "In every great performance there is a sense of *danger,* a feeling that the performer may destroy the role and violate the artistic structure that is being created." In both Shakespeare and opera, the most breathtaking performances are those that have the effect of Olivier's Titus Andronicus: "As usual," exclaimed a reviewer, "he raises one's hair with the risks he takes."

Not surprisingly, in the event, the image of a high-wire artist has proved a common fallback. The actor Tony Church, for instance, insists on the necessary danger when he comments, "Each time you do a speech it goes too far one way or the other. It's like walking on a tight-rope all the time. Yet if it's not, then there's something wrong: you're not playing dangerously enough." Bert States makes not only this point but also the related one that heroic performance is possible only in live as opposed to filmed conditions: "Virtuosity, in theater as in athletics, is not simply skill but skill displayed against odds. . . . One might liken the film actor to the aerialist who works with a net. His performance is no less a thing of beauty than that of the stage actor, but it is relieved of all possibility of disaster except that of poor acting. But the theater offers the actor no net; the play is one long danger." Live opera is, even more strikingly, "one long danger," as Amelita Galli-Curci made very clear: "It is the destiny of a coloratura soprano to be like an acrobat on a steel wire. The public waits anxiously to see if you are going to crack on a certain note or be unable to finish a certain filatura, just as the public in a circus follows attentively to see if the poor wretch is going to miss a leap and fall."

Starring roles in Shakespearean plays and standard operas create a climate of risk for another obvious reason: so many of them have become cultural icons. Their scripts and scores are profoundly embedded in the common aural memory. *Pace* Mark Twain, who defined a classic as a book everyone talks about but nobody reads, the classic protagonists of the Top Bard and of Mozart, Verdi, Wagner, Puccini, and Strauss are cynosures of very much talk indeed and are also constantly being "re-read" before the public. The regular playgoer or operagoer is, in fact, sometimes inclined to feel a little sorry for these characters and the relentless commentary and reincarnation to which they are subjected. Debussy once had occasion to express a similar sympathy for Leonardo's famous portrait: "Being a masterpiece is not always an enviable profession. As proof of this one need merely consider the dreary plight of Mona Lisa and her smile. Because of too much notoriety and scrutiny over the years, it has become rather strange. If only on seeing her for the umpteenth time the art critics had confessed that she really wasn't smiling at all! If only they had admitted that she was actually pouting, perhaps a bit voluptuously in some private Lombardian way." (The "masterpiece" that elicited Debussy's whimsy, by the way, was *Parsifal*.)

The analogy with the most famously enigmatic, Hamlet-like of all paintings is pertinent. For it is a main attraction of the cynosure role that, being hard to "locate" or "fix" interpretively once and for all, it can embrace numerous possibilities. Margaret Webster's assertion that "Shakespeare allows his actors a greater margin of interpretation than can possibly be pinioned by any single mind" applies as well to the most charismatic operatic roles. Most of us have often experienced in performance quite different, or even mutually exclusive, yet still quite convincing versions of Cleopatra, Lady Macbeth, Carmen, Iago, Tosca, Prospero, Salome, or whomever . . . all emanating astonishingly from the same script or score. The risk for the performer is whether or not his or her version will emanate credibly from the core of the character. Bentley has written that "the 'great' characters—Hamlet, Phaedra, Faust, Don Juan—have something enigmatic about them," and he adds that "a mysterious character is one with an open definition . . . open as, say, a circle is open when most of the circumference has been drawn." While Bentley complains that critics are always "closing the circle which Shakespeare left open," he would surely not deny that it is the actor's task to make this risky, daunting choice and, during a few hours' traffic of the stage, close the circle: only one of the several possible Lears or Hamlets can take the stage at any given performance.

What makes the choice even more daunting is the fact that a great classic role will speak challengingly to the mediocre or merely competent performer rather as Hamlet does to Guildenstern: "You would play upon me, you would seem to know my stops, you would pluck out the heart of my mystery, you would sound me from my lowest note to the top of my compass." The executant not possessed of the full range of expression Hamlet refers to will fare exactly as he promises Guildenstern: "though you fret me, you cannot play me."

A corollary risk of the all-too-familiar classic role is this: the performer must not only compete with the character itself but also with the listener's own imagined ideal version of the role, as well as the versions of one's predecessors and contemporaries. An audience will come to *La Traviata* or *Romeo and Juliet,* for the most part, familiar from top to bottom with the roles of Violetta and the two lovers. The stars will have to measure up. Agate once commented in a review that "Bulwer's old title—*What Will He Do With It?*—must be in the mind of everybody who goes to see a new Othello," and there are likewise usually few aural neophytes who attend *La Bohème* or *Aida* without bringing a personal aural template to the proceedings.

William Archer, an eminent critic and playwright, articulated this challenging situation succinctly with reference to Shakespeare, but what he says applies as well to the world of opera: "We have each our own private ideals of Macbeth, Hamlet, Othello, Lear; we have all of us read of, if we have not seen, great performances of these parts; so that every actor who undertakes them has to pass through a triple ordeal, encountering, first, our imagination, kindled by Shakespeare; second, our idealized memory of performers which used to please our perhaps unripe judgment; third, our conceptions of the great actors of the past, gathered from the often extravagant panegyrics of contemporaries." The chief responsibility amid all this baggage of ideal projection and historical memory—and only those with star quality will be able to sweep such baggage away—is quite literally to renovate the role . . . forge it anew and create the illusion of "the first time." Agate laid it down as a rule that "every good performance of any Shakespearean play is recognizable by the fact that it arouses fresh shock at something which one knows so well that one takes it for granted—to each spectator his own shock, of course." Most memorable individual vocal performances in opera succeed in this fashion.

The final risk worth noting can be simply stated: great roles demand that one *give,* and there is an ever-present temptation to give too much—a temptation that comes not only from the wonderful verse and music but also from the audience, which is eager to have greatness thrust upon it. The satisfaction of succeeding is obvious: "there is nothing for me in all the world," Leonie Rysanek has said, "compared to a good performance, when I can go home and go to bed and think, 'You did a fine job.' It is a wonderful feeling." As for failure . . . one often wonders how a performer can bear up under the pall of disappointment, or worse, resentment. Conrad Osborne has summarized the mood among standees at the old Metropolitan Opera: "If you gave, all was forgiven; if not, no fate was too cruel." And on the subject of great roles' revealing executants to be impostors, Shaw was more than usually sulphuric: "I *hate* performers who debase great works of art: I long for their annihilation; if my criticisms were flaming thunderbolts, no prudent Life or Fire Insurance Company would entertain a proposal from any singer within my range, or from the lessee of any opera house or concert room within my circuit."

The cruel fact is that, on these virtuosic heights, the means of pleasing the standees and the Shaws mingle with the means of self-destruction. This is seen in Tebaldi's remark about endangering the voice in order to "do something real on the stage," or in the fairly commonplace gossip that a particular singer has compromised and shortened a career by exertions too wonderful for his or her own good. Thus, for instance, the *Metropolitan Opera Encyclopedia*'s entry for Ljuba Welitsch: "her fiery temperament seemed tireless in her first Met season, but her prodigal use of these exceptional resources surely shortened her career, though producing some of the most memorable operatic impersonations of the immediate postwar years." Her colleague Irmgard Seefried reminisced about Welitsch: "When she sang Donna Anna with me, I was overwhelmed by the velvet sound. But she gave too much of herself, and nature is fierce, for the price to be paid is high."

The careers of stellar Shakespeareans are rarely as dramatically meteoric as those of singers, not only because the vocal stress is not nearly so drastic but also because actors have far more opportunities to vacation from the heroics of Shakespearean service. Singers are doomed to a more relentless outpouring of vocal generosity. Shaw, we have seen, singled out Verdi for tempting his singers to "unnatural and dangerous assumptions and exertions." Expanding on this view, he wrote that Verdi's "worst sins as a composer have been sins against the human voice. His habit of taking the upper fifth of the compass of an exceptionally high voice, and treating that fifth as the normal range, has a great deal to do with the fact that the Italian singer is now the worst singer in the world." This, in 1893, was no news to Verdi's countrymen. In 1847, before he hit his stride with *Rigoletto, Trovatore,* and *Traviata,* a journalist noted, "there is no lack of those who blame him . . . as a waster of singing voices because of the *tessitura* of his parts and because of the nature of the *cantabiles,* which are too showy in their declamation." Still, while other composers may be more "grateful" to their heroes and heroines, I think an insistence upon immoderate exertions is common to most of the repertory's "good parts." Most of the great roles fall in one way or another into the category of "too showy." This is their reason for existence.

What do these parts require? Shaw wrote that "to kindle art to the whitest heat there must always be some fanaticism behind it."* White heat, not lukewarmness or mere competence, is necessary. Fanaticism: perhaps this is Shaw's unwitting translation of Verdi's "il diavolo adosso." Both notions, of course, are vague and impressionistic, as perhaps befits any attempt to define star quality. I am, in fact, tempted to close the present meditation on this quality with an impressionistic *jeu d'esprit* of my own and introduce a character from fiction who seems to me to capture some of the salient qualities of the typical Shakespearean and operatic protagonist, and of the star performer as well. He is Henderson, the title character of Saul Bellow's novel, *Henderson the Rain King.* (I was first tempted to choose a figure from Bellow's *Humboldt's Gift,* the author's embodiment of the life force, who is named—perfectly for my purposes—Rinaldo Cantabile.)

Henderson: a wealthy man in his mid-forties. One might guess he is operatic just by looking at him: six-foot-four, weighing 230 pounds, and known for his "blustery ways." But the true marks of opera and Shakespeare are inside—in the vastness of his emotional energy (he admits he is "full of excess"), in his awesome ability to suffer ("I am to suffering what Gary is to smoke"), and in the voice inside him that is constantly crying, "I want, I want, I want, oh, I want." He is not merely stirred by life but "horribly stirred." When he lies awake at night it is because he is "bathed in high feeling." He is set apart from the normal people, the *eirons* of life: "I admire rational people and envy their clear heads." Responding to Billings' "sumthing outrageous" inside himself, Henderson, soon after the novel begins, flees from his wife, children, a vast hog farm, and a small fortune to the wilds of Africa. His is the goal of many a Shakespearean and operatic protagonist: "My purpose was to see essentials, only essentials, nothing but essentials." His exuberance is astonishing, and it comes as no surprise when he starts quoting from Whitman: "Enough merely to be! Enough to breathe! Joy! Joy! All over joy!"

Henderson goes to Africa in search of an epiphany. Opera, in its impetus toward concentrated and striking expressivity, is an epiphanic art form, and its world is one of high relief, magnification, and escalation.* Henderson's cata- pulting spirit captures just these operatic (and, I hasten to add, Shakespearean) qualities:

> I think the noble will have its turn in the world . . .
>
> I am a high-spirited kind of guy . . . I am unkillable . . .
>
> I am the inspirational type and not the systematic type . . .
>
> I wanted to raise myself into another world. My life and deeds were a prison . . .

Bellow arranges for Henderson's epiphany to occur among an African tribe, its ruler acting as a soul-liberating guru. From him, Henderson learns what we also learn from the great characters of the stage: "the noble self-conception is everything." This climactic scene, which has a distinctly Whitmanesque elevation to it, takes a marvelous, explicitly operatic turn when the ruler exhorts Hender- son, "You can have a new poise, which will be your own poise. It will resemble the voice of Caruso, which I have heard on records. . . ." And the novel ends (as so many operas end) with a brilliant, ecstatic gesture from the protagonist: "I felt it was my turn to move, and so went running—leaping, leaping, pounding, and tingling over the pure white lining of the gray Arctic silence."*

18

Technique and the Gemini Factor

One must try to do coolly the things that are most fiery.

HECTOR BERLIOZ

All "staged" or "played" human action ultimately involves a separation between the role and the actor who assumes it. Hamlet refers to this separation in his famous rhetorical question:

> Is it not monstrous that this player here,
> But in a fiction, in a dream of passion,
> Could force his soul so to his own conceit
> That from her working all the visage wann'd,
> Tears in his eyes, distraction in his aspect,
> A broken voice, an' his whole function suiting
> With forms to his conceit?   (2.2.551–557)

In the theater we are constantly aware, to a greater or lesser degree, that the actor is bending his "whole function" (his body, technique of projection, and artistic personality) to his own "conceit" (his imaginative conception of the role). "In the audience, we too enact a double performance," William Worthen observes in *The Idea of the Actor;* we "grasp both reality and artifice, actor and character." Worthen's—and Hamlet's—point was also made, I think, by John Webster in his 1615 sketch of the ideal actor: "what we see him *personate,* we thinke *truly* done before us." Strictly speaking, personation and truth are mutually exclusive. What reconciles them, as Webster suggests, is our ability, willingness, and even pleasure to *think* that the "fiction" of what is done before us is, for the time being, "true."*

Hamlet, being naturally in a very foul mood on the subject of personation and truth, views the actor's dissembling in the light of Elsinore's hypocrisy and calls it "monstrous." But of course on the stage it is unavoidable. David Belasco, whose play *Madame Butterfly* Puccini turned into an opera, made the point when he observed, "In acting there can never be, in the very nature of things, any real feeling."* Samuel Johnson drove the same point home with his usual bluntness in a conversation with the great Kemble reported by Boswell: "Johnson, indeed, had thought more upon the subject of acting than might be generally supposed. Talking of it one day to Mr. Kemble, he said, 'Are you, Sir, one of those enthusiasts who believe yourself transformed into the very character you represent?' Upon Mr. Kemble's answering that he had never felt so strong a persuasion himself: 'To be sure not, Sir (said Johnson); the thing is impossible. And if Garrick really believed himself to be that monster, Richard the Third, he deserved to be hanged every time he performed it.'" Real emotion, rather than emotion "suiting/With forms" of artistic intention, will invariably destroy a performance . . . just as unfeigned emotion on stage wrecks the *commedia* at the end of *Pagliacci*.

But some thespian dreams of passion, like dreams we have in real life, are more verisimilar than others. The most naturalistic, or representational, styles of acting—notably those associated with Stanislavsky and the so-called Method—minimize (or, ideally, erase) the ironic, self-consciousness-filled distance between actor and role by seeking the illusion of complete reality . . . or what Wesley Balk calls "inner honesty." Balk also observes that Stanislavsky's influence in America has been strongest in cinematic acting, where the probing and intimate eye of the camera encourages "non-technical emotional performances." The essence of this acting style, Balk adds, is "the idea that if you actually try to do something with your voice, you will be phony; and its corollary, that if you simply 'feel it,' the voice will do whatever needs to be done without conscious exercise of technique." The set designer's equivalent of this ideal, of course, is the stage picture that seeks to mimic an actual place by recreating super-realistically all the geology, flora (often fauna, too), architecture, and appropriate props. It is small wonder that the film and television camera, so amenable to deployment in real locations, encourages this "non-technical" acting.

Vincent Canby, the film critic for the *New York Times,* reviewed a movie in 1987 that handily proved this rule by being a remarkable exception. We are already in familiar territory when Canby writes that *Orphans,* directed by Alan Pakula, "provides a field day for its three remarkable stars and, in turn, for the audience that beholds them." Even more pertinent is his description of an acting style that "is extremely risky and rare in movies." This is the style of virtuosity sallying forth that, as we have seen, is essential to any fine operatic or Shakespearean impersonation: "*Orphans* could be subtitled *Actors Acting.* The performances of Mr. Finney, Mr. Modine, and Mr. Anderson aren't self-effacing. That is, the actors don't entirely disappear inside their roles. One watches them with the same sort of wonder with which one attends a performance by an opera singer hitting notes never heard before. The actors don't vanish, but seem to stand just a little to the side of the characters, as delighted as we are by what they're able to do."

Actors Acting, Singers Singing: it is quite impossible for our two dramaturgies to escape from their own respective subtitles. Though we often hear players and singers mouthing, in so many words, the professional conceit of this proto-Methodical couplet dating from 1760—

> . . . Fool or Monarch, happy or distrest,
> No Actor pleases that is not *possess'd*. *

—the hard fact is that it is as impossible to perform blank verse or a melodic line "non-technically" as it is for a plane to take off and land without movable wing flaps. Typically, the noted verismo soprano Magda Olivero pleasures herself in the conceit as she describes her Tosca: "I live every part completely, and I become Floria Tosca." But in her very next sentence she concedes that "another side of me, another part of me, is standing beside me and watching, and controlling the sound." Shakespeare encourages a similar thespian self-consciousness by demanding that the actor confronted with a challenging speech or "performance situation" keep his wits about him and exert technical control. Thus, the actor Michael Pennington is but paraphrasing Olivero when he speaks, apropos one of Hamlet's soliloquies, of the need to "control the floodtide of emotion and to discipline it mentally and technically."

The discrepancy—perhaps even the antagonism—between the "fiction" of impersonation and the "truth" of virtuoso technique would seem to align Shakespearean and operatic dramaturgy with the acting style at the opposite end of the spectrum from naturalism. And this is very much the case. This style is reflected in the principles of Japanese Kabuki theater and Brechtian theater, where the stage is transparently a platform for acting; it may be decorated, but it is not *disguised* as another place. Several generalizations made in Earle Ernst's study of Kabuki thus seem apt to the premises of a typical Shakespearean or operatic performance: "the whole . . . technique of acting is directed outward toward the audience"; "the aura surrounding the Kabuki actor can best be described as one of detachment . . . [the] line of detachment is clearly drawn between actor and character"; the audience is most appreciative of "passages which show brilliant technique and execution." Brecht, too, was highly aggressive in demarcating the two realities of impersonation (acting) and fiction (the experience enacted). He refused to shy away from, but rather exploited, the fact "that the actor appears on the stage in a double role, as Laughton and as Galileo." And he insisted, for his "epic theater," that the actor "do all he can to make himself observed standing between the spectator and the event." It was thus the essence of the "alienation effects" that Brecht used to subvert the illusionism of theater that "the actor must invest what he has to show with a definite *gest* of showing." All of which allows Worthen to summarize that the Brechtian and Stanislavskian styles are indeed antithetical: "The cardinal principle of Brechtian acting is . . . the actor's evident estrangement from his own emotions and from the role he is playing. The actor's frank acceptance of the public nature of the theater, and the inevitable artificiality of theatrical performance, marks his furthest remove from the Stanislavskian actor."*

Shakespeare and opera—which rest firmly on those supreme alienation effects, blank verse and sung speech—will always press toward the Brechtian end of the continuum, and toward a frank acceptance of the artificiality of performance. Clifford Leech practically recites Brecht's menu at the end of an essay on the acting of Elizabethan plays: "We shall do . . . best if we forget the picture frame. We need a theater where the actors are throughout manifestly on a stage . . . a theater, moreover, where gesture has to be elaborated to be seen around the encompassing auditorium, where speech must be deliberate and fully articulate . . . where illusion is simultaneously created and defied." And Leech is surely seconding Brecht's eagerness to upset, disconcert, and engage the audience (too prone to sink into inert passivity) when he asserts that "To stage Marlowe and Shakespeare properly . . . we must have faith in the audience's flexibility of mind." Marlowe's *Edward II* was Brecht's first directorial effort, and it seems clear that "the theatrical self-consciousness of Renaissance drama" mentioned by Worthen powerfully influenced Brecht. One Brecht scholar has ventured, in fact, that he "was never more modern than when he was borrowing formal, structural, and stylistic elements from Shakespeare."

It is also easy to imagine that the Elizabethan era's forthright embrace of the artificiality of the dramatic event influenced Brecht. The phenomenon of play-presenting and organized, commercial playgoing was still something very new under the sun when Shakespeare first came to London: the first London playhouse, James Burbage's Theatre, had opened its doors scarcely a decade earlier, in 1576. Elizabethan playwrights (and, it would appear, audiences, too) were intrigued by the techniques and pretensions of this new form of entertainment. They were fascinated, in Worthen's phrase, with "the histrionic dynamics of drama" and reveled in reflexive allusions and devices that drew attention to the "fiction" of the events taking place on the raised platform. And no one doted on these precursors of Brecht's "alienation effects" more than Shakespeare. Michael Goldman, for instance, finds *Hamlet* a supreme example of this reflexivity: "*Hamlet* is the most interesting play ever written—and one reason for this is that it is immediately concerned with the nature of theatrical interest itself, with the relation of the actor's art to life."

We have touched already upon several passages in *Hamlet* that led Goldman to his conclusion, but many other memorable moments of thespian self-consciousness are scattered throughout the canon. Duke Senior's comment in *As You Like It* (which gives Jaques his cue for "All the world's a stage . . .") is such a one:

> This wide and universal theater
> Presents more woeful pageants than the scene
> Wherein we play . . .

Another is Berowne's comment, in the last scene of *Love's Labour's Lost,* that

> Our wooing doth not end like an old play:
> Jack hath not Gill. These ladies' courtesy
> Might well have made our sport a comedy.

After Malvolio is baited outrageously in *Twelfth Night,* Fabian observes, "If this were play'd upon a stage now, I could condemn it as an improbable fiction." A boy actor, we must remember today, was given these lines to speak when

Cleopatra imagines how she will be treated if Octavius takes her back to Rome: "The quick comedians / Extemporally will stage us . . . and I shall see / Some squeaking Cleopatra boy my greatness." And of course *The Tempest* is impossible to experience without an awareness—climaxing with Prospero's "Our revels now are ended"—that the play is among other things a meditation on the presentation of theatrical "projects" and perhaps even Shakespeare's farewell to the profession.

Ostentatiously reflexive reminders of artificiality also occur in opera. Sometimes a character will follow the habit of a Shakespearean soliloquist and address the audience directly—as in Figaro's misogynistic outburst in *Le Nozze,* "Aprite un po quegl'occhi"; in Berta's complaint about the uproar *chez* Dr. Bartolo in *Il Barbiere;* or in Loge's sarcastic remarks, uttered conversationally from the forestage, as he watches the gods enter Valhalla at the end of *Das Rheingold*. Shakespeare's several prologues and epilogues—which exist precisely to call attention to artifice—are paralleled in the famous prologues of *Pagliacci* and *Lulu* and in the spoken epilogue of *Gianni Schicchi* or the *scena ultima* of *Don Giovanni,* which—like the fugal free-for-all at the end of Verdi's *Falstaff*—can be very effective when treated by the director as a bit of wise moralizing addressed to the audience.

Other self-conscious touches in opera can be very droll in the Shakespearean vein. Mozart quotes *Figaro's* "Non più andrai" as part of the table music at Giovanni's last meal, and Leporello remarks, "That tune I know only too well." (Also delicious is a reference to "that infernal nonsense *Pinafore*" in Gilbert and Sullivan's *Pirates of Penzance*.) Berg turned Wedekind's Alwa into a composer in *Lulu* and has him say of the heroine at one point, "One could truly write an interesting opera about her." Such touches, as in Shakespeare, are invariably ingratiating, as Strauss suggested in a letter to Hofmannsthal about the role of the Composer in *Ariadne auf Naxos:* "Give your sense of humor its head, drop in a few malicious remarks about the 'composer'—that sort of thing always amuses the audience and every piece of self-persiflage takes the wind out of the critics' sails." Which is often what Shakespeare intended to do by inserting his references to plays and playing. The most striking and elaborate bit of "self-persiflage"—a more charming phrase than "alienation effect"—in the entire operatic repertory, by the way, is Strauss' career-ending *Capriccio*. As a meditation on the nature and artifice of sung theater, it occupies the same position in the Strauss canon as *The Tempest* does for Shakespeare. Each is a self-conscious vanity of its creator's art; each is enriched by its creator's bittersweet mood of artistic valediction.

Another way Shakespeare and opera call attention to the reality of "actors acting" and "singers singing" we have already explored: the tendency to create characters who themselves become performers (role-players, chameleons, jugglers of masks) or who are simply thrust into several performance situations in the course of their stage lives. Consider Prospero, for instance. He calls for "some vanity of mine art" and produces the splendid hymeneal masque featuring Ceres and Juno, but Shakespeare also cast Prospero himself as a "vanity" of the actor's art, requiring him to pass through a kaleidoscope of "interior" roles as he interacts with all the other characters: strict father to Miranda, disgusted master to Caliban, stern but benign master to Ariel, vengeful enemy to the usurper of his dukedom, compassionate master-of-ceremonies in asides. The two realities of fictional "truth" and artistic impersonation are hard at work here in tandem—as they have been many times before in Shakespeare's plays.

Prospero is but the last of many characters who are not only avatars of thespian vanity and virtuosity in their fictional worlds but whose actual roles—the lines themselves, with all their challenges to heroic elocution—are premised upon the artistic vanity of a Burbage, Condell, Armin, Heminge, or any of several nameless boys. In such roles, as with many operatic roles of this ilk, emotional trauma or heroics issue inevitably in heroic technical impositions. The consequence is obvious: "Any striking instance of technical virtuosity," Goldman writes, "in a play's verse will, if the actor can master it, inevitably present itself as technical virtuosity in performance." So it is that in opera as in Shakespeare the great roles are premised upon what can be called the Gemini factor: the inescapable twinning of the role and the performer.

The supreme goal of the naturalistic or Method actor is to "live" rather than merely "play" the part; in other words, to banish the Gemini factor as completely as possible and achieve a completely consistent effect on stage. He can achieve this effect only as long as he is given speech and actions approximating those of the normal extra-theatrical world. What prohibits the Shakespearean or operatic actor from achieving such a goal is the unavoidable presence of technique. Consistency of effect is out of the question, because the two dramaturgies are magnificently *in*consistent and *un*natural in their pressure on performers to run the gamut from utterly prosaic normalcy to the heights of purple-patched blank verse and coloratura fireworks . . . or pressure to shift suddenly from *ppp* to *fff*, negotiate vertiginous changes of tempo, and express themselves *dolcissimo* one moment and *con slancio* or *avec explosion* the next.

The main challenge of this kind of pressure is perhaps best expressed by Lady Macbeth's doctor: "Unnatural deeds / Do breed unnatural troubles." The elaborate rhetoric of verse and music breeds the "unnatural troubles" of elocution, breath control, projection, and the husbanding of resources. The Gemini factor is born of the dual tasks that derive from this inescapable corollary of performance: one must master these troubles through technical facility and resourcefulness while making Lady Macbeth's or Rosina's or Siegfried's "unnatural deeds" humanly convincing. John Barton's remark about the way to succeed with Shakespeare's set speeches, quoted earlier, is pertinent to opera: "The problem is how to do justice to the rhetoric and yet to keep it human."

Hamlet (and Shakespeare) acknowledges this dual challenge of the virtuoso rhetoric of a big role when he warns the players: "in the very torrent, tempest, and, as I may say, whirlwind of your passion, you must acquire and beget a temperance that may give it smoothness." Hamlet's image and its implications are fascinating and worth a close look. He seems to grant that "unnatural deeds" are the player's proper *métier:* for he must ride out the "torrent" of larger-than-life passions. But the nub is that a player must become not passion's slave but its artful moderator. *Temper* and *temperance* Shakespeare usually uses to suggest moderation or balance. Hamlet's phrase "modesty of nature" reiterates the point. The player, he says, must "acquire" (by training and practice) and "beget" (deliver, through his own artistic ingenuity) the technique necessary to control his effects. There must be, to paraphrase Hamlet's warning, a calm artistic self-discipline at the center of the passionate whirlwind. Now, *smoothness* is a tantalizingly ambiguous word. What does it really mean? I think it is not too farfetched—given the gist of his other advice—to guess that Hamlet is thinking of "credibility"; or, in Barton's simple phrase, "keeping it human."

Boito began his essay on the *personaggi* for the *Otello* production book (published in 1887) by quoting almost all of Hamlet's advice in a fairly free translation. He introduces it by saying, "All artists of the theater, even the finest ones, ought to have impressed in their memory the following words, written three centuries ago, which are today still the most perfect and most modern lesson in theatrical delivery that has ever been conceived." Boito, clearly, felt this advice suited to Elizabethan drama was also germane on the opera stage. Hamlet's admonition under scrutiny here comes out as follows when translated literally back into English from Boito's Italian: "Always, even when the whirlwind of passion is most violent, you must keep your wits about you and be masters of your selves."* It appears that Boito interpreted Hamlet's advice as I am inclined to do: as an explicit recognition of the Gemini factor.

This advice is most pertinent (and the Gemini factor most insistently present) when a character's passions are torrential, when the conflicting challenges of heroic emotive "roughness" and shrewd technical "smoothness" are greatest. This occurs mainly in the arduous, all-too-famous "Mona Lisa" speeches, arias, and ensemble pieces of the two repertories. Here, in these familiar obstacle courses, a kind of performative "ecstasy" is most likely to occur (*ecstasy* derives from the Greek for "to stand outside"). We have noted it in Nilsson's observation that "Sometimes you must separate the voice from the emotions"; in the occasions Olivero found "another part of me . . . standing beside me and watching, controlling the sound"; and in Bumbry's flat reiteration of Hamlet's warning: "keep your wits about you." John Barton and the actors he worked with in *Playing Shakespeare* advert often, in so many words, to the Gemini factor and the need to embrace, even revel in, a conscious artistic distancing from the character itself. Unsurprisingly, these remarks are usually about how to play speeches and interior performances that are highly operatic. In a chapter aptly titled "Passion and Coolness," for instance, Michael Pennington makes Nilsson's point as he discusses the role of Hamlet; indeed, he describes the way a singer must preserve the vocal line even under the pressure of full emotion: "the language of the play is terrifically wrought and elaborate. And it is possible through an excess of feeling to distort that language instead of working through it." The task for the performer in either case is to discipline the artificially heightened expression, as Pennington puts it, "mentally and technically."

Death scenes, to take another example, are a notorious operatic staple, and Barton's discussion of a few death scenes in Shakespeare echoes what an operatic stage director might tell a tenor rehearsing Edgardo's "Fra poco a me ricovero" at the end of *Lucia,* or a soprano preparing Manon Lescaut's "Sola, perduta, abbandonata." Barton, for example, rejects the idea of doing Hotspur's death speech in *1 Henry IV* ("O Harry, thou hast robb'd me of my youth . . .") as racked with realistic pain, saying very sensibly that "Hotspur's pain is in the mind rather than in his body." He insists the speech is "partly choric"—Barton's term for speeches where "language is more important than character and . . . takes the driving seat." Hotspur's final words must be approached more self-consciously and ecstatically if its "poetic juices" are to be fully released. "Hotspur in part stands outside his own dying," Barton explains, "what he says about thoughts and life and time seems at first blush out of character. But it is exciting how a man who has lived so much in the moment should see clearly beyond that moment." As with virtually every operatic death aria ever written, the final effect, Barton urges, should be, not one of actual pain, but one of the "wonder of dying." Another Shakespearean death scene is merely reported: Quickly's marvelously affecting account of Falstaff's last moments babbling of green fields is manifestly an interior "performance" in *Henry V*. Barton urges that its choric function be fully embraced as such, and not subverted, as it often is, by waves of realistic grief from the good Mistress. "The moral," says Barton (and this is the moral of many an operatic aria), "is that it's sometimes more important to make a text resonate than to be moved oneself."

Interior performances, incidentally, offer a perfect way to draw attention to the Gemini factor, for they require not only the actor but the character himself to indulge in a conscious stepping outside the self. Take Richard II, one of Shakespeare's most persistent play-acting characters. He is always showing off, always demonstrating his virtuoso "poetical" agility. One of his several blank-verse arias is the thirty-three-line speech to Aumerle, Carlisle, and Scroop on "the death of kings" (3.2.144–177). It contains the famous sentence that begins, "within the hollow crown / That rounds the mortal temples of a king / Keeps Death his court. . . ." Barton urges the actor to make this speech, not a realistic effusion of self-pity and grief, but rather an ironic, self-conscious *tour de force*. Richard (*not* the actor) must put the speech in inverted commas, Barton insists: "Say to the audience, 'This is going to be great, telling the story of the death of *kings*' . . . if you go totally into your grief you'll lose the idea of standing outside yourself and yet luxuriating in the situation." Barton is very pleased when the actor Richard Pasco recites the speech in this vein, and the way he expresses his pleasure points directly at the Gemini factor: "You managed both to be inside the character and yet to *stand outside yourself* at the same time. That's the double vision that lies at the heart of irony. It's not easy to do both at once." The same double vision, of course, is called upon during the interior performances in opera.

Though Barton does not make the connection as strenuously as he might, he lays down virtually the same premises for the "outer" performances of actors themselves. The most obvious premise is that heightened verbal (or, let me add, musical) utterance exists to be savored, exploited, and celebrated to the hilt. The actor, like Richard, must be of a mind to luxuriate in the *artistic* as well as the dramatic situation. Barton thinks Richard should utter the first lines of his speech—

> For God's sake let us sit upon the ground
> And tell sad stories of the death of kings.

—as if he were telling his audience, "It's delicious, it's going to be glorious. We're going to have fun." Superlative impersonations in both dramaturgies always display performers who attack the challenging moments of their roles with the same optimistic panache. Barton reiterates his point about savoring artifice when he discusses that most artificial element of Shakespearean verse: rhymes. Should the actor play them or, by imposing naturalistic speech-rhythms, try to pretend they do not exist (rather as a singer might ignore a composer's ornamentation of a melodic line)? Barton is emphatic: "he should play them because they are there in the text. To dodge them is a cop-out and a textual distortion. They need to be relished consciously."

The other premise implicit in Richard's speech—which follows from the priority of relishing heightened utterance—is that the actor must sometimes stand outside his or her character in order to achieve the ideal performance. This becomes necessary when language or notes take the driver's seat and, hence, when emotion must share the stage with technique. Barton takes as a typical example of such an event the speech of Hamlet's mother describing Ophelia's suicide ("There is a willow grows askaunt the brook" 4.7.166–183). This is Gertrude's longest speech in the play, her one verse-aria in which to shine. Like Hotspur's death speech, it is uncharacteristically "poetical" for her and has the distinct feel of being artificially "indented" in the action. "Clearly," says Barton, "the choric function is dominant in the sense that the thing described matters rather more than the feelings of the speaker." Consequently, Barton urges that an actress reciting this speech relish its heightened language rather than attempt to make it, quite inappropriately, merely a conduit for Gertrude's emotional distress at the loss of a promising daughter-in-law. Barton generalizes from this example: "Shakespeare very often requires an actor to show emotions and at the same time to stand outside those emotions. He asks for passion and a degree of detachment, if not coolness, at the same time."

"Choric" events like Gertrude's speech are fundamental to operatic drama, and one obvious source of the detachment mentioned by Barton is also very familiar to the singer. This is preoccupation with technique: clean elocution, salient projection, justness of pronunciation and accent, deftness of phrasing and nuance, all of which, as Michael Pennington has noted, can be so easily distorted by excessively realistic emotion. Most memorable passages in Shakespeare and opera occur when characters become spectacularly articulate under extreme emotional pressure. When a Shakespearean character is found in this situation, Barton concludes, "we can be pretty sure he is to some extent standing outside his own emotions." And with this conclusion Barton rings in Hamlet's advice about temperance and smoothness, which he paraphrases in this way: "an actor needs to be sure when [emotion] predominates and when it's mixed with the intellectual capacity to stand outside and handle those emotions."

All the remarks on acting made by Hamlet, Shakespeare's most intellectual hero, finally come down to intellectual capacity, the capacity to acquire and beget the technical means of handling the huge challenges that our theaters of virtuosity thrust upon their executants. Those who do not possess that capacity are bound to die miserably and unmourned on the slopes of Mount Lear, Mount Othello, or some other well-known thespian Matterhorn. In the theater of virtuosity the Gemini factor is thrown into highest relief, for the audience comes, not only to see a fictional character live out a dream of passion, but also to experience a skilled technician unfeignedly at work. Indeed, with some of the most familiar works, the audience may approach the performance decidedly more intrigued to experience the technician's than the character's stage life. Interest will be focused on how the artist skirts the technical crises lying in wait, like so many pockets of quicksand, between the first and final curtain.

It is a commonplace bit of theatrical cant—like the cant of saying so-and-so "becomes" or "is" Cleopatra or Carmen—to praise an artist by saying he or she makes the difficult appear easy. In the theater of virtuosity, I think, this is rather faint praise. The pleasure of virtuoso performances is rather in recognizing and celebrating the difficulty of the achievement, in sensing the outpouring of technical and caloric energies . . . the sweat and the heaving of the diaphragm as Donna Anna sings "Or sai chi l'onore" or the massive swings of a Lear moving from the rage of "Pour on, I will endure" to the pathetic whimper of "In such a night as this." We want to applaud, as Peter Hall says, the allegory of man's life, "full of effort and then exhaustion," that any heroic performance will present to an audience.

Because technical skill is an ever-present element of performance in our two dramaturgies, we can perhaps arrive at another consequence of the Gemini factor: the Aristotelian impetus to *empathize* passively with the imaginary character's experience is always balanced, in Shakespeare and opera, by the Brechtian encouragement to *sympathize* or collaborate with the performer *qua* performer as he or she traverses a role's technical challenges. An unavoidable double vision will present us with Laughton and Galileo, Callas and Tosca, McCracken and Otello, Jacobi and Richard II, Vickers and Peter Grimes. Shakespeare would surely have objected to Brecht's combative notion of an actor's doing his best to stand between the spectator and the action, but there is still an important emphasis on a special, self-conscious intimacy between performer and audience in his plays.

This same intimacy occurs in opera. A recent description of Elizabethan acting style, in fact, might well do service for operatic acting: "This kind of acting makes particular demands upon the player: he cannot be naturalistic but must do two things, play the character and define the role; that is, comment by gesture or inflection on what he is doing as he does it. In this he could draw upon the audience's knowledge of himself."* The performer must "define the role," of course, by technical means, and it is in fulfilling this part of the dual task that he or she will move toward a special intimacy with the audience. The palpable excitement (when all is going well) of the performer's executive successes and the audience's enthusiastic appreciation is a manifestation of this intimacy. Casca, in *Julius Caesar,* alludes to this volatile and sometimes noisy intimacy between Elizabethan actor and audience when he describes Caesar theatrically refusing the crown before the Roman populace: "If the tag-rag people did not clap him and hiss him, according as he pleas'd and displeas'd them, as they use to do the players in the theatre, I am no true man."

Intimacy leads one to think of foreplay, and it is no wonder that Shakespeare's star roles naturally impel the actor toward the forestage. Nor is it surprising that references throughout theatrical history to this interposition of executant personality between spectator and events on stage should often remind one of the behavior of opera singers. We have Cocke's complaint in 1615 about the player who does not look "directly at his fellow's face" but aims his "voice into the assembly for applause-sake." We have the classical actor in the Kemble-Siddons tradition, who "shows less regard for what is going on on the stage and speaks, almost as from a rostrum, to the spectators." And we have, too, among memorable Shakespearean and operatic performers, the constant awareness of the voice as a personal *instrument* played with technical brilliance. Thus, Peter Brook of John Gielgud: "a magician . . . his tongue, his vocal cords, his feeling of rhythm compose an instrument . . . [that] stands half-way between the music and the hearer."

The technical demands upon the vocal instrument in Shakespeare and opera are in several ways strikingly alike. "Serve God and act clearly," admonished one Elizabethan playwright in a prologue, but when one considers all of these technical preoccupations that must accompany impersonation one begins to recognize how difficult that simple task really is. The most obvious demand is for extraordinary range. Shakespeare's theater, says J. L. Styan, "could permit both the rant and the whisper," and the remarkable vocal compass this implies is suggested, I think, in an amusing observation made by the nineteenth-century actor and playwright, Dion Boucicault: "the great English tragedians before Kean used their treble voice—the tea-pot style. They did it as if they played the flute. Then came the period when the tragedian played his part on the double bass. . . . Now we perform that part in the present age in what is called the medium voice." The implication, I venture, is that there is a call for *all* the registers—treble, middle, and bass—in Shakespeare's most imposing roles. The operatic comparison would be with those roles which either require remarkably different kinds of voices at various points in the music (notably Norma, Leonora in *Trovatore,* and Violetta) or which can be effectively assumed by, say, a lyric or dramatic soprano. The extremities of Shakespeare's vocal demands were exampled in the way Olivier sought out a "violet velvet" timbre for his Othello. Sounding like a singer approaching a role slightly beyond his range, Olivier recalls, "I didn't think that I had the voice for it. But I did go through a long period of vocal training especially for it, to increase the depth of my voice, and I actually managed to attain about six more notes in the bass. I never used to be able to sing below D, but now, after a little exercising, I can get down to A, through all the semi-tones."

As regards range of declamatory style, Wesley Balk insists the complete singing actor be able to "move along the full range of the continuum" between naturalistic and formal or ceremonious delivery. Shakespeare requires exactly the same breadth, for his speech runs the full gamut between realistic dialogue that could have been heard in the street outside the Globe (see, for example, the conversation between two teamsters in *1 Henry IV,* 2.1.1–31) and the gaudiest blank-verse bravura speech. Another obvious demand burdening the actor and singer alike is for painstaking articulation. The actor Paul Scofield has remarked that "the plays of Shakespeare . . . deal with articulate people," and—opera being, as Meredith has said, "An image of articulateness"—operatic executants must likewise be highly articulate: there can be no true eloquence without superior elocution. Harbage has noted that all of history's eminent Shakespeareans "were great *elocutionists*" . . . the word deriving from the Latin for "to speak out." If there were a like word in our vocabulary for "to sing out," we could apply it as categorically to the great singers of history.

The similar tasks of articulation can perhaps be underscored here by drawing attention to a charming little book, *The Art of Singing,* that was translated and published by the lutenist-composer John Dowland when Shakespeare was still active in London. Its eighth chapter opens with an amusing sketch of the "divers fashions" of singing in "divers Nations": "Hence it is that the English doe carroll; the French sing; the Spaniards weepe; the Italians which dwell about the Coasts of Janua caper with their Voyces, the others barke; but the Germans (which I am ashamed to utter) doe howle like Wolves." The author, Andreas Ornithoparcus, then moves on to discuss briefly "the Ten Precepts necessary for every Singer." What is remarkable about several of these precepts is that they virtually echo Hamlet's advice to the players.

His request to "suit the action to the word" is matched by the third precept: "Let every Singer conforme his voyce to the words, that as much as he can he make the *Concent* sad when the words are sad; & merry when they are merry." Hamlet's concern that nothing "come tardy off" finds its equivalent in Ornithoparcus' warning, "Above all things keepe the equalitie of measure. For to sing without law and measure is an offence to God himselfe, who hath made all things well in number, weight, and measure." Shakespeare, in his verse, also made all things well in number, weight, and measure, and performers of this verse must, like singers, obey the "law" of meter, rhythm, and accent or risk committing the most audible gaffes. They thus feel something of the tyranny of "measure" that Kiri Te Kanawa describes. In opera, she says, "the music is always closing in on you. You have to accomplish something in six bars, and they will not be stretched out or shortened. It's inevitable, like a clock ticking."

Hamlet makes his main plea for articulation when he insists the players pronounce their speeches "trippingly on the tongue." Those who "mouth it" bombastically he scorns. Ornithoparcus treats inarticulate singers with the same ridicule. Performances in which "neither one Voyce can be distinguished from another, nor one sillable from another," he says, should "be punished with the severest correction." And there is not a little of Hamlet's sarcasm in this rhetorical question: "Think you that God is pleased with such howling, such noise, such mumbling in which there is no devotion, no expressing of words, no articulating of syllables?" Verdi, incidentally, paraphrased both Hamlet and Ornithoparcus pretty closely when he complained, apropos his demand for a *Falstaff* cast that could sing the words very trippingly: "In general our singers always use a fat tone; they neither have elasticity in their voices, nor clear and easy pronunciation, and they often lack [proper] accentuation and breath." That singers in general must pay careful attention to all these matters is obvious enough, but it may be less obvious that Shakespeareans feel similar obligations. For them, too, the momentary loss of concentration can lead to disaster. "If you forget your lines," says the actress Sheila Hancock, "and substitute a one-syllable word where there should be a two-syllable word, it's like an electric shock. It jars and feels awful."

A third technical preoccupation of the singer and actor is with the effort—always in conflict with clear articulation—of projecting the voice adequately into the auditorium. Since time immemorial, it appears, vocalists have been tempted to assume that brute decibel force rather than technical finesse can carry the day. Thus, Hamlet despises the player whose pleasure it is to "spleet the ears of the groundlings," and Ornithoparcus warns the singer to "take heed, lest he begin too loud braying like an Asse." He even wonders if the Saxons along the Baltic coast "delight in such clamouring" because "they have a deafe God." His conclusion is that "God is not pleased with loud cryes, but with lovely sounds: it is not (saith our Erasmus) the noyse of the lips but the ardent desire of Art, which like the lowdest voice doth pierce Gods eares." At the death of the tenor Gayarré, Shaw had occasion to bring this distinction between decibels and "art" up to date: "Actors and singers who have small voices should remember that the problem for them is to make themselves *heard,* and by no means to make themselves *loud.* Loudness is the worst defect of quality that any voice, large or small, can have." The point of all these dehortations, of course, is that laryngeal over-exertion is bound to subvert clear and natural articulation. Actors who "mouth it," says Hamlet, are as tiresome in the theater as town-criers. And he adds, later on, that whatever is "overdone" will inevitably force actors into such strutting and bellowing that they will have neither "th' accent of Christians nor the gait of Christians." The result: an abominable imitation of humanity. Ornithoparcus is more succinct: "The uncomely gaping of the mouth, and ungracefull motion of the body, is a signe of a mad Singer."

Boito, it might be added, follows Hamlet (and Ornithoparcus) very closely. In the *Otello* production book he advises future sopranos impersonating Desdemona "not to make bug-eyes, not to agitate themselves with their body or their arms, not to walk with strides a yard long, not to seek after so-called *effects. . . .* It ought to be possible to express every sadness and every joy (except in rare cases where horror verges on excess) without contorting the face, making one's eyes bulge, or exaggerating one's accentuation."

To sum up the effect of all this constant, painstaking technical preoccupation upon the performer (and upon the audience, perfectly aware of this preoccupation), one is likely to turn to images of deceptive simplicity: the fly-wheels and gears behind the face of a clock or the intricately engineered infrastructure of a clean-lined skyscraper. The feats of time-keeping sonic engineering that a Shakespearean or operatic actor must perform will result ideally, to borrow Hamlet's word, in a "smooth" effect. But we in the audience cannot help but be aware that very much scrupulously rehearsed, utterly calculated effort—technique, in other words—has produced this effect. What is more, in this theater premised on virtuosity we enjoy this awareness; it heightens our pleasure. We like to feel about a performer as Shaw did about the actor Louis Calvert. Thinking of Calvert's technical acuteness, he praised "his perfect intelligence that finds the nail in every phrase and hits it on the head unerringly," which is also a good recipe for superior opera singing. Our consciousness of and pleasure in the "perfect intelligence" of a performer, quite apart from the action on stage, is but another manifestation of the Gemini factor. This factor recalls to mind Shaw's advice to a would-be Othello to "shout cautiously." The paradox is perfectly suited to the Gemini factor for the simple reason that Othello's great shouts of passion—like those of operatic heroes and heroines—cannot be compassed without cautious technique. We admire the shouting and the caution simultaneously.

Such double awareness is present everywhere in the theater of virtuosity. The author or composer is conscious of his own creative agility; he is conscious, too, of the virtuoso talents of his performers, which he will exploit, and of the audience that will come in expectation of virtuoso text and performance. The player or singer, on the other hand, will be triply conscious of his character's challenging "dream of passion," the role's challenge to his technical skills, and the audience's challenge to him to match his own reputation as an artist. And, finally, the audience sits fully conscious of the disparity between the imagined "real" human actions on stage and artificial bravura feats of highly trained elocutionists and vocalists.

It is easy to see how such theater can collapse into ghastly artistic self-indulgence. Olivier bluntly said that "acting is just one big bag of tricks" (perhaps the most common pejorative definition we have for "technique"), and the theater of virtuosity can, in its dreariest incarnations, seem to be merely a pretext for stars to dip ostentatiously into their bag of tricks. Or into a singer's bag of tricks, which ideally includes the Sutherland trill, the Caballé pianissimo, the Tucker sob, the Bergonzi or Kraus phrase-sculpting, the Schwarzkopf nuance, the Price velvet, and the fill-in-a-name High C. Reduced to absurdity, this thespian self-indulgence will produce what one critic, for example, called in a review of *The Tempest,* "the prima donna approach to Prospero's big speeches as if they were so many arias." The point of such star turns, he adds acidly, is "to give the peak moments their full resonance and gloss over the rest with a smooth veneer." A depressingly high percentage of Shakespeare and opera performances pursue this approach and sink beneath contempt. Approached thus, no great Shakespearean play or opera can escape being viewed, as Thomson viewed *La Gioconda,* as "hokum . . . written for tiptop show-off."

That said, however, we should perhaps draw back a little and honestly admit that hokum properly so-called and tiptop show-off are essential elements here. Indeed, the reviewer of that *Tempest* production, faced with the results of a director's attempt to eschew "rhetoric" and "extravagance," finally ended up thinking that the "prima donna" approach "may be the only way to play it"! Thomson, after all, admits that hokum can be exciting when it is "properly performed," and he goes on to say that *La Gioconda* "is capable of producing shivers no end when the musical execution is a bang-up one." It may even be that Thomson was using the word *hokum* in the surprisingly neutral sense of its first definition in *Webster's:* "a device found to elicit a display of . . . emotion from an audience and therefore deliberately used to impel persons to a desired action."

If so, "hokum" fairly accurately describes those obvious and very calculated "devices" in Shakespeare and opera that impel characters and artists to their respective emotional and executive virtuosities. When performed in a bang-up way, these devices will elicit displays of emotion from an audience. Such tricks hath hokum . . . and superior technique. Thus, it is small wonder that praise for a recent bang-up impersonation at Stratford-on-Avon is couched in exactly the terms one might use to praise a brilliant operatic performance: "John Wood's craggy, passionate, nervy Prospero, reveling in the verse, exploiting, ostentatiously at times, his impressive vocal range, is firmly at the center and in control of the production."

It is worth asking, in conclusion, if there is a sensible way to embrace the inevitability of technique in drama being tied to heroic utterance. One "trick" by which to do so is suggested by the actor Mike Gwilym: "A director often asks an actor to deal with a particularly heightened piece of language by putting it into inverted commas. The danger with that is that it sounds as if the actor is being very self-conscious about what he is saying. The trick is, I think, for the character rather than the actor to put the words into inverted commas. In this way he acknowledges to the audience that the language he is using is not common parlance." The trick is one for the audience, too: it must try to imagine that the heroic utterance derives from the imaginary human experience on stage. The audience must think what it sees is, in John Webster's phrase, "truly done" before it. But of course with blank verse and sung speech we cannot banish our awareness of the inverted commas themselves; our ears must "quote" such utterance as something very far from the commonplace. There is a pretense of deception on both sides of the apron, but nobody is fooled.

John Barton feels the dilemma in rather different terms. "Shakespeare's language should be made to work on an audience as powerfully as an actor's emotions can. A heightened, poetic text has the power to do that, but the playing of emotions can sometimes strangulate that power." Barton hastens to add that he does not mean that the actor "shouldn't have emotions, but that they need to be channeled and controlled like the rest of his performance." He then points to Hamlet as the supreme example of Shakespeare's insistence, in an actor's performance, upon "a mixture of passionate feeling and intense intellectual thought at the same time." The great thespian battle, according to Barton, must take place along "the borderline between passion and coolness" ("coolness" is Barton's synonym for Hamlet's "smoothness").

Passion and coolness: Barton says they are "intense and locked together" in Hamlet. Perhaps this is why Hamlet stands apart from Shakespeare's other tragic figures, a perfect archetype (possessed of his own Gemini factor, which makes him a virtuoso ironist and master of "ambiguous giving out") of the Shakespearean and operatic *performer*. Perhaps this is why Verdi never seriously thought of turning him into an operatic *character* . . . the overwhelmingly engaged protagonist being what he and other composers have tended to prefer. And, finally, perhaps Hamlet may suggest that the ideal Shakespearean and operatic acting ought not be conceived as a diplomatic compromise between the "possessed" and Stanislavskian or Method acting style and the "alienated" Brechtian style, but as a volatile, *yin*-and-*yang* coexistence of both styles.*

Maynard Mack may have been getting at the essence of the Gemini factor when he wrote of Shakespeare's art as a whole in an essay pertinently titled "Engagement and Detachment in Shakespeare's Plays." In this essay Mack sought to apply to Shakespeare this observation made by the art historian Kenneth Clark on Titian: "he could maintain that balance between intense participation and absolute detachment which distinguishes art from other forms of human activity." The application to Shakespeare is utterly plausible, especially in respect of thespian style. That participation and detachment should be seen as equivalents to passion and coolness seems to me natural. And since Shakespeare manifestly evinced both attitudes toward his profession, his plays, and his characters, perhaps the same exciting, razor's-edge balance should be the supreme goal of artists in live performance.*

The risk will always be that a superb technician's detachment may ruin the performance, but with a powerfully engaged artist an ideal balance is possible. Arthur Symons expressed this ideal (and what happens when it is not achieved) when, in a book on Eleonora Duse, he recorded his "impressions" of her great rival, Sarah Bernhardt. Symons' thumbnail sketch of Bernhardt's performance, I think, has much to do with the nature of the finest operatic acting, as Symons' own musical analogy suggests. His sketch is also a fitting last word on the Gemini factor.

> The first thing one notices in her acting, when one is free to watch it coolly, is the way she subordinates effects to effect. . . . Bernhardt's acting reminds me of a musical performance. Her voice is itself an instrument of music, and she plays it as a conductor plays upon an orchestra. . . . The pleasure we get from seeing her as Francesca or as Marguerite Gautier is doubled by that other pleasure, never completely out of our minds, that she is also Sarah Bernhardt. . . . [She] is always the actress as well as the part; when she is at her best, she is both equally, and our consciousness of the one does not disturb our possession of the other. When she is not at her best, we see only the actress, the incomparable craftswoman openly laboring at her work.*

19

The Right Voices

All waits for the right voices;
Where is the practis'd and perfect organ?
WALT WHITMAN

Byron once explained his antipathy to the theater of his time in these words of unconcealed disdain: "Who would condescend to the drudgery of the stage, and enslave himself to the humors, the caprices, the taste or tastelessness, of the age? Besides, one must write for particular actors, have them continually in one's eyes, sacrifice character to the personating of it, cringe to some favorite of the public, neither give him too many or too few lines to spout, think how he would mouth such and such a sentence, look such and such a passion, strut such and such a scene. Who, I say, would submit to this?"

The answer to this question is clear: Shakespeare, most opera composers, and virtually any writer who hopes for success, not in his closet but on the actual public stage. Can it be any surprise that Byron, who did write several plays, had only *Marino Faliero* produced in his lifetime? And can it be surprising that the unanimous opinion of posterity has been that Byron's plays read far better than they act? Indeed, his haughty attitude toward the collaboration between dramatist and actors may explain why the history of English Romantic poets and the stage was a "short, sad" one. As one student of this period has wittily concluded, "the Romantics were ventriloquists rather than chameleons." They refused to condescend to those chameleons—the actors—who alone can give a play life.

Working in theaters of supreme expressive virtuosity, Shakespeare and opera composers could least afford to ignore, as Byron would have liked to have done, the talents and limitations of their actors. The burden of many preceding pages has been to give full prominence to the executive heroism inherent in the two dramaturgies. All is premised ultimately upon the superlative impersonator. Granville-Barker splendidly emphasized this point when he offered an excuse for analyzing in detail a scene from *Antony and Cleopatra:* "this close analysis of every turn in the showing of a character and composing of a scene . . . must go to *giving a play the simple due of its acting.* As a reader he cannot lose by knowing what demands the play's art makes on the actor's. The greater the play, the more manifold the demands! When he sees them fulfilled in the theater his enjoyment will be doubled. If they are not, he will a little know why, and so much the worse for the actor; but, at long last, so much the better."

As with the chicken and the egg, it is idle to debate whether Elizabethan playwriting or Elizabethan play-acting came first. What is clear, though, is that they could not have existed separately. This symbiotic relationship was one of the enviable elements of drama of the period, as Richard Flecknoe suggested in 1664, in one of the very first retrospectives on this golden age: "It was the happiness of the Actors of those Times to have such Poets as [Shakespeare, Jonson, Beaumont, and Fletcher] to instruct them and write for them; and no less of those Poets, to have such docile and excellent Actors to Act their Playes, as a *Field* or *Burbidge.*" Whether the actors of the day were indeed "docile" we shall never know; but excellent they must surely have been. The "happiness" of actors and stage poets, as Flecknoe emphasizes, was mutual. It derived from mutual exploitation.

A similar mutual happiness and exploitation has been central in opera through-out its history, and it will be my purpose in the following pages to explore a little how the two dramaturgies depend, for their essential character and remarkable similarities, upon the same *modus operandi:* an insistence upon "the right voices" and the tailoring of roles to the personality and technical resources of a specific "practis'd and perfect organ." It could scarcely be otherwise in this theater of virtuosity. For here a postulate laid down by Michael Goldman in his theory of drama weighs especially heavily: "The powers of the actor determine the play-wright's art." This theater routinely asks its principals to take risks, go to extremes, and assert their resources beyond normal limits. Finally, though, limits must be ascertained and respected. An author or composer for the actual stage, as opposed to the closeted Byronic one, clearly cannot ask the impossible, and the most natural way to avoid such folly is to tailor executive demands to the abilities of familiar artists.

This is what Shakespeare and the most successful composers have done. They appear to have been expert at working with what they were given—challenging their artists to the utmost but not asking them to destroy themselves or look ridiculous. Pragmatic sanity has been their hallmark. No great creators for the stage have lacked a keen sense of the line that separates the possible from the impossible. Their greatness often resides in the fact that their sense of where the line lies will differ radically from that of their contemporaries. "When someone says, 'It can't be done,'" Verdi wrote to Ricordi, "you can be sure he is an ass." But of course Verdi was constantly explaining to all and sundry what, to *his* mind, was artistically possible and not. Backstage at the Globe, Shakespeare was probably told "it can't be done" on numerous occasions . . . and went his merry way. Still, in all this, the great determining feature of Shakespeare and opera is the executive powers of principle artists who once lived and breathed. So it is worth examining some of the evidence that masterpieces have often been profoundly affected by the performers who were available at their inception.

Shakespeare was the creature of a very special time. Indeed, it is a little frightening to think how small a "window of opportunity" there was in English theater history for a Shakespeare to appear, find a stable working environment, and write the kinds of plays he did. First, the ascent of acting style and commercial production in London from crudity to sophistication occurred practically over-night, in the mid to late 1580s. It is a breathtaking coincidence that this was very probably when Shakespeare arrived in the city. Second, his active London years coincided with a period of considerable popularity for and stability in the acting companies and in the acting profession. During this period the companies became less dependent on wearying provincial tours and could settle comfortably into their own familiar acting-places. Shakespeare was especially fortunate in his association with the Lord Chamberlain's (later the King's) Men, a company that remained stable for the sixteen years or more that he was a sharer, house dramatist, actor, and (presumably) director there. Finally, Shakespeare was able, by coming to London when he did, to cultivate his characteristic playwriting style and bring it to its remarkable apex just before the relatively bare thrust-stage of the Globe Theatre was affected by a revolution in stage technology that commenced in 1604, with the coming of the Stuarts and their elaborate court masques (this revolution is discussed in the following chapter). In sum, if Shakespeare's presence in London had been very much earlier or later than the decades on each side of 1600, his canon would surely have been very different from the one the First Folio now presents to us.

Though we know next to nothing certainly about Shakespeare within his close-knit company, it is plausible to picture a life of great familiarity with his physical surroundings and his acting colleagues. Several parallels from operatic history leap to mind: Handel's association from 1720 to 1728 with the Royal Academy of Music at the King's Theatre, which produced such masterpieces as *Giulio Cesare,* *Rodelinda,* and *Tamerlano,* written to display the talents of Cuzzoni and Senesino; Rossini's spectacular Neapolitan period of 1815 to 1822, when he wrote for the Teatro San Carlo's fine orchestra and such brilliant fioraturists as Isabella Colbran and Giovanni David; Donizetti's two-decades-long association with various Neapolitan theaters; Strauss' long association with the Dresden Opera; and Britten's cultivation of a well-known cadre of English singers (Janet Baker, Peter Pears, Jennifer Vyvyan, John Shirley-Quirk, and others) and his work within the limitations of auditoriums near his home at Aldeburgh. As these composers prepared their scores, contemplating in advance the familiar vocal personalities, agilities, and quiddities of stage presence of their casts, so must Shakespeare have written his plays. "The membership of his company was remarkably stable during the entire span of his professional career," William Ringler has summarized, "so that when he wrote a play he knew ahead of time . . . the individual actors who would perform each major part." John Russell Brown, in *Discovering Shakespeare,* is even more explicit in his re-creation of the ambience backstage at the Globe: "The performers of his play would be men whom he knew very well from daily and creative contact. Their individual rhythms and voices, the pitch, pressures, excitements and tricks of their charac-teristic performances . . . would all be somewhere in his consciousness as he wrote." Brown adds the very plausible speculation that "actors were seldom far from his mind as he worked."

The indications that Shakespeare wrote *to* his actors and his company's resources are ubiquitous and remarkable. All of his plays are constructed so that they can be acted, with some doubling, by no more than a dozen men and a few boys. And perhaps nothing about Shakespeare's plays is more striking than the effect on his dramaturgy of the convention of using boy actors. By any standard of our own day—when rote memorization, unfortunately in great discredit, has all but vanished in schools—the thespian feats required of these boys are astounding. And they were very young. In 1602 one of them, Salomon Pavy, died. Jonson, for whom Pavy had played several roles, wrote an "Epitaph" on the occasion that contains these lines:

> Years he numbered scarce thirteen
> When Fates turned cruel,
> Yet three filled zodiacs had he been
> The stage's jewel.

The notion of a twelve-year-old of uncracked voice (Hamlet notes that the players' boy has grown and worries, "Pray God your voice . . . be not crack'd") credibly handling the well-over-500 lines of a Juliet or a Portia, or even the 300-plus lines of a Viola, Isabella, or Desdemona must leave us full of admiration.

Several facts, however, show that Shakespeare was generally quite careful to minimize the boys' burden and to allow the performances of the adults—capable of greater vocal range, virtuosity, and study—to dominate the stage. Only once, for instance, did he write a play containing more than three female roles of over a hundred lines: *Richard III*. Perhaps this experience, early in his career, taught him never to tempt fate like that again. Fifteen of the plays have only one female role over a hundred lines long, and nine others have but two. And while Shakespeare's female characters come to spectacular life on stage—in the comedies they usually gain the upper hand—this is often achieved by economical eloquence rather than by extensive speeches.

One can see the careful husbanding of the boys' powers in the quite disparate role-lengths of several famous pairs of characters: Petruchio and Katharine in *The Taming of the Shrew* (580 and 220 lines, respectively); Berowne and Rosaline in *Love's Labour's Lost* (620, 180); Benedick and Beatrice (490, 320); Troilus and Cressida (520, 270); Macbeth and his wife (700, 270); Lear and Cordelia (770, 120). It is perhaps the most marvelous of all Shakespeare's stage illusions that his women hold the stage as brilliantly as they do while under several stark disadvantages: the average length of the leading male and female roles in the thirty-five plays is 660 lines and 310 lines, respectively; only two female roles are among the twenty longest in the canon (Rosalind at 750 lines, Cleopatra at 670); and female characters speak *a mere seventeen percent* of the total lines in all the plays.

With a few notable exceptions, we have no knowledge of which actors played which roles in Shakespeare's company. Yet there are some hints that his habit was to conceive roles with specific actors in mind. Enough has already been said of

Burbage: obviously, the playwright's task was to fit out a dramatic action that would give his star and perhaps three or four of his ablest principals, substantial roles in which to shine. A performance without a Burbage was surely tantamount to a *Hamlet* without the Prince; and, as we have already noticed, a large number of his plays and those of his contemporaries put the burden of a performance on a single pair of shoulders.

Shakespeare's attentiveness to type seems evident in the case of Will Kempe. Kempe was the main comedian in the company until about 1600; he apparently had rather a coarse presence and style and was also given to extemporal additions and mugging. Perhaps with Kempe no longer in the company as Shakespeare wrote *Hamlet,* Shakespeare felt he could let his hero insist that "your clowns speak no more than is set down for them." At any rate, we do know that Kempe was the original Dogberry, the bluff, bumbling constable of *Much Ado*. One Richard Cowley was Verges, Dogberry's feckless deputy and straight man. This information is tantalizing.* For one notices in a series of plays written between 1594 and 1600 the recurring feature in the *dramatis personae* of a comic duo made up of a garrulous, self-important (and possibly rotund) party and his retiring (possibly thin) sidekick. Even in the distribution of lines these pairs seem to shadow Dogberry and Verges:

> Dogberry and Verges (202 lines, 30 lines, respectively)
> Costard and Dull in *Love's Labour's Lost* (205, 32)
> Peter and Sampson in *Romeo and Juliet* (63, 38; Peter a known Kempe role)
> Bottom and Quince in *Midsummer Night's Dream* (270, 133)
> Launcelot and Gobbo in *Merchant of Venice* (189, 44)
> Shallow and Silence in *2 Henry IV* (222, 43)

It is plausible to imagine these roles were written for the same pair of actors, and to imagine Shakespeare searching for ways to employ his undynamic but highly comic duo.* If the braggart and his reticent straight man were still working as a team when Shakespeare came to write *Twelfth Night,* it would be pleasant to think of Kempe's Sir Toby Belch to Cowley's Sir Andrew Aguecheek as a glorious finale to their collaboration. Toby would thus have been the clown's splendid farewell role for the Chamberlain's Men—just before he left to indulge in the highly successful publicity stunt of dancing the morris from London to Norwich. It is worth noting too that, if these roles were indeed Kempe's, Shakespeare became very clever at isolating his comic turns from the main action of his plays. He was thus able to prevent his clown from, as Hamlet phrases it, setting "some quantity of barren spectators to laugh too, though in the mean time some necessary question of the play be then to be consider'd."

Shortly after Kempe's long-distance caper to Norwich, another clown, Robert Armin, joined the company. Here is another likely example of a series of roles written to type. Armin was apparently more in the chameleonic than the swaggering comic tradition, and he could sing. It is thought that Shakespeare's clowns soon became more witty, graceful, learned, and songful in response to Armin's abilities . . . witness Touchstone in *As You Like It* and Feste in *Twelfth*

Night. Perhaps Armin (as Feste) and Kempe (as Sir Toby) shared the stage on at least one occasion.

Baldwin speculates that Armin took the clown roles, even the very minor ones, in the rest of Shakespeare's works. But Ringler thinks that, after seeing Armin as Touchstone and Feste (both big roles of over 300 lines), the playwright began to exploit his considerably greater thespian dexterity. Ringler focuses on *King Lear* and argues plausibly that Armin did not play the Fool as many have supposed (Ringler urges the familiar notion that Cordelia and the Fool were a doubled role for an experienced boy actor). Rather, Ringler suggests Armin played Edgar.* Edgar's is the second largest role in *Lear,* and he also gets special billing on the quarto title-page: "With the unfortunate life of Edgar, sonne to the Earle of Gloster, and his sullen and assumed humor of Tom of Bedlam." The role requires great chameleonic talent, for the actor must assume six different interior characters during the course of the performance; one of them, poor Tom, requires considerable foolery. Ringler imagines that Armin was the man for this part, which requires an actor "of exceptional versatility, adept at rapid costume changes, able to speak in dialect and to modulate the tone and rhythm of his voice even more subtly, and with the ability to shift from one character to another in an instant in full view of the audience."

If his guess is correct, one is encouraged to think Shakespeare might have created the role of Iago for him a year or two before. Granville-Barker says of Iago: "he has abounding vitality, a glib tongue, and a remarkable faculty of adapting himself to his company." Boito also makes this a central aspect of Iago's character in the *Otello* production book: "One of his arts is the faculty he possesses of changing his aspect according to the persons with which he finds himself, in order to trick them and dominate them the better." These are, of course, hallmarks of the ingratiating comedian who is highly conscious of his various "audiences." It is very tempting to think of Shakespeare's going to the company's prime clown for his Iago. It has, in fact, been my experience that the most successful Iagos exude a quicksilver gaiety.* We know, incidentally, that Burbage and Armin played together in Jonson's *The Alchemist* in 1610 for the King's Men. If they had indeed succeeded as Othello and Iago, perhaps they were well up to the demanding roles of Subtle (the alchemist) and Face (his quick-witted housekeeper). And maybe, a year later, they undertook the even more rancorous duet of Prospero and Caliban in *The Tempest*. If so, we see Shakespeare—with two of his star actors in mind—working several successive and brilliant variations on the tense, often rancorous relationship between master and servant-jester that is a commonplace of so many Elizabethan plays.

As in any repertory company, certain highly versatile actors like Burbage and Armin encouraged the creation of multiple-personality roles . . . or roles in succeeding plays that displayed their talent for portraying utterly different character types. Obviously, other actors were, though admirably skilled, limited by temperament, quality of voice, or stage demeanor to one particular "line," or *Fach*.* While we have no idea which actors settled into them, several obvious types appear regularly in the Shakespearean dramatis personae. One is the graybeard: irascible or congenial, paternal or avuncular. Nearly every play boasts a role of this sort, and Baldwin assigns most of them to John Heminge: Boyet the "old love-monger" of *Love's Labour's Lost*, Egeus in *Midsummer Night's Dream*, the Chief Justice of *2 Henry IV*, Capulet, Polonius, Brabantio, Kent, and Camillo in *Winter's Tale*. The guess is plausible, since Heminge was a relatively old man of thirty-seven when he began his long collaboration with Shakespeare in the early 1590s. One wonders a little if the playwright's affection for his old friend rubbed off on that picture of autumnal goodness and wit, Gonzalo, in *The Tempest*— perhaps his last role for Heminge. A witty epigram from the period promises the company's star what will happen when he dies and stands "at the judgement throne."

> Then fear not, Burbage, heaven's angry rod,
> When thy fellows are angels & old Heminge is God.

Another recurring type in the later plays is the voluble, proud, unsympathetic social outcast. Baldwin assigns several of these roles to John Lowin, who came to the company in about 1603: Falstaff in *Merry Wives*, the con-man Parolles in *All's Well*, treacherous Aufidius in *Coriolanus*, foul-mouthed Apemantus in *Timon*, and the extortionate pick-pocket Autolycus in *Winter's Tale*. The report that appeared many decades later saying Lowin acted Falstaff and Jonson's Morose and Volpone (all figures of social isolation) "with mighty applause" gives some substance to Baldwin's speculation.

A more specialized type was the adult man who was adept at playing superannuated female characters like Margaret, the mannish "she-wolf of France" in *Richard III*, Juliet's Nurse, and Mistress Quickly. And inevitably, toward the bottom of the players' roster, must have come the amiable fellow of small gifts and innocuous stage presence who could be trusted not to run amuck with the Sea Captain's 30 lines in *Twelfth Night*, Peter's 35 lines in *Measure for Measure*, or the Duke's mere 20 lines in *All's Well* . . . all of which roles Baldwin assigns (for no convincing reason) to Shakespeare himself! Finally, there is the typical small female role—like Lady Percy in *1 Henry IV* (60 lines) or Anne Page in *Merry Wives* (30 lines)—in which a promising neophyte boy actor might be allowed to cut his thespian teeth.

In sum, it is hard to avoid feeling that Shakespeare had greatness thrust upon him to a significant extent by the circumstances of his company's personnel and his stage. Hints of this are both very small and large. Among the former one might place the jousting over Helena's "tall personage" and Hermia's "dwarfish" stature in *A Midsummer Night's Dream*, which surely was evoked by the height of the two boys who played these roles. And could the first Sir Andrew Aguecheek or Cassius of the lean and hungry look have been other than slender in profile? We can also see Shakespeare going with the flow in a small detail of *Othello*: he must have had, at one time, a boy in Desdemona's role who could not sing, for in the 1622 quarto the willow-song passage and a later one in which Emilia recalls the song are absent. We might not now have Rossini's and Verdi's splendid willow-songs if Heminge and Condell had not been able to include in the Folio a performing version for a boy able to sing. A broader hint of Shakespeare's facing constraints with his usual common sense is the fact that, in this drama relying on boys-as-women, chaste kisses are occasionally offered but there is only one scene in the entire canon that could conceivably be called a "necking" scene. This is where Falstaff is with the whore Doll Tearsheet in *2 Henry IV;* Shakespeare obviously wanted here the effect of ludicrousness that he was otherwise extremely careful to avoid when sex reared its head on stage between men and boy actors.

One might, finally, guess at Shakespeare's habit of accommodating his dramas to his players from his manifest desire, throughout his career, to do what he could to please his audience. Samuel Johnson's chiding *obiter dictum* that Shakespeare "was much more careful to please than to instruct" cannot be denied. As Harbage summarized in a book aptly entitled *As They Liked It,* Shakespeare "thinks of himself as a caterer, and shows a solicitude for the digestions of his audience." This caterer's instinct is evident in all of the plays, but especially in those that show him working overtime to please. In *Merry Wives,* for instance, he appears to have brought Falstaff back from the dead to delight not only Queen Elizabeth but also an audience that had mightily applauded the fat knight in two earlier plays. The commercial cynicism of the venture—worthy of Hollywood's sequel-itis—is perhaps also to be noted in the playwright's apparent willingness to countenance a cut-to-shreds version of *Merry Wives* (extant in a quarto edition) for "an audience not aristocratic and not primarily intellectual." *Macbeth,* to take another example, has been shown to be a play carefully and elaborately conceived to appeal to the vanity and interests of the new Stuart king just arrived from Scotland, as would befit the first new play from a member of the just-patented *King's* Men company. Many believe, though it has not been proven certainly, that *Troilus and Cressida* shows Shakespeare seeking through much fine oratory, irony, and forensics to please, not the "general" public, but an audience of lawyers at an Inn of Court, which, it is supposed, commissioned the play. If so, he succeeded all too well with this heap of dramatic caviar, for there is no record of a public performance of *Troilus* prior to 1898. Finally, *The Tempest* shows Shakespeare brilliantly seeking to please his courtly and his public audiences, both of which were completely agog in 1610 and 1611 over the spectacular scenic effects introduced in lavish court masques. Indeed, Prospero's remark in the epilogue that his "project" was "to please" can be applied to the dramatist's entire canon.

In these last pages I have sought to emphasize Shakespeare's practical and pragmatic *modus operandi.* He appears to have recognized that the path to commercial *success* (as opposed to that airy "fame" which the foolish King in *Love's Labour's Lost* says "all hunt after in their lives") lay in exploiting his players' abilities to the fullest and catering to the tastes of his audience. He may have existed on a plane well above his more "barren spectators," but he clearly embraced their presence among his audience and found many ways to entertain them. Verdi was of a similar bent—yet another reason to think of him as the most Shakespearean of operatic composers. He scorned rule-mongers in art and insisters upon the theoretically "proper" way of doing things as much as Shakespeare scorned the pedantry displayed by his comic butt Holofernes and the snobbish arrogance paraded by his puritanical Malvolio. A letter Verdi wrote to Ricordi as *Aida* was in progress typifies his stance: "I have substituted a chorus and a *romanza* for Aïda in place of a chorus in four voices worked out in the style of Palestrina. This chorus could have gotten me a 'bravo' from the bigwigs, and with it I could have hoped for a position as a contrapuntalist in any old academy. But I had some scruples about imposing *Palestrina on the harmony* of Egyptian music. . . . After all, it's my destiny! I'll never be a *savant* in music; I'll always be a *hack* [*guastamestiere*]." One can picture Shakespeare—criticized for ignoring the Aristotelian dramatic "unities" and borrowing hackneyed plots from hither and yon ("mouldy tales," Jonson sniffed)—retorting in precisely the words of Verdi's last sentence.

Julian Budden casts Verdi in a thoroughly Shakespearean light, too, when he has occasion to compare him with the most famous operatic savant of all: "He was not an artist like Wagner, who wished to mould the public to his own image. . . . He did not plough his own furrow regardless of obstacles and without stooping to compromise, as some of the rugged individualists of our own century have done. . . . Like Mozart he wrote for the present; he was even ready to make concessions to the taste of a foreign public, as his French versions of *Il Trovatore* and *La Forza del Destino* demonstrate, though he would only carry the resulting alterations into the definitive scores if they satisfied his artistic conscience. In other words, Verdi's creations are the outcome of a dialogue between composer and public."* They are also, as in Shakespeare, the outcome of a dialogue between the dramatist and the personality and resources of his executants. Verdi, in writing "for the present," wrote with present (and ever more carefully chosen) voices in mind. For he well recognized—as Hofmannsthal was to remind Strauss—that one of the crucial necessities in a theater premised on virtuosity was "good parts." And good parts required tailoring, not to an all-purpose dummy, but to individual artistic dimensions. This awareness led Verdi, as soon as his reputation allowed him to do so, to assure his control of casting contractually. It also led him to make changes to accommodate individual vocal abilities, often up until the final dress rehearsal.

Verdi's habits are well documented in his preparations for his first Shakespearean opera, *Macbeth*. His Burbage was the baritone Felice Varesi, and from the project's outset Verdi made him a *sine qua non:* "Varesi is the only artist in Italy today who is able to do the part I have in mind, both because of his style of singing and his feeling—even because of his appearance. All other artists, even those better than he, couldn't do that part for me as I'd like." A week later he was writing directly to *"Carissimo"* Varesi, "Do you want to come to Florence during Lent? If you do, I'll write *Macbeth* for you!" Amid the uncertainty, Verdi was poised to launch into one of two possible projects: he was waiting solely for the right voices. His pupil Muzio wrote to Verdi's father-in-law and patron, Antonio Barezzi: "Either there'll be [the tenor] Fraschini and then he'll do *Masnadieri,* or there won't be, and then he'll do *Macbeth* with Varesi."

Varesi was finally engaged, and the result proved historic. For, as Budden writes, this was "the first time Verdi attempted the portrayal of an entirely individual character. This was possible only with the collaboration of the singer concerned . . . at a very early stage. . . . Verdi sent him the music number by number in draft form, with detailed instructions for its interpretation and occasionally asking his advice about scoring." The composer was just as attentive to his Lady Macbeth, Marianna Barbieri-Nini. He could be firm with her on large dramatic matters ("I saw how much you would have wanted a cantabile of the type in *Fausta*. But observe well the character of this role and you will see that this could not be done without betraying it, and declaring open warfare on good sense"), but in more technical local matters of execution he could be very accommodating. In the same letter, just weeks before the premiere, he promises the soprano "a *cabaletta*, which I shall write for you in Florence so that it will suit your voice perfectly and be sure to make an effect." In the *brindisi*, Verdi notes, "there is a passage written in two ways; tell me which suits you better." And he points out a chromatic scale in this duet that "must be sung rallentando and end in a pianissimo; if this proves difficult for you, let me know." The passage apparently did make Barbieri-Nini uneasy, for it was eventually replaced. In an earlier letter accompanying the second-act finale Verdi discusses ornamentation, adding "I do not well recall whether you trill with ease; I have included a trill, but if necessary it can be taken out at once." Happily, it appears the soprano had a trill.

Broadcast throughout Verdi's correspondence are eloquent indications, large and small, that he could scarcely function but with a satisfactory and familiar cast firmly in mind. The most explicit reason for the demise of his long-cherished *Lear* opera, for instance, was failure to secure a suitable cast. At the top of his list of demands to the Teatro San Carlo management, in a letter of 22 April 1856, was "a company of singers I am satisfied with." And he is very specific: a splendid baritone for the title role; a leading soprano "not dramatic but highly affecting"; two good secondary artists; a very good contralto; a dramatic tenor with a good voice for a smaller role. The letters of the following months make droll if frustrating reading today. On 11 November: "The company selected is still not adequate to perform *Re Lear*. . . . Penco, although she is an excellent artist, could not play Cordelia the way I see her." For Mica (i.e., the Fool), only the contralto Giuseppina Brambilla would suit Verdi. On 7 December: "It is not my custom to allow artists to be foisted on me, not even if Malibran were to come back into the world. Not all the money on earth would make me relinquish this principle." On 15 January 1857: "It's impossible to compose *Re Lear* for a contralto and a mezzo I don't much care for, a soprano not suitable to the subject. For this opera I would have to reduce myself to having [the baritone Filippo] Coletti alone. It's paltry!!" On 17 June: "With the exception of Coletti, absolutely no one in the cast is set"; and Verdi warns, "Believe me, it would be a great mistake to risk *Re Lear* with a cast of singers who are, though very fine, not so to speak stamped with their parts." Finally, on 13 October, Verdi agrees to start a year later, "provided you have the right singers, which you know is imperative." And so, though a complete libretto had been hammered out by Verdi and Antonio Somma, *Re Lear* began its long languishing on the shelf.

La Forza del Destino, whose dramaturgy has been likened to Shakespeare's, also shows Verdi tailoring his musical and dramatic characterization to specific singers. Indeed, one imagines many a juicy Shakespearean comprimario role issuing from the kind of artistic intimacy that existed between Verdi and the baritone Achille De Bassini, whom the composer had set his heart on for the *buffo* role of Fra Melitone. To De Bassini—who had created the *primo baritono* roles in *Foscari, Corsaro,* and *Luisa Miller* over a decade earlier—Verdi wrote: "I have a part for you, if you would be willing to accept it—comic, very charming—it's that of Fra Melitone. It will fit you to a T, and I've almost identified it with you personally." Verdi thinks better of this and quickly adds, "Not that you are a buffoon, but you have a certain humorous vein which squares perfectly with the character that I've intended for you." A Shakespearean practicality is also to be noted in Verdi's writing the *cabaletta* "S'incontro la morte" in the excruciating key of C major because the tenor of the St. Petersburg premiere, Enrico Tamberlik, was equal to the tessitura. But at the Madrid premiere, when his old colleague Gaetano Fraschini (the creator of five Verdi tenor roles) sang Alvaro, he transposed it to B-flat and lightened the orchestration for him. He then wrote to his publisher, saying he had done this "since no one will be able to perform what was written for Tamberlik. Please put [these changes] in all the scores which you hire out." And thus it stands in the published score.

Many of the most difficult notes, passages, and arias in the repertory exist now because composers have had such special talents like Tamberlik to work with . . . and because they did not have the compromising second thoughts that Verdi had on this occasion. On another occasion—the 1869 premiere of the revised *Forza*—Verdi warned that no transpositions would be countenanced, but in the end he relented and moved Carlo's double aria in act 3 down from F to E for the baritone's comfort and Preziosilla's "È bella la guerra" up from B-flat to B. Shakespeare's habit was to lavish care on his very colorful comprimario characters, and it was in fact with the casting of *Forza*'s peripheral roles that Verdi was most concerned. Apropos the 1869 premiere he wrote to Ricordi, "Don't forget that in *Forza* you need three artists who are completely at ease on stage to do Preziosilla, Melitone, and Trabuco. Their scenes are comedy, pure comedy. [How reminiscent of Shakespeare's comic interludes for Kempe and Armin!] Therefore good diction and an easy stage manner. See to that."*

A last example of Verdi's writing to specific talents concerns one of his most delightful Shakespearean characterizations, Mistress Quickly in *Falstaff*. Verdi treasured the role and was distressed to learn that a mezzo with fine agility but meager stage presence—Guerrina Fabbri—was contemplated for her. "La Fabbri with her fine voice could be successful with melodies based on agility, as in *Cenerentola,* etc. But the part of Quickly is another matter. You need singing and acting, much ease on stage, and the proper stress on the main syllable. She hasn't these qualities, and so we run the risk of sacrificing a part which is the most colorful and original of the four [women]." And he adds, trying to make himself utterly clear, "this is a comedy: music, notes, and words; no cantabiles; mobility on stage and plenty of energy."*

Then Giuseppina Pasqua came on the scene. She had made her debut in 1869 as the page Oscar in *Un Ballo in Maschera* at the incredible age of fourteen, and had gone on to distinguish herself as Azucena, Amneris, and as Eboli in the 1884 premiere of the four-act *Don Carlo*. After meeting with her, Verdi wrote, in delight, "she's intelligent and she understands what it's all about; she will be happy to do this part and should do it well." Verdi must have been very pleased with what he saw and heard during Pasqua's visit to Sant' Agata, for he adds, "I myself have noticed that at certain points . . . Quickly is on stage too long without having anything to say, and I think that we could take the odd word or sentence here and there from Alice and Meg and give them to Quickly without spoiling the comedy." James Hepokoski has studied the alterations Verdi made with Pasqua in mind—among them a new high G, G-flat, and F—and concludes that she exercised "a profound influence on the composition of Quickly's part." Perhaps we can permit ourselves to imagine Shakespeare at his writing desk, thinking of Kempe or Armin as he laid out a comic scene, when we read Budden's conclusion about Pasqua: "So it seems that even in an opera written uncompromisingly to please no one but himself, Verdi was not above making concessions to the qualities of a great executive artist; and we may be sure that it was to Pasqua's prowess as a comedienne that we owe Quickly's substantial solo at the start of act 2, scene 2."

A general similarity between the Shakespearean and Verdian *dramatis personae* is worth commenting on here, for it underscores their characteristic writing to specific talents. I have noted earlier that roles of a similar cast are discernible in several plays from the same time, the inference being that Shakespeare wrote them for a specific actor. In Verdi's case, of course, no inference of a similar habit of writing for specific voices is necessary, since we know all their identities. Though we can now only guess how most of them sounded in real life, we can still safely assume, for instance, that the tenor role's similar cantilena and range in *Alzira, Il Corsaro, La Battaglia di Legnano, Stiffelio,* and *Ballo* owe something to the fact that Fraschini was the tenor at all of their premieres. The rip-roaring soprano vocalism, on the other hand, of some of the operas of the 1840s—*Ernani, Corsaro, Lombardi,* and *Macbeth* among them—must say something of the virtuoso gifts of Sofia Loewe and Barbieri-Nini. And the keen intelligence and potent upper register of Felice Varesi must answer to a considerable degree for the similar vocal personality of the three great roles he created: Macbeth, Rigoletto, and the elder Germont. Similarly, the development of the baritone category in general should be seen as a tribute to Verdi's favorites of his early years: De Bassini, Coletti, and Filippo Colini (they created nine primo baritono roles for him).

In 1916 Strauss wrote Hofmannsthal concerning *Ariadne auf Naxos,* informing him that "I shall give the part of the Composer to Mlle. Artôt, since the tenors [in Vienna] are so terrible." Lola Artôt had been Strauss's first Berlin Octavian, and he knew the role, as it was, would not be meaty enough to attract her. So he urged his librettist to consider how "we might further furnish the part for her with, say, a little vocal number, or perhaps you could write an additional pretty little solo scene for the Composer at the end . . . wistfully poetical." He adds at the end, "Please let me have this soon: because I can only win Mlle. Artôt for our piece if I can offer her a kind of small star part." Hofmannsthal was utterly aghast and shot back, "I fear your opportunism in theatrical matters has in this case thoroughly led you down the garden path. . . . [I]n Berlin they happen to have Mlle. Artôt, but who is to sing the part elsewhere, in Vienna, for instance? And what is more: if you do adopt this irrational idea of giving the part to a woman, you must not, for heaven's sake, cut the part *to fit the performer.*"

Strauss hastened to calm Hofmannsthal. He stood firm, however, and emphasized his practical considerations in typically Straussian (and, I dare say, Shakespearean) fashion. "A tenor is impossible . . . because he would cost the management too much, and the part would be too small for him. A leading baritone won't sing the Composer: so what is left to me except the only genre of singer not yet represented in *Ariadne,* my Rofrano [i.e., Octavian, a mezzo travesty role], for whom an intelligent female singer is available anywhere: Artôt in Berlin, Sanden in Leipzig, Schoder in Vienna, Krueger in Munich, and a score of others. As a rule this is the most talented woman singer in a theater, who will look forward to the 'little cabinet' part and will make something of it." As his final shot, Strauss points out with wry humor, "Don't forget that the best singers aren't available for the operas of living German composers, or only in exceptional cases, but are kept for Verdi, Meyerbeer and Flotow! So we stick to Artôt, and it's got to be a delightful part! That's final!" We can be grateful now for the composer's "opportunism," for this "small star part" as it finally emerged is, as he predicted, a delight. The "little vocal number" that Strauss wanted and finally put near the end of *Ariadne*'s prologue often brings down the house. Lotte Lehmann, as it turned out, was the Composer at the Vienna premiere, while Artôt introduced the role to Berlin.

I draw attention to this anecdote by way of observing that throughout operatic history the opportunism of Strauss has fairly regularly triumphed over Hofmannsthal's horror at fitting a role to a specific performer. The kind of opportunistic exploitation that led Shakespeare to chance writing a 750-line role—Rosalind in *As You Like It*—for an extremely promising boy actor or the 1,100-plus lines of Iago for one of his skilled comic players is to be found everywhere in the operatic repertory: heroic abilities excited the creation of heroic tasks.

Mozart's canon is full of examples of this. Many a bass today could easily sing the role of Osmin in *Die Entführung aus dem Serail* if only Mozart had thought practically of who would be singing it in Geneva, Milan, and Santa Fe in later centuries. But in fact he was writing for the foremost German bass of the day, Ludwig Fischer, an excellent actor who was capable of extremely wide leaps, excellent coloratura, and sepulchral low notes. So Mozart saw to it that Osmin's role was much enlarged from his source-story and created a virtuoso spree for Fischer, including a low D that, nowadays, must often be imagined by an audience. Mozart was so taken by Fischer that he even contemplated rewriting the title role of *Idomeneo* for him. His opportunism also extended to the soprano engaged for *Entführung:* "I have sacrificed Konstanze's aria a little," he wrote, "to the flexible throat of Mlle. Cavalieri." The loveliest number in *Figaro,* "Deh vieni, non tardar," was written expressly for one of Mozart's favorite sopranos, Nancy Storace. The first Don Giovanni, Luigi Bassi, insisted on a substantial new solo, and Mozart obliged by writing—several times recasting—"Deh! vieni alla finestra." He was also familiar with all six members of his first cast for *Così fan tutte:* among them, the Fiordiligi was Da Ponte's current mistress, the Despina had been his first Cherubino, and the Guglielmo had been his first Figaro. Their roles were surely tailor-made. And *Die Zauberflöte* must have been just as carefully tailored to its premiere cast, for Mozart's sister-in-law was the Queen of the Night, the Pamina had been his first Barbarina five years earlier, and his Sarastro was familiar as the foremost bass at the Theater auf der Wieden (Mozart had already written a concert aria for him). Mozart actually lived with Emanuel Schikaneder—his librettist and Papageno—while working on the opera. Finally, it is worth noting that Mozart took obvious advantage of his Tamino, Benedikt Schack, who was also an expert flutist.

Other major opera composers have also shown a keen attentiveness to "the practis'd and perfect organ"—Rossini, for instance. Philip Gossett tells us that Rossini's finest operatic writing came from his settled years in Naples, surrounded by familiar singers: "In his Neapolitan works Rossini rarely failed to exploit the characteristic strengths of these voices." And the tenor Chris Merritt, who has performed seventeen Rossini roles, summarizes that the composer was "extremely intuitive about the specific voices for which he wrote. He always tried to wrap the music around the voice of the individual singer. If you study all the parts that he wrote for the same singer, you begin to get a second sense of what that particular voice was like. You can see all sorts of patterns, such as parts of the voice he favored and those he avoided, the kinds of ornament in which that singer excelled."

An exchange of letters between Vincenzo Bellini and the tenor Domenico Donzelli shows how carefully he tailored his music to his principals. Contracted to undertake Pollione in *Norma,* Donzelli wrote from Paris in detail about his abilities: "The extension of my voice is almost two octaves; that is, from the bass D up to the high C. Chest tones, then, up to the G; and it is in that range that I can declaim with equal strength and sustain all the force of the declamation. From the upper G to the high C, I can avail myself of a falsetto which, used with artistry and strength, is a resource for ornamentation. I have sufficient agility, but I find it very much easier when descending than when ascending. . . ." Bellini replied cordially, "You have merely anticipated my wish to write you in order to learn all the precise details that you have given me about the nature and flexibility of your voice. . . . I shall rack my brain to make you as pleased with me as I am with you." Bellini was just as eager to please his Norma, Giuditta Pasta, even offering to change "Casta diva" because she found it "ill-adapted" to her abilities, if, after a week of living with it, she still felt this way. Fortunately, her scruples eventually gave way.

William Ashbrook and Julian Budden conclude that Gaetano Donizetti was "particularly responsive to the individual qualities of the singers for whom he wrote." Thus, the mark of the highly skilled, florid singer Tacchinardi-Persiani was left on several operas written for her, among them *Lucia di Lammermoor.* The more powerful declamatory style of Pasta and Ronzi de Begnis, on the other hand, obviously influenced the vocal character of the prima donna roles in Donizetti's *Anna Bolena, Maria Stuarda,* and *Roberto Devereux.* For a mezzo with very little agility at all, Rosine Stoltz, Donizetti relinquished ornamentation in favor of long-lined simplicity as he composed the role of Léonore in *La Favorite.* And the birth of the baritone as a distinct vocal category may be discerned in the several later operas Donizetti wrote for Giorgio Ronconi, who, incidentally, created Verdi's first great baritone role, Nabucco, and became a favorite of the composer. More recent composers whose attentions to individual singers have been notable of course include Strauss (with Lehmann) and Britten (with Janet Baker and Peter Pears).

By way of conclusion I want to recall Granville-Barker's remark that Elizabethan drama was "built upon vigor and beauty of speech." He goes on to say, "We may suppose that at its best the mere speaking of the plays was a brilliant thing, comparable to *bel canto*." It is for this reason that justice to Shakespeare's art must be, for the producer and director, above all a matter of seeking out Whitman's perfect and practiced organs for the plays' brilliant *bel canto* roles. Unfortunately, *bel canto* is a displeasing notion to some people. Some years ago, the composer Ned Rorem had occasion to define *bel canto*. His definition is far from friendly, but it will help me emphasize the central point of the present chapter: "*Bel canto* is to opera what pole-vaulting is to ballet, the glorification of a performer's prowess and not a creator's imagination."

One is tempted to pause and give pole-vaulting its due. Both pole-vaulting and ballet are utterly gratuitous forms of bodily transport; the vault and an *entrechat à six* bear roughly the same relationship to the simple act of hopping. One might also suggest that, as the ultimate *grand jeté* of the sporting world, the pole-vault entails something of ballet's heroic effort and can surely be accounted beautiful when superbly performed. But the important point is that in our two dramaturgies of virtuosic utterance there can be no antithesis, such as Rorem's insult to *bel canto* assumes, between the performer's prowess and the creator's imagination. If there is to be true glory—as opposed to mere glorification—it must come simultaneously, like (to return to the image once again) Falstaff and Ford exiting together *sotto braccio* from Verdi's Garter Inn. I am also tempted to add that any view of Shakespearean or operatic performance that, as Rorem would appear to desire, seeks to minimize the element of heroic execution seems to me quixotic. The "performer's prowess" is an ineradicable part of both dramaturgies.

In this I am reminded of J. F. E. Halévy's grand opera in a prologue and three acts based on *The Tempest,* to a libretto by Scribe. Curiously, this *La Tempesta* was commissioned by an Englishman, Benjamin Lumley, for Her Majesty's Theatre in London. The cast was stunning. Prospero was Filippo Coletti, who had created Francesco opposite Jenny Lind in *I Masnadieri* and who was to have been Verdi's ill-fated Lear. Miranda was the famous soprano Henriette Sontag; Ariel, the eminent dancer Carlotta Grisi (cousin of Giuditta and Giulia). But the *pièce de résistance* was clearly Luigi Lablache, reigning bass of the day, as Caliban. There is much to be witty about at what happens to poor Shakespeare here. The storm scene ends with a choral *preghiera*. The "uninhabited island" is overrun with "sylphides," doubtless for choreographic purposes. There is no masque, and most of the great speeches, including the "revels" speech, are nowhere to be found. Obviously, the centerpiece was Lablache, for Prospero recedes drastically in favor of Caliban. The opera ends, not with the mage dominating the stage, but with Caliban singing a bravura aria called the "Miranda *variazioni*."

Still, one should very much like to have been present to hear the practiced and perfect voice of Lablache. For Lumley reports a "tremendous success" at the premiere in 1850 and says that Lablache's performance became the "town talk": "It was the last, as it was the best 'creation' of this rare artist." Perhaps, one wonders, *The Tempest* might have been a very different play if Burbage had, in a whimsy, decided he wanted to play Shakespeare's "salvage and deformed slave" instead of Prospero.

20

Staging Shakespeare and Opera;
or, Big Tiny Little

Whatʼs in a name? When I was growing up, the Lawrence Welk Show was a constitutional part of the family scene. Filial duty and a desire for Saturday dinner—without Welk on the television it was unlikely to be served—behooved me to watch it. My favorite part of the program came to be the ragtime piano numbers. I assume my abiding interest in ragtime began with this exposure—though I know now that Scott Joplin and his colleagues would have been horrified at the coloratura additions and hell-bent tempos ragtime suffered on the program. My favorite Welk ragtimer was a large, exuberant man billed as Big Tiny Little. His full name, it turns out, enhances the drollery: Big Tiny Little, Junior.

When I begin to think about how Shakespeare and opera should ideally be staged, I find Mr. Littleʼs paradoxical name often comes to mind. It also frequently comes to mind as I leave a performance. Sometimes I am left mulling over his name simply by an awareness of the strange business of shoe-horning very big traumas, masses of humanity, even very big edifices like *Tosca*ʼs Castel Sant' Angelo into (relatively) tiny little stages. This paradox was particularly keen in Shakespeareʼs mind when he wrote his epic-size *Henry V*. In its prologue he wondered out loud how he could possibly "cram / Within this wooden O" all the "vasty fields" of France and all the armor that would "affright the air at Agincourt." His solution was—as it must be in any theater where technology, funds, and, hence, scenic effects are limited—to ask the audience to "piece out our imperfections with your thoughts."

Sometimes, on the other hand, I am left contemplating Big Tiny Littleʼs paradox in reverse: for example, when I experience a big, eye-popping production that dwarfs an essentially intimate or introspective human drama. I recall, for example, a recent production of Massenetʼs *Manon*, with the two fugitive loversʼ cozy Parisian apartment roughly the size of a jai alai court. With such scenic elephantiasis Shaw gradually lost patience: "There was a time when I did not greatly mind seeing Violetta laid up with consumption in a small bed in the middle of an apartment rather larger than Trafalgar Square." One becomes particularly mindful of the uneasy cohabitation of the Big and the Little when one attends a play or opera that consists primarily of intimate scenes but has one crowd scene that somehow causes everything else to inflate to its level of grandiosity. In this distractingly lavish visual environment, even the bigger-than-life characters of Shakespeare and opera can seem tiny indeed. One is apt to leave the theater, in the familiar phrase, humming the scenery.

Shakespeare and opera, we observed very early on, present a fundamentally *aural* theatrical experience. The prologue in *Henry V,* it will be recalled, closes by conjuring the audience, "Gently to *hear,* kindly to judge, our play." This shared aurality gives us perhaps the most obvious and most compelling explanation for the fact that the two dramaturgies (and their executants) suffer in a remarkably similar fashion when they are subjected to lavish production values. When, that is, the effort to disguise the acting platform as a fictional place is intensified *à la* David Belasco at the beginning of the century and Franco Zeffirelli toward the end of it. Or when, in yet other words (those of a Renaissance architect), "the *Royaltie* of *Sight*" begins to dominate the theatrical event in the form of careful use of geometrical perspective, scene-painting, historically accurate costumery, elaborate props, and verisimilar lighting effects. The history of Shakespeare and opera subverted by "big" productions is curiously intertwined and, in fact, began when Shakespeare was still writing his plays. Certain highlights of this long history are worth considering, for it is a history whose lessons, it seems, are periodically forgotten and must be periodically relearned.

We can, in retrospect, be grateful that Shakespeare's active London years did not come ten years later than they did. We should also be grateful that Queen Elizabeth's death and James' ascent to the throne did not occur several years earlier than 1603. For in either event he would probably have written his greatest works in a style quite different from that with which we now associate him. As it is, Shakespeare is thought of principally as an "Elizabethan" dramatist and, therefore, as a writer for a mostly bare platform-stage thrust out into a smallish "wooden O." Though not as severely stylized as Greek and Kabuki theater with their elevator shoes and masks (but sharing the convention of male actors' playing female roles), Elizabethan theater was also primarily ceremonious and presenta-tional ("stylized"), rather than naturalistic and representational ("realistic"). Illusionism was an honor it dreamed not of. Reflexive self-consciousness was the norm for author, actor, and audience alike. Incapable of the merest effects of real life—like that most obvious and oft-needed one, the darkness of night—Shakespeare depended emphatically upon the audience's ability to "piece out" the obviously stagey "imperfections" of the physical performance through its powers of imagination. Implicit in all of his plays are the numerous explicit pleas made by his very talkative presenter, Gower, in *Pericles:*

In your imagination hold   . . . never could I so convey,
This stage a ship . . .   Unless your thoughts went on my way . . .

Imagine Pericles arriv'd at Tyre . . .   In your supposing once more put your sight . . .

More to the point, not merely for *A Midsummer Night's Dream* but for all plays in the canon, is Puck's simple, "Think but this, and all is mended."

Obliged to give "local habitation" only a *name* rather than a physical reality, and able to travel lightly with baggage mainly of costumes and a few multiple-use props, Shakespeare was able to conceive of his actions without intermission and with the very minimum of dead time spent moving those props. Simultaneous exits and entrances from opposite sides of the stage—allowing no time for the eye to focus upon vacancy—was his standard procedure. Quarto editions of Elizabethan plays, which probably reflect actual acting versions, routinely give no act or scene divisions, and even in the Folio six of the plays—after an initial *Actus primus. Scœna prima*—also fail to make any divisions. Relentless speed of performance was clearly of the essence for several practical reasons, not least being winter's limited daylight hours after 2 P.M. and a reluctance to impose too much upon the large part of the audience standing in the pit. And on several occasions one is inclined to think that Shakespeare cynically but rightly expected the haste of his action to conceal loose plot-threads and inconsistencies.

The great demand upon the Elizabethan audience was, rather, aural and mental: to comprehend as much as possible from the breathless pace of the blank verse and let it work to the fullest on the imagination. It has already been suggested that in order to properly stage the plays of Shakespeare, Marlowe, and other Elizabethans, we "must have faith in the audience's flexibility of mind." This is simply because these dramatists had this faith themselves. Flexibility of mind is the supreme premise of this theatrical golden age. It is by now a commonplace of praise for Shakespeare in particular that, unable to rely complacently upon the illusionism and spectacle of later stage technologies, he created his fictions upon a purely verbal-poetic plane. Hence all those evocative intimations of night-time scattered through these plays that were performed in broad daylight.

Granville-Barker, to take another astonishing example of this verbal power, eloquently describes how Shakespeare turned to advantage the most notoriously constricting of all the conventions he inherited, the impersonation of women by boys:

> Feminine charm . . . was a medium denied him. So his men and women encounter upon a plane where their relation is made rarer and intenser by poetry, or enfranchised in a humor which surpasses more primitive love-making. And thus, perhaps, he was helped to discover that the true stuff of tragedy and of the liveliest comedy lies beyond the sensual bounds. . . . Much could be said for the restoring of the celibate stage!*

And Granville-Barker perhaps goes some way toward explaining why, in Shakespeare, actresses notably lacking in conventional feminine beauty or charm can nevertheless evoke extraordinary beauty in performance. He observes of the typical Shakespearean heroine and the "usurping" actress who now assumes her role: "Shakespeare . . . has left no blank space for her to fill with her charm. He asks instead for self-forgetful clarity of perception, and for a sensitive, spirited, athletic beauty of speech and conduct, which will leave prettiness and its lures at a loss, and the crudities of more Circean appeal looking very crude indeed."* Beauty, in other words, must emanate from the vocal cords.

It can be no coincidence, since both dramaturgies operate on high expressive planes, that we make the same demands of an operatic heroine: no merely physical beauty will suffice if the singer's "conceptual" and artistic beauty is lacking. Antonio Ghislanzoni, the librettist for *Aida*, had occasion to make just Granville-Barker's point when Verdi, in a surprising lapse of theatrical common sense, worried that an unhandsome soprano might render laughable a libretto reference to Aïda as "*bella*." Ghislanzoni wrote to Giulio Ricordi, saying, "See if you can persuade Maestro Verdi to leave Radames' lines 'Morir! si pura e bella, etc.' as they are. Apart from the fact that in the theater all women are beautiful, or at least are made beautiful by musical idealism, I feel that any substitution of words whatever in this place would diminish the effect of the supremely beautiful musical phrase the Maestro has found. Even if we had a monster from Lapland onstage, the public would go into ecstasies." Depend, in other words, upon the artist's execution and the audience's flexibility of mind. Thus it is that many a soprano has been admired, for instance, as Grace Moore was: "She was not beautiful, but she created the illusion of beauty."

Just before the rustics' "Pyramus and Thisby" makes its heavy demands on the mental flexibility of the Athenian court, Theseus makes his famous reference to the "tricks" of "strong imagination." Shakespeare himself clearly depended on such "tricks" to transform the "base and vile" props of his stage and the stage itself into a place of "form and dignity." The crucial point to be made here is that it is a very risky and perhaps even an ultimately futile business to crowd out the audience's imagination with the *fait accompli* imagination of the set designer. Any such attempt to make the acting-area visually represent what the playwright expected to be seen by the mind's eye is bound to result in a kind of double or blurred vision . . . and result in a subversive competition with the verse itself. As many have observed, this is a competition into which even angels might fear to enter. "The poetry of *The Tempest* is so magical," said Shaw, for example, that it will "make the scenery of a modern theater ridiculous," and Bert States has acutely observed, more recently, "If there is one way to make Shakespeare verbose, it is to speak his poetry in a milieu that usurps its descriptive function. There is no dawn that can do justice to Horatio's morn 'in russet mantle clad,' walking 'o'er the dew of yon high eastward hill.'" States goes on to make a related point that emphasizes how beside the point of the main premises of Shakespearean dramaturgy are elaborate production values and *trompe l'œil* realism: "it is interesting to note how little attention Shakespeare's people pay to their given environment; there is, in other words, very little reference, gestural or vocal, to the practical furnishings of the world through which they move. Objects are of little interest to them." How well this observation applies to most characters in opera.

So much, then, for the Shakespeare who was. We can turn now to the Shakespeare who might have been, had his theatrical reign begun several years later, or had

James' reign begun several years earlier than it did. If either of these events had occurred, the playwright might well have reacted differently than he did to the revolution in stage technology—indeed, the most spectacular and complete revolution in all English theatrical history—that took place when James became king in 1603. (This was the same revolution, incidentally, that brought opera to England for the first time.)

This very expensive revolution could not have occurred under Elizabeth. For she was a famous pinch-penny, except in matters of her own clothing and jewelry, and she did not encourage great outlays by her Master of the Revels for courtly entertainments. If her chief courtiers wished to risk bankruptcy entertaining her, of course, she was less inclined to fret. When James and his Queen Anne came down from Scotland to London, however, the several decades of Stuart extrava-gance commenced in swift and awesome fashion, most notably with the opulent court masques so favored by Anne. These masques set the Revels Office back by the then-enormous sums of between £1,000 and £3,000 (court performances of plays were worth a mere £10 to a company like Shakespeare's). Introduced from the wealthier courts of Europe, especially in France and Italy, these masques were performed but one time only, before the gathered aristocracy at Whitehall, usually on a festive occasion such as Twelfth Night. The first such masque, Samuel Daniel's *Vision of the Twelve Goddesses,* was mounted on 8 January 1604; and the next year, on 5 January 1605, *The Masque of Blacknesse* inaugurated a series of collaborations between the greatest English masque designer (England's first major native architect, Inigo Jones) and the greatest English masque poet (Ben Jonson). The King's Men were paid to assist at some of these masques, so it seems likely that Shakespeare was familiar with them at first hand.

How thoroughly these masques must have upstaged the paltry goings-on at the Globe Theatre! It is now impossible for us, so accustomed to the basics of modern stage-furnishing, to conceive of the excitement and astonishment felt by the first English masque audiences (and the less fortunate, who had to read the descrip-tions of them that were hastily printed afterward). For with these masques England was introduced, in one fell swoop, not only to the concept of the proscenium arch and geometrically organized stage perspective, but also to painted and moveable scenery, cloud-cars, *periaktoi* (forerunner of the revolving stage), complex lighting effects (by candles), very rich costumery (worn by noble participants), elaborately choreographed dances, and musical numbers arranged for a large consort. Of the many admiring descriptions of these masques published at the time, one can serve to give a flavor of the eye-stealing delight they must have evoked. Here Jonson speaks of his *Hymenaei,* dating from 1606: "Such was the exquisite performance, as (besides the *pompe, splendor,* or what we may call *apparelling* of such *Presentments*) that alone (had all else beene absent) was of power to surprize with delight, and steale away the *spectators* from themselves. Nor was there wanting . . . *riches,* or strangenesse of the *habites,* delicacie of *daunces,* magnificence of *scene,* or divine rapture of *musique.*" It is worth noting that Jonson uses the word *spectators* rather than *auditors.* His point here is that the masque would have been a great success even in the absence of his own contribution, the plot and poetry ("had all else beene absent . . .").

This dawn of Baroque stage design in England, then, began just a few years past the midpoint of Shakespeare's career; several of his greatest works were yet unwritten: *Macbeth, Antony and Cleopatra, Coriolanus, The Winter's Tale,* and *The Tempest*. These masques, in effect, gave him an excellent foretaste of the dawn of English opera—though the first masque to be performed throughout in what Jonson called *"stilo recitativo"* and musical numbers—*Lovers Made Men*—did not appear until 1617, a year after Shakespeare's death. With the advent of all this stunning technology, we can easily imagine the playwright torn between the stage of his earlier years—the one reflected in Macbeth's farewell (the "poor player" who frets his hour upon a relatively bare stage)—and the more sumptuously decorated stage of the masquing world. This stage was to give Shakespeare his imagery for that last and most poignant of all farewells, Prospero's "revels" speech:

> The cloud-capp'd towers, the gorgeous palaces,
> The solemn temples, the great globe itself,
> Yea, all which it inherit, shall dissolve,
> And like an insubstantial pageant faded
> Leave not a rack behind . . .

It would have gone very much against the Shakespearean grain to hang on sentimentally to a waning theatrical fashion, even one that he had exploited so happily. Being thoroughly pragmatic, fashion-conscious, commercially astute, and eclectic, Shakespeare did what he could to make use of some of the production values that gave the masques their enormous fascination—within, of course, the limitations imposed by his company's primary commitment to its Globe and Blackfriars public rather than now-and-again "command" performances at court.

We can see the results of his efforts to borrow some of the masque's sex appeal in several of his plays written after the masque vogue was in full swing. There is the masque in *Timon of Athens,* the various panoply and dumb shows of *Pericles,* the apparition scene with its descent of Jupiter "in thunder and lightning, sitting upon an eagle" in *Cymbeline,* the dances and *coup de théâtre* of the living statue in *The Winter's Tale,* and the vision (involving music, dance, and mime) and several elaborate ritual events in *Henry VIII*. The culmination of these efforts, of course, is *The Tempest;* it is a masterpiece of compromise between play and masque elements. In some of these later plays Shakespeare obviously altered his usual style and habits to make his accommodations to the more visual aesthetic. *The Tempest* is the second shortest of his plays, surely in part because the several special effects, songs, and dances took up considerable time.

It is tempting to think Shakespeare, having given the new stage technology his best shot with *The Tempest,* decided that this way madness lay, and that 1611—with a quarter-century in the playhouse and three dozen plays behind him—was just the right time to retire to Stratford. And, though Jonson tells us Shakespeare was a man of "gentle expressions," it is also tempting to imagine him casting a few rancorous aspersions at this new technology and its subversive effects upon the

theater of his salad days. If he had been able to see into the future, though, he
certainly would have had ample cause for disgust. For the technology whose
English debut he experienced and which he embraced as best he could on his
commercial stage was, ironically, to become the bane of Shakespearean production for centuries to come. From *The Tempest* onward, the battle between
Shakespeare's aural, poetic riches and the designer's visual panoply was joined.
Increasingly, the latter was triumphant.

Jonson, who could explode with expressions far from gentle, lived long enough
to see the disastrous effects of masque stage technology upon words, poetry,
human passions, and, quite simply, the *meaning* of the action presented. He soon
fell out of love with his colleague Inigo Jones and all his fancy stage-furnishings.
Sometime after 1630, he wrote a bitter "Expostulation with Inigo Jones" that
makes him sound for all the world like certain critics today, who are appalled by
the latest scenery-stuffed stage of a Zeffirelli opera or the latest show on Broadway
that is very thin in book, music, and human interest but filled with spectacular
visual effects. This mordant poem is perhaps the first great whoop of disgust—
and still very pertinent for us today—at theater cynically presented to the eye
rather than to the ear and the mind:

> O Showes! Showes! Mighty Showes!
> The Eloquence of Masques! What need of prose
> Or verse, or sense to express immortal you?
> You are yᵉ Spectacles of State! . . .
> You aske noe more than certaine politique Eyes,
> Eyes that can pierce into yᵉ Misteries
> Of many Colors! read them! & reveale
> Mythology there painted on slit deale!
> Oh, to make Boardes to speake! There is a taske!
> Painting & Carpentry are yᵉ soul of masque!

Jonson calls Jones a purveyor of mere "painted cloth" and "deal boards" and
derides his ambition to be not only costume designer but also the "maker of the
properties," the "music master, fabler too," and the "main dominus-do-/All in
the work!" Several other Jonsonian phrases ring remarkably true today. He speaks
of "This money-get, mechanic age" and attacks those shallow folk who "cry up the
[stage] machine, and the shows!" And in lines relevant to the perils suffered by
Shakespeare on the seventeenth-, eighteenth-, and nineteenth-century stage,
Jonson attacks what we would call "designer" theater. "Design," he says, is "a
specious fine/Term of the architects," but in fact, it "in the practiced truth
destruction is/Of any art" other than the designer's own.

The history of Shakespearean production after the Restoration was, as Jonson
unwittingly prophesied, one of increasing subjection to what Granville-Barker
nicely termed "the fal-lals of illusion" . . . and to increasing co-optation into
operatic modes of presentation. Even before William D'Avenant made his
notorious semi-operatic "improvements" on Shakespeare, Richard Flecknoe
expressed his awareness of the risks courted by producers who are preoccupied
with visual effects. In his "Short Discourse of the English Stage" (1664), he wrote,

"The difference betwixt our Theatres and those of former times" was that the earlier theaters "were but plain and simple, with no other Scenes nor Decorations of the Stage, but only old Tapestry and the Stage strew'd with Rushes, with their Habits [costumes] accordingly, whereas ours now for cost and ornament are arriv'd to the height of Magnificence." Flecknoe then offers his perceptive assessment of the consequences of this trend: "That which makes our Stage the better makes our Playes the worse perhaps, [we] striving now to make them more for sight than hearing, whence that solid joy of the interior is lost." Solid joy of the interior—a happy phrase for what is often lost when a Shakespearean play or opera sinks under the weight of a lavish production.

From a much longer historical perspective, Granville-Barker comes close to suggesting that the decay of the English golden age can be laid upon the "mighty shows" despised by Jonson and criticized by Flecknoe: "The Elizabethan drama made an amazingly quick advance from crudity to an excellence which was often technically most elaborate. The advance . . . may be ascribed to the simplicity of the machinery it employed. That its decadence was precipitated by the influence of the Mask and the shifting of its center of interest from the barer public stage to the candle-lit private [and indoor] theater, where the machinery of the Mask became effective, it would be rash to assert; but the occurrences are suspiciously related." Granville-Barker adds a "postulate" that resonates throughout the history of operatic as well as Shakespearean staging: "Man and machine . . . are false allies in the theater, secretly at odds; and when man gets the worst of it, drama is impoverished; and the struggle, we may add, is perennial. No great drama depends upon pageantry. All great drama tends to concentrate upon character."

The long history of great drama impoverished—made tinier and littler—by big, pageantry-filled productions began during the Restoration. *The Tempest* was an obvious prey, and in 1667 D'Avenant and Dryden brought it forth as *The Enchanted Island* to great success. Samuel Pepys saw it eight times in the next two years. Among its "improvements" were a man "that never saw Woman" to balance Miranda, and a sister for her, too. The first scene offered to view a real "Tempestuous Sea in perpetual agitation," a sinking ship, airborne horrid spirits, and a shower of fire. Young Love appears to have commandeered center stage, for Ferdinand and Miranda, that sweetly insipid pair, are given about a *thousand* new lines! And no doubt the twenty-four violins, "Harpsicals and Theorbos" placed "between the Pit and the Stage" subverted the intimacy and interaction between actor and audience that must have been usual at the Globe Theatre.

And what might Shakespeare, let alone Jonson, have said of the operatic version of *A Midsummer Night's Dream* that appeared in 1692 as *The Fairy Queen*? It offered two and a half hours of Purcellian music (not a single Shakspearean line set to it, however) plus about 1,600 lines of spoken text. So much for two hours' traffic of the stage. Constant Lambert guessed it must have run seven hours, but surely it could not have been compassed in less than four. As for the fate of the play's original lines, 750 were used, 950 were cut, 400 modified, 200 new lines added to be spoken, and 250 added to be sung. But it was clearly the scenery and special effects that starred here. They cost a total of £3,000. Among the stage pictures were a "Prospect of Grottos, Arbors, and delightful Walks," a "Walk of Cypress Trees . . . [with] a very large Fountain, where the Water rises about twelve Foot," and a "Chinese Garden . . . quite different from what we have in this part of the World." Among numerous dances and theatrical coups were these: "Two great Dragons make a Bridge over the River; their bodies form two Arches, through which two Swannes are seen"; "While a Symphony's Playing, the two Swans come swimming . . . [and] turn themselves into Fairies"; "Six Pedestals of *China*-work rise from the Stage . . . in which [are] six *China*-Orange-Trees." At the end, "the Grand Dance begins of Twenty-four Persons." Mighty shows, indeed! Within less than a century the minnow of 1604 had grown into a whale and swallowed Shakespeare whole. It was reported of *The Fairy Queen* that "the Court and Town were wonderfully satisfy'd with it." But the "Expences in setting it out being so great," we learn not too surprisingly, "the Company got very little by it."

From the Restoration onward, "mighty shows" was the password for the production of operas and of Shakespeare's plays, and they shared for the most part equally in the eagerness to exploit new technology. No ingenuity was spared in satisfying the audience's thirst for ever more picturesque scenery and astonishing special effects. John Addison has fun with this scenic mania in his *Spectator* of 6 March 1711, where he reports "great designs on foot for the improvement of the opera" at the Haymarket Theatre. It has been proposed, he says,

> to break down a part of the Wall, and to surprize the Audience with a Party of an hundred Horse, and that there was actually a Project of bringing the *New-River* into the House, to be employed in Jetteaus and Water-works. This Project, as I have since heard, is postponed 'till the Summer-Season, when it is thought the Coolness that proceeds from Fountains and Cascades will be more acceptable and refreshing to People of Quality. In the mean time, to find out a more agreeable Entertainment for the Winter-Season, the opera of *Rinaldo* [by Handel] is filled with Thunder and Lightning, Illuminations, and Fireworks; which the Audience may look upon without catching Cold, and indeed without much Danger of being burnt; for there are several Engines filled with Water, and ready to play at a Minute's Warning, in case any such Accident should happen. However, as I have a very great Friendship for the Owner of this Theater, I hope that he has been wise enough to *insure* his House before he would let this Opera be acted in it.*

The steady elaboration of scenic elements in the eighteenth century, though perhaps never so extravagant as in opera of the period, surely caused Shakespearean performances to become more and more lugubrious and led, in consequence, to ever more drastic condensations. Cibber's *Richard III* of 1700, it will be recalled, reduced the original 3,600 lines to about 2,050.

We may guess from Byron's assertion early in the nineteenth century—"a good acting play should not exceed 1,500 to 1,800" lines—that ever more time was being taken up with scene changes and special effects (the average Elizabethan play was somewhere between 2,500 and 2,800 lines). Evidence comes from all sides that, in this century, production values kept well apace with ever larger commercial stages and halls and with the progress of technology, especially in gaslight, limelight, and finally, electric arc-lighting. A few lonely voices, too, could be heard in the wilderness, regretting gaudy trimming of the stage at the expense of the work itself. Hazlitt attended *A Midsummer Night's Dream* in 1816 and reported acidly: "The spirit was evaporated, the genius was fled; but the spectacle was fine: it was that which saved the play." He then adds a sarcastic little fillip that could apply to many a theatrical production of recent years: "Oh, ye scene-shifters, ye scene-painters, ye manufacturers of moon and stars that give no light, ye musical composers, ye men in the orchestra, fiddlers and trumpeters and players on the double drums and loud bassoon, rejoice! This triumph is yours!" Sir Walter Scott, in an 1819 "Essay on the Drama," drew the obvious conclusion from this state of affairs: "The first inconvenience arises from the great size of the theatres . . . nothing short of mask and buskin can render [actors] distinctly visible to the audience. Show and machinery have therefore usurped the place of tragic poetry."

But the juggernaut lumbered on. In the Victorian period, scene-changes came to average ten or twelve during an evening. "As the nineteenth century wore on, the tendency to more and more elaborate productions posed serious problems to the actor," Michael Booth has summarized, "the very size and coloring of the [stage] picture could obscure and even obliterate his own performance." The Victorian taste for grandiosity was not confined to thousand-voiced Handel choirs, either, as an 1857 production of *Richard III* suggests. Its playbill listed 121 persons and advertised a funeral procession of monks with torches, priests with a golden cross, 59 banner-men, an equally large coronation procession, and an army for Richard of 58 men dressed in historically accurate costumes. For such a spree, of course, only Cibber's drastic condensation could be used. This period of "heavy upholstery and mountainous realism" was continued well into the twentieth century by the likes of Sir Herbert Beerbohm Tree, subsiding somewhat only during the expressionist 1920s. Even now the heavy-upholstery style is fitfully with us. Bernard Levin visited a Zeffirelli *Othello* at Stratford-on-Avon in 1961, and his review almost exactly echoes Hazlitt: "I have never seen so many *objects* on any stage in my life. . . . For the eye, too much; for the ear, too little; for the mind, nothing at all."

The same trend toward productions that dwarfed the principal artists occurred on the nineteenth-century opera stage, most flamboyantly at the Paris Opéra. Berlioz, in his *Evenings with the Orchestra* (1853), satirically exaggerated the menu there in a catalogue of the ways its management sought to keep an increasingly jaded public engaged: "Drastic methods followed in hopes of shaking it into wakefulness: . . . bells, cannon, horses, cardinals under a canopy, emperors covered with gold, queens wearing tiaras, funerals, fêtes, weddings, and again the canopy, always the canopy, the canopy beplumed and splendiferous, borne by four officers as in *Malbrouck,* jugglers, skaters, choirboys, censers, monstrances, crosses, banners, processions, orgies of priests and naked women, the bull Apis, and masses of oxen, screech-owls, bats, the five hundred fiends of hell, and what have you—the rocking of the heavens and the end of the world, interspersed with a few dull cavatinas here and there." This spirit of scene-stuffing was shared by most other opera houses of lesser means, but the Opéra stood supreme as the epitome of the fastuous style. And so it took the brunt of Shaw's wrath in the postscript of a letter he wrote in 1917 to the eminent musicologist Ernest Newman. The point he makes brilliantly here about the "size" of operas and the often disparate "size" of the productions they receive ought to be an obvious one, but it is amazing how often it is ignored by impresarios and producers:

> For a whole century that disgusting overgrowth, the Grand Opera of Paris, has been compelling quite charming composers to try to be too big for their boots, and to turn out dainty operettas and ballets in Handelian-Wagnerian armor. Do Meyerbeer's *Dinorah* on the scale of [Boughton's] *The Immortal Hour* at Glastonbury, or the third act of *Les Huguenots* without cuts at the Birmingham Repertory, or Gounod's *Mireille* on somebody's lawn, and everybody will be enchanted to discover that they are not gigantic obsolescent bores, but just as charming and dramatic in their romantic little way as Auber's *Fra Diavolo*. Such a thing as a really big opera is very rare: what we have had is little operas dressed up like big ones, with intolerably tedious results. Let us now have a period of *soi-disant* big operas treated as little ones, and they will come right as if by magic. We want chamber opera: music is an *intimate* thing.

Verdi's *Otello* might seem to be one of those rare big operas, but in fact the composer was very much of Shaw's mind about it. In a conversation in Paris he anticipated Shaw's point about the Opéra too: "My opera is a drama of passion, not a spectacle; it is almost an intimate drama. I even intend to reduce the size of the Scala stage for the last act. I fear that the Opéra stage will be too vast for *Otello*."*

Several years later *Falstaff* evoked from the composer further thoughts about production values. The fancy preparations contemplated by his publisher, Ricordi, began to worry him, and he wrote a letter worthy of Jonson or Hazlitt: "You talk to me of scenery, of sending painters to London (to do what?), of costumes, of machinery [*macchinismi*], of lighting? In order to do sets for the theater, one must have painters for the theater. Painters who would not have the vanity to value their own technical *bravura* but rather serve the Drama. For the love of God, let's not do as we did for *Otello*. In trying to do things well, it was overdone."* And it is a subtle but splendid instance of Verdi's essential loyalty to the dramaturgy of Shakespeare's pre-masque "early manner" that the mechanical demands of *Falstaff* are minimal and thus worthy of the "wooden O." "As for the mechanical aspect," Verdi wrote to Ricordi, "there's little to do, apart from the scene of the basket, which doesn't present difficulties as long as the backstage hands don't make pointless complications. Regarding lighting effects, there's need only of a bit of darkness in the forest scene, but let it be well understood, a darkness that allows the faces of the actors to be seen." Then Verdi adds a cautionary example to underscore his distaste for *macchinismo* for its own sake. He warns that he does not want "lighting effects like those in the last act of *La Wally:* very beautiful, if you wish, but they completely ruined the entire dramatic effect and ended the opera coldly!"

What cautions can be drawn from this brief chronicle of the temptation to allow "ocular proof" to validate our two dramaturgies supremely intended to create worlds of sound? One can first observe that any zealous attempt to put Shakespeare or opera into environments of *trompe l'œil* realism is likely to seem superfluous: for the terrain their characters traverse is conceived and experienced emotionally, not topographically. One might add, too, that remarkably few Shakespearean plays—*Shrew* and *Merry Wives,* perhaps—and few operas truly cry out for picturesque exertions and busy prop-building. Granados' *Goyescas,* obviously, and perhaps one might put *Werther,* Britten's *Gloriana,* Vaughan Williams' *Sir John in Love,* or Bartók's *Bluebeard's Castle,* Stravinsky's *Rake's Progress,* and Janáček's *Cunning Little Vixen* in the odd little group of truly picturesque operas. Shaw nominated Mascagni's *L'Amico Fritz* as well, when he said that it is "more an idyllic picture than a story. The cherry-tree duet ought really to be hung in the Royal Academy." Probably the greatest of the picturesque operas is *Turandot.* Puccini got just what he asked for when he wrote his librettist, Giuseppe Adami, "Be more sparing of words and try to make the incidents clear and brilliant to the eye rather than the ear."*

It might also be observed that there is something a bit arrogant and otiose in lavish efforts to provide a picturesque, realistic foil for an acting style that must be, by the nature of blank verse and sung speech, artificial and stylized. William Archer made something like this point when he objected to Henry Irving's highly touted effort to add "realistic" touches to his Shakespearean impersonations: "Realism, by its very nature, gravitates toward the ridiculous, and it is only the firmest artistic sense and the rarest execution that can keep [theater and realism] apart." The obvious corollary is that the proliferation of realistic effects on the stage is bound to undermine the power of the blank verse or musical line to elicit the audience's own imaginative responses. Speaking of the heavy demands made on the audience by drama premised upon an empty space within a sunlit wooden O, Granville-Barker suggested, "It is possible that the more we are asked to imagine the easier we find it to do so." (Those who experienced the years of radio's heyday often speak of how much more exciting that medium was than television, allowing as it did the individual listener the full freedom of visual imagination.) A very full stage, however, gives the audience little to do but receive the wealth of visual data. Nothing is to be done with it, and so the audience is bound to become—as Berlioz said it did at the Paris Opéra—inert, passive, and easily bored. The less the audience is required to imagine, surely, the harder it will be to do so when it is necessary.

Ghislanzoni assured Verdi that a *brutta* soprano would become *bella* through the effect of "musical idealism." The crucial transformations on the Shakespearean stage are also achieved by aural means. The "poetic idealism" of his verse, as it were, gives his "shadows" and his local habitations convincing existence. Stage realism cannot help but compete disconcertingly—and sometimes quite ludicrously—with the "reality" that is projected through the actors' voices. An image Verdi used in a different context perfectly describes the consequence of big productions' making these voices seem even tinier constituents of a very large stage picture: "The order is reversed. The frame has turned into the painting!!!"

Big-frame–small-picture productions are common on the Shakespearean and operatic stages, perhaps because production values are perceived as the one obvious way to create novel approaches to works that are for the most part familiar to the public. The lines and notes cannot be changed (much), but their visual environments certainly can be subjected to virtuoso manipulation. Thus, Beckerman's warning about staging Shakespeare—"not to overvalue the necessity or even the desireability of novelty in staging"—will most often fall, one imagines, on deaf ears. But any approach that shows a preoccupation with what Mencken called the "carnal husk" of opera and not the music (or the "carnal husk" of Shakespeare's plays and not their poetry) is bound to have a deadly effect in the long run. Perhaps this is one reason why Peter Brook, in his book titled (aptly enough for my purposes) *The Empty Space,* associated Shakespeare and opera with the least attractive of his four kinds of theater: the Holy, the Rough, the Immediate, and the Deadly. "Nowhere does the Deadly Theater install itself so securely, so comfortably, and so slyly as in the works of William Shakespeare," Brook writes; and "Grand Opera," he says, is "the Deadly Theater carried to absurdity." One thinks, too, of the deadly boredom etched in many faces at a routine performance of Shakespeare or opera when, Brook remarks, some Deadly Theater succeeds merely because "just the right degree of boringness is a reassuring guarantee of a worthwhile event."

It will always be a temptation to assuage the ever-incipient boredom of Shakespeare and opera—which are by no means easily or immediately accessible forms of entertainment—through the easy if expensive course of visual ingenuity and busyness. (There is also the sad fact that it is often easier to offer the public a $1,000,000 *Aida* than a spectacular Aïda.) But are the inevitable side effects of this approach—dwarfing the heroic performers and weighing down and slowing down the performance with scenery and props—worth it? These side effects, in turn, often lead to objectionable "condensations" of the works themselves. Brook remarks that the absence of scenery was one of Shakespeare's "greatest freedoms," allowing him to speed the spectator through a succession of illusions. And Beckerman has pointed out how much his plays depend on "vigorous movement of the actors coming on and off the stage. The actors themselves, rather than the stage equipment, provided the impetus of the play's progression." Much opera from the standard repertory—thoroughly dependent on the cultivation of rhythmic momentum—functions in the same way. Verdi, at mid-career, when scenic *mutazioni* were becoming ever more elaborate, sensed the danger of "mighty shows" to the cumulative momentum of a performance: "If in opera there were neither cavatinas, duets, trios, choruses, finales, et cetera, and the whole work consisted, let's say, of a single number, I should find that all the more reasonable and proper. . . . [All] these numbers isolated like this by change of scene for every number have for me more the quality of concert pieces than an opera." What Verdi was getting at, without knowing it, was an approximation of a Globe Theatre performance of a play as a "single number"—without intermission for scene changes.

Mencken, that cynical enemy of opera, grumped that "an opera may have plenty of good music in it and fail, but if it has a good enough show it will succeed."* The *horror vacui* style of stage-dressing owes much to this article of faith, and not uncommonly this style has considerable success with the public—as witness Zeffirelli's current *Tosca, Bohème, Turandot,* and *Traviata* productions at the Metropolitan Opera. This is because a fair portion of an opera audience is present, as Mencken also observed, "not to hear music, not even to hear bad music, but merely to see a more or less obscene circus." Working hard to please these "barren *spectators*" is a paltry business. Hamlet has just the right squelch for those who indulge in such pandering: "That's villainous, and shows a most pitiful ambition in the fool that uses it." The proper ambition in Shakespeare and opera, of course, is to fill the empty space first and foremost with the appropriate and splendid touches of poetry's and music's sweet harmony. Only then, in Flecknoe's lovely phrase, can we possibly seize upon the solid joy of their interior.

Operas from Shakespeare

21

Squeezing the Orange

In the following brief chapters I propose to range rather serendipitously through the vast repertory of Shakespearean operas with an eye to arias, local details of plotting or characterization, and aspects of musical coloring that throw into high relief (occasionally into high camp) important issues raised by composers and librettists when they "read" Shakespeare, then "translate" him into their own expressive media. This repertory is a huge one, of well over two hundred operas. Included within it are perhaps fewer than a half-dozen masterpieces, a dozen or more that are estimable and stageworthy, several peculiar works that one would happily attend out of pure curiosity, a few dozen more that either boast brilliant moments or raise interesting historical or aesthetic issues, and, last, that cherishably batty group of operas in which the Shakespearean source is so insouciantly disfigured as to create real hilarity and mirth. Given this vast feast, I have no illusion about tasting comprehensively or even deeply in the potpourri that follows; sharpening the palate and perhaps tempting further savoring is what I hope for.*

If a single caveat or credo could be said to govern these excursions, it is one that has been well expressed by the same man who felt such trepidation at squeezing the great Shakespearean orange into an operatic container: Arrigo Boito. It is a convenient commonplace to speak of the "translation" of Shakespeare into opera, but this word is not entirely satisfactory, as Boito had occasion to point out to Verdi during their *Otello* collaboration. While pondering a few new lines sent to him by his librettist, Verdi had noticed—and this is a tribute to his artistic scruples where "Signor Guglielmo" was involved—that Boito made use of Carlo Rusconi's very free paraphrase of Iago's line, "Were I the Moor, I would not be Iago" (Rusconi makes this "I would not wish to see an Iago near me"). Verdi then consulted the French translation by Victor Hugo and another Italian one, by Andrea Maffei, and found both of them more nearly literal. While admitting he did not dislike Rusconi's version, Verdi fretted about such an infidelity and queried Boito about it in a letter. Boito speedily replied:

What I am about to write seems a blasphemy. I prefer Rusconi's phrase. It expresses greater things, which the text does not express, reveals the evil soul of Jago and good faith of Otello, and announces to those who hear it a whole tragedy of deceits. For us, who have had to give up the admirable Venetian scenes, where these sentiments are uttered, Rusconi's phrase is most useful. My opinion is to keep it the way he has it.

Perhaps feeling this might not be sufficient to satisfy Verdi, Boito added an insightful coda in which he rejects the notion of translation and eloquently insists on their own artistic freedom:

This is not to deny that Rusconi was wrong to adulterate a thought of Shakespeare's. A translator's fidelity ought to be very scrupulous, but the fidelity of one who illustrates a work from another art with his own art, it seems to me, can be less scrupulous. He who translates has the duty not to change the letter; he who illustrates has the mission of interpreting the spirit. One is a slave, the other is free. . . . Proceeding with that rationale we arrive at the following conclusion: *We are right in adopting Rusconi's wrong.**

Boito's refusal to stand on high artistic principles or to indulge in a self-defeating mania for "fidelity" where two very different dramaturgies are concerned shows admirable sanity and self-possession. Six years earlier he had staked out the same artistic breathing room in fewer words (also addressed to Verdi): "An opera is not a play; our art lives by elements unknown to spoken tragedy."

The challenge of Boito's position, obviously, is to avoid leaping to derisive or dismissive conclusions about this or that "adulteration" of Shakespeare's text without considering very carefully the prior constraints that are imposed by that greatest element "unknown" and unimagined by the playwright: music. It is inevitable, we should remember, that virtually every lyric version of Shakespeare will have to entail the sort of apology that was made for an opera based on *The Tempest* that appeared in 1756: "It is hoped that the Reader will excuse the omission of many passages of the first Merit, which are in the Play . . . it being impossible to introduce them in the plan of this Opera." Besides, Shakespeare was himself utterly shameless in the ways he changed his dramatic and non-dramatic sources for his own purposes. Samuel Johnson, in fact, surmised that the reason for the relative failure of *Julius Caesar* was an atypical slavishness: "adherence to the real story . . . seems to have impeded the natural vigor of his genius." We, too, must be wary of cherishing a slavish adherence to Shakespeare and thus impeding the "natural vigor" of music's unique ways of creating life on stage.

Boito's remark to Verdi that their task was one of "interpreting the spirit" of their source perhaps points to a second challenge. Shakespeare's plays are famous for disclosing—according to who is interpreting them—numerous, often mutually exclusive "spirits." As with all the great operas or operatic roles, there is no one privileged or "correct" way to execute them. Shakespeare gave remarkable, even daunting artistic license to his interpreters; and our challenge, as we judge musical versions of his works, is to grant this license to composers and librettists as well . . . that is, to situate an operatic "reading" of a play, a character, or even a speech within the capacious Shakespearean realm of multiple possibilities before we rule it a failed interpretation of some monolithic "spirit" of one's own choice. In other words (Verdi's): "If you wish to set Shakespeare you must not hesitate, as the French say, to take the bull by the horns."

The first essay that follows (appropriately, it is also the longest) addresses one of the most notorious alleged failures of interpretation in our repertory. Ironically, the accused party is Boito himself. I believe a detailed examination of the circumstances surrounding his interpretation and the criticisms leveled against him will suggest how important it is—and yet how difficult!—to meet the challenges I have just described and render unto both Shakespeare and the logic of music, not in a partisan spirit but with something approaching equanimity.

22

Incredible Credo?

Arrigo Boito, it would appear, has much to answer for in inventing Jago's evil fulmination in the second act of *Otello,* "Credo in un Dio crudel." Joseph Kerman, while calling this soliloquy "tremendous," chides its "somewhat muddled" theology and concludes that "Shakespeare's very complicated human being" is degraded by it "into that perennial operatic standby Mephistopheles." Peter Conrad asserts that Boito makes Iago into "a romantic satanist . . . by equipping him with a sardonic Credo" and finds this transposition "dramatically ruinous." The Credo is surely responsible for Frits Noske's view of Jago as "a born villain with not half the motivation of his Shakespearean namesake," while Emrys Jones counts the "slightly embarrassing Credo" among the unfortunate libretto's "massive intrusions of alien, or non-Shakespearean, material from other traditions." Julian Budden praises the Credo as "one of the most powerful depictions of evil in the operatic repertoire," but nevertheless calls Boito's text "a piece of high-flown nonsense which has its entire justification in the musical setting." Most recently, the Credo has been called "a profession of belief having apparently little to do with Shakespeare," and another commentator, noting the "acute discomfort" of English and American critics over the text, summarizes in a similar vein: "As is well known, there is no clear Shakespearean equivalent" for the Credo.

Verdi thought otherwise. He responded delightedly to Boito's text in an 1884 letter: "Most beautiful, this Credo, most powerful and Shakespearean through and through."

Who is right in this thoroughly polar contradiction of views? I believe that in all the important respects Boito and Verdi were, and thereby must hang a detailed look into the nature of Shakespeare's villain and the genesis of the opera's Credo—since the charges against the latter are several and at first blush compelling.

The obvious place to begin is with the question of Iago's motivation, long the focus of a hot debate that has produced much smoke and a little light. Noske, we have seen, thinks Jago has "not half" the motivation of his counterpart, while Kerman finds Boito's "simplification of Iago's personality actually relieved him of the whole vexatious matter of motivation." Pinpointing Iago's motivation can be a very complex or very simple matter, depending upon how one chooses to view him. Judging him from the outside—his cunning, quick-change demeanor, his often ironic and convoluted language—one might arrive at Kerman's conclusion that he is a "very complicated human being." Other critics judge him from the inside—his sang-froid spirit of negation—and conclude that he is an utterly simple humanoid creature, hermetically sealed off from genuine emotion or humane instinct, and therefore lacking all the complexity of a "human being." Thus, Conrad finds, not complexity, but merely a "prosaic vacancy" in Iago, and Budden finds that his soliloquies reveal a "spiritual void." Such radically opposed responses are perfectly appropriate to a character who actually utters the oath "By Janus." Kerman, Conrad, and Budden are each precisely half right: combine their views of Iago and one moves closer to his paradoxical, puzzling nature.

It should be noted, too, that much of the "vexatious matter" clouding Iago's motivation is the carefully devised sham creation of Iago himself. Othello justly praises his ensign at one point for knowing "all qualities, with a learned spirit, / Of human dealings" (3.3.259–260), and one absolutely vital bit of knowledge in human dealings Iago possesses is an awareness that the average pathetically naïve human being expects actions to be *motivated*. Given a plausible motive, such a person will be satisfied and not search more deeply. Iago is a master at broadcasting bogus reasons for his actions—like his "fury" at being passed over for promotion, or his "revenge" for Emilia's supposed adultery with Othello—just as he is brilliant at larding his speech with attractive words like *honest* and *love*.*

Indeed, it is possible that Shakespeare wished to frustrate members of his audience who naïvely expected explicit motivation for villainy when, toward the end, he allowed his profoundly naïve hero to ask: "Will you, I pray, demand that demi-devil / Why he hath thus ensnar'd my soul and body?" (5.2.301–302). Othello, like many another tragic protagonist, has not been paying careful attention, and thus his pathetic question is sincerely asked. But Iago's reply is noncommittal: "Demand me nothing; what you know you know." Coleridge's familiar idea of Iago's "motiveless malignity" chiefly hangs on this response, which is also responsible for James Agate's praise of one actor playing Iago because he left "the motive for villainy very much where Shakespeare left it, that is in the dark."

Such views, however, seem altogether too respectful of Iago's truculent reticence at his last appearance. For when an audience at *Othello* hears Iago say "what you know you know," it will surely register that it "knows" very much more than the hapless Moor, not merely from watching Iago interact with all the other major characters but also from seven highly revealing soliloquies and several crucial asides spoken exclusively for the audience's edification. This seems more than sufficient acquaintance from which to draw some plausible conclusions about the motives behind Iago's malignity. Indeed, I think Shakespeare would have been nonplussed to learn that he had left Iago's primary motivation in the dark, for there are many instances of very explicit pointing and morality-play–like simplification in *Othello* that suggest he went out of his way to make the main constituents of this motivation shiningly clear.

But it is Boito's assumptions about Iago's motivation that concern us here, and the best way to assess them is by looking at his first version of the Credo, a monologue he prepared for Verdi in 1879. It was cast in four neat quatrains of *doppi quinari* (lines of two balanced five-syllable phrases, rhymed *a a b b*). In English, it runs as follows:

> The snare is taut, I have the deceits in hand; you are swelling, Envy, let you corrode me! I spill your gall in my path, you furious goad in my mind.
> The idea reigning here (*touching his brow*)—firm and hidden—is set to the task in means and end. Evil has put there such a cruel mark that mortal force cannot foil it.
> I guide and appoint the fate of Otello; *I am wicked because I am a man* because I have the dregs of hatred in my heart, while he lives amid glory and love.
> But the patient one, assiduously ripening time, brings joy and misfortune. O triumph, my menacing proud genius! Now the general weeps, and the ensign laughs!!*

This is a remarkably comprehensive synopsis of the Shakespearean villain's motives and methods (only the italicized line survived in the final 1884 Credo). If Verdi had accepted it, critics would have had little cause for disgruntlement. Most obviously, there is no "theology" here. Boito has concentrated on the various secular aspects of Iago's evil nature: his cunning calculation, his furious intensity, and the extraordinary priority of pure hatred in his behavior ("I do hate him as I do hell-pains . . ."; "I have told thee often, and I retell thee again and again, I hate the Moor"; "I hate the Moor . . ." says Shakespeare's villain).

Boito also emphasizes, in the third stanza, one of the two most important driving forces behind what Iago calls his "peculiar end"; namely, delight in exerting control over others. Iago is a monstrous rebel against "the curse of service"; clearly, it is genuine pleasure for him to exercise manipulative power for its own sake. That his means is an end in itself is one reason for his mysterious motivation. Boito captures well the will to power that is expressed, for instance, when Iago tells Roderigo, "I follow [Othello] to serve my turn upon him," or when he boasts about Roderigo himself, "Thus do I ever make my fool my purse." But Othello is Iago's supreme object of scornful manipulation. Iago exults to think that "The Moor . . . will as tenderly be led by th' nose / As asses are," and his excitement is perhaps never more electric than when he explains in soliloquy his plan to

> Make the Moor thank me, love me, and reward me,
> For making him egregiously an ass,
> And practicing upon his peace and quiet
> Even to madness.   (2.1.308–311)

Some of Boito's details are well culled, too: the opening image of the snare should remind us of Iago's true prophecy that Desdemona's goodness will "make the net / That shall enmesh them all," and of Othello admitting he was "ensnar'd" by Iago; the reference to "time ripening" reminds us of Iago's own perfect sense of timing; and the mention of the patient man echoes his frequent reiterations of the word ("How poor they are that have not patience!"; "confine yourself but in a patient list").

The most intriguing part of Boito's first version is his isolation (and capitalization) of the other driving force in Iago's character: Envy ("Invidia"). That the word *envy* occurs in two dozen Shakespearean plays but never in *Othello* need not, of course, lead to a conclusion that the thing itself is absent from the action. Iago makes enough invidious comparisons to justify such a generalization: of Othello he says, "The Moor (howbeit that I endure him not) / Is of a constant, loving, noble nature"; he says Desdemona is "fram'd as fruitful / As the free elements" and seconds later vows to "turn her virtue into pitch"; and his observation that Cassio "hath a daily beauty in his life / That makes me ugly" is the quintessential Shakespearean expression of envy.

It is possible that Boito decided on his own to "interpret the spirit" of Iago and change the Shakespearean letter by bringing Envy into his text. But James Hepokoski has argued persuasively that he probably took his cue from Victor Hugo's commentary on his French translation of *Othello*. In the crucial passage on Iago, marked in Boito's copy with several penciled lines and a caret, Hugo observes:

> The principal cause, the true cause of Iago's hatred must be sought in his own nature [i.e., not in the actual lines]. Iago is a man who is unable to accept or endure any kind of superiority. . . . Iago envies Othello for being everything that he is not. He resents him for being honest . . . heroic . . . loved by the people . . . adored by Desdemona. And that is why he wants revenge. Ah! Othello is genius! Well then, let him be careful! for Iago is envy.

Hugo's synopsis of Iago's character—combining the two pre-eminent characteristics of obsessive eagerness to exert his superiority and his envy of all forms of goodness because (like Mount Everest) they are there—is well within the bounds of interpretive plausibility. No wonder Boito was taken by it.

This preceding version of the Credo has been worth examining because Boito did not significantly change his view of Iago's psychological motivation when he created the much-ridiculed 1884 version we now have. Indeed, Boito's very first words of advice on playing Jago, in the 1887 production book, remain almost exactly those of Hugo: "Jago is Envy. Jago is a scoundrel. Jago is a critic." What was explicit in the first version is simply rendered implicit in the Credo we now know. *Invidia,* in happy fidelity to Shakespeare, evanesces back into the play's grim atmosphere unverbalized. But Envy's chief consequence—a keen desire to negate the finest human ideals—remains a central element of the final version as well. Though translated into different words, the crucial expressions of contempt, hatred, evil, firmness of intent, and envy (now of "tears, kisses, [loving] glances, sacrifice, and honor" rather than "glory and love") all survive fully in the final Credo:

Credo in un Dio crudel che m'ha creato	I believe in a cruel God, who has created me
Simile a sè e che nell'ira io nomo.	in his image and whom, in hate, I summon.
Dalla viltà d'un germe o d'un atòmo	From the baseness of a germ or atom
Vile son nato.	Base I was born.
Son scellerato	I am evil
Perchè son uomo;	because I am a man;
E sento il fango originario in me.	and I feel in me the primeval filth.
Sì! questa è la mia fè!	Yes! this is my faith!
Credo con fermo cuor, siccome crede	I believe with a firm heart, like
La vedovella al Tempio,	the young widow in church,
Che il mal ch'io penso e che da me procede	that the evil I conceive and produce
Per mio destino adempio.	I accomplish by my fate.
Credo che il giusto è un istrion beffardo	I believe the honest man a mocking actor
E nel viso e nel cuor,	and both in face and heart
Credo tutto è in lui bugiardo,	everything in him is a lie:
Lagrima, bacio, sguardo,	tears, kisses, glances,
Sacrificio ed onor.	sacrifices and honor.
E credo l'uom giuoco d'iniqua sorte	And I believe man the toy of wicked fate
Dal germe della culla	from the germ of the cradle
Al verme dell'avel.	to the worm of the grave.
Vien dopo tanta irrision la Morte.	After so much mockery comes Death!
E poi?	And then?
La Morte è il Nulla,	After Death is Nothing,
È vecchia fola il Ciel.	And heaven an old wive's tale.

This similarity should not be surprising, since it appears that Verdi objected, not to the *content,* but to the *poetic form* of the first version. For in the letter accompanying his revision, Boito wrote: "I have been remembering that you were not content with a second-act scene for Jago in *doppo quinari* and desired a more broken, less lyric form. I proposed to make a sort of 'wicked Credo' and have tried to write one in broken meter and without symmetry." And so Boito produced his ragged series of five-, seven-, and eleven-syllable lines more in the fashion of a *recitativo accompagnato* than a strophic aria. One of the two important changes in the final Credo, then, was purely a matter of prosodic—and ultimately musical—form.

Verdi's desire for such a change raises a fascinating possibility; namely, that his view of Iago was influenced by his prior acquaintance with another Shakespearean villain: Edmund in *King Lear.* In fact, a transaction very similar to that involving the 1879 and 1884 versions of the Credo took place as Verdi worked from 1853 to 1855, with Antonio Somma on a *Re Lear* libretto.* Verdi's view of Edmund was perceptive. He recognized that Edmund was not a conventional stage blackguard, but rather a mercurial, ingratiating fellow. "He is not a repulsive villain like Francesco in Schiller's *The Robbers* [the source for Verdi's *I Masnadieri*]," Verdi admonished Somma, "but rather one who laughs and jokes all the time and commits the most atrocious crimes with utter indifference."

Verdi was to see Iago in much the same way. To a painter who suggested that Iago have the face of an honest man, Verdi replied, "Extremely good! Iago with the face of a gentleman. You've hit it. Oh, I knew it, I was sure of it." And over a year later he elaborated his view of Iago's manner: "distracted, nonchalant, indifferent to everything, incredulous, sparkingly tossing off good and evil thoughts as if he were thinking of something quite other than what he actually says." Such buoyant wit is also reflected in Boito's main caveat for playing Iago in the *Otello* production book: "The coarsest, most vulgar error an artist interpreting this role can commit is to make him a kind of man-devil! —and screw into his face a mephistophelean smirk and allow his eyes to bug out satanically. . . . He ought to be handsome and seem jovial, frank, and rather good-natured . . . [and display] a great charm of pleasantness in his persona."

This, needless to say, is splendid advice for playing Iago. Edmund Kean's "admirable" Iago was, according to Hazlitt, exactly in accord with the vision of Verdi and Boito: "The accomplished hypocrite was never so finely, so adroitly portrayed—a gay, light-hearted monster, a careless, cordial, comfortable villain . . . quite equal to anything we have seen in the best comic acting." Dickens admired another actor because his Iago was "not in the least picturesque according to the conventional ways of frowning, sneering, diabolically grinning, and elaborately doing everything else that would induce Othello to run him through the body very early in the play." And Olivier took Boito's point to its extreme when he played to Ralph Richardson's Othello: "I played Iago entirely for laughs . . . terribly sweet, and as charming as can be."*

Verdi's wish that his Edmondo be an agile, blithe villain had an interesting result. Somma had initially taken Shakespeare's main speech disclosing Edmund's evil character (his "Thou, Nature, art my goddess" soliloquy, a kind of "credo of the illegitimate") and turned it into a conventional *scena* consisting of a recitative, slow aria, then fast aria. Verdi objected, finding the regular prosody and old-fashioned musical form at odds with Edmondo's peculiar character. He requested instead something with a "variety of colors: irony, contempt, wrath" and pointed out that he could not possibly give "a cantabile to such a *personaggio.*" In conclusion he urged Somma, "Try to avoid an aria that is a decoration." Verdi was, in effect, asking from Somma precisely what he was to ask thirty years later from Boito: something "more broken, less lyrical" for a mercurial Shakespearean villain. The revision Somma eventually provided—a fine multifaceted and asymmetrical scene—was no mere "decoration." It is easy to see in the pat, orderly prosody of the 1879 Credo precisely the same danger of a musical and theatrical "decoration" that Verdi was increasingly eager to avoid in his later years.

Verdi's view that materialistic, reductive *personaggi* like Edmondo and Jago ought not be allowed much room for lyric expansiveness was particularly insightful: idealists like Lear, Cordelia, Othello, and Desdemona—not earth-treaders—are more likely to leap upward into lyric flight. Perhaps there is something of Verdi's view in Peter Conrad's remark on the futility of making Iago operatic. Conrad feels the Credo "unbalances the Shakespearean action by filling in the envious, motiveless, prosaic vacancy 'of the character. Otello's natural medium is music, but Shakespeare's Iago cannot exist in it, because, not knowing what he is, distrusting and ironically disowning all his emotions, a shifty and selfdetesting negator, he has nothing to sing about." Though, as he often does, Conrad exaggerates his point for effect, there is a kernel of truth here that I think governed Verdi's creation of Jago's role.

For in his later years Verdi became ever more disposed to insist on the priority of dramatic over purely vocal values, and Jago in particular finds him adamant about this priority. Several times he makes Conrad's point that Jago has nothing to *sing* about:

A curious thing! The role of Iago, apart from a few *éclats* could all be sung *mezza voce*!

Iago cannot be performed, and is not possible, without extraordinarily fine pronunciation. . . . In this role it is necessary neither to sing, nor to raise one's voice (with few exceptions). If, for example, I were a singing actor, I would do everything with lips pursed, *mezza voce*.

Just as Jago must only declaim and sneer [*ricaner*] . . . just as Othello . . . must sing and howl, so Desdemona must always sing.

Here again we find Verdi repeating an admonition he had made before about Edmondo: "I have said to make Edmondo a mocking character because in music he will prove more varied: making him otherwise, he would have to sing *heavy* phrases, with shouts. Mockery, irony are portrayed (and it is more novel) *mezza voce*: it becomes more terrible." That Verdi should have been so insistent upon *mezza voce* for Iago, Shakespeare's supremely mocking and ironic villain, should really come as no surprise.*

Conrad's assertion that Iago has "nothing to sing about," of course, is finally specious and need not rule him out of opera. For he is a great pretender: he pretends to be human, a loyal friend, an honest ensign, and so on. He may have nothing to sing about *sincerely,* but this does not prevent him from pretending to engage in those human experiences that lead into song. He pretends, as Hazlitt said a good Iago should, to be an "excellent good fellow, and lively bottle-companion," and so sings a rousing drinking song. By way of faking bosom loyalty he utters a tremendous oath with Othello ("Si pel ciel"). Even a horribly unlulling lullaby is in his repertoire ("Era la notte"). These pretenses at vocalism, like all his other pretenses, are brilliantly carried off. Several great operatic characters equally void at their core—Don Giovanni, Mephistophele himself, and Lulu chief among them—have, like him, nothing to sing about . . . and sing it very well indeed.

In addition to the very different prosodic form, Boito made one other important change in his final version—a change almost inevitable, as he transformed his earlier monologue into a "wicked" version of the Church's *Credo in unum Deum:* he gave the whole a more satanic and religiously subversive cast. This has elicited some of the harshest criticism of the Credo—Kerman chiding its "muddled" theology and regretting that Jago lapses into "that perennial operatic standby Mephistopheles," Conrad complaining that Jago becomes a caricature "romantic satanist."

This line of criticism—indeed, all criticism that suggests there is (in

Hepokoski's phrase) "no clear equivalent" for the satanism of the Credo in the play—is very odd. For I do not think Shakespeare worked harder or more elaborately in his entire career than he did to make it pristinely clear that Iago is a Luciferian agent—not least in his boastful reversal of Yahweh's "I am what I am." W. H. Auden, in fact, has asserted that "Iago's statement, 'I am not what I am,' is given its *proper explanation* in the Credo which Boito wrote for him in . . . Verdi's opera."

Othello presents, in essence, a morality-play allegory in which "bless'd" goodness (Desdemona) and "hellish" evil (Iago) struggle for the soul of Everyman (Othello), and so the play is impregnated with the religious language that must attend such a cosmic struggle. *Devil* and its derivatives occur more often (twenty-seven times) than in any other Shakespearean play, and the occurrence of *heaven* forty-nine times also sets the Shakespearean record. There is no genuine "theology" in the play, but simply a constant emphasis upon the monumental contest between the divine and the diabolical in human nature. And there is no theology, muddled or otherwise, in the Credo—just an explicit expression of this antithesis.

Verdi doubtless found the satanism of the Credo "wholly Shakespearean" in part because he was attuned to the allegory submerged in the action: "This Iago is Shakespeare, humanity, that is to say a part of humanity—evil." He and Boito were clearly alive to how lavishly the dramatist orchestrated his efforts to give Iago's evil a Christian context.* The very first word he utters—"'Sblood" (God's blood)—is a sacrilegious oath. Indeed, a *sacrilegus* was a stealer of sacred things, and Iago is an expert thief and manipulator of sacred language itself: *divinity, heaven, heavenly, baptism, holy writ, bless'd, redeemed, sin, confess,* and *grace* all fall from his lips. Such words—like *honest*—are for him mere counters to use to ingratiate himself with his victims.

Especially in the final scene, Iago's satanic identity is relentlessly underscored. The word *damned* carried a more specifically theological connotation in Shakespeare's time, and Iago is called "a damned slave" both by Montano and Lodovico. Lodovico also calls him "this damned villain," and Roderigo's dying words are "O damned Iago!" Lodovico also refers to "this hellish villain," and Othello, we have noted, calls him a "demi-devil."

There are also several moments in the play when Shakespeare practically hints in neon that Iago is a henchman of Satan. By the time Iago describes Cassio to Roderigo as "a slipper and subtle knave, a finder-out of occasion; that has an eye can stamp and counterfeit advantages . . . a devilish knave" (2.1.241–245), the audience can relish the irony that Iago is in fact describing himself. When Cassio says to him of wine, "if thou hast no name to be known by, let us call thee devil!" (2.2.282–283), we turn our eyes to the real devil onstage. Similarly, it is hard to avoid the obvious irony of Othello's calling Desdemona a "fair devil" (3.3.479) at just the moment he names the real "white devil" (the Renaissance term for a hypocrite) near him his new lieutenant. The satanic irony is enormous when, a short time later, Othello "explains" to Iago, "They that mean virtuously . . . The devil their virtue tempts" (4.1.7–8). And just seconds later a frenzied Othello shouts "O devil!" and falls in a trance. Iago stands over him and exults as the perennial standby Mephistopheles often does: "Work on, / My medicine, work! Thus credulous fools are caught. . . ." At the play's end all of these ironies are ratified when Othello alludes to the devil's cloven feet—

> I look down towards his feet; but that's a fable.
> If that thou be'st a devil, I cannot kill thee   (5.2.286–287)

—and when he concedes that a "demi-devil" has ensnared his "soul."

All of this led Boito to what one would think was an unexceptionable conclusion: "Othello is like a man moving in circles beneath an incubus, and under the fatal domination of that incubus he thinks, acts, suffers, and commits this dreadful crime." The Credo—introduced with Jago's boast flung at a departing Cassio, "Your evil demon drives you on, and I am your evil demon" ("Ti spinge il tuo dimone, e il tuo dimone son io")—is a perfectly appropriate, one might even say inevitable dramaturgical consequence of this view. The blame for creating a "romantic satanist" should thus be directed not at Boito but at Shakespeare. And we should also keep well in mind, apropos Kerman's objection, that Mephistopheles was not only a "perennial operatic standby" but also a commonplace figure on the Elizabethan stage. Iago is merely the most brilliant of the species.

In content, then, the Credo has very clear equivalents in *Othello*. It also fills a dramaturgical function that clearly parallels Shakespeare's use of the soliloquy. Both the earlier monologue and the later Credo were obviously conceived by Boito as a kind of synopsis of the seven solos Shakespeare gave Iago (Othello has only one). The librettist was to perform precisely the same operation in *Falstaff*, when he condensed Ford's three jealousy-soaked soliloquies into the opera's only big, old-fashioned solo aria, "È sogno? o realtà?" At the end of his cover letter for the Credo's final version, Boito wittily added, "Look how many ghastly things I've made him say." This is precisely Shakespeare's occupation in Iago's solo speeches. Granville-Barker says about them, "Of Iago's opinions and desires and of what he means to do they will tell us truly." Boito was duty-bound by the spirit of these seven speeches to invent the Credo; if anything, it shows his genius for the theatrical *brevità* that the composer relentlessly demanded from his librettists.

The playwright clearly wished to create an intimate (and discomforting) bond between Iago and the audience . . . to allow us to see, in other words, what no one onstage can see until the very end: Iago's true nature. When he is alone, "ghastly things" pour from him. None are more so (and both emphasize his satanism) than such outbursts as—

> I have 't. It is engend'red. Hell and night
> Must bring this monstrous birth to the world's light . . . (1.3.403–404)

> *or*
>
>   Divinity of hell!
> When devils will the blackest sins put on,
> They do suggest [tempt] at first with heavenly shows,
> As I do now.   (2.3.350–352)

Budden thinks the Credo "a conglomeration of sentiments which Shakespeare's villain would not have thought worth formulating," but the plain fact is, they are very similar to these two and several other sentiments Iago dispenses in his soliloquies. These soliloquies were vital to Shakespeare as he spun out the elaborate web of ironies that make *Othello* such a painful experience for an audience. It is thus impossible to conceive of the play as an opera without at least one big solo aria like the Credo, full (in Kerman's apt phrase) of "blasphemous bravado."

It should be pointed out, too, that the soliloquies and the Credo are vital in throwing into spectacular relief Iago's many disguises. Both Shakespeare and Boito were careful to circumscribe the villain's eruptions of satanism; most of the time he is observed as the brilliant actor of various utterly charming roles. Granville-Barker writes that Iago has "a remarkable faculty of adapting himself to his company," and this is the heart of Boito's and Verdi's conception of the role. The production book explains (p.5):

> One of [Jago's] gifts is the faculty he possesses of changing his demeanor according to the persons with whom he finds himself, the better to deceive and dominate them. Easy and jovial with Cassio, ironic with Roderigo, seeming good-natured, respectful, devotedly meek with Otello, brutal and threatening toward Emilia, obsequious with Desdemona and Lodovico. Here is the basis, here is the pretence, here are the various aspects of this man.

Without the soliloquies or the Credo to play against, this display of thespian dexterity—and much of the action's poignant irony—would fail of their tremendous impact. *Paradise Lost* would have suffered similarly, had Milton not given his villain such Iago-like utterances in soliloquy as these before he charms Eve and turns her virtue into pitch:

> Which way I fly is Hell; myself am Hell . . .

> Farewell remorse! All good to me is lost;
> Evil, be thou my good . . .

> For only in destroying find I ease
> To my relentless thoughts . . .

In the Credo, Jago calls the honest man an "istrion beffardo," or mocking actor. But this is precisely what Shakespeare's and Boito's villain is for most of his time on stage. The worst possible error an actor or singer in these roles can commit—as Boito pointed out—is to allow the satanic Iago to overwhelm the thespian Iago. This was, curiously enough, the crux of Shaw's criticism of the baritone who created the role for Verdi, Victor Maurel: "There is too much of Lucifer, the fallen angel, about it—and this, be it remarked, by no means through the fault of Verdi, who has in several places given a quite Shakespearean tone to the part by *nuances* which Maurel refuses to execute." Shaw's praise of Verdi's musical conception of Jago is just, I think: when we leave a performance of *Otello* thinking of Jago, along with Kerman, as a typical operatic Mephistopheles, we should be more inclined to blame the baritone or the director who guided him than the composer and librettist.

A final criticism of the Credo remains to be dealt with—Budden's assertion that its verbal text is "high-flown nonsense." This is not literally true, of course: every sentence is crystal clear; there are no hard words or vexingly ambiguous phrases. Even if the text *were* nonsense in purely verbal and logical respects, it would have been faithful to Shakespeare. For on several occasions he allows Iago to utter such portentous, tautological nonsense as "I am not what I am" or "What you know, you know." For example, when Othello says men should be what they seem, Iago bafflingly retorts: "Men should be what they seem, / Or those that be not, would they might seem none!" (note how the vigorous exclamation point carries the day). When Othello mentions Desdemona's honor, Iago opines, "Her honor is an essence that's not seen; / They have it very oft that have it not." Nonsense. Soon Othello presses him for what he knows, and Iago responds, "No, let me know, / And knowing what I am, I know what she shall be." More nonsense; to which the distraught general stupidly replies, "O, thou art wise." And when a puzzled Lodovico asks of the noble Moor, "Are his wits safe? Is he not light of brain?" Iago utters this masterpiece of nonsense:

> He's that he is; I may not breathe my censure
> What he might be. If what he might he is not,
> I would to heaven he were! (4.1.270–272)

Again, the confidently assertive exclamation!

Shakespeare's point in all this, of course, is that Iago is capable of the full gamut of nonsense: verbal, moral, and religious. Evil *must* be nonsense if one posits a Christian view of the world. Milton's Satan talks nothing but logical, moral, and theological nonsense in *Paradise Lost* when he rationalizes his miserable expulsion from heaven and when he persuades Eve to partake of the apple. Such is Iago's *métier*. He knows very well that "Knavery's plain face is never seen till us'd," and that the plain face of "blackest sins" must be disguised in "heavenly shows." And what better disguise than the impressive nonsense he purveys so often with such appalling and overmastering conviction? Shakespeare fully expected his audience to find every nonsensical sentiment that Iago utters to be—in both the theological and non-theological senses of the word—damning. Budden's assertion of the nonsense of the Credo should therefore be considered high praise indeed. For even this nonsense has its clear equivalent in Shakespeare's play.

23
Two Lears

*K*ing *Lear* is a frightening play. It frightened Nahum Tate into concocting his notorious happy ending and so affected crusty Samuel Johnson that he could not bear to reread its last scenes until, as Shakespeare's first editor, he was forced to do so. It intimidated Charles Lamb into saying it was "too large" for the actual stage and could be consummately appreciated only in the reader's imagination. As is well known, it frightened Verdi, who began planning a *Re Lear* in 1849 and left a virtually complete libretto abandoned at his death fifty years later. Benjamin Britten was tempted, too, but he also turned away from the challenge at the draft-libretto stage.

Aribert Reimann was not frightened.* Verdi often complained that he could not find the right singers to inspire him to begin composing, but this was not Reimann's problem. The pre-eminent baritone, Dietrich Fischer-Dieskau, is in fact acknowledged in the score as the source of the composer's "initiative and courage" to scale this theatrical Everest. Fischer-Dieskau's offer to assume the title role proved irresistible, and so in 1978, *Lear*, an "opera in two parts after Shakespeare," received its Munich premiere. Since then productions have been mounted elsewhere, and a full-length recording has appeared.

It is worth looking at this *King Lear* opera that *was* written from the perspective of that most tantalizing work in the might-have-been repertory, Verdi's *Re Lear*. For in these two "readings" of Shakespeare the unique and separate splendors of the nineteenth- and twentieth-century operatic idioms are vividly displayed. Verdi's *Lear* would have differed utterly from Reimann's *Lear*, and thereby hangs a tale.

Verdi left his views on *King Lear* in numerous letters and in a thoughtful synopsis of the action he made in 1850 for his first librettist on the project, Cammarano. The full and revealing libretto emerged several years later from his second collaborator, Somma (who also created the *Ballo in Maschera* libretto). The synopsis shows the basic contours Verdi had in mind. Predictably, he gravitated toward passages suitable for set pieces: Lear's swearing of vengeance upon his daughters; a "magnificent quartet" for Lear, Edgar, Kent, and the Fool in the storm; a mad-scene for Lear; a duet for Cordelia and Lear at the end; and a finale "in which Lear must lead." He pictured five principals (Lear "a big baritone," a Cordelia "not necessarily with a big voice, but deep feeling," a "very good contralto" as the Fool, Edgar, and Edmund) and four comprimarios (Goneril, Regan, Kent, Gloucester).

It is fascinating to speculate on *Lear's* attractions for Verdi. The most obvious needs no elaboration: Verdi was steeped in the work of what he called a "master of the human heart." He must have sensed (as Verdi scholars have also sensed) that his first Shakespearean opera—*Macbeth*—had represented the first full-fledged flight of the "Verdi style." He was pleased with the artistic effects of the opera (he wrote his father-in-law that it was his best to date) and doubtless began to cast about for another likely Shakespearean subject. The completeness of his *Lear* scenario suggests that he gave it much thought between the *Macbeth* premiere in March 1847 and February 1850, when he sent it to Cammarano.

Verdi might have been attracted to *Lear* for a number of conventional reasons. Next to *Macbeth,* it is the shortest of the great tragedies. Verdi—willing to excise the blinding of Gloucester and the cliff scene and abbreviate the machinations of the villains—perhaps thought the violence of operatic compression could be successfully borne. He must also have sensed that the pattern of malediction—distress—explanation—reconciliation and despair that was so common in early nineteenth-century Italian opera was archetypically reflected in *King Lear*.

Other aspects of the play may have appealed to Verdi's personal sensibilities. The kind of heroine Cordelia represents was particularly attractive to Verdi. Gerald Mendelsohn, a professor of psychology, has argued that Verdi heroines of the middle period are "essentially innocents who are corrupted or destroyed by the world, a world ruled by the passions of men and the inflexible order created by men."* Delia (the operatic name) would have been precisely such a victim. Another theme supremely evident in the play proved rich for Verdi in his middle period: the relationship between father and child. Macduff's heartfelt aria on his murdered children in *Macbeth* is an inkling of the powerful interactions of father and child to follow in *Luisa Miller* (1849), *Rigoletto* (1851), *La Traviata* (1853), and *Simon Boccanegra* (1857). The final scene of Lear, as Verdi envisioned it, is almost exactly paralleled by Rigoletto's discovery of his dying Gilda. And not a few Verdians suspect that the poignant reunion of father and daughter in *Boccanegra* grew from Verdi's thinking about *Lear*. Perhaps Delia and Amelia are related by more than their similar-sounding names.

Finally, one might add that, with roles like Nabucco, Macbeth, and Rigoletto, Verdi had almost singlehandedly created the vocal category of the heroic baritone. It was a vocal range and timbre he gloried in, and he was perhaps excited by the idea of writing the role of the tragic father-king for "a big baritone." The only singer contemplated for the first cast that Verdi approved of was Filippo Coletti, the first Germont *père* in the revised *Traviata* and a specialist in several of Verdi's "big baritone" roles.

We will never know precisely why Verdi abandoned *Re Lear*. Mascagni said that Verdi told him, *à la* Lamb, "The scene when King Lear finds himself on the heath terrified me." Vincent Godefroy comments wittily on this excuse: "This really savors of the secure romancing of an octogenarian lion couchant and regardant." Most likely, a combination of reasons caused the inaction: chiefly, the lack of appropriate singers and the inexperience of Somma—a lawyer—at libretto-making. Perhaps time and effort began to weigh on the composer: so he relaxed into the better part of discretion, the "too cold breath" of fussing over the libretto finally cooling his musical imagination. Or perhaps as Verdi grew from the fatherly to the grandfatherly age, and as his childless liaison and eventual marriage to Giuseppina Strepponi began to occupy him, the themes of *King Lear* loosened their hold. A play centering on the painful and fatal folly of old age—not the witty folly of a Falstaff—may have become less and less attractive.

Even after he had gained his reputation and a measure of artistic control, Verdi was ever and rightly fearful that the premieres of his operas would be compromised by managerial crassness and the half-efforts or worse of singers and musicians. Reimann was more fortunate. His premiere occurred under festival conditions, with the supreme German baritone of the day at center stage. The production was conceived by one of the most resourceful designer-directors, Jean-Pierre Ponnelle; the assumption of the other roles, conducting, and orchestral playing were excellent, as the complete recording shows; and the reaction of the Munich public was described by one reviewer as "outspokenly, almost stormily positive."* What is the work like, and how does it "read" Shakespeare? A short explanatory note by Reimann's librettist, Claus Henneberg, provides two interesting comments with which we can attempt to answer these questions.

The first one is inevitable and echoes Boito: "Transforming a Shakespearean play into an opera libretto requires reduction, tightening, and lyric-dramatic intensification. I have cut from Shakespeare's text what the music can express." More intriguing is Henneberg's remark that he found the courage to make a German version of *Lear* when he came upon a copy of the very first German translation by Johann Eschenburg (1777). He found this translation "on the whole harder, more clear, and more theatrical" than its better-known nineteenth-century successors. Rather inauspicious epithets of praise, one might think, for Shakespeare's rich banquet of words. *Hardness* and *clarity* do not really fit one's memory of the play's great moments, and today the word *theatrical* has the connotation of streamlining and "less is more," which is un-Shakespearean. Likewise inauspicious: Eschenburg's was a *prose* translation.

Henneberg's libretto is indeed hard and clear. It is also (intentionally, I think) colorless and in important respects empty—colorless and empty after the fashion in which Peter Brook rendered the world of his powerful film of *Lear* a stark black and white. The first of three general aspects of the original that the librettist chose not to translate into musical form, then, is the "poetry."

There is of course no question of retaining all or even most of the original's memorable lines, as the play's 25,000 or so words were compressed into 6,000 or 7,000. Even then, with a kind of music quite apart from the word-music of Shakespeare's poetry ringing in his ear, Henneberg shows a systematic tendency to sharpen and shorten by de-poeticizing. Here are a few examples of what I mean; in each case a literal translation of the German in parentheses follows Shakespeare's version:

> Peace, Kent!
> Come not between the dragon and his wrath.
> (Kent, step not between her and me.)

>   Out, vile jelly!
> Where is thy luster now?
> (Out with it!)

>   I am bound
> Upon a wheel of fire, that mine own tears
> Do scald like molten lead.
> (But I am bound upon a wheel of fire.)

The libretto consistently shows this desire to excise metaphorical phrasing, remove colorful adjectival qualifiers, and vitiate Shakespeare's remarkable imagistic associations. In search of hardness and clarity Henneberg thus often achieves a verbal flatness worthy more of Beckett than of Shakespeare.

Several of the play's memorable high-flying utterances are therefore kept remarkably earthbound. Lear's great "O, reason not the need" speech—which convenes most of the action's crucial themes—is radically cut; while his moving prayer, "Poor naked wretches," disappears entirely. The poignant "Is man no more than this?" speech, arguably the play's centerpiece, is much shortened and rendered unintelligible by being placed in one of Reimann's few ensembles. Even when a major speech survives, there is still considerable stripping down.

Take, for example, Lear's volcanic speeches in the storm ("Blow, winds, and crack your cheeks!" 3.2.1–24). Much poetic energy is lost here. Vanished are the superb complex adjectives like *thought-executing, all-shaking,* and *high-engendered* (very odd, this, since complex adjectives are a hallmark of German syntax). "Nature's moulds" are left uncracked, oaks uncleaved by "vaunt-courier" thunderbolts, and the "thick rotundity" of the world is unflattened. Crucial abstractions are, as throughout the libretto, also lost: *elements, unkindness, ingrateful.* Vanished also are perhaps the two most important lines in these speeches, which capture the conflict between Lear's "royal" will to endure and his "human" need to capitulate: "Then let fall / Your horrible pleasure. Here I stand your slave, / A poor, infirm, weak, and despis'd old man."

A second general aspect of the source was also not of much interest to the librettist: the play of ideas and philosophical/moral abstractions. The tremendous over-meaning is but haphazardly conveyed, and there is little concern to evoke the central concepts of fellowship, friendship, justice, authority, service, fault, madness, love, sacrifice. The libretto does not show an interest in integrating the ideals by which men live or do not live (as, say, Verdi and Boito integrate the concepts of honor, sacrifice, love, duty, and honesty in *Otello*). A startling number of Shakespeare's most resonant phrases are not part of the Reimann opera. Among them: "I did her wrong"; "I would not be mad"; "this tempest in my mind"; Take physic, pomp"; "they told me I was every thing"; "I should e'en die of pity / To see another thus"; "Would I were assur'd / Of my condition!"

The third and last general characteristic of Shakespearean drama that inevitably suffered an operatic sea-change was the interplay of themes and images. Many of the most important ones—for example, those relating to blindness, weeping, inner and outer storm, clothing, and violent animals—are absent from the opera. And a few vital themes are emphasized, but at wild variance from Shakespeare's timing. Lear's urge to sleep is expressed in the play, but not—as in the opera—as early as Lear's very first speech ("Ah, ah, dieses Verlangen nach Schlaf"—"ah, ah, this urge toward sleep"—is apt only for the Lear of act 4).*

If all this sounds like the carping of a Shakespearean purist, it is because I have been talking about the libretto apart from the music. When we turn to Reimann's music the full power of his achievement—and something of the opera's limitations—become more apparent. The reviewer for the *Neue Zürcher Zeitung* (12 July 1978) was generally pleased with the world premiere, but made this important qualification: "Though Riemann possesses little vocabulary for the maturing process that Lear must undergo—'ripeness is all'—he still manages never to fail in the labyrinths of evil insinuations, hideous injustice, and acts of brutality." Here, in sum, is Reimann's artistic forte: the power to express the barrenness of evil and the alienation of the human being . . . not the power to grow and to love.

Reimann's opera is worthy of Shakespeare's play in its evocation of humanity and civilization left "darkling." Its music powerfully vitalizes those phrases cast throughout *King Lear* that suggest entropy and chaos: nature on the "very verge" of her confines; "this tough world"; "this dreadful pudder"; "this great decay." The composer accomplishes aurally what Brook achieved visually in his film's vast, blank spaces that seem to sustain no plant life whatsoever. The very amorphousness of Reimann's tone-row–dominated music evokes the movement from form to formlessness, order to anarchy, which is part of the original play's structure. The imponderable atmospherics of pessimism are brilliantly conveyed by the score. In fact, Reimann tells us that a preceding composition—a "Variations for Orchestra"—could stand as a prelude to his opera because it expresses "the isolation of humanity in complete solitariness, the brutality and dubious quality of life." It will come as no surprise, therefore, that the main musical effect even at the richly emotional act 4 reunion of Cordelia and Lear is one of stasis and oppression.

This is a plausible way to read the play, but it is not the only one. One cannot help feeling that Verdi would have focused on the "maturing process" that the Zurich reviewer found missing in Reimann's opera. Lear's irrational renunciation of love and the subsequent revival of his power to love were clearly central to Verdi's plans. He was eager to describe a paradise regained (but tragically too late), but there is nothing of paradise in Reimann's view of Lear. He passes over that one last moment of repose and joy in the last scene ("Come let's away to prison"—5.3.8–19) not with lyricism but rather a solemn dirge. One can tell clearly from the finished libretto that Verdi's *Re Lear* would have succeeded in a Romantic, idealistic, upward-thrusting vein. Reimann's *Lear,* on the hand, seems to take life from the Shakespearean (and Hobbesian) notion of humanity preying upon itself, as his Duke of Albany says, "like monsters of the deep." Reimann interprets the play very darkly; the light of Cordelia's goodness flickers very faintly indeed.

Reimann achieves his grim effects through techniques introduced by Richard Strauss, Arnold Schoenberg, and Alban Berg: intricate formalistic patterning, the marshaling of tone rows and tone clusters, extreme variation in sound mass, extreme variation of vocal utterance from spoken word to coloratura, a tendency of the orchestra to sound in separate string, brass-and-wind, and percussion groups, and a willingness to forego all regular metrical organization. The two primary elements of the Verdian style—melody and emphatic rhythms—are abandoned by Reimann.

His vocal line is angular, often more declaimed than sung. Voice and orchestra seldom parallel each other, and in important passages the music tends to act as punctuation between phrases. It does not support and "lift" emotional gestures as with Verdi, but rather surrounds and pervades them. The Reimann style is epitomized in the storm scene, with its alternation of vocal and orchestral comment, the frequent pulse of a dirge, and a tendency to set important phrases to a single note that is simply repeated.

One might summarize Reimann's musical ethos as relying on an undifferentiated string sound-mass arranged into tone clusters rather than chords. At some points every one of the forty-eight string players has his or her own stave in the orchestral score; sometimes they are instructed to maintain a droning tone cluster. All this helps to evoke the primordial randomness of Lear's world. On the other hand, the passionate exertions that play against this sound-mass are usually reflected by the woodwind, brass, and percussion sections—sounding in unpredictable, nervously skittering outbursts.

The compositional techniques of the Second Viennese School, so abundant in the score of *Lear,* require enormous labor. The composer must spend much energy rationalizing and relating the various formal devices imposed by the tonerow system. He spends much of his time not on the Verdian level of scene-length structuring or vast key-relation designs, but on the level of musical microincidents. Those who are not friendly to such methods will perhaps find justice in the hellish agony Reimann suffered as he wrote the storm scene: "Three weeks I lived in this chaos. At nights I always found myself in this vortex of sound; musical figures and harmonies expanded beyond their normal dimensions, turned into abstract forms, which tormented, menaced, surrounded and crushed me"—an atonal composer's nightmarish version of "The Sorcerer's Apprentice"!

Reimann's unremittingly dark view of the play perhaps forced him to avoid any striking differentiation in vocal characterization. His casting shows this: three sets of characters show an odd lack of distinction in range and tessitura: Lear and Gloucester (baritone and bass-baritone); Edmund and Edgar (tenor, countertenor); Goneril, Regan, Cordelia (dramatic soprano, soprano, soprano). In each case the similarity of vocal category blurs spectacular differences in character— most obviously the far greater charisma of Lear in comparison with Gloucester and the truly lyric sweetness one might have expected to distinguish Cordelia from her sisters. In fact, the vocal lines of the three sisters are remarkably similar.

That Reimann cast the Fool as a speaking role constitutes one interesting parallel with Verdi, who wanted a contralto in the role. Both composers clearly saw the Fool as a radically "different" participant in *King Lear*'s world. This was perceptive, for the Fool is certainly the best combination of wisdom, observation, and tact in the play. He is as astute as Kent, but with one crucial difference: he does not participate in the moral scheme of the play. The play's world is too profoundly "foolish" to be affected by the truth-saying of a Fool. The Fool might as well be a Martian, for all the good he does. To convey this apartness, Reimann ingeniously distinguishes his cerebral Fool musically from everyone else on stage by giving him throughout that most cerebral of accompaniments, the solo string quartet.

Reimann was also not interested in achieving any sense of variety of locale. In the second part, for example, the scenes in Albany's palace and the Dover camp occur simultaneously on stage. The moods of these scenes in the play are vividly contrasting, but in the opera the music underlying them is not. There is rather a terrible unity of effect: for Reimann, Lear's *whole world* is inclement, darksome, and aleatory. The music, whether noisily frantic or eerily calm, is thus always oppressively aloof and seemingly random.

Shakespeare's play exists in part for its grand poetic and philosphical gestures. These did not interest Henneberg and Reimann; their real interest and power lay in the expression of effects of decomposition and negation that are undeniably part of the play. As one might expect, Reimann's setting of Lear's five annihilating *never*'s in the last scene is tranfixing (they were of no interest to Verdi and Somma!). His style, which runs along the continuum from static, unmeasured calm to a boisterous anarchy of sound, is not the kind of style that could rest *on* Shakespeare's great lines that propel the upward movement of the play (Lear discovering that his "true need" is to love Cordelia). I find it easy to imagine Verdian music doing just that. Reimann's style, however, allowed him to fall into the terrible human abyss *between* Shakespeare's lines, and he proved that a powerful stage work lay in that abyss. Verdi, for all his genius, could not have done that.

24

The *Measure* of Wagner

The libretto of *Die Meistersinger,* it was ventured recently, "may well be the most humanely prodigal—the most Shakespearean—comedy in modern literature." When one thinks of the playwright's many comic figures who perfectly fit Hans Sachs' remark about Beckmesser—

> The hour of weakness comes to every man;
> then he becomes ridiculous . . .

—one is tempted to give this spectacular praise some credit. Wagner himself would surely have been pleased by it, for he once wrote that Shakespeare was "the mightiest Poet of all time" and associated him on several occasions with his supreme musical idol. He referred to "these twain Prometheuses—Shakespeare and Beethoven," and asserted once that by "applying Beethovenian Music to the Shakespearean Drama might lead . . . to the utmost perfecting of musical Form."*

Wagner's admiration—like Verdi's—appears to have been one of very early onset. One of his biographers tells us he threw himself into the study of English "merely to make a thorough acquaintance with Shakespeare," and we have hilarious evidence from Wagner himself that the plays fired a young boy's imagination. At the age of fourteen he busied himself with a grandiose tragedy called *Leubald und Adelaide,* which he described as "nothing but a medley of *Hamlet* and *King Lear."* The plan of it, he wrote, was "gigantic in the extreme; forty-two beings perished in the course of this piece, and in its working out I saw myself compelled to call the greater number back as ghosts, since I should otherwise have had no characters left for its latter acts." One is reminded here of Shakespeare's own bloody piece of juvenilia, *Titus Andronicus,* in which a modest thirteen members of the *dramatis personae* suffer a violent demise.

About seven years later, Wagner carried through his first and only attempt at a Shakespearean opera: *Das Liebesverbot (The Ban on Love).* This "grand comic opera in two acts" based on *Measure for Measure*—it followed *Die Feen* but was the first Wagner work to reach the stage, in Magdeburg on 29 March 1836—is an exuberant, elephantine beast. Only dyed-in-the-wool lovers of Bellini, whose vocal style it often mimics, would argue with Martin Gregor-Dellin's conclusion that it is the "least successful of all his works."* Even the preface to the 1922 piano-vocal score (a massive 594 pages) makes bold to suggest that, in composing the opera, Wagner was able to exorcize reactionary artistic tendencies that were hostile "to a better self he had not yet discerned." Dean devastatingly concludes that "a first glance at the score suggests a mixture of *Zampa* and *The Pirates of Penzance.*" He also concludes—echoing Andrew Porter's *bon mot* about Berlioz' *"Much Ado* without the ado"—that "there is no 'measure' of any kind in the libretto."

And yet . . . *Das Liebesverbot* is very much worth examining here, not least because the juvenilia of great and highly characteristic artists are often prophetic of future achievements. We would not now mourn the early death of, say, Jane Austen, if she had left behind only the charming miniature "novels" she wrote as a child, any more than we would mourn a Shakespeare who died with only *A Comedy of Errors, Titus,* and *Henry VI* to his credit, or a Wagner boasting only a *Feen* and *Liebesverbot.* Yet, because in each case much splendor followed, these apprentice works take on an importance that extends far beyond their intrinsic artistic worth. The tree growing as the twig is bent, they put this later splendor in illuminating perspective. As Dean hints, *Liebesverbot* offers but a meager foretaste of the familiar Wagnerian musical style; it is in fact Wagner's one *echt bel canto* opera. But it is remarkably prophetic of certain more general contours of Wagner's art. More to the point, this work also exemplifies several of the ever-present conflicts and compromises that attend the transposition of Shakespeare into music.

It is a commonplace that what a particular actor, director, scholar, or composer finds in a Shakespearean play will often tell more about the finder than about the play itself. This is especially true when the interpreter falls, as Wagner does, into that general category of artists famously defined by Schiller as "sentimental" (in opposition to "naïve"). The "naïve" poet, according to Schiller, disappears in his work: "The object possesses him entirely. His heart does not, like a cheap metal, lie on the very surface, but, like gold, must be sought in the depths." In a brilliant essay, Isaiah Berlin sought to put Verdi in this class, along with Shakespeare, Homer, and Cervantes. On the other hand, the soul of the self-absorbed "sentimental" artist, Schiller writes, is "fascinated by its own play . . . he never sees the object, only its transformation by his own reflective thought." Berlin summarizes: "the effect of the sentimental artist is . . . tension, conflict with nature or society, insatiable craving, the notorious neuroses of the modern age, with its troubled spirits, its martyrs, fanatics, and rebels, and its angry, bullying preachers [like] Rousseau, Byron, Schopenhauer, Carlyle, Dostoevsky, Flaubert, Wagner, Marx, Nietzsche." And sentimental music as Berlin describes it—"self-conscious, self-regarding, doctrine-influenced, accompanied by theories and manifestos"—is supremely represented by the Wagner canon. What makes *Das Liebesverbot* interesting is that Wagner's alterations of *Measure for Measure* unabashedly show the emergence of the opera repertory's greatest "sentimental" composer. The sentimental artist's habit of investing his own preoccupations and life experiences none too subtly in his work is clearly apparent in those parts of the play that Wagner chose to throw into high relief.

Before looking at his choices, let us consider the play that in Germany is called *Masz für Masz*. It will be recalled that Shakespeare combined two time-honored plot devices—the Corrupt Governor and the Disguised Ruler—in his story of the Duke of Vienna, who leaves his power in the hands of a rigorous deputy (Angelo) in order to observe, under the guise of a friar, the administration of his duchy's laws. Shortly afterward, Angelo hears the eloquent plea for mercy of Isabella, a novice in the order of Saint Clare, on behalf of her brother (Claudio), who is condemned to death for impregnating his fiancée before the marriage is actually performed. Suddenly overwhelmed by lust for Isabella, Angelo—like Puccini's Baron Scarpia—offers a reprieve in exchange for sex with her, and from this offer the play's various events develop.

Upon this plot Shakespeare raised one of his most complexly and penetratingly intellectualized actions. It proceeds just as its master-manipulator Duke proceeds, "By cold gradation and well-balanced form." J. L. Styan has laid down this rule of playwriting, "At its peril does a play begin with ideas, abstractions, themes, intellection"; but in *Measure* Shakespeare merrily flings this rule to the ground and tramples on it: in the first twenty-three lines there are no fewer than twenty-one abstract terms (*justice, worth, power, grace, honor,* etc.)! The play is, perhaps more than any other in the canon, about ideas. The dramatist almost seems, for once, more interested in exploring the finely balanced antitheses of justice and mercy, courage and zeal, integrity and compromise that his plot encompasses than in creating credible human emotions and characters. This may be why Hazlitt felt "in general a want of passion" in *Measure for Measure,* and why others have found it "dark and bitter"—a decided member of the "problem play" category.

Since music is notably inept at describing abstract concepts or intellectual processes, it is remarkable that Wagner should have been drawn to the greatest "drama of ideas" in the canon: no other composer has ever been tempted by it. And what on earth, one wonders, drew the scarcely pious youth to what has been called the "most theological" and "most specifically Christian" of all Shakespeare's dramas? It is also surprising that the twenty-one-year-old composer gravitated toward a work of such sober maturity; it was, after all, Shakespeare's twenty-seventh play, written when he was forty (in the Renaissance, the beginning of old age).

But our composer, who was eventually to turn his mistress, Mathilde Wesendonck, into Isolde and his arch-enemy critic, Hanslick, into Beckmesser, knew precisely what elements of *Measure for Measure* would allow him to express "sentimentally" his own mood, preoccupations, and political inclinations, and he went after them with a single-mindedness that was to mark all his subsequent projects. Wagner was not the first—or last—to ransack a Shakespearean source for precisely what appealed to him, and it is at least to his credit he did so consciously and with candor.

When we learn Wagner's state of mind as he fleshed out the *Liebesverbot* libretto on a "wonderful summer trip" through the spas of Bohemia, we begin to see how he achieved the incredible feat of transforming *Measure for Measure* into what has been called "a vivacious celebration of free love, free indulgence, and free everything else that engages the senses." Many years afterward Wagner recalled, "I was then . . . full of youthful zest and armed with a positive outlook on life. . . . I had left abstract mysticism behind and became endeared to materiality. The beauty of substance, wit, and spirit were prodigious revelations to me." Then he continues, "The fruit of these impressions and moods was *Das Liebesverbot.* . . . I drew the subject from Shakespeare's *Measure for Measure* and, in following my state of mind, reworked it very freely into a libretto."

Swept up in the early stages of his passionate affair with Minna Planer and already restive with the stuffiness of German bourgeois society, Wagner naturally identified with Shakespeare's Claudio and Claudio's incorrigible, amoral, free-spirited friend Lucio (Wagner's Luzio becomes the opera's primo tenore). From this utterly self-absorbed bias sprang Wagner's manipulation of the play. He explains, forthrightly: "the keynote of my conception . . . was particularly directed against puritanical hypocrisy, which thus led to a provocative exaltation of the 'liberated sensuality.' I took care to understand the grave Shakespearean theme only in this sense. I could see only the gloomy, austere deputy, his heart aflame with most passionate love for the beautiful novice, who, while she beseeches him to pardon her brother . . . at the same time kindles the most dangerous fire in the inflexible Puritan's breast."

And so Wagner retained the Corrupt Magistrate and jettisoned the Disguised Ruler. It is astonishing but not surprising, then, that Shakespeare's central figure—the Duke, who has over thirty percent of the play's lines—vanishes completely. Wagner was obviously in no mood to devote his energies to a hero who says he has

> a purpose
> More grave and wrinkled than the aims and ends
> Of burning youth. (1.3.4–6)

More to the composer's taste was Lucio's "I had as lief have the foppery of freedom as the morality of imprisonment" (1.2.133–134). His Angelo figure thus becomes the opera's focus and principal male voice . . . a villainous bass, of course, named Friedrich. "I transposed the scene to the capital of glowing Sicily, where a German viceroy, aghast at the incomprehensible laxness of its populace, attempts to carry out a puritanical reform." At the opera's happy end his King returns. Though he is given no words, Wagner says that the King "will surely be pleased to see how ill the sour puritanism of Germany becomes hotblooded Sicily." That is to say: how ill German puritanism suited the passionate young Wagner.

Aside from reflecting Wagner's abandonment at this time of German in favor of Franco-Italian music, this major change was prophetic in two other respects. First, Friedrich is made an overtly repulsive tyrant, worthy of Lucio's witty remark in the play that when Angelo "makes water his urine is congeal'd ice." He's called in the opera a "man of lies," a clown, hypocrite, heinous man, and monster, and his subjects are consistently and unanimously satirical about "the German fool" ("der deutsche Narr") who rules them. This sets up an un-Shakespearean "revolutionary" theme; Wagner in fact has his hero Luzio shouting to the citizens of Palermo, "Down with the despicable tyrant," just before the King's entourage arrives to placate matters (shades of the *Fidelio* climax). As C. F. Glasenapp observed, the people of Palermo are "made active interveners in a manner that would never have occurred to the politically conservative Shakespeare." Surely there is a premonition here of Wagner's partisanship just a decade later, when libertarian revolutions swept the Continent.

Second, Wagner was eager, as he said, to have his hypocritical viceroy "brought to justice only by the avenging power of love." This looked back to German opera's greatest heroine, Beethoven's Leonore; especially reminiscent is the moment Isabella uses Leonore's word for Pizarro and calls Friedrich "Abscheulicher!" More important, this liberating love looked forward to the actions of several loving redemptresses Wagner was to create later on, notably Senta, Isolde, and Brünnhilde.

Shakespeare's carefully balanced conflict between human nature's "wanton stings and motions of the sense" (1.4.59) and the "temper'd judgment" (5.1.473) of law charitably applied thus fell by the wayside. Wagner candidly admitted his operatic license: "The fact that these powerful features are so richly developed in Shakespeare's creation only in order that, in the end, they may be weighed all the more gravely in the scales of justice, was no concern of mine: all I cared about was to expose the sin of hypocrisy and the unnaturalness of a ruthless code of morals." Thus, he concludes, "I discarded *Measure for Measure* completely."

It should be hastily added that Wagner's *idée fixe* did bring to life an authentic and vital aspect of Shakespeare's play. For *Measure for Measure* in a sense *is* an argument against rigidly legislated controls of human conduct. Wagner's typecasting of Angelo-Friedrich as an "inflexible Puritan" was also trenchant and just. Though laws condemning adulterers to death did not exist in Shakespeare's day, there was much pressure from the growing Puritan sector of English society for just such a draconian statute (Puritans were also the most vocal enemies of the theater). For instance, George Stubbes wrote in his popular *Anatomy of Abuses,* several editions of which appeared during Shakespeare's life: "I would wish that the Man or Woman who are certainly known . . . to have committed the horrible fact of whoredom, adultery, incest, or fornication . . . should drink a full draught of Moses' cup, that is, taste of present death as God's sword doth command, and good policy allow." In 1650, with the Commonwealth, the Puritans finally got their way, and capital punishment for adultery was established . . . and public theaters of course forbidden.*

Measure for Measure mounts a pre-emptive attack on such cruel punishment. Shakespeare knew where his audience's sympathy would lie: the Globe, after all, was smack in the middle of London's most notorious "tenderloin" area. He knew they would side with the part-time pimp, Pompey, who warns his judge that one might as well "geld and splay all the youth of the city" as pass a law against prostitution. Make it a capital crime, and Pompey wittily predicts the depopulation of the city: "If you head and hang all that offend that way . . . in Vienna ten year, I'll rent the fairest house in it after threepence a bay" (2.1.238–242). And it was perhaps Lucio's sensibly realistic view of human sexuality that so enamored Wagner of him: "A little more lenity to lechery would do no harm . . . it is impossible to extirp it quite, till eating and drinking be put down." We are surely all of Lucio's party when he exclaims, "Why, what a ruthless thing is this in him, for the rebellion of a codpiece to take away the life of a man!" (3.2.97–103, 114–116). This "countercultural" view embedded in Shakespeare's play becomes the battle cry of *Das Liebesverbot.*

Most of the other noteworthy changes made in the play derive from the excision of the Duke. Perhaps most important, Isabella takes over some of his duties as benevolent manipulator of the action. Thus, she—not the Duke disguised as a friar—has the idea of substituting Mariana (Angelo-Friedrich's former betrothed, rejected for lack of dowry) for herself in the sexual part of the bargain. That she has already hatched this plot when she has her memorable interview with her brother spoils the keen conflict in the play between her righteous insistence upon her chastity and his understandable fear of death. Wagner forces her, rather nastily, to leave Claudio in fear of execution as punishment for his "faint-hearted vacillation." Her exultation near the end also makes it clear she has taken over the Duke's *deus ex machina* duties: "Triumph! Triumph! The plan is consummated! I play the game with death and with laughter, and in cunning retaliation win the victory!"

But, oh, what a fall there is for Shakespeare's heroine at the opera's end! Wagner deprives her of her magnanimous plea to the Duke for mercy to Angelo (everyone on stage makes the gesture in unison). And, with no ducal figure on hand to make her a fit husband, Wagner has her suddenly admit that she has loved Luzio "from the first" and then exit arm-in-arm with him.

In fairness, a few instances of fidelity to Shakespeare deserve to be noted. The extended duet in which Friedrich is suddenly inflamed by Isabella's eloquent goodness nicely captures the flavor of the original. The sweet and attractive vocal personalities of Mariana and Claudio do justice to their Shakespearean counterparts. And perfectly natural was the transformation of the play's two clown roles—the "simple constable" Elbow and Pompey—into a basso *buffo* captain of the watch, Brighella (Italian for "clown"), and a tenor *buffo* charmingly named Pontio Pilato. Finally, it has been observed that the point of the brief clowning scenes in the play is to display "the comic ebullience of the private individual undaunted by institutions." The exuberant, carousing vitality of the "suburb of pleasure" and carnival scenes in *Das Liebesverbot*—though their music may be clichéd—is worthy in spirit at least of the bawdy, noisy nightlife of Shakespeare's own Bankside environs, which lay safely outside the jurisdiction of the City of London's sober governors.

Speaking of which, it is an amusing irony that the sober governors of Magdeburg considered Wagner's title objectionable for an Easter-season production. "The Novice of Palermo" proved much more appealing to them. Wagner also found himself assuring the police—rightly enough, but, given all his changes, surely with his fingers crossed—that his opera was "modeled on a very serious play by Shakespeare."

The first (and last) performance clearly fell into the category of Delicious Debacles. The management could not afford to print the libretto, so the audience was left, Wagner recalled, "in a complete haze." All the male principals were unsatisfactory; especially the tenor, whose memory was "most defective." The women were only fitfully up to par, while the orchestra "poured forth its inexplicable effusions, often with exaggerated noise." Just before the second performance was to begin, the tenor singing Claudio, who was having an affair with the Isabella, was assaulted by the latter's husband. This led to minor bloodshed and "general fisticuffs" backstage. When Wagner peeked through the curtain, he could see only three persons in the auditorium—an intrepid Herr and Frau Gottschalk and a Polish Jew in full evening dress. Wisely, the performance was canceled, and *Das Liebesverbot* sank into oblivion. After the disaster Wagner permitted himself, in a letter, a comment on the "powers that be" in Magdeburg that was perhaps worthy of Shakespeare's "lewd fellow," Lucio: "They are all shitheads here!"

25

Taming *The Shrew*

Near the beginning of Shakespeare's exuberant domestic farce, *The Taming of the Shrew,* the servant Tranio offers his master this sensible advice:

> No profit grows where is no pleasure ta'en.
> In brief, sir, study what you most affect.

This sentiment applies to the canon itself: so great is its variety that we are bound to seek our pleasure in those plays we have the most affection for—and leave the others alone.

It is a matter of personal taste. These last several years, for instance, the seldom-performed *Love's Labour's Lost* has been my favorite of all the plays (it is Shakespeare's most profound, searching comment on the powers and pitfalls of language in general and poetry in particular). It is always on my syllabus, while *Hamlet,* a play that has never appealed to me, recently made its first appearance there. Margaret Drabble has said that *Antony and Cleopatra* is "the most sublime work ever written . . . whenever I think of it I know that art is timeless, that the human spirit is immortal and that joy is forever." But Verdi said he felt "the loves of the chief characters, their personalities, and even their misfortunes arouse little sympathy." Though he had a tin ear, Samuel Johnson's very decided personal tastes have often been ratified by opera composers. His view of *Julius Caesar,* for instance, pretty well explains its lack of success in music: "I have never been strongly agitated in perusing it, and think it somewhat cold and unaffecting." But Falstaff, Shakespeare's most successful operatic *buffo* role, left Johnson uncharacteristically at a loss to explain his delight: "unimitated, inimitable Falstaff, how shall I describe thee?" And his summary of *Romeo and Juliet* explains why it has been the most popular of the tragedies with composers; it is, he says, "one of the most pleasing of our author's performances. The scenes are busy and various, the incidents numerous and important, the catastrophe irresistibly affecting, and the process of the action carried on with such probability . . . as tragedy requires." To each, in other words, his own.

An obvious corollary to Tranio's advice is that those premises and personal affections we bring to a play will determine where and how we eke our "profit" from it. *The Taming of the Shrew,* as it happens, offers one of the starkest examples of this fact in the entire canon. I am referring, of course, to the notorious speech uttered by Kate just before the end of the play. The radically different kinds of dramatic profit to be reaped here can be seen by focusing solely on this remarkable performance, for everything else in the action depends on how it is interpreted by a director, or a composer and librettist.

Kate's speech would seem custom-made for operatic exploitation. First, it is without question the play's denouement: the action flies toward it like an arrow to a bull's-eye. At forty-four lines—twice the length of her tamer's one big solo—it is also the play's only knock-'em-dead set speech . . . eloquent, elaborate, and poised. What better pretext for an aria? Since Shakespeare nowhere else placed a climactic speech so near a play's end (just ten lines follow), it would seem perfectly perched for a rousing finale by a triumphant soprano. Furthermore, composers dote on moments when a character must give an "interior" performance on stage, and Kate's speech is indeed a command performance, with the entire cast assembled as her audience. And yet . . . in two of the *Shrew* operas we shall consider here, the composers, incredibly, chose to ignore the speech *entirely.* Before looking at why and how this main liberty was taken, however, it is necesssary to brush up on some of the Shakespeare criticism that has roiled around the tamed shrew's *tour de force.*

The great problem is that what Petruchio (and Shakespeare) oblige the heroine to say in this speech is hard to stomach. Shaw called the speech "degrading" and said it "might have passed with an audience of bullies" in "an age when woman was a mere chattel." And in a letter to the editor of the *Pall Mall Gazette* under the pseudonym of Horatia Ribbonson he expressed the sanguine hope that "all men and women who respect one another will boycott *The Taming of the Shrew* until it is driven off the boards." A recent commentator has correctly summarized that the speech "has embarrassed generations of critics," and even its warmest defenders feel obliged to advert to its "self-extinguishing" import and "superficial tidiness" or grant that the inevitably dehumanizing farce plot leaves Kate at the end displaying "some vestiges of a trained bear." One critic simply throws up her arms and confesses such a denouement is unhappily typical: "Shakespeare's comic endings often dissatisfy us . . . because they sweep much dust under the rug."

The context and gist of the speech can be briefly rehearsed. Soon after Petruchio has succeeded in taming Kate and they have had a brief scene alone proving they are now a genuine "item" (5.2.142–150), Petruchio and two other newly married husbands wager a hundred crowns on whose wife will come most promptly when bidden. Hortensio's refuses flatly; then Lucentio is turned down by Kate's sister Bianca, hitherto of "mild behavior and sobriety"! This sets the stage for Kate to stun everyone with, as Petruchio boasts, her "new-built virtue and obedience." She appears at once, bringing the "froward wives" with her. Petruchio says, "Katherine, I charge thee, tell these headstrong women / What duty they do owe their lords and husbands," and so follows her masterpiece of "womanly persuasion":

Fie, fie, unknit that threat'ning unkind brow,
And dart not scornful glances from those eyes,
To wound thy lord, thy king, thy governor . . .
Such duty as the subject owes the prince,
Even such a woman oweth to her husband;
And when she is froward, peevish, sullen, sour,
And not obedient to his honest will,
What is she but a foul contending rebel,
And graceless traitor to her loving lord?

As the visual climax for her lecture she leads the two in the ritual gesture of placing her hand under Petruchio's foot: "In token of which duty, if he please, / My hand is ready, may it do him ease."

Obviously, Shakespeare thought he had created a breathtaking theatrical coup for the end of his farce. But now, in the era of women's liberation and equal rights, it is—at first blush—breathtaking in a way he could not have foreseen. The speech is at least discomfiting and at worst, in Shaw's words, "altogether disgusting to modern sensibility." Even a coarse Restoration version of *The Shrew* called *Sauny the Scott* (circa 1670) dropped the speech, and Garrick gave the last six lines quoted above to Petruchio to soften Kate's odious self-abnegation in his *Catharine and Petruchio* (or merely to fatten his role), which elbowed Shakespeare's version into the wings for over a hundred years.

In this century considerable ingenuity has been expended in trying to arrive at a comfortable way to play the speech. The most conservative faction asserts that Shakespeare must be taken as a man of his age, for which male dominance within the marriage bond was a biblically ratified, time-honored fact of life. To wish the play otherwise is our own time-bound and sentimental self-indulgence.* They also say that Kate at the beginning is not a pretty picture—"rough," "wonderful froward," "intolerable curst," "sullen," and "stark mad" with vengeful, selfish instincts—whatever means can transform this "wild Kate" into a "household Kate" are justified.

A more diplomatic faction urges careful attention to the actual language and dramatic context of the speech. They say that reciprocity, not unilateral domination, is the keynote. Ideally, the husband

cares for thee,
And for thy maintenance; commits his body
To painful labor, both by sea and land;
To watch the night in storms, the day in cold,
Whilst thou li'st warm at home, secure and safe;
And craves no other tribute at thy hands
But love, fair looks, and true obedience.

Then they quote from marriage manuals of the day, which actually tried to extinguish the bad old wife-as-chattel belief. Six editions of Robert Cleaver's *Godly Form of Householde Governemente* appeared during Shakespeare's career, for instance; and he wrote, "surely they are but very fooles, that judge and thinke matrimony to bee a dominion" of the husband over the wife. They also point out that wifely obedience is required, not to chauvinistic willfulness, but to a husband's *"honest* will." They also suggest that another popular treatise, *The Instruction of a Christen Woman,* may hit on the effect Shakespeare was after: "A good woman by lowely obeysaunce ruleth her husbande." This faction thus emphasizes the positive and views the denouement, more subtly, as a paean to "marriage as a rich, shared sanity. . . . Marriage is addition, not subtraction: it is a sad let-down if the dazzling action of the play produces only a female wimp."

This faction also points out how the context softens the apparent chauvinism of the event. They say the little dialogue in the preceding scene—which ends

> *Katharine:* Nay, I will give thee a kiss; now pray thee, love, stay.
> *Petruchio:* Is not this well? Come, my sweet Kate.
> Better once than never, for never too late

—makes it clear that the two are now in cahoots, and thus the behavior of both in the wager scene is "performed" with self- (and audience-) delighting irony. Petruchio is, indeed, showing his trust in Kate by giving her the power to make a public fool of him. As one partisan of this faction concludes, "The wager has already been won, and husband and wife are playing a game whose object is to demonstrate their superiority as a couple to their scornful relatives."

A third and last major faction takes this "ironic" reading and runs with it for a feminist touchdown. This faction cannot imagine that when Shakespeare wrote such stuff for Petruchio as—

> I will be master of what is mine own.
> She is my goods, my chattels, she is my house,
> My household stuff, my field, my barn,
> My horse, my ox, my ass, my any thing . . . (3.2.229–232)

—he was not subversively reducing the tenets of male dominance to repulsive absurdity. Thus, they view Kate's speech as spoken with spirited tongue-in-cheek irony. Margaret Webster first urged this approach in 1942, and Harold Goddard, in 1951, concluded similarly that "the play is an early version of *What Every Woman Knows*—what every woman knows being, of course, that the woman can lord it over the man so long as she allows him to think he is lording it over her."

Subsequent feminist readers have, of course, refined and advanced this position (called "Katolatry" by its critics). Coppélia Kahn takes pleasure, for instance, in how Kate's "spirit remains mischievously free" as she "steals the scene from her husband." Kate "treats us to a pompous, wordy, holier-than-thou sermon which delicately mocks the sermons her husband has delivered to her." Remaining "outwardly compliant but inwardly independent," she pretends to complete the fantasy of male dominance while, with a wink to the audience, she is in fact mocking it as a mere fantasy. And so we are left pitying Petruchio, both for being duped and for authenticating his manhood in such a boorish way.

And so we have three starkly contrasted—and mutually exclusive—ways to translate *The Shrew*'s climactic speech into stage action. I am inclined toward the middle position between the "historical" and "ironic" extremes, but at the same time suspect that, were Shakespeare writing the play today, he would find the feminist slant appealing. In all his other comedies, after all, the women seem to get the upper hand. As for music, one can plausibly imagine certain of the great composers leaping characteristically—and successfully—to one of the three choices. Wagner's expansive male ego would have led him to the literal reading, redemptive female self-extinction being a hallmark of several of his plots. One can imagine Mozart or Verdi admirably steering the nuanced middle course, while the insouciant subversions of the ironic interpretation would seem just right for a Prokofiev or Shostakovich.

Unfortunately, no composer of the first rank has taken up the challenge. The composers who have set the "peremptory" Petruchio and "proud-minded" Kate at musical loggerheads have, also unfortunately, evaded the challenge at the heart of the play. Still, there is almost always a grain of interpretive truth even in the most appalling infidelity . . . or, at the very least, fascination in how cunningly the infidelity is achieved. So a brief look at a few *Shrew* operas will, I think, prove worthwhile.

The first, "freely fashioned" after Shakespeare, was composed by Hermann Goetz (1840–1876); its German title is *Der Widerspenstigen Zähmung*. It premiered in 1874 and came to London in an English version a few years later. The opera had a considerable vogue, and, perhaps in part because Goetz dispensed with the "disgusting" speech, Shaw was much taken with it. In 1886 he called it "the greatest comic opera of the century, except *Die Meistersinger,*" and a few years later he put it in august company indeed, referring to "Goetz's *Taming of the Shrew* and Wagner's *Die Walküre,* both of them beautiful and popular works." In 1947, at the age of ninety, he wondered out loud, "why are such masterpieces as Goetz's *Shrew,* Mozartean in its melody . . . shelved for operas by Rossini which our vocalists cannot sing?" Dean, on the other hand, is fairly hard on Goetz, saying he softens "the bite of the story" and emasculates the anti-love affair (Petruchio "repeatedly declares his love in conventional terms, while [Kate] is eager to surrender almost from the start").

The main difficulty of the Goetz *Shrew* is that it responds to a single—but important—element of the play: romantic love. The Arden editor makes it clear that "romantic love is a presence in the play from beginning to end" and that "it functions as a kind of prolonged and delicious illusion." The trouble is that Shakespeare subjects this illusion to subversion and satire. Goetz and his librettist, J. V. Widmann, however, became, like several of the play's comic butts, prisoners of the delicious illusion. They take the romantic illusion seriously, and so their couple sinks into a rather soupy "Romantische Strasse" relationship that lacks all the wit and din of the play. Even the wooing scene—fit for a boxing ring—turns out, as Dean observes, "tender as any romantic love-duet." Henry Krehbiel complained about all this many years ago when he wrote that the Goetz opera "was too much like good Rhine wine, and too little like champagne to pass as a comic opera," though surely a robust Bardolino would be more suitable to the venue. The vocal casting, too, hints at the difficulty: having Petruchio a baritone and Kate a soprano evokes more than a little Wagner's Dutchman and Senta, especially when the soprano gets a big solo *scena* well before the end in which she all too nineteenth-century-operatically divulges, "I'd give my life for him."

Expectedly, Shakespeare's tiny nine-line hint of true-love-at-last balloons into a full-fledged love duet, on which the entire cast eventually "looks in." There being no wager, it is only necessary for the soprano to utter a two-line synopsis of The Speech: "All my stubbornness and pride, and more, have I sacrificed to him, whom I must love with all my heart." Needless to add, there was no place in this version for Petruchio's down-to-earth, brilliantly anticlimactic line, the play's most wonderful and famous: "Why, there's a wench! Come on, and kiss me, Kate."

The *Shrew* of American composer Vittorio Giannini (1903–1966) received its premiere in Cincinnati in 1953. When the New York City Opera mounted it, the *New York Times* greeted it as a "capital show" and noted in it what would seem to be all the right qualities—"crispness . . . vivacity . . . boisterousness" and an orchestra "alive with witty comment." Written mainly in the idiom of Puccini and Menotti, the opera begins with auspicious fidelity: Bianca gets all three of her suitors, and the marking *ruvidissimo* ("very coarsely") occurs with apt regularity. But, something goes bizarrely, amusingly haywire as the opera nears Kate's speech.

For, as with Goetz, the original play is given up as hostage to the illusion of romantic love: Giannini and his co-librettist Dorothy Fee were determined to take seriously this illusion at which Shakespeare takes so many witty potshots. This caused a problem, however, since the play boasts no flowery poetry to support the lyricism they had in mind. So they went shopping in the most obvious Shakespearean malls for what they needed: the Sonnets, "Venus and Adonis," and *Romeo and Juliet*. One can perhaps see from the act 2 scene for the secondary lovers, Bianca and Lucentio, that The Speech is doomed. It is thoroughly conventional (*andante sostenuto,* with violin obbligato and a climactic high B for the two singers), and its text is shamelessly patched out of Juliet's "I have no joy of this contract to-night" in the balcony scene and the famous love-at-first-sight sonnet at the Capulet feast ("If I profane with my unworthier hand . . .").

The opera's climactic scene pulls out the same stops, but on a grander scale (there is no wager business here, either). After a noisy concertato involving the marriage of Lucentio and Bianca, Katharine and Petruchio are left alone. The chamber is flooded with moonlight and a lush *Lento—assai espressivo.* The hero, mischievously described by Shaw elsewhere as a "coarse, thick-skinned money-hunter," is here asked to sing "with infinite love" and then, with Kate, to sing some very odoriferous poetry to each other:

> *Katharine:* Is it thy will thy image should keep open
>   My heavy eyelids to the weary night? (Sonnet 61)
> *Petruchio:* How sweet and lovely dost thou make the shame. (Sonnet 95)
> *Katharine:* Such is my love. . . . That for thy right I shall bear all wrong.
>   (Sonnet 88)

> *Both:* Touch but my lips with those fair lips of thine
> (*lost in rapture*) And the kiss shall be thine own as well as mine.
> ("Venus and Adonis")

This magpie ingenuity has to be admired, not least because the very first published reference to Shakespeare was as an "upstart crow" who beautified himself with others' feathers. But if Shakespeare were to attend this opera he might be forgiven for wondering how on earth his bawdy, bold farce came to be confused with the heavily perfumed fantasy world of Romeo and Juliet. Petruchio's "Well, there's a wench," etc.—not exactly in keeping with the ambience of, say, "what light through yonder window breaks"—is dropped from this opera, too.

One other *Shrew* opera deserves mention, especially since Dean thinks it "unquestionably the best," boasting "humor and high spirits, a light touch and tunes that haunt the ear." This is by V. Y. Shebalin (1902–1963); it premiered in 1957.* Its final scene cleverly—perhaps too cleverly—attempts to ingratiate all three factions at once. There is no climactic wager scene (a different bet occurs in the first act), and hence no need for Kate's performance. Instead, she and Petruchio are given a duet in which contradictory sentiments of obedience-over-all and what-every-woman-knows, of love romanticized and love satirized, and even the moderate faction's ironic detachment are all shuffled together:

> *Petruchio:* I did not tame Kate; no, love's powerful force enflamed my fiery blood. Love can tame even a male. Where the light of love is shining there is no master or slave. I am a conquered conqueror. I am at your feet in love with you, but you are silent?
> *Katharine:* I was silent but my heart and soul spoke to yours. Oh yes, Petruchio, you are right: it is you who tamed my wild temper—your love tamed it. I fell deeply in love with you; every moment I love you more and more. You are my husband and my master. You are my governor . . . still, in our weakness we women are stronger than you men and you cannot escape the captivity of tender embraces, my precious friend!

Like any "all of the above" solution to a problem of interpreting Shakespearean multiplicity, this one is bound to leave no one completely happy. At any rate, Shebalin brings the curtain down very quickly after these speeches, giving his audience no time to parse the incongruities.

The coast is still clear, then, for an operatic reading of *The Shrew* that accepts the full challenge of Kate's crucial speech. And, because music is such an able means of conveying an ironic disparity between *what* and *how* something is said, it is easy to imagine an ironic-feminist curtain aria being very effective in the theater. An attempt by a woman composer to tame *The Shrew* would seem, in any case, to be long overdue.

26

Some Ado about Berlioz

If Verdi was—in terms of theatrical art (his practical savvy, the enormous canon, the three operas)—the most Shakespearean of all composers, Hector Berlioz was surely the most Shakespearean in the harder-to-pin-down sense of affinity of "spirit" or "soul." On the last page of his memoirs he bittersweetly confessed he was still unreconciled to never having known Shakespeare, "who might perhaps have loved me."* Coming from any other composer the remark would sound fatuous; coming from Berlioz, it is ineffably touching and sugges-tive . . . not least because both men took such seminal and persuasive liberties with their respective art forms. Berlioz was in effect speaking of himself and the liberating part his works would play in nineteenth-century music when he offered this praise of Shakespeare in a letter to the editor of London's *Morning Post:* "music is one of the forms of poetry . . . and on the same freedom which is characteristic of Shakespeare's immortal conceptions depends the development of the music of the future."

Certainly, no other composer's life and compositions were so passionately or extensively involved with Shakespeare's plays in one way and another. Like Verdi, Berlioz called him Papa Shakespeare, but there is a special élan in the Frenchman's bursts of enthusiasm. Though he could forthrightly announce that "Shakespeare is the supreme creator, after the Almighty," he was often subtle in his apprecia-tions: "It is much harder for a Frenchman to sound the depths of Shakespeare's style than it is for an Englishman to catch the individual flavor and subtlety of La Fontaine or Molière. They are continents, Shakespeare is the world." Late in life, he apostrophized the man he wished so keenly to have known: "Shakespeare, the great dispassionate genius, impassive as a mirror, reflecting what it sees. Yet what ineffable compassion he must have had for all things!"

It is also doubtful that any other composer has known the canon as well as Berlioz.* He admitted that his first infatuation occurred when he could "only glimpse Shakespeare through the mists of . . . translation," but he eventually became a fluent reader and decent speaker of English. He came to be a Shakespearean in his circle of friends, and once read to them all of *Hamlet* (a play he knew by heart in French). His knowledge extended well off the beaten path, too: "I am quite transported by some words of Old Nestor in . . . *Troilus and Cressida*. I have just reread this amazing parody of the *Iliad,* where nevertheless Shakespeare makes Hector even greater than Homer did." On Nestor's line describing Hector raising his sword—"Lo, Jupiter is yonder, dealing life!"— Berlioz comments: "God in heaven but it's beautiful. I feel my heart will burst when I come across lines like that. . . ." It is clear he achieved a high level of connoisseurship. He wrote from London, for instance, "I saw *Hamlet* recently. The new Hamlet, Brooke, is superb, much better than Macready or Kemble." To a critic who said that the composer of the *Romeo and Juliet* symphony failed to understand Shakespeare, Berlioz responded, with justifiable heat, "Toad, swollen with wind and crapulous stupidity! Let him prove it!"*

Berlioz came to know the plays well enough to be disgusted at cut-and-paste stagings, too. He fairly shouted at the "miserable poetaster" who invented the happy ending for *Lear* (quoting *Richard III* at the end): "You fool! You have committed the most odious of crimes, for you are guilty of an offence against . . . genius. May you be accursed! 'Despair and die!'" Perhaps the most accomplished writer among all the great composers and the most devoted to Shakespeare, he was moved to conclude: "*Hamlet* and *Romeo* doctored, new characters added to *The Tempest*—so it goes on. This is where Garrick's example has led. Everyone is telling Shakespeare how he should have written." This keen sense of the dramatist's artistic integrity led Berlioz to what is perhaps the most poignant expression we have of the burden and challenge of adding music to the equation of great literature:

> It is beautiful because it is Virgil; it is striking because it is Shakespeare. I know it. I am only an interloper: I have ransacked the gardens of two geniuses, and cut a swathe of flowers to make a couch for music, where God grant she may not perish, overcome by the fragrance.

More impressionistically, it should be observed that Shakespeare several times figured in bittersweet moments—both major and minor—in Berlioz's life. The Ophelia in his first *Hamlet* was Harriet Smithson, whom he fell tumultuously in love with and eventually married. While waiting for the Opéra to give him a version of *Hamlet* "admirably arranged as an opera," he commenced work on *Benvenuto Cellini,* a splendid work that displays not a little of the exuberance of Shakespeare's romantic comedies, as well as the Romantics' newfound ardor for Shakespeare's times. In the moving pages on the death of his wife, he quotes from *Hamlet* and cries, "Shakespeare! Shakespeare! I feel the flood rising, my grief sweeps over me, I drown! Let my cry come unto thee!" In a happier moment, after the premiere of *Les Troyens à Carthage,* he told of one congratulation "which begins with this quotation from Shakespeare [*Dream,* act 5]: 'Well roared, lion!' Isn't that delightful!" His memoirs end with this terrible valediction: "For you, morons, maniacs, dogs, and you, my Guildensterns and Rosencrantzes, Iagos, Osrics, gadflies, crawling worms of every kind: farewell, my . . . friends. I scorn you, and hope not to die before I have forgotten you." And in a letter to his dear friend the Princess Sayn-Wittgenstein he wrote, in 1862: "Yesterday I put down the last orchestral note with which I shall ever blot a sheet of paper. 'Now no more of that. Othello's occupation's gone.'" This bitter late mood—all the more terrible in its contrast with his enthusiastic youth—is finally to be noted in the four-and-a-half lines from *Macbeth* ("Life's but a walking shadow . . .") that are placed at both the beginning and the end of the *Memoirs*.

Shakespeare, of course, figures very prominently in the Berlioz canon. There is the "dramatic fantasy" on *The Tempest,* a *King Lear* overture, a piece on the death of Ophelia, and a march for *Hamlet*. Each of the *Huit scènes de Faust* carries a Shakespearean motto culled from *Romeo* or *Hamlet*. Berlioz said his "greatest symphony" was based on *Romeo and Juliet*. And he confided to a friend in 1856 that he had "conceived of a vast opera on the Shakespearean plan, based on the second and fourth books of the *Aeneid.*" Ironically, one might well call *Les Troyens,* along with *Boris Godunov,* the two and only great operas based on the general "plan" of Shakespeare's ten chronicle plays.* And there is, of course, its magnificent love duet for Didon and Énée, boldly purloined from Lorenzo and Jessica in *The Merchant of Venice*.

Finally came Berlioz' last opus, *Béatrice et Bénédict*. Like *Falstaff,* it was a buoyant, valedictory *jeu d'esprit* that seemed to appear miraculously amid the gloom of old age, growing misanthropy, and pain. (He composed it while suffering from "my infernal neuralgia" and conducted the premiere at Baden on 9 August 1862 in "excruciating pain.") Since it is in several respects brilliantly characteristic of Berlioz—and in several more, true to the spirit of *Much Ado About Nothing*—this opéra-comique in two acts "Imité de Shakespeare" deserves our attention here.

Many Shakespearean operas carry very un-Shakespearean titles that strongly hint where a composer's theatrical interest will be focused. Sometimes the change is unsurprising. The two *As You Like It* operas naturally take *Rosalind* as their title; after all, this was, at about 750 lines, the largest female role Shakespeare ever created. Five operas retain Shakespeare's title of *King Lear,* but in an art form where the prima donna often dominates, it is no shock to find three works named after Cordelia. Two *Winter's Tale* operas are named after Perdita, another after Hermione. In a few cases, the variety of titles is amusing: five operas set on Shakespeare's Venetian Rialto are named after the merchant, but three others are named after Shylock, two after Jessica, and one after Portia. Dean lists two bona fide *Twelfth Nights* (and two that take Shakespeare's subtitle, *What you Will*), but also three *Violas,* two *Cesarios,* and two *Malvolios*. As for *Much Ado* . . . aside from Berlioz' opera and one that is, incredibly, named after Hero, it has mainly been *Viel Lärm um Nichts* or *Beaucoup de bruit pour rien*.

But of all operas that carry tendentious titles, *Béatrice et Bénédict* must surely stand next to *Falstaff* as—within the limitation the title implies—a splendid evocation of Shakespeare. From the beginning of his *Much Ado* project, Berlioz consciously circumscribed his ambitions, most obviously by conceiving his opera as a one-act divertissement. When his son asked how he could possibly boil five acts down to one, Berlioz replied: "I have only taken one theme from the play (the rest is of my own devising)—that in which Béatrice and Bénédict, who detest each other, are mutually persuaded of each other's love, whereby they are inspired with true passion. The idea is really comic." A year later (and one year before the premiere) he assured his good friend Humbert Ferrand, "I am getting on with a one-act opera for Baden. . . . I promise there will not be much Ado in the form of noise in it." Only when Berlioz enlarged the opera after the premiere did it reach its present evening-length form.

Given that Berlioz had radical pruning in mind, it is hard to object to his priorities. *Much Ado* is comprised of four main elements: romantic love comedy (focusing on Claudio and Hero), witty comedy (Beatrice and Benedick), plebeian comedy (Dogberry and the Watch), and malevolent intrusion (Don John's attempted ruin of Hero's reputation). Some critics have concluded that this theatrical recipe is, finally, a botch. Shaw referred to it as Much Adoodle-do and said, with his usual razzing hyperbole, it "is a shocking bad play" that can "only be saved by Dogberry picking it up at the end." Even its most eager admirers admit the huge difficulty of getting the interpretive balance right in performance . . . or in one's mind.

Berlioz, in any case, chose to use but two of the four elements: the witty and the plebeian comedy. (I will discuss later an important new element added by Berlioz that is, if not strictly speaking derived from *Much Ado,* Shakespearean in a more general way.) And who would deny he made the right choice? Beatrice and Benedick—who have far and away the play's longest female and male roles—steal the show. They are, as a recent editor summarizes, "the play's outstanding interest and the one for which it is primarily known." Shaw must be the only spectator in history who has refused to be charmed by them. And Dogberry—a juicy role tailored for Will Kempe—is one of Shakespeare's most delightful pride-puffed malapropists. As for the Hero–Claudio affair, not much really "happens" in it, and Claudio has with some reason been called a "miserable specimen" and the "least amiable lover in Shakespeare." As personalities, Hero and Claudio seem like wallflowers in contrast to the two merry warriors. Don John's Iago-esque villainy is baldly shoehorned into the action of what must be considered, finally, one of the playwright's happier comedies, and the artless way he is dispensed with at the end represents something of an apex in Shakespearean expedience. So one cannot complain too bitterly—if other rewards are forthcoming—that Berlioz dispenses with him entirely.

In what remains the rewards are indeed great. Aside from making much nearly verbatim use of crucial passages for Béatrice and Bénédict in his spoken scenes, Berlioz' music, I think, captures their spirit splendidly. He was more accurate than most composers in assessing his own music and wrote of his work-in-progress: "It is gay, keen-witted, and now and then poetic; it makes the eyes and the lips smile." This is exactly what an actor who plays the "evermore tattling" hero and the actress who plays "my Lady Tongue" must be and do. And the vocabulary that admirers of these characters exhaust in describing them—*sprightly, vivid, animated, vital, vivacious,* and so on—is perfectly appropriate to describe the effects Berlioz achieves with the jazzy syncopations, veering melodic lines, trills and arpeggios, dotted- and cross-rhythms, and lilting triplet figures of the music for his title characters. Beatrice is praised for her ability to "dance out the answer" in repartée, and Benedick is "from the crown of his head to the sole of his foot . . . all mirth": Berlioz' debonaire and prominent piccolos, flutes, and violins dance out this mirth, while their "self-endeared" pride and the disdain and scorn that "ride sparkling" in their eyes resound in blaring brass and a thumping bass drum. The two lovers are given very accurate advance publicity in the play:

> There is a kind of merry war betwixt Signior Benedick and her: they never meet but there's a skirmish of wit between them.*

Berlioz' accurate advance publicity, of course, is the insouciant *allegretto scherzando* of his overture. Surely he was thinking of this element of the opera when he described *Béatrice et Bénédict* as "a caprice written with the point of a needle and [which] thus needs extremely delicate playing."

Though Berlioz retained *Much Ado*'s plebeian comedy, Master Constable Dogberry vanishes. Some critics, naturally disappointed at such a loss, tend to vent their frustration on his replacement, the parodic composer and conductor at Don Pedro's court, Somarone. But in several respects he is loyal to the letter and spirit of Shakespeare's lowbrow hijinks. *Somarone* is Italian for "big ass," a lovely reminder of Dogberry's great exit line: "O that I had been writ down an ass!" Furthermore, it was a common Shakespearean practice to ring in characters for the sole purpose of offering *moments musicaux;* in *Much Ado* we have Balthasar, who sings "Sigh no more, ladies" and "Pardon, goddess of the night." Somarone performs a similar function with his "Épithalame grotesque." Benedick has some fun ridiculing Balthasar's vocalism—"And he had been a dog that should have howled thus, they would have hanged him, and I pray God his bad voice bode no mischief"—and Berlioz aims this speech verbatim at the "voix discordantes" of Somarone's chorus. There is also much satire of fancy "poetical" language in *Much Ado,* and this, too, is carried over in the fun of the "haute poesie" of Somarone's hymeneal text. The sentiments of this text, incidentally, are horribly malapropos ("Mourez, tendres epoux . . ."), as is the funereal music itself (Berlioz wanted it played "lourd," or ponderously). These are nice equivalents, it seems to me, of Dogberry's malapropisms. Balthasar warns before he sings, "There's not a note of mine that's worth the noting," and that's a good description of the music Berlioz concocted for his musical ass.

A last, more general similarity between Dogberry and Somarone, though, is perhaps the most resonant. The constable is one of Shakespeare's many comic sendups of (to borrow Isabella's phrase in *Measure for Measure*) "man, proud man, / Dress'd in a little brief authority." One can safely assume that his complacency, self-inflation, and thickness were observed by Shakespeare in real life . . . and that Shakespeare was delighted to aim his palpable hits at "the powers that be" of his London environment that harassed him and his profession. Perhaps there had been a Dogberry or Constable Dull in the Stratford of his youth. Somarone, just like Wagner's Beckmesser, performs the same function: he allowed Berlioz a final Bronx cheer at all the self-enamored, talentless, hubristic musical fools he had suffered throughout his life. Somarone's combination of high self-esteem (he orders his "ebony-and-gold baton #37," reserved for "les personnes de qualité," brought to him when Don Pedro enters!) and utter lack of artistic sense or "soul" produces the kind of humor that Shakespeare indulged in frequently, notably with Bottom and Malvolio. (In a scenario from about 1852, Berlioz turned Dogberry into a gardener named Pippo; his final gambit produced an essentially more Shakespearean result.)

Though Berlioz' libretto is most loyal to the play's original lines in the spoken scenes, several of the fifteen numbers perform Shakespearean functions in piquant ways. Hero's status as the conventionally "fair" and "gentle" romantic lover, for instance, is wittily (and beautifully) established by an utterly conventional entrance aria consisting of a dreamy, lovesick *larghetto cavatina*, then a rousing *allegro con fuoco cabaletta* complete with coloratura cadenza (no. 3, "Je vais le voir"). Berlioz exactly translates Benedick's superb line, "What, my dear Lady Disdain! Are you yet living?" and uses it to introduce the one duet (no. 4, "Comment le Dédain") that epitomizes the merry war. The lively trio in which Don Pedro and Claudio vent their "matrimoniomanie" and Bénédict his "matri-moniophobie" (no. 5, "Me marier!") gathers many of the play's witty passages in which the debate between married and single life is pursued. The drinking chorus (no. 9, "Le vin de Syracuse") led by Somarone, and its surrounding "downstairs" slapstick action, follows Shakespeare's familiar habit of relaxing from high-flown or sophisticated action and indulging in good, clean vulgarity. One thinks, for example, of the scene with Peter and the musicians just after Juliet is found dead on her wedding morning (4.5.102–146).

Perhaps Benedick's most memorable line, "I will be horribly in love with her," is also exactly translated and leads into the tenor hero's big aria, a rondo (no. 7, "Ah! je vais l'aimer"). It is naturally derived from his biggest solo speech in the play, though Berlioz had to elaborate on the single line of outright praise—"By this day, she's a fair lady!"—to eke out enough text for a vocal display piece. Berlioz also chose the most obvious speech of Beatrice's to turn into his mezzo's one big bravura number (no. 10, "Dieu, que viens-je"). This is when she emerges alone after overhearing Hero and Ursula gossip about Benedick's love for her. This is the only speech of more than a few lines that Shakespeare gave her which is not in prose, and to underline Beatrice's sudden burst of amorous lyricism he very unusually rhymes it so as to fall but a quatrain short of making a sonnet (*abab cdcd ee*, 3.1.107–116). Since Beatrice, even in love, is not given to "roman-tic" gushing, Berlioz had to invent his text for the *andante* section, but the following *allegro agitato*—beginning with her declaration, "Oui, Bénédict, je t'aime!" sung "avec explosion"—is true to the spirit of Shakespeare's

> . . . Benedick, love on, I will requite thee,
> Taming my wild heart to thy loving hand.

In one respect Berlioz improved on Shakespeare. At the end of the play a messenger announces the capture of Don John, and Benedick lamely and prosaically concludes, "Think not on him till tomorrow; I'll devise thee brave punishments for him. Strike up, pipers!" Berlioz' codas are among the most consistently thrilling in music; he knew a lame ending when he saw one. He saw where the emphasis should have been: upon his Béatrice and Bénédict. He therefore translated their last tart dialogue nearly verbatim, had the chorus enter with four large signs announcing, "Ici . . . l'on voit . . . Bénédict . . . l'homme marie!" and then launched his couple into a fleet *scherzo duettino* that matches the effect of the swirling dance that ends *Much Ado*. Its final lines, though invented by Berlioz, are perfectly in character and make an exuberant coda: "Yes, for today the truce is signed . . . we'll become enemies tomorrow!"

I mentioned above that Berlioz included a significant musical element in *Béatrice et Bénédict* that is not in *Much Ado*. This is the evocation of powerful, romantic passion—of which there is very little in Shakespeare's text. Since the play's main sphere, as the Arden editor writes, is "that of strategic challenge, little room remains for the lyrical enrichments so central to most of the other comedies." It is obvious that Berlioz—having removed the "serious" business of Don John's plot—recognized that he needed something else more sustained, lyric, and emotionally weighty—in short, more genuinely beautiful—to balance all the "coy and wild" spirits and broad humor of his other characters. And so there is a lambent *andante un poco sostenuto* in the midst of the busy overture and several other slow movements interspersed in the following action . . . in most of which Hero takes part. Berlioz was in effect seeking to balance the wit of a Mercutio by adding something of the passionate intensity of a Juliet; the sophisticated, witty love of a Portia with the youthful exuberance of a Jessica and Lorenzo. Thus, it is perhaps no coincidence that the most radiant (and famous) number in the opera—the *duo-nocturne* for Hero and Ursule (no. 8, "Nuit paisible et sereine!")—ranks with the music of Berlioz' love scene in *Romeo and Juliet* and the love duet in *Les Troyens*. The effect, in fact, is of a refreshing aural oasis amid the bustle. Mozart sought this same effect toward the end of *Figaro* with the *andante* of Susanna's peaceful and serene "Deh vieni, non tardar."

A poignant, unintended irony surrounds this duet, by the way. In it Berlioz wanted to express in music the opposite of Beatrice's character, as Shakespeare's Hero describes it: "She cannot love, / Nor take no shape nor project of affection." Worn out by his long Parisian musical wars, the composer was tempted to a similar withdrawal from human intercourse; disdain and scorn often rode sparkling his writings. No wonder, then, that Béatrice's musical personality derived so naturally from the characteristic Berlioz style. But what music did he think fit for Hero, his epitome of the woman truly capable of "un tendre sentiment l'un pour l'autre"? The irony is that for this purpose Berlioz, in musical terms, returned to his youth by casting the duet in a very old-fashioned bel canto style. It is rather as if Shakespeare had found a place in his last play, *The Tempest,* for a full-fledged sonnet reminiscent of his Elizabethan youth, as he had done in *Romeo and Juliet* almost two decades earlier. Why did Berlioz do this? After all, he had written in 1854 that "Since those days [the early 1830s], duets in unison have been worked to death." It is tempting to think of the duet as a remembrance of things past: in order to introduce a beautiful romantic reverie into Shakespeare's play he chose to write one last unison duet, complete with parallel thirds and interlaced melodic lines, worthy of Norma and Adalgisa. Its traces of melancholy are perhaps in part due to Berlioz' memories of a time when his first Ophelia and Juliet, Harriet Smithson, drew from him "un tendre sentiment l'un pour l'autre."

It is perhaps worth asking why Berlioz produced such an exuberant swan song. One reason seems clear: this "petit" opera must have been a recreational diversion from the rigors of his "grand" opera *Les Troyens* and its protracted gestation. (Verdi had a similar diversion in mind when, weary of his *King Lear* project, he asked his librettist to think about a new project in the vein of Bellini's charming, genteel *La Sonnambula*.) Another possible reason is more speculative. In January 1833—his first reference to such a project—the composer promised, "I am going to write a very merry [*fort gai*] Italian opera on Shakespeare's comedy *Much Ado*." Why the thirty-year delay? Perhaps because this is a decidedly "premarital" comedy, and in October 1833 he married Harriet Smithson. After a few years the marriage began to deteriorate; he took a mistress in 1842; and in 1844 Berlioz separated from his wife. She sank into alcoholism and became (to borrow a few phrases describing Beatrice) "curst" and "shrewd of . . . tongue." At her death in 1854 he recalled not only the "radiant beauty and genius" of the early years, but also her "ungovernable jealousy" and their "bitter domestic strife." It is easy to imagine that all these tribulations would discourage a composer from setting a play that made fun of husbands who set their "neck into a yoke, wear the print of it and sigh away Sundays" . . . or that gaily ridicules its hero at the end as "Benedick, the married man."*

Finally, one is forced to conclude that *Béatrice et Bénédict* is a Shakespearean *succès d'estime* not so much in comprehensive dramaturgical substance as in its vitality of spirit. The characteristic Berlioz style was, as he had long before hinted in his Queen Mab scherzo, perfectly suited to the verbal style of Shakespearean comedy. One can see this in how easily praise of *Much Ado* can be applied to the Berlioz score: "Nothing about the play's style is casual or characterless. Falling on the ear with spirited assurance, it is always so fresh as to seem spontaneous. . . . The prose . . . is extraordinarily vivacious. It is lively in rhythm, expressive for the voice, never heavy-footed, flaccid, or haphazard. It is happily self-conscious but unaffected."* Conversely, it would be hard to find more apt epithets for the style of *Much Ado* than those Berlioz chose for himself in an 1858 postscript to his memoirs: "The predominant features of my music are passionate expression, inward intensity, rhythmic impetus, and a quality of unexpectedness."

27

The Tempest *Tossed*

No other Shakespearean play has tossed as often and as vehemently on the high seas of music as *The Tempest*. Winton Dean discovered over thirty versions in 1966, and the number continues to rise: John Eaton's *Tempest* of 1985 was coldly received by the critics, while Lee Hoiby's version introduced the very next year drew considerable praise.* As we shall see, even the list of *Tempest* operas contemplated but never composed is intriguing. All told, however, the history of musical *Tempest*s is a long, melancholy one, punctuated by bursts of unintended hilarity. If Hoiby's work gains a toehold in the operatic repertory, it will be the first time Prospero and his "project" have done so. For the most part, it seems, composers and librettists have been happy to fare forth without a mainsail, mainmast, or sometimes even a rudder . . . that is, some part of Shakespeare's play one would have thought indispensable. And because *The Tempest* is perhaps Shakespeare's profoundest, most ineffable drama, it is a hard play to grasp, let alone fully encompass. For all its manifestly operatic qualities—the spectacle, starkly simplified characters, melodramatic clash of good and evil, and several set speeches crying to be arias—*The Tempest* presents, as Dean says, "many traps for the unwary," and he grimly concludes that "no composer has yet solved any" of them.

This paradox, that Shakespeare's most musical play has proved so intractable to composers, is perhaps explained, at bottom, simply by the cavalierly exploitative way it has usually been approached. Emmanuel Chabrier gives us a flavor of such effrontery in a letter he wrote to his publisher in 1886:

> I now know *The Tempest* by heart; there are many good things in it, but Blau has reason to want some expanding of the love story, otherwise papa Prospero would get to be a bore. As for the drama properly so called, where is it? Are they genuinely dramatic, those conspiracies of the old fogeys Alonzo, Antonio, Gonzalo, Stefano? And those two debauchées Caliban and Interpocula [Trinculo]—who cares about their peregrinations around the island? So the interest will have to be spiced up elsewhere: first, in the idyll between Ferdinand and Miralda [*sic*] which is of the finest

order; second, in the whole nuptial atmosphere and the spirits of air in act 4; third, in the buffoonery with the drunkards. Is that enough to make an opera?

Given Chabrier's meager interest in Prospero, eagerness to inflate the most "tasteful" love affair in all Shakespeare, and generally bumptious attitude, we can thank the good Lord that he apparently ended up answering his question "no" and shelving the project. Many other similarly bold composers and librettists, however, have had their way with the play, and the sea changes it has suffered as a result are, to say the least, breathtaking. We need not loiter here over the opera in which Ferdinand was a mezzo and the same singer doubled as Prospero and Trinculo, the version where "Our revels now are ended" *precedes* the revels, the one in which Miranda finds herself temporarily in the arms of Trinculo, or the one that has Prospero arrive in a chariot drawn by Caliban and three dragons. But I think it is worth looking at certain other *Tempest* operas that illuminate the exigencies of musical translation in particular historical periods.

The first, Purcell's *The Tempest, or the Enchanted Isle* (1695), is surely the most frustrating. How pleasing it would be to have a full-fledged version of the play from England's first great theatrical composer and in a musical idiom closest to that of Shakespeare's own time. Unfortunately, his text was derived by the notorious Thomas Shadwell (butt of Dryden's scathing poem *Mac Flecknoe*) from the already radically un-Shakespearean Dryden–D'Avenant *Tempest:* little genuine Shakespeare survived. Also, in so-called operas of the time, huge chunks of spoken dialogue were interspersed with an occasional song and then a few massive masquelike vocal and orchestral scenes (there is no music whatever in the first act). Thus two casts, one of actors and one of singers, go their separate ways. And it is awful to contemplate the fact that the 265 lines Shakespeare gave his annoyingly well-behaved lovers is increased to over 1,200! Purcell conformed his music completely to this dramaturgy of his time. Only two characters, Ariel and Dorinda (Miranda's new sister), get to sing, while several airs are given to such newcomers as Aeolus, Amphitrite, Neptune, two devils, and various Nereids. What we have, then, is some very fine but decidedly incidental music.

J. C. Smith's 1756 *Tempest* warns that "many passages of the first Merit" in the play had to be jettisoned, obviously because Smith was mainly concerned with providing the two dozen airs, three duets, and chorus that were pretty much the standard recipe for baroque opera. The original poetry is kneaded into rhymed iambic pentameter or doggerel like this for Prospero:

> Let magick sounds affright no more,
>   While horrors shake the main;
> Nor spell-bred storms deface the shore,
>   Let sacred nature reign!

The texts for many of the airs, by the way, are gathered with magpie abandon from Jonson and lesser poets.

More laboriously odd, a century later, was the translation into Italian of an opera by a Frenchman for a London premiere at the Haymarket Theatre. This was

Halévy's *La Tempesta* of 1850, which I had occasion to discuss at the end of Chapter 20. The impresario who commissioned the opera, Benjamin Lumley, wrote quite plausibly, "it seems probable that this play was the first which, in the Elizabethan age, assumed in some degree the form of an opera"; but then he goes on to observe that it "lacked (be it said with all reverence) the stage 'situations' rendered necessary to lyrical works by modern requirements." Every age has its "modern requirements," and in this case they were mainly for big numbers for stellar executants—Lablache as Caliban and Sontag as Miranda.

A brief look at the Scribe libretto (rejected earlier by Mendelssohn) shows how the industrious hack produced the usual number-opera "situations" like sausage from a grinder. We can bet that Miranda and Ferdinand are made for each other, for instance, because they both get entrance-arias on virtually the same hoary text. She: "Ah! who will give my heart peace?" He: "Ah! give to my soul the peace of my youth!" The first act ends with a love-plighting duet phrased in time-honored libretto clichés. Given the two stars on hand, it was inevitable that Scribe would make a big scene of Caliban's seduction and attempted rape of Miranda, which are only briefly mentioned in the play. No self-respecting nineteenth-century librettist could have resisted Shakespeare's obvious openings for a drinking-song; Scribe gives one to Stephano and throws in a bacchanal for good measure to fill out the act. Here Caliban shows his true character: no "salvage and deformed slave" but merely a blowsy, aging *opera buffa* roué. No surprise in this: a few years earlier Lablache had created the title role in Donizetti's *buffa* masterpiece, *Don Pasquale.*

Prospero does get a somewhat faithful browbeating aria addressed to Antonio and Alonso at the beginning of act 3. But then, incredibly, he almost vanishes; in the denouement he has but *four* lines! Instead, Scribe rings in another love duet ("À moi seul, oui, à moi seul, O mon âme! etc."), gives Ariel a few spectacular coups (he produces "Un palais aérien resplendissant de mille feux" and "un vaisseau magnifiquement pavoisé"), and gives Miranda the last word: "Nature and love have conquered. . . ." Lablache appears to have brought down the curtain (or stepped in front of it) for the display of his wares in a number called the "Miranda *variazioni.*" All this—but no Gonzalo, no plot on Prospero's life, no masque, no "revels" speech, no "Ye elves of hills," and Ariel rendered mute for the sake of Carlotta Grisi's pirouettes. Lumley spoke of "reverence" for Shakespeare, but as was often true in the history of Shakespearean opera this is mere lip service. The true reverence displayed here by composer, librettist, and impresario was for the expectations of both the artists and the audience of their time.

The "modern requirements" of Frank Martin's time made his *Der Sturm* (1956) as different an opera as one could imagine. Far from searching for or inventing the pretexts for arias and concerted numbers as Scribe did, the Swiss composer went far—in Dean's view much too far—in trying to avoid these *Effektstücke* that Wagner had complained about so loudly. Martin's heavy reliance on serialist compositional methods and on declamation produced an opera, Dean concludes, that is "almost devoid of contrast" and "austere to a fault." Martin's libretto (his own) is very faithful to Shakespeare's letter: he sets an almost literal, slightly abridged version of Schlegel's pre-eminent German translation. His infidelity lay elsewhere: the musical ethos of the Schoenberg school, of which Martin had become an exponent, was simply not congenial to Shakespeare's heroic utterance and *Effektstücke*. There are great moments and tempting set pieces throughout Shakespeare's play, and the modernist allergy to expressive exuberance was bound to mute their effect. Thus, Martin replaces the busy, festive, and "most majestic" masque of act 4 with a grave, stately pavane, and he does everything he can to make the "revels" speech into a lullaby: the tempo is *andantino,* the rhythm is gently rocking, and it is marked *sempre pianissimo*. Scarcely what one would expect from a man who has just been described as "touch'd with anger, so distemper'd." The opera's unsatisfying effect is finally attributable to this consistent lack of interest, to borrow Verdi's phrase, in taking the bull by the horns.* Also to the point is Agate's remark that Shakespeare shows quietism the door.

Two *Tempest* projects that never bore fruit are especially tantalizing: Verdi's and Mozart's. Sometime in the late 1840's Verdi jotted down a list of attractive opera subjects; it began: *"Re Lear, Amleto, Tempesta, Caino* [i.e., Byron's *Cain*], *Rois'amuse. . . ."* In the Verdi archives, it turns out, there is a finished libretto for a *"Melodramma in tre atti"* based on the play; it is dated 1866 and the *argomento* is presumably signed by its author, one Andrea Panizza. Though we have no way of knowing whether Verdi approved of it or not, this libretto does show very plausibly how *The Tempest* would have been modified for the nineteenth-century Italian opera stage and is worth a brief description.

The constriction of the play becomes immediately apparent. Panizza writes, "The meeting of Miranda and Fernando, the moral and physical agitation of the guilty ones, and other related scenes form the argument of one of the most marvelous fantasy plays of Shakespeare . . . except for slight modifications, we have attempted to reproduce this argument in the present opera." Yet again, Shakespeare's low-profile love interest is inflated, presumably for a stellar soprano and tenor. And it is astonishing for Panizza to speak of the play's fantastic element when, in fact, Ariel and Caliban vanish entirely. This may have been due to the reluctance of censors to allow witchcraft on stage. In 1857, for instance, the Neapolitan censors told Verdi he had to transfer his *Ballo in Maschera* action to a pre-Christian age when witchcraft and the summoning of spirits was believed in. The low comedy of Trinculo and Stephano is also completely cut, perhaps because of the very long sleep of Verdi's comic muse between *Un Giorno di Regno* (1840) and *Falstaff* (1893).

Unavoidably, the plot devolves into matters of love and political conspiracy . . . and into a successful search for pretexts for numbers. Prospero and Miranda watch the storm in a standard duet; then in a big concerted scene Gonzalo and a chorus of mariners console Alonso; this is followed by a "plotting" duet for Antonio and Sebastiano. The rest of act 1 is given over to Fernando, in "a delightful part of the island," remembering his father to the tune of a *"melodia melanconica."* Soon Miranda enters, and he utters the all-too-true line, "Ecco la Diva!" With Prospero's arrival this love duet turns into a typical trio finale, its formula strophes rhymed *abbcac*. Act 2 opens with a stock *scena* for the tenor: after a slow aria in the farewell-to-my-homeland vein, the voice of Miranda ignites a *cabaletta* ("Oh! if it were granted to see that divine angel"). She appears for another love duet, and the next scene, where Prospero recounts some past history and Miranda pleads for mercy ("Revenge causes sadness, forgiveness makes divine the hand that offers it"), was obviously intended to give Verdi an opening for a *gran scena concertato a due*. The still-obligatory *largo*-of-astonishment finale is elicited as, amid storm and thunder, Prospero saves Alonso and Gonzalo from the plot of Antonio and Sebastiano. A chorus of mariners and island spirits joins the soloists to express fear and astonishment.

Act 3 is the most loyal to Shakespeare, for it is given over entirely to the denouement of the play's last scene. Prospero appears in ducal robes and "with a cheerful look in his face." "Movingly, he raises his eyes to heaven" and blesses Fernando and Miranda. The scene changes to a "delightful garden" in the center of which is a statue of Amor. Here we get the equivalent of Shakespeare's masque: a four-quatrain chorus of spirits that accompanies a ballet danced by nymphs, reapers, and "various other forms of apparitions." Finally, "the singing recedes and the apparitions vanish amid confusion and noise." The couple is ushered into the garden, and Prospero is left alone for a big solo abjuring "this art that made me so powerful" and announcing that "the merciful voice of pardon ought to sound from my lips." Alonso's party enters, the two villains (unlike Shakespeare's) turn penitent, the lovers are discovered in a cave (but *not* playing chess!), and the opera ends with a sextet and chorus.

All in all, one can say that Panizza succeeded in turning the play into what looks, on paper at least, like a typical vehicle for middle-period Verdian musical dramaturgy. Perhaps, if the 1866 date is correct, it (like the *Re Lear* libretto) came to seem to the composer too clichéd, old-fashioned, and far removed from Shakespeare . . . and landed permanently on the shelf at Sant' Agata.

Even more tantalizing—and mysterious—is the possibility that Mozart, in his last months, might have worked on a *Tempest* opera. On 6 September 1791 his *La Clemenza di Tito* premiered; then came *Die Zauberflöte* on 30 September. By 31 October one Gottfried Bürger was writing to Schlegel, "Gotter has written a magnificent free adaptation of Shakespeare's *Tempest* under the title *The Enchanted Island*. The ladies cannot praise it enough. Mozart is composing the music." Truth? or idle gossip on the Viennese musical scene?

The Friedrich Gotter referred to in the letter had in early 1791 adapted a version of Shakespeare's play called *Die Geisterinsel,* which had been fashioned by Friedrich von Einsiedel. The correspondence between these two men certainly suggests that Mozart was part of the picture and that they were in touch with him. On 24 March Gotter wrote to Einsiedel to say that, after "having heard about the new works of this modern Amphion," he, too, was agreed on Mozart as their composer. By 3 November, he writes, "the edifice is quite finished and ready for Mozart to embrace it with his heavenly choirs [*Himmelschören*]." A few days later Gotter writes that, "after a joint revision of the piece a letter along with the first act will go to Vienna, and who knows whether Mozart might not in exchange bestow upon us an aria . . . ?" On 15 December, Gotter sends to Einsiedel the second and third act, which he calls "rich material for Mozart's unparalleled humor." Shortly after this, however, news arrived in Thuringia that Mozart had died in

Vienna on 5 December. He left no known music for the project nor any reference to it in his own correspondence.

Other factors make a Mozartean *Tempest* project seem plausible. *Die Zauberflöte* was obviously derived in part from another "magic" opera—*Der Stein des Weisen, oder die Zauberinsel*—that Schikaneder had produced the year before. An opera on Shakespeare's play would have been an obvious further entry in this then-popular sub-genre of opera. Also pertinent is Edward Dent's observation that Schikaneder's reputation as an actor and manager (he likens him to Dickens' Vincent Crummles in *Nicholas Nickleby*) had largely been made with productions of Shakespeare. Dent even goes as far as to suggest that Schikaneder presented "a Falstaffian figure both in appearance and personality" on the German theatrical scene. It is easy to imagine Mozart's librettist and first Papageno egging him on to repeat the huge success of *Die Zauberflöte* by venturing onto Prospero's magic island.

The irony of all this is that the Gotter libretto—eventually set to music by four composers in the banner year of 1798—is "singularly remote from Shakespeare in language and spirit" (Dean) and "a magic opera and nothing more, without the deeper charm and meaning" of the play (Einstein). The whole royal party except for Ferdinand vanishes, and, since enchantment was to be the chief part of the menu, two dancing roles are added: Maja, a good spirit, and Sycorax, the mother of Caliban only mentioned by Shakespeare. Ariel becomes very bossy, and the main plot interest is the struggle between Prospero, potent only in daylight hours, and Sycorax, the—well—queen of the night. As a version of the play, this farrago would have amounted to a cynically slavish sequel to *Die Zauberflöte*. Though we would certainly cherish more music such as Mozart's valedictory opera contains, we should perhaps be grateful that he devoted himself in his last weeks on earth not to Gotter's travesty but to a more serious project: his great, unfinished *Requiem*.

The choicest irony of all, though, is that *Die Zauberflöte* itself may unwittingly be—in its ethical and dramaturgical soul if not in the names of its characters—the best and most loyal *Tempest* opera we have. Dean says that it "is only the most famous of many operas that combine magic (good and evil), spectacle, ethical humanism, and low buffoonery." These are exactly the four major ingredients of Shakespeare's play. The similarities between the two casts have often been noticed as well: Prospero and Sarastro presiding benevolently with their arcane arts, Caliban and Monostatos both sex-crazed but feckless social subversives, and Ariel perhaps briefly appearing in the three genii. The noble lineage of Pamina and Tamino and the youthful naïveté of Papagena and Papageno are both reflected in Miranda and Ferdinand. And then there is the lesson at the center of the opera as Egon Komorzynski describes it in his book on Schikaneder: "*Die Zauberflöte* is an artwork of the highest pedagogical value, for it makes possible 'The Education of Humanity.' Whoever truly grasps this work . . . will learn that we can raise ourselves above the ordinary, the commonplace, learn utterly to despise the abysmal meanness of Monostatos, [and] learn that not selfishness but magnanimity must guide us."

Certainly no other *Tempest* opera discussed above shadows with such fidelity the significance of Prospero's Monostatos, Caliban, and the resounding magnanimity of Prospero's promise at the end:

> Though with their high wrongs I am strook to th' quick,
> Yet, with my nobler reason, 'gainst my fury
> Do I take part. The rarer action is
> In virtue than in vengeance. They being penitent,
> The sole drift of my purpose doth extend
> Not a frown further.   (5.1.25–30)

Leigh Hunt pointed out the other obvious similarity of the two works as early as 1819: "*Die Zauberflöte* . . . is to Mozart's other works what *The Tempest* is to the most popular of Shakespeare's comedies."

28
Britten's *Dream*

Igor Stravinsky once observed about Benjamin Britten, rather sourly, "He's not a composer—he's a kleptomaniac." Perhaps this was Stravinsky's response to Britten's remark that he had liked the Russian composer's only opera, *The Rake's Progress*, very much—"everything but the music." In any event, there is truth in the aspersion. Indeed, it is an aspersion any sensible theatrical creator might wear with pride. History suggests that most great playwrights and composers who prove consistently successful with a substantial number of works eventually become, in one or more important ways, kleptomaniacs of genius. Having this knack for purloining what is useful from one's artistic inheritance and one's own environment would seem a virtual prerequisite for any composer contemplating a Shakespearean opera. After all, Shakespeare began his career condemned as a thieving magpie by Robert Greene and went on to become the theater's supreme kleptomaniac, cobbling the plots for his plays from a huge variety of pre-existing sources.

And *A Midsummer Night's Dream* bids fair to be his masterpiece of kleptomania. Bullough, in his massive study of Shakespeare's sources, writes of the "great ingenuity" he exercized in unifying "a hotch-potch of material" stolen from other authors. Bullough reproduces the texts of *eleven* certain or likely sources: the Theseus–Hippolyta plot taken from Chaucer and Plutarch; some of his fairy lore from a popular book on witchcraft; the ass's head from *The Golden Ass* of Apuleius; the plot of "Pyramus and Thisby" from Ovid, to name the most important. The play's many styles also show Shakespeare's genius for borrowing from his own past successes. From his earlier satire of sonneteering in *Love's Labour's Lost* and *Romeo and Juliet* comes such flamboyant poetizing as Demetrius' outburst as he awakens madly in love with Helena:

> O Helen, goddess, nymph, perfect, divine!
> To what, my love, shall I compare thine eyne?
> Crystal is muddy. O, how ripe in show
> Thy lips, those kissing cherries, tempting grow!

The slapstick clowning of the rustics was an obvious spinoff from similar hijinks in several preceding comedies, while the gimmick of a play-within-a-play during which aristocratic couples take witty potshots at inept actors repeats the hilarity of the play mounted by the comic butts Holofernes and Nathaniel toward the end of *Love's Labour's Lost*. The only decided novelty for Shakespeare, in fact, was his *Dream*'s "fantasy" or "fairy" style.

The miracle of the play, as Bullough suggested, is that its four radically different "worlds" (the real one of Athenian aristocracy, the lovers' unreal one of fiercely vexed passions, the lovably dull-witted world of the "hard-handed men" who would be actors, and the fantasy realm of Oberon and Titania) are magically blended, unjarringly and coherently, into a theatrical whole. That Britten and his co-librettist Peter Pears (also his first Flute/Thisby) were able to work approximately the same miracle must surely be the main reason why their opera has been called a work of "almost formidable authenticity" . . . and why Winton Dean has flatly asserted that it is "the most successful Shakespearean opera since Verdi." In 1929 James Agate said of music for this happiest of Shakespeare's comedies, "I must have Mendelssohn . . . since in this connection all is dross that is not Bartholdy." But I think few would argue that, with the premiere of Britten's opera at Aldeburgh in 1960, Mendelssohn's familiar and charming incidental music finally received some serious competition.

One reason for the opera's sense of authenticity is its complete reliance on the original language. While Purcell in his 1692 operatic version of this play, *The Fairy Queen,* managed to avoid setting a single Shakespearean line to music, Britten and Pears made use of about 1,100 of them. Indeed, only one line in the entire libretto is non-Shakespearean.* This *Dream* is, in fact, the only important opera in the repertory based entirely on Shakespeare's language, though much careful culling for lines suitable to music drama was necessary. In the end, some set speeches like Titania's "These are the forgeries of jealousy," two minor characters (Hermia's father and Philostrate, Theseus' Master of the Revels), and approximately half of the 2,150 original lines were excized. Given how much more time-consuming it is to sing rather than speak words, the cutting of fifty percent of Shakespeare's lines nevertheless produces a running time of 145 minutes, which is about the same time the play usually runs.

Perhaps even more crucial to the success of the opera is the verve and ingenuity with which Britten was able to create distinctive aural equivalents for the various verbal styles in *A Midsummer Night's Dream*. This is obvious at first hearing and has been commented on hitherto by everyone who has looked into the score. Just a few touches of Britten's skill in this regard need be mentioned here. The generally four-square meter and prosaic quasi-folksong melodies and cadences are a perfect accompaniment for the rustics, and Britten fittingly accompanies Bottom's egomaniacal exertions with a gregarious trombone. Aptly, the vocal lines are rather angular and undifferentiated for the four conflicted lovers (until, of course, the harmonious end), and Britten shrewdly isolated Puck's agility and fleet-rhymed poetry by keeping his role declaimed rather than sung and identifying his escapades with a jazzy side-drum and boisterous solo trumpet.

Particularly poignant and relevant to Shakespearean and operatic emphasis on set pieces was the musical style Britten employed to evoke the fairy world. Though he was not very loyal to the play's most famous set speeches, he was—in giving the fairy realm a local habitation and a sound—loyal to their lyric equivalent: the distinct "number" or aria of baroque opera. He was a great admirer of Purcell (he edited a concert version of *The Fairy Queen* for his Aldeburgh Festival), and Purcell haunts Britten's fairy scenes. Clearly inspired by Purcell are the dreamy coloratura of Oberon's aromatic "I know a bank where the wild thyme blows," Tytania's enchanting "Come, come, now a roundel," and the bouncy aria which tells the fairies how to treat her new lover Bottom, "Be kind and courteous." Britten even instructs in the score that the stately little number in which Bottom and the fairies exchange courtesies be played "Purcellian style." Casting Tytania (Britten's spelling) as a coloratura and Oberon as a countertenor—the score also permits a contralto—itself alludes to baroque performance tradition. As the Britten specialist Philip Brett summarizes, "the fairy denizens . . . are much more likely to sing in rounded set-pieces than are the Rustics or Lovers who invade their territory. Theirs is a world of formality, of decorum, of a certain innocent perfection, and ultimately of nostalgia as well." These rounded set pieces take the place—and fulfill the same eloquently riveting functions—of the set speeches that Shakespeare carefully deployed in his play. Britten's nostalgic love of Purcell's music thus proved perfectly apt to Shakespeare's way of structuring of his action.

But perhaps nothing in the entire Shakespearean opera repertory underscores the similarities of the two dramaturgies as clearly and amusingly as do the two versions of "Pyramus and Thisby." The rustics' tedious, brief play is Shakespeare's compendium of the myriad ways a play of his time could be awful. It could have an overly informative prologue: Quince's amounts to nearly a third of the script. The blank verse could be leadfooted of meter, repetitious, barren of happy conceits, littered with filler-words and old-fashioned alliteration, and dreadfully punctuated: the prologue Shakespeare invents for the rustics commits these and a half-dozen other stylistic gaffes. A bad playwright could forget to rely on the audience's power of imagination ("I, one Snout by name, present a wall . . ."), overdo the ecphonesis ("O wall, O sweet, O lovely wall . . ."), or fall on his face in the purple patches ("These lily lips, / This cherry nose, / These yellow cowslip cheeks . . . are gone!"). More generally, he could allow the plot to become a transparent pretext for a series of thespian feats for various actors to gain applause. Everyone in "Pyramus and Thisby" gets his moment to shine, from slow-of-study Snug ("Well roared, Lion!" says Demetrius of his growling aria) to that epitome of the egomaniacal star actor, Bottom. Shakespearean drama riveted attention on passion's slaves, and so does "Pyramus and Thisby": each of the title characters gets a "passion" with which to end the play. The irony of all this satire of the ludicrous potentialities in Elizabethan drama, of course, is that a few years before concocting "Pyramus and Thisby" Shakespeare had taken a very similar plot and got right most (but by no means all) of the tricks of good playwriting . . . and produced *Romeo and Juliet*.

"Pyramus and Thisby" is a mere 124 lines in all, and Britten knew precisely what to do with Shakespeare's miniature send-up of Elizabethan drama: transform it into a miniature send-up of "number" opera. He made his satirical intentions clear by switching from English to melodramatic Italian directions when the rustic actors enter Theseus' palace. Thus, Wall discharges his lugubrious part in a *lento lamentoso* and Bottom's suicidal "passion" is an *allegro disperato*. That Britten's parody of opera should fit Shakepeare's parody of his own dramatic methods reiterates the theme of synonymy in "Shakespearean" and "operatic."

The average opera's running time, we have seen, is about 140 minutes (Britten's score says the first production lasted 144 minutes), with about 18 numbers, which last, on average, a little over seven minutes each.* Britten's Lilliputian opera-within-an-opera is perfectly in scale: it lasts 17 minutes and consists of 14 numbers that last about 75 seconds each. The music of his prologue, with its rapid-fire semiquavers on the same note, proves the rustics haven't a clue what melody or "phrasing" means, let alone the niceties of iambic pentameter. It has already been hinted by the composer that the rustics don't have the operatic knack, for in the rehearsal scene, words like "love" and "lofty" are given very low rather than high notes. The word-setting of their play shows such ineptitude on numerous occasions. But what is most vividly caricatured is the progression from aria to aria, with chit-chat and brickbats from the aristocratic audience in between. Wall gets his all-too-earthy entrance aria (in Schoenbergian *Sprechgesang* and notation, a wicked bit of Britten satire), then comes Bottom with a recitative ("O grim-look'd night," marked *moderato ma tenebroso!*) and an aria ("And thou, O wall") that ends as most nineteenth-century Italian arias do, *con espansione*. The heroine's first appearance is accompanied, predictably, by harp and flute in a prim *allegretto grazioso*. Seconds later, love is naturally being plighted in an *allegro brillante*.

Shortly afterward, Snug presents his Lion and, fearful that genuine fright might sweep through his audience, sings his part at first *leggiero,* then *dolce* and *intimo*. Still, Britten gives him a *presto feroce* to drive Thisby offstage. At this point the audience is in a very generous mood; even Starveling's Moon gets a bravo ("Well shone, Moon!" says Hippolyta). The catastrophe comes with admirable haste; Bottom spies Thisby's bloody mantle and dispatches himself: "Thus die I, thus, thus, thus, thus, thus." But, like certain operatic heroes and heroines, he rises for one last repetitive lyric outburst that milks the passionate moment for all it is worth:

> Now am I dead,
> Now am I fled,
> My soul is in the sky,
> Tongue lose thy light!
> Moon take thy flight!
> Now die, die, die, die, die.

Thisby soon appears to converse, as Donizetti's Lucia does, with an imitative flute, then shifts into an *adagio lamentoso* for the aria that brings her to stab herself and end the opera: "Thus Thisby ends, Adieu, adieu, adieu. . . ." In lieu of an epilogue comes a ballet divertissement in the form of bergomask dances. It is telling that Britten uses every part of Shakespeare's parody of his own style and dramaturgy to parody the methods of composers from the heyday of grand opera.

It remains to note two changes made by Britten and Pears that I cannot help thinking regrettable. The first was their decision to cut the play's first scene, which introduces us to Theseus, Hippolyta, and their imminent nuptial celebration. Shakespeare was careful to "frame" the mysterious, magical world of the lovers and the fairies at the beginning and end of his play with the real world of Athens, and the loss of this frame upsets this symmetry. It should be added, though, that the play's splendid opening speeches for Theseus and Hippolyta do occur when we arrive at Athens for the first time in the opera's last scene.

It is also frustrating that the play's most resonant, crowning—indeed most famous—speech was entirely cut. This is Theseus' very reasonable satire upon the life of the imagination. It begins,

> I never may believe
> These antic fables, nor these fairy toys.
> Lovers and madmen have such seething brains,
> Such shaping fantasies, that apprehend
> More than cool reason ever comprehends.
> The lunatic, the lover, and the poet
> Are of imagination all compact . . .   (5.1.2–8)

Though Britten had the example of Verdi's cutting *Othello*'s crowning speech ("Soft you; a word or two before you go"), I wish for several reasons that he had set it to music, not least because there seems to be a place for it in the action. If any original element of the play is shortchanged in the opera it is the Athenian court, and this speech would have bodied forth more satisfyingly the "real" world against which the splendid poetry (and music) of the lovers' and fairies' worlds are to be appreciated. This speech, making light of the "tricks of strong imagination," is also the perfect introduction to a silly play put on by actors with not an ounce of imagination among them. It also puts us in the audience firmly on the side of "shaping fantasies," for we *know* that "cool reason" is here completely wrong: we ourselves have delighted in the "fairy toys" of the Athenian wood. Finally, though, one can simply say that this is one of those Shakespearean speeches (Bottom puzzling out his dream is another) that set the mind reeling at the numerous tantalizing questions they raise: can human existence of any value exclude either "cool reason" or "imagination"? is Shakespeare satirizing the kind of real-life, dull-witted aristocrat who would attend a command performance by the Globe players and not lend his own imagination to the event? does the speech reveal some Shakespearean thoughts about his own "poet's pen"? Theseus' speech, in sum, *deepens* the play in a remarkable way. I wish Britten had retained it . . . perhaps by setting it in a brief oasis of the very dry recitative that a voice of "cool" (and utterly unoperatic) reason deserves and surrounding it with the tricks of strong musical imagination that are apparent everywhere else in the *Midsummer Night's Dream* score.

29

Wherefore *Romeo?*

Everybody remembers Juliet questioning aloud from her window (Shakespeare mentions no balcony) why the boy she loves bears that fatal name: "O Romeo, Romeo, wherefore art thou Romeo?" It is surely the play's most oft-repeated line, though very few reciters correctly stress *thou* or *Romeo,* not *art* (*wherefore* always meant *why* to Shakespeare). About two dozen operas have been based on *Romeo and Juliet,* and over the centuries, lovers of Shakespeare's play have had numerous occasions to ask Juliet's question—in tones of bemusement and sometimes outrage—when confronted by the liberties taken with the original plot and *dramatis personae.* Wherefore did the composer and librettist bother to call their work *Romeo e Giulietta, Romeo und Julia,* or even *Les Amantes de Verone* and *I Capuleti e i Montecchi?* so far flung from Shakespeare they all seem to be.

Take, for example, what is probably the best verismo-style version of the play, by Riccardo Zandonai. His *Giulietta e Romeo* (1922), heavy with the musical aromas of Puccini and Mascagni (especially *Cavalleria Rusticana*), manages the story very nicely without the services of Capulet, Paris, Mercutio, or Friar Lawrence. He even commences his story with the lovers already married. There is a balcony scene here, but it takes place *during* the Capulet ball. Several operatic love-duet clichés are here laid end to end, and the climax is dual high A's as the rope ladder touches the ground and Romeo climbs up. As one would expect from a veristic approach to the story, the two lovers betray nothing of their childish exuberance in Shakespeare's play, and an air of danger hangs constantly in the music. Tybalt is turned into a mafioso villain, and the heroine has good reason to sing, "I am fearful. . . . My soul is swollen with tears."

The verismo penchant for picturesque divertissements is also served in Zandonai's opera: a torch dance in the Capulet courtyard, a drinking song in Mantua, a rousing "vento furibondo e clamore" with lightning and thunder to usher Romeo back to Verona, and a lovely offstage chorus ("Alba di Dio!") to accompany the final moments. An opera composer's eye for in-set "performance" situations led to at least one clever invention: the fugitive Romeo learns of Juliet's death when "Il Cantatore" sings a ballad based on the latest news from Verona. Dean comments, "As in other verismo operas the fringe characters, such as the town-crier and the ballad-singer, enjoy better music than the principals." The death scene is laid on with a Puccinian trowel, with Giulietta ending her life "leaning toward Romeo, like a dying lily."

Such conspicuous failure to encompass fully the various elements of Shakespeare's tragedy of star-crossed love has been the unfailing rule since the competition began in 1776 with a Leipzig *Romeo e Giulia*. This is not a little strange, since (as two great nineteenth-century composers, Berlioz and Verdi, excitedly suggested) the play seems heaven-sent as an operatic vehicle. When visiting Florence in 1831, Berlioz heard rumors of a very good *Romeo* opera being performed there—Bellini's, which had premiered in Venice the year before. He was already familiar with several "lamentable attempts" to transform the play into opera, and so went to the Pergola Theater full of hope:

> What a subject for an opera! How it lent itself to music! To begin with, the dazzling ball at the Capulets, where amid a whirling cloud of beauties the young Montague first sets eyes on "sweet Juliet" . . . then those furious pitched battles in the streets of Verona, with the fiery Tybalt presiding like the personification of Revenge; the glorious night scene on Juliet's balcony, the lovers' voices "like softest music to attending ears" . . . the dashing Mercutio and his sharp-tongued, fantastical humor; the cackling nurse; the stately hermit . . . and then the catastrophe, extremes of joy and despair drained to the dregs in the same instant . . . and, at last, the solemn oath sworn by the warring houses, too late, on the bodies of their children. . . . Yes! I hurried to the Pergola. . . .

He came away, as we have already seen, in "bitter disappointment!" (Wagner was hard on Bellini's opera, too: "downright threadbare music . . . hung upon an opera-book of indigent grotesqueness.") It is a choice irony that Berlioz' "dramatic symphony" based on the play, which appeared eight years later, captured almost every one of the significant aspects of Shakespeare's action he had specified (there is no cackling nurse). It stands now as the musical work most faithful to the spirit, if not the dramaturgy, of *Romeo and Juliet*.

Rossini lived just long enough to sample Charles-François Gounod's *Roméo et Juliette* (1867) and utter his sardonic synopsis of the opera: "It is a duet in three parts: one *before,* one *during,* and one *after.*" Verdi felt the same way. As with Berlioz, the play enthralled him, and he, too, was willing to criticize an operatic colleague who failed to make anything of its kaleidoscopic contrasts. Two months after his own *Otello* premiered, Verdi confided to a visitor:

> And yet *Romeo!* What a tempting subject! I envisage the work, I live this work. It is there in my head. Background: hatred, bloody strife between Montagues and Capulets. Foreground: the tragic love of the two children. Then there's the entire comic side, which Gounod ignored. I would have wanted to create a more spirited work with greater contrasts, not a long duet.*

Another lamentable Shakespearean lacuna in the Verdi canon.

Notwithstanding the brickbats flung at them by proprietary Shakespeareans, the Bellini and Gounod operas are by far the most important versions of *Romeo and Juliet* we have. Bellini's opus, especially its final scene, is considered by most experts to mark his entry into the short period of mastery that produced *Norma, I Puritani,* and *La Sonnambula,* while Gounod's opera has become a repertory mainstay in recent years, on the boards even more frequently than his great war-horse, *Faust.* Each deserves a few comments that isolate some of the remarkable infidelities and fidelities to Shakespeare's familiar story. And the ways the fatally ironic death scene is handled are also worth examining briefly.

Especially when considering an Italian opera like *I Montecchi,* it is only fair to remember that the Romeo and Juliet story was a very familiar one decades before it arrived on Shakespeare's desk in the form of Arthur Brooke's 3,000-line poem, *The Tragicall Historye of Romeus and Juliet* (1562). Shakespeare's may be the most famous telling, but it is by no means the only "spin" on the boy-meets-and-loses-girl story. Indeed, it seems plausible that the supreme librettist of the bel canto era, Felice Romani, did not think of Shakespeare at all when he concocted his action, but burrowed instead into the preceding native versions by Luigi Da Porto (*Istoria . . . di due Nobili Amanti,* 1530) and Matteo Bandello (*Le Novelle,* 1554), which in several respects are closer to the opera's plotting. The main point, of course, is that, with any essentially simple but powerfully affecting story like this one, several possible and attractive variations in plot or characterization are imaginable. The corollary to this is that not all of Shakespeare's narrative decisions need be accounted the best.

Even bearing these caveats in mind, Bellini's opera administers several shocks to expectation. Most notably, there is nothing of Shakespeare's persistent emphasis on youthful exuberance versus adult propriety . . . perhaps because seasoned operatic principals must perforce be, physiologically, thoroughly adult personages. Shakespeare takes pains to let us know Juliet is thirteen, but nothing is said of her extreme youth in the libretto. And we are introduced to Romeo, not (as in the play) as a laughable victim of puppy-love for one Rosaline, but rather as "the most abhorred, evil, and violent" leader of the Ghibelline forces (Romani, setting the action in thirteenth-century Verona, translated the Capulet–Montague rivalry into the context of the historical Guelph–Ghibelline feud).* What is more, as the action begins, he has already slain a Capulet son. Given Italian *opera seria*'s allergy to comic elements, it is no surprise that Mercutio and the Nurse vanish, while Tybalt is transformed into a stock villain, as well as the rival nominated by Capulet to marry Juliet.

But the real shock comes in the first two Romeo and Giulietta duets, in both of which the heroine, incredibly, manages to summon the courage to prefer her father's wishes to Romeo's. When she refuses to depart with him in the first duet, Bellini's stupefied Romeo asks, "What do I hear? what power is greater to you than love?" Most un-Shakespeareanly but very Italianately, she replies: "That of duty, law, and honor." And thus eighteen minutes of Bellinian melodic abundance end with the two lovers leaving separately! The peculiarly strong appeal of father–daughter relationships in Italian opera is further demonstrated when, at the beginning of act 2, Giulietta is given a big set piece in which she pleads, "with a voice that weeps," for Capulet's forgiveness. While Shakespeare's father turns ferociously upon his reluctant daughter, Bellini's Capulet is merely firm.

On the other hand, the denouement of the opera in some respects improves on Shakespeare. It begins, for instance, with one of the few passages truly reminiscent of the play. Romeo's

> O del sepolcro profonda oscurità,
> cedi un istante, cedi al lume del giorno,
> e mi rivela per poco la tua preda . . .

is a close pass at Shakespeare's lines:

> Thou detestable maw, thou womb of death
> Gorg'd with the dearest morsel of the earth.
> Thus I enforce thy rotten jaws to open . . .

One can see in these Shakespearean lines, though, the tendency toward the too-clever-by-half that aggravated Samuel Johnson. Of the play's characters he wrote, their "pathetick strains are always polluted with some unexpected depravations. His persons, however distressed, 'have a conceit left them in their misery, a miserable conceit.'" Just such a conceit is all too apparent in the big forty-seven-line "aria" Shakespeare gave Romeo just before he sips his fatal beverage in the Capulet tomb. Contractual and marine imagery always came easily to the young Shakespeare:

> and, lips, O you
> The doors of breath, seal with a righteous kiss
> A dateless bargain to engrossing death!
> Come, bitter conduct, come, unsavory guide!
> Thou desperate pilot, now at once run on
> The dashing rocks thy sea-sick weary bark!

In contrast with this incessant poetical cleverness, the simplicity of the operatic Romeo's equivalent speech—in concert with a typically gorgeous Bellini melody—is refreshing:

> I beg you, fair spirit,
> who ascend to Heaven,
> turn to me, take me with you:
> you cannot forget me, leave me thus
> in my suffering.

And one must also say the thirty-second denouement that follows the death of Giulietta is a great improvement over the 180 lines and dozen minutes that follow the lovers' death in the play . . . easily the flattest and most anticlimactic ending in all Shakespeare.

To judge by its cast page, Gounod's opera is much more faithful to Shakespeare. Bellini and Romani managed with just five characters who figure in the play, while the librettists Jules Barbier and Michel Carré bring in eleven and even make use of the Prologue's sonnet (sung by the chorus). Indeed, *Roméo et Juliette* was happily the first opera to reinstate Mercutio and make an aria out of his virtuoso Queen Mab speech. Though the scorching summer sun of Italy scarcely shines in Gounod's sophisticatedly "Frenchified" music (Dean writes that he placed the story in a "voluptuous and sugary envelope"), the success of the lyrical duets for the lovers cannot be denied. As with the Zeffirelli film version, the two lovers' romantic energies are magnified at the expense of all Shakespeare's satiric and moralizing commentary on his childish lovers.

Otherwise, the opera is only fitfully faithful. It was a happy touch to take the famous "sonnet" dialogue for Romeo and Juliet at the Capulet ball and turn it into a similarly artificial "madrigal à deux voix," but one is a little disappointed that the clichéd text and waltz–rhythm of Juliet's ariette at the ball ("Ah! je veux vivre") is the closest equivalent in the opera to Shakespeare's marvelous "Gallop apace, you fiery-footed steeds" speech. The catastrophic sword-fight scene unfolds Shakespeareanly, but the rather pious and stiff Gounod hadn't a prayer of making the good-willed Friar an appealing character or of capturing the dizzy volubility of the Nurse (unfortunately given the name Gertrude in the libretto).

The various ways the lovers "shake the yoke of inauspicious stars" in Shakespeare, in the earlier narratives, and in the two operas are intriguing. Shakespeare, following Brooke's poem, would seem to have opted for a less theatrical and anguishing denouement: Romeo being already dead when Juliet awakes, there is no opportunity for a last passionate, irony-laden duet. One wonders if Shakespeare paused to consider the alternatives here, or whether, as with many details in the Brooke poem, he simply pressed slavishly forward with what his source gave him.

Opera composers have (understandably, I think) been reluctant to give up such a riveting theatrical moment for *both* their principals. In this they did not follow Shakespeare, but rather the two main Italian sources of the story. In Da Porto's version Romeo embraces Juliet after taking the poison and she awakens. They have a last interview and then, after the Friar fails to persuade her to enter a convent, she "drew in her breath and held it a long time, and then, uttering a great cry, fell dead on the corpse of Romeo." This novel (and of course impossible) form of suicide may be repeated in Bellini's opera, for his Giulietta simply falls dead over her lover's body without benefit of poison or knife. That Bellini's hero urges Giulietta to live on after him suggests that Bandello's version of the ending may have played a part, too. In Bandello, Romeo takes the poison, then Juliet awakens in fright at his disguise; they lament their fate, and Romeo urges her to live on after him. She also refuses the notion of a nunnery, embraces Romeo, and dies without any specified cause.

Gounod ends with the more extended duet that Verdi disliked, and it is larded with more usual nineteenth-century opera libretto clichés:

> Come! let us flee to the ends of the earth!
> Come! let us be happy, let us flee together, come!

God of goodness! God of mercy!
Be Thou blessed by two happy hearts!

But as in Shakespeare, she first sips from the empty flask of poison, then stabs herself. However un-Shakespearean their last unison words are—"Lord, Lord, forgive us!"—they certainly capture the exalted, immemorial notion of sacrifice-for-love at the end that is the supreme payoff for all tellers of this story: "O infinite, supreme joy of dying with you!" Juliette sings with her last breath; "Come! One kiss! I love you!" This, surely, is the *image* one carries in one's mind of the climax of the play . . . even if Shakespeare did not actually write it that way.

30

Barber's "Pair So Famous"

With its stupendously drawn central characters, exotic terrain, and countless soaring poetic gestures inviting bravura vocalism, *Antony and Cleopatra* would seem a natural play for operatic treatment. As James Agate pronounced, "If ever a Shakespearean play calls for music, processions, and Tadema-like excesses in bathroom marble, *Antony and Cleopatra* is that play."

However, a closer look begins to suggest why it is surely the trickiest of the tragedies to stage and act . . . and why it might be a daunting vehicle for music. To begin with, the play sprawls: there are over thirty speaking roles and a dizzying succession of forty-two scenes (the record for the canon). It has been observed that the play was based on a story that "is not in its essence dramatic at all," and the great Bradley bluntly concluded, "It is, no doubt, in the third and fourth acts, very defective in construction." Finally, as regards the tone struck by its hero and heroine, this play might be called the kinkiest of the tragedies: somehow Shakespeare managed to tie farce, nobility, raunchiness, solemnity, witty flippancy, idealism, and reductive satire together into one "knot intrinsicate." Rare is the director—or the actors playing the lovers—who can slip this knot with artful dexterity rather than damaging scissors of some sort.

That the title characters are a far from admirable pair may also help explain why they have not proved attractive to composers.* Shakespeare, I believe, expected his audience to associate Cleopatra and her Alexandria with the sleazy taverns and bordellos hard by the Globe Theatre (several times she's called a prostitute), and he perhaps expected Antony to be thought of as a high-ranking "lord" who has ferried across the Thames from the very proper City or Court for a Bankside debauch.* As Samuel Johnson suggested, some of Cleopatra's "feminine arts" are "too low" . . . and were certainly not the stuff of nineteenth-century operatic heroines. Thus, Verdi rather predictably indicated the love, personalities, and misfortunes of Antony and Cleopatra aroused "little sympathy" in him. And Berlioz, daydreaming about an *Antony and Cleopatra* opera, made it clear that precisely what he liked in the play was what would send his contemporaries running in the opposite direction:

> Of all the subjects that could inspire me, this one is the least accessible to French taste. Do you think I would have the impertinence to disfigure Shakespeare's creation by fashioning an academic [read Boieldieu-esque!*] Cleopatra? . . . Oh no! It's just because that blazing, mercurial Egyptian lass is the very antithesis of such clods that she would captivate me. I adore the giddy creature . . . the imperial grisette, who plays hop-scotch in the streets of Alexandria, whose mood changes twenty times in a quarter of an hour at the will of fantasy, who brazenly questions Mardian the eunuch about his sexual predilections, and finally the silly, weak woman, fleeing the battle of Actium without knowing why. What a character for musical fantasy!

And yet . . . there is but a single important opera for us to consider here, and even this one—perhaps because we do not readily associate its composer with mercurial musical fantasy—raises disconcertingly numerous puzzles and qualms.

The premiere of Samuel Barber's *Antony and Cleopatra* on 16 September 1966, inaugurating the Metropolitan Opera House at Lincoln Center, will in all likelihood stand for some time to come as the most famously dreadful night at the opera in the history of the American (if not the Roman) republic. The next day the *New York Times* described the opera—directed by Franco Zeffirelli and with a libretto credited to him—as "a Swinburnian mélange of sad, bad, mad, glad." Even *Variety*—under the flavorful headline "Tres Chi Chi Bow of $45 Million Metop Faces Very Chilly *Antony and Cleopatra*"—chimed in, calling the work an "almost uninterrupted bore, of little musical and no sexual conviction." Several years later, another critic recalled it as a "haircurlingly awful production." Leontyne Price, the Cleopatra, was traumatized by the debacle, and Barber himself sailed the next day into several years of gloomy European reclusion.

Battered but unbowed, the composer a few years later undertook revisions, depending principally on the stage-directing and libretto-making expertise of Gian Carlo Menotti. Though, when the revised score was published in 1976, Zefirelli's name as librettist was oddly replaced by "William Shakespeare." Liberated from the hype of opening night for a huge new theater (then the world's largest)—and from the scenic gigantism that its fancy new stage machinery encouraged—the opera was revived at the Juilliard School in 1975, to generally more favorable critical response.* Two years after Barber's death, performances in 1983 at Spoleto led to a commercial recording of this revised version. Several issues raised by Barber's changes in the course of revision are intriguing; they can help pinpoint some of the difficulties involved in squeezing the Shakespearean orange into a form appropriate to the musical stage. As well, the opera, even in its final form, offers a few, mostly lamentable object lessons in the difficulties of "reading" a Shakespearean script musically.

With one glaring exception, Barber's revisions made good theatrical sense. The original cast of thirty singers was shrunk to twenty-two. The disappearance of Lepidus, the feckless third triumvir, does no harm, and the eunuch who amused Berlioz vanishes, along with an interpolated non-Shakespearean song for a countertenor that needlessly halted the action. The thankless role of Antony's proper Roman wife, Octavia, becomes effortless as well: she is demoted from lyric soprano to mute (in any event, she should have been a mezzo, for both play and libretto describe her as "low-voiced"). A few divertissements, like a stick dance on Pompey's galley, were deleted, and the obligatory drinking song that Shakespeare himself invited with these rousing lines—

> Come, thou monarch of the vine,
> Plumpy Bacchus with pink eyne!
> In thy vats our cares be drown'd,
> With thy grapes our hairs be crown'd!
>   Cup us till the world go round . . .   (2.7.113–117)

—was also removed. Especially in the first act, many isolated lines without dramatic point were cut; in all, this reduced the 342-page score by about fifty pages and something under an hour of running time.

Aside from a prelude inserted between the two monument scenes in the last act, the only major addition was the interpolation of a two-stanza song from a Beaumont and Fletcher play at the beginning of the big act 2 love duet.* The music for this, marked *tranquillo e sereno* and dominated in the patented Barber vein by languorous string writing, is beautiful . . . highly reminiscent, in fact, of his haunting scena for soprano, *Knoxville: Summer of 1913*. And the lines are undeniably handsome:

> Take, oh take those lips away
>   That so sweetly were forsworn,
> And those eyes like break of day,
>   Oh lights that do mislead the morn,
> But my kisses bring again,
> Seals of love, though seal'd in vain.
>
> Hide, oh hide those hills of snow
>   Which thy frozen blossom bears,
> On whose tops the pinks that grow
>   Are those that April wears,
> But first set my poor heart free,
> Bound in those ivy chains by thee.

The purpose of this addition is easy to see: Barber and Menotti wanted at least one luxuriant oasis of conventional idyllic amorousness (and beautiful music) just before the battle of Actium sends the lovers careering to disaster. And they succeeded.

But at quite a cost: that of seriously changing the nature and even the verbal style of the torrid, violent, and self-destructive love affair that Shakespeare actually portrays. This interpolated song is everything that Antony's and Cleopatra's poetry is not: sweet-toned, elegant, and polished. Only interpolating Ben Jonson's similar but more famous "Drink to me only with thine eyes" could have been wider of the mark. "Take, oh take" masks the sexual battle, very much out in the open in the play, behind the Petrarchan formula of the ice-princess (her breasts are "hills of snow") enthralling her male lover ("those ivy chains") that Shakespeare had made fun of on numerous previous occasions. A song like this might have given pleasure in a London palace, where bawdy "levity" was frowned on and discreet sophistication was the norm. But I think we can be pretty sure that the renegade couple described so bluntly in the play's first speech (tellingly absent from the libretto) would have found something more exciting to do than recite such "tasteful" stuff to each other:

> His captain's heart . . . reneges all temper,
> And is become the bellows and the fan
> To cool a gipsy's lust . . . you shall see in him
> The triple pillar of the world transform'd
> Into a strumpet's fool.   (1.1.6–13)

If anything, the song is redolent of the self-consciously "poetical" style of *Romeo and Juliet* or Sonnet 116, which so famously defines love as "an ever-fixed mark." In *Antony and Cleopatra,* his last great lovers' play, Shakespeare opens up a harsher, more adult, more cynical world of sexual politics, as befits a "mutual pair" well past their salad days. This is the world of the Sonnets' Dark Lady (Cleopatra says she is "with Phoebus' amorous pinches black") and of that greatest of all poems about lust, Sonnet 129:

> Th' expense of spirit in a waste of shame
> Is lust in action, and till action, lust
> Is perjur'd, murd'rous, bloody, full of blame,
> Savage, extreme, rude, cruel, not to trust . . .

Shakespeare's heroine, but not Barber's, is several if not all of the above. She, like the barge in which Antony first sees her, is more stately in the opera, also more regally diva-like (the casting of Price, a not particularly agile or resourceful actress, doubtless influenced composition). In this respect, then, the interpolated song—though subversive of Shakespeare—was actually appropriate to the character of the opera's Cleopatra.

The challenge of the Egyptian queen is to convey her quick-changing histrionic skills in what Berlioz called musical "fantasy." The challenge of Antony is quite different: that is, to show what he himself calls the "miserable change" from "the greatest prince o' the world" into a seriocomic failure even at committing suicide. Barber did not succeed here, either, simply because the vocal writing for his bass-baritone remains rather stiff, bluff, and more or less the same throughout . . . decidedly more a soldier's than a lover's cantilena.* Antony's effeminizing "dotage" over Cleopatra—often remarked upon in the play—is, in fact, scarcely projected at all. The singer is thus given little interpretative "room" to project his character's disintegration into a "noble ruin"—as, for instance, a Wotan can do over an admittedly vaster time-span. Also, Antony is a creature of impulse, and this requires emotional and vocal agility that would seem to make the choice of a low baritone an odd one, as I think the recording unfortunately demonstrates.

Other problems also arose from the way lesser principals were musicked. Part of Shakespeare's strategy was to invite his audience to side with the two magnificent derelicts simply because the alternatives are so uninteresting or repulsive: tediously virtuous Octavia (a Micaela role—with no big arias) and cold-bloodedly machiavellian Caesar. But the opera's libretto and music do little to capture any of the numerous little touches that urge us to think the "Roman" way is inhumanly unpassionate and ungenerous. To use the distinction of the messenger, we naturally prefer to spend our time with "breathers" like Cleopatra rather than "statues" like Octavia. Barber said he was happy to cut "all that Rotary Club talk by the Romans," and indeed many Romans vanish and Caesar's role is severely cut. But what remains is too close to neutral to make him the odious "politician" Shakespeare intended. The failure to recreate this character one loves to hate is starkly shown in how Caesar's eulogy for Antony ("The breaking of so great a thing should make / A greater crack . . .") is handled. Shakespeare clearly intended it to be perceived by the audience as mere lip-service performed to enhance Caesar's own public image. The tears he refers to in it are surely of the crocodile variety, and, pertinently, he cuts the eulogy short as soon as he sees a messenger with "business" in his looks. Disastrously, Barber takes the eulogy at face value and transforms it into a genuinely emotional aria—marked *con passione, espressivo,* and *appassionato.* The playwright would have been amazed to find his epitome of politic cool turned momentarily into one of passion's slaves! (He might have sympathized a bit more with the idea of a mute Octavia, having biased the competition so clearly in Cleopatra's favor. Enobarbus' remark about Octavia's "holy, cold, and still conversation" scarcely recommends her for musical excitement. But perhaps an inspired operatic adventurer might have gone in an *opposite* un-Shakespearean direction, as Berlioz had a mind to do: "Perhaps there could be some way of bringing the frigid Octavia and the wild Cleopatra face to face. What a confrontation that would make." Agreed.)

When Shakespeare wholly invented a character not in his sources, he usually had very good reasons for doing so. Enobarbus is the only such character in *Antony and Cleopatra,* and clearly, to borrow his own phrase, he "earns a place i' th' story" as a witty, agile, cynical commenter on all the pretences around him, especially those of the two lovers. He is the play's Mercutio figure and, like Mercutio, the playwright does away with him when he no longer wants satire in his plot. The famous description of Cleopatra's barge, in glowing blank verse, is very unlike him, but Barber seized on it as central to the musical identity of Enobarbus, whom he cast as a lyric and far-from-agile bass. Nothing of the character's sharp tongue and intelligence survives in the opera. Shakespeare banked on our pleasure at these qualities to affect us when his suicide speech arrives. But since we are only slimly acquainted with Enobarbus in the opera, it is curious that Barber chose to give the suicide of this rather sketchy character the full lyric, *con espansione* treatment. This is one of the evening's longest solos, but since we don't care about Enobarbus it doesn't generate much of an effect.

A small but crucial ingredient of the success of Shakespeare and the great stage composers is the ability to create credible and memorable comprimario roles. One thinks of Lepidus and the rustic fellow who delivers the asps, and of such operatic bull's-eyes as Mozart's Monostatos and Barbarina, Verdi's Sparafucile and Padre Melitone, or Puccini's Sacristan and Spoletta. Barber's inability to give any of his minor roles a distinctive aural presence and personality is perhaps the clearest hint that he was, finally, not destined to be remembered as a composer for the stage.

Finally, though many of the play's familiar lines do appear in the libretto, one is frequently conscious of Barber's shying away from the brilliant passages that would seem to cry out for vigorous, lavish vocal gestures. Consider the play's first great speech, which comes but moments into the action:

> Let Rome in Tiber melt, and the wide arch
> Of the rang'd empire fall! Here is my space,
> Kingdoms are clay; our dungy earth alike
> Feeds beast as man; the nobleness of life
> Is to do thus—when such a mutual pair
> And such a twain can do't, in which I bind,
> On pain of punishment, the world to weet
> We stand up peerless.　　(1.1.33–40)

Opera thrives on elaborate, stance-taking oaths like this, and one wonders why Barber did not accept the challenge to establish his hero's character emphatically here, as Verdi did for Radames in that other Egyptian opera with "Celeste Aida." Instead, the first line-and-a-half appears fleetingly on page 100 of Barber's score. Elsewhere, for the coda of the act 2 love duet, he has both lovers sing "O thou day o' the world" several times over. Yet he leaves the tremendous *con slancio* lines that follow resting in peace:

> *Antony:*　Chain mine arm'd neck, leap thou, attire and all,
> 　　　　　　Through proof of harness to my heart, and there
> 　　　　　　Ride on the pants triumphing!
> *Cleopatra:*　　　　　　　　　　　Lord of Lords!
> 　　　　　　O infinite virtue, com'st thou smiling from
> 　　　　　　The world's great snare uncaught?　　(4.8.14–18)

From so many memorable lines like the following, in which Antony blisters Cleopatra—

> O this false soul of Egypt! this grave charm,
> Whose eye beck'd forth my wars and call'd them home,
> Whose bosom was my crownet, my chief end,
> Like a right gipsy, hath at fast and loose
> Beguil'd me to the very heart of loss.　　(4.12.25–29)

> You have been a boggler ever . . .　　(3.13.110)

> I found you as a morsel, cold upon
> Dead Ceasar's trencher . . .　　(3.13.116–117)

—Barber chose to set only the pale "To the young Roman boy she hath sold me." As Antony is heaved up into the monument, Cleopatra has these lines showing both her penchant for the royal verbal flourish and her "I want what I want when I want it" philosophy of life:

Had I great Juno's power,
The strong-wing'd Mercury should fetch thee up,
And set thee by Jove's side. Yet come a little—
Wishers were ever fools—O, come, come, come.   (4.15.34–37)

Barber sets only the last four words. And completely absent is perhaps Antony's most affecting speech, in which he contemplates the implications of the changing shapes of clouds and the woman "Whose heart I thought I had, for she had mine" (4.14.3–22).

One last example of a resounding Shakespearean aria or arietta debilitatingly whittled into nonexistence is the play's real eulogy for Antony (as opposed to Caesar's fake one). The moment does not have the wished-for effect in the opera, in part because only the italicized sentences survived in the libretto. Cleopatra fondly recalls,

> *His legs bestrid the ocean, his rear'd arm*
> *Crested the world,* his voice was propertied
> As all the tuned spheres, and that to friends;
> But when he meant to quail and shake the orb,
> He was as rattling thunder. For his bounty,
> There was no winter in't; an autumn it was
> That grew the more by reaping. *His delights*
> *Were dolphin-like, they show'd his back above*
> *The element they liv'd in.* In his livery
> Walk'd crowns and crownets; realms and islands were
> As plates dropp'd from his pocket.   (5.2.82–92)

Shorten the runway thus, and what lyric flights are possible?

That the line in this speech holding the key to Antony's character—"For his bounty / There was no winter in't"—was not used, points to what may be the most serious and puzzling ailment of the opera's libretto; namely, an almost perverse avoidance of memorable lines that define the two lovers' personalities, make their characters coherent and three-dimensional. Antony's exuberant impulsiveness ("Things that are past are done"), his acute knowledge of his shame ("Ten thousand harms. . . . My idleness doth hatch"), the mixture of hostility ("The witch shall die") and adoration ("My heart was to thy rudder tied by th' strings, / And thou shouldst tow me after") are too gingerly conveyed in the opera. And, while Barber and Zefirelli and Menotti strangely used up valuable stage time with a few scenes of no great value to the plot (the bawdy soothsaying scene and Antony's soldiers hearing strange music "under the earth"), many a juicy passage one would have thought indispensable in evoking Cleopatra's magnetic but violent spirit is abandoned. For instance, Antony's "Fie, wrangling queen! / Whom every thing becomes—to chide, to laugh, / To weep," and her "I will give thee bloody teeth" and "Ram thy fruitful tidings in mine ears," as well as her admission that she is "Wrinkled deep in time" and past her "salad days." And why deny us the most famous praise of sexual charisma in all theater, Enobarbus'

> Age cannot wither her, nor custom stale
> Her infinite variety. Other women cloy
> The appetites they feed, but she makes hungry
> Where most she satisfies; for vilest things
> Become themselves in her.   (2.2.234–238)

Even Barber's elaborately conceived final scene passes over several superb details that would have produced a more infinitely various heroine. For instance, her jealous thought when Iras dies before her: "If she first meet the curled Antony, / He'll make demand of her, and spend that kiss / Which is my heaven to have." Or her (and our) vindictive pleasure at the thought of foiling cunning Caesar: "O, couldst thou [the asp] speak, / That I might hear thee call great Caesar ass / Unpolicied!" And the obvious desire to leave the queen (and the soprano) the visual focus as the curtain falls deprived us, too, of the poignant lines from Caesar that make a perfect coda to all the passionate sound and fury:

> . . . she looks like sleep,
> As she would catch another Antony
> In her strong toil of grace.

31

The Other *Otello*

Stendhal was not inclined to let friendship interfere with aesthetic judgment. Consider poor Francesco Berio, author of the libretto for Giaochino Rossini's *Otello* and a good friend of the volatile Frenchman. In the pages on the opera in his *Life of Rossini,* Stendhal wrote of "the marchese Berio, a man who is as charming a companion in society as he is unfortunate and abominable as a poet!" He also speaks of "the unmentionable literary hack" whose "unspeakable ineptitude" led to the libretto's "orgy of blunders."

Stendhal was able to let all this pass, however. He loved the opera, on this side of idolatry, and he was not about to let a mere travesty of Shakespeare spoil his pleasure. In fact, his praise in the following quotation could well be applied to that disconcerting class of operas that offer dubious Shakespeare but irresistible music: "One factor alone contrives to redeem *Otello,* and that is *our* reminiscences of Shakespeare's *Othello* . . . we are so electrified by the magnificent musical quality . . . so spellbound, so overwhelmed by the incomparable beauty of the theme, that we invent our own libretto to match." He also writes of the score's "outstanding quality" and "fiery urgency," and ecstatically recalls Pasta's performances as Giulietta (he preferred her to the role's creator, Isabella Colbran).

It is easy to understand—and share—such mixed feelings about *Otello.* Byron, in an oft-quoted letter reporting that Italians were "crucifying *Othello* into an opera," took exactly Stendhal's position after he attended a Venetian performance in 1818. He called the music "very good," but . . . "as for the words! All the real scenes with Iago cut out—& the greatest nonsense instead."* (Byron, as we will see, was wrong about Iago's scenes.) But this cheek-by-jowl existence of much "nonsense" alongside several passages and one whole act that are potently Shakespearean is what makes this opera so interesting for our purposes. Rossini's opera, like Shakespeare's own plays, was a creature of its time, place, and the circumstances of its first performances, and much of the "nonsense" (a common way of saying "different from Shakespeare") made very plausible sense backstage at the Teatro del Fondo in Naples in the fall of 1816.

Take, for example, the opera's largest dollop of nonsense, the handling of Shakespeare's Roderigo. It will be recalled that he is the pathetic little dupe, absurdly hopeless in his pursuit of Desdemona, whom Shakespeare contrasts with his larger-than-life dupe, Othello. Iago calls him a "snipe" and, with his usual vicious condescension, a pimple annoyed to bursting: "I have rubb'd this young quat. . . . And he grows angry." Berio makes him no less than the Doge's son and a serious love rival for Otello; Rossini promotes him to the status of the opera's second *primo tenore*. Indeed, during much of the action of the first act it seems that Berio was inspired by a different Shakespearean play: Desdemona's father Elmiro (= Capulet) gives her (= Juliet) in marriage to the well-connected Rodrigo (= Paris, son of the Prince of Verona), rather than Otello, a potential political rival (= Romeo). Why? And why, too, is Iago's role shrunk so drastically in the process (in the play Iago has 230 lines more than Othello)?

One obvious explanation would be that Italian opera of the time doted on love triangles and contretemps they generate. In fact, Berio displayed his devotion to this cliché by creating in his libretto a love rectangle. He returned to the story by Giraldi Cinthio that Shakespeare had used as his source and changed Iago's motivation from sheer envy and delight in manipulation to disappointed love for Desdemona.*

But I think there is a better explanation for Rodrigo's ascendancy over Iago. Iago, in both play and opera, is a far-from-heroic character. He works by indirection and innuendo; he's not a stance-taker, but a behind-the-scenes operator. This was not the sort of character to inspire or sustain the vocal virtuosities that were the heart of Rossinian opera. I think that what Berio and Rossini wanted was a plot that would position two prominent, aristocratic, heroic characters as rivals, for the simple reason that at the time there were two fine tenors in the stable of the great Neapolitan impresario Domenico Barbaia: Giovanni David and Andrea Nozzari. (Indeed, the peculiarity of *Otello*'s requiring five tenors must be explained by the fact that the Barbaia roster was knee-deep in that category.) A crucial concern in preparing the opera, then, must have been an eagerness to exploit the talent at hand. At that time, when artists capable of coloratura virtuosity were at the height of their dominion, no one would have thought to let the mere felicities of a source story stand in the way, however esteemed the author. It was thus the height of hypocrisy for Stendhal—after casting such aspersions at Berio—to admit to the real attraction in early bel canto opera; namely, heroic vocal competition: "to hear a performance barren of the glorious rivalry of some Nozzari and David striving to outdo each other in perfection, goaded higher and ever higher in achievement by the spurs of emulation, is to hear nothing." It is tempting to imagine that a similar thespian competition between the actors playing the Moor and his ensign was a part of the play's first performances . . . and that Shakespeare may have tailored the two roles to their specific talents. Indeed, one wonders how different *Othello* might have been if Shakespeare had had at his disposal *two* actors of heroic cast like Burbage and no one (Lowin? Armin?) capable of Iago's chameleonic agility.

Often, it cannot be denied, Berio jettisons elements of the story one would have thought vital. Many keen ironies are muted, for instance, by making Desdemona aware from the outset that Iago is a nasty character. There is, astonishingly, not a single moment of happy love or tenderness for Otello and Desdemona together. And the scene is set entirely in Venice, which perhaps caused the Republic's notorious internecine politics to loom larger in the opera. Several other changes were obviously designed to create occasions for favored musical and dramatic clichés of bel canto opera. The nuptial celebration for Rodrigo and Desdemona is dramatically interrupted by Otello, and this leads to a typical *larghetto* of dramatic shock and a rousing *allegro* finale (Donizetti's famous sextet in *Lucia* derives from exactly the same situation). Bel canto composers doted on scenes in which the heroine, surrounded by a big crowd, is given a heartrending plea; such a finale occurs at the end of the second act (Verdi followed this tradition memorably at the end of his *Otello*'s third act). It is harder to get upset about certain other changes. As many have remarked, the Emilia–Iago marriage is hard to believe (in Cinthio there is even a three-year-old daughter); Berio drops it. It is hard, too, to share Byron's fury over the handkerchief's being replaced by an intercepted letter and lock of hair. Either way, it is a cheap device, as Shaw was happy to point out. *Othello*'s plot, he wrote, "is a pure farce plot: that is to say, it is supported on an artificially manufactured and desperately precarious trick with a handkerchief which a chance word might upset at any moment."

Amid all these alterations and transpositions, however, are several effects that nevertheless capture the Shakespearean spirit. A plausible equivalent, for instance, of Iago's terrible aside as he watches Othello and Desdemona express their love are the asides for Rodrigo and Iago during the general's triumphal first scene ("Vincemmo, o prodi!"): "Restrain yourself, we must cautiously conceal our revenge," says Iago. And the next scene—in which Iago pretends friendship for Rodrigo, urges trust, and promises quick and happy fulfillment of his desires—perfectly catches the ingratiatingly "honest" seductions that Iago performs with all his "friends" in the play. Rossini's honeyed and graceful vocal writing for Iago here, as well as the pleasant scoring for woodwinds and strings, nicely conveys the villain's charming sociability. (Cinthio's wicked ensign is, in fact, described as having a "handsome presence" and, even in the play, should not be cast as a conventional ogre-ish fellow.) Finally, I think it can be said the vocal writing for Desdemona as a whole is outstandingly faithful to Shakespeare's "most exquisite lady." In her first scene, Emilia says to her, "Sempre è con te sincero" ("It is ever sincere with you"), and Rossini produced several displays of sincere and sympathetic emotion for Giulietta. Even the cornucopia of Rossinian embellishments is most suitable to a heroine who is intended to exude the essence of generosity, and who is described as "bounteous" and "fram'd as fruitful / As the free elements." The stamina of Desdemona's virtue is thus aurally symbolized in the heroics Rossini first required from Madame Colbran (whom he later married).

Two scenes in the opera—the finest, I think—owe much indeed to Shakespeare's poetry and dramaturgy. The first is the duet for Otello and Iago in act 2. Byron, mystifyingly, says all the "real scenes" for Iago are cut from the libretto, but this duet in fact brilliantly conflates the play's two memorable Othello–Iago scenes (3.3 and 4.1). While structured as the usual recitative–slow movement–fast movement, the duet still embraces most of the aspects of character that Shakespeare emphasized. We meet Otello alone and dejected that he has sacrificed "glory" and "honor" in his love for Desdemona, but like Shakespeare's hero he is still attracted to her. Iago enters, notices he is upset (equivalent to Shakespeare's "My lord, I see y' are mov'd"). Otello asks, "What must I do?" and Iago urges him to "control" himself (= "Marry, patience / Or I shall say y' are all in all in spleen"). This failure to give a more specific answer frustrates the general: "Your stammering, your doubts, your irresolute countenance, have plunged my poor heart in such anxieties" (= "Would I were satisfied!"). Just as in the play, the villain's coy obliqueness is excruciating ("Ah! do not keep me in this dreadful uncertainty," says Berio's hero) and encourages Otello to leap to a conclusion: "Ah, perhaps unfaithful. . . ." Iago chides, "Ah, calm yourself" (= "I see, sir, you are eaten up with passion"). Finally, urged by "friendship," Iago makes it clear that the terrible supposition is correct and fuels Otello's desire for "vendetta." Suddenly Otello warns Iago, "betrayal [by you] now would also be a crime" (= "If thou do'st slander her and torture me, / Never pray more"). Shocked at the doubt of his honesty, Iago divulges the ocular proof: the intercepted love letter Desdemona had written to Otello, which Otello assumes was intended for Rodrigo.

Here the recitative ends. In a wistful *adagio,* Otello's misery pours out while Iago felicitates himself, "Already fierce jealousy has poured out all its poison; already it inundates his heart completely and leads to my triumph." This splendidly echoes many of the villain's most famous lines: "O, beware, my lord, of jealousy!"; "The Moor already changes with my poison . . ."; "Work on, / My medicine, work!" Otello is brought out of his grief when Iago produces the lock of hair that accompanied the letter. This has the same effect as the tale of Cassio's dream in the play, sending Otello into a fit of vindictive rage. Rossini therefore shifts into a thrilling *allegro* that is very similar to Verdi's "Si vendetta" in *Rigoletto.* What follows is pure Shakespeare. Otello swears avenging fury (= "Arise, black vengeance, from the hollow hell!"). Iago tells him to be calm (= "Patience, I say; your mind may change"). Otello swears he will have courage to avenge her (= "my bloody thoughts . . . Shall nev'r look back . . . Till that a capable and wide revenge / Swallow them up"). Finally, Iago swears he will help the general to "overthrow [his] enemies" (= Iago giving his "wit, hands, heart, / To wrong'd Othello's service"). This vaulting finale is wonderful and bears comparison with the duet "Si pel ciel" at the end of act 2 in Verdi's opera. In one detail at the end—Iago's last line addressed to the audience, "I shall triumph over him"—Berio remained more Shakespearean than Boito. Indeed, though perhaps six hundred lines of dialogue were in this scene boiled down to 130 or so, one still can conclude that the spirit of Iago's seduction scenes was re-created remarkably well by Berio, especially given the constricting, three-part formula of the bel canto *scena* he was dealing with.

It is curious, and one would like to think not entirely coincidental, that three great nineteenth-century Italian composers experienced remarkable artistic growth in the course of working on Shakespearean subjects: Rossini with *Otello,* Bellini with *Capuleti,* and Verdi with *Macbeth.* To underscore this fact with regard to Rossini, it is often pointed out that *Otello* was the first opera he wrote entirely for orchestral (as opposed to harpsichord) recitatives, as well as one of the first Italian operas to carry a tragic ending.* But what really proves the point is a look at the opera's *pièce de résistance,* its third act, which Rossini conceived as one large scene moving from recitative to aria, prayer, duet, and brief choral climax.

The chorus of praise for this scene has been virtually unanimous. Stendhal said the plot of this act "is contrived with far greater skill than in either of the other two." Meyerbeer thought it "truly divine . . . splendid declamation, always impassioned recitative, mysterious accompaniments full of local color, and, especially, the style of the old romances in its highest perfection." In his study of Rossini, Francis Toye calls it "a masterpiece with only one defect: the introduction of the crescendo from the calumny song in *Il Barbiere,* which effectually ruins the otherwise magnificent vigor of the scene between Otello and Desdemona when he comes to kill her." More recently, the pre-eminent American Rossini scholar, Philip Gossett, has concluded very impressively that "if one were to choose a single moment as the watershed between the worlds of eighteenth- and nineteenth-century Italian opera, it would have to be the third act of *Otello.*" Rossini, it appears, might have accepted the high praise. When he was asked late in life what in his canon would survive, he replied: "The last act of *Otello,* the second act of [*Guillaume*] *Tell,* and *Il Barbiere di Siviglia.*"

Gossett's belief that "Rossini came of age as a musical dramatist" with this scene may suggest in part why it is the opera's scene most loyal to the spirit (if not, in many respects, the letter) of Shakespeare's action . . . and also the scene most like its counterpart in Verdi's opera. Its many skillful effects are readily apparent, so it is necessary only to draw attention to a few telling parallels with and diversions from the play. The compelling and vigorous music that Rossini gives to Desdemona both in her recitatives and set pieces should be emphasized at the outset. However "exquisite" her goodness, Shakespeare made her a spirited, confident heroine (though she is not always acted this way): there is no doubt that she is such a woman in this scene. Also in the Shakespearean spirit was Rossini's pragmatically eclectic idea of having Desdemona and Emilia hear a distant gondolier singing the very apt and famous lines from the Francesca and Paolo episode in Dante's *Inferno:*

> Nessun maggior dolore
> che ricordarsi del tempo felice
> nella miseria.*

This puts Desdemona in a melancholy mood, which leads her into the harp-accompanied *canzone,* "Assisa a piè d'un salice" (the Willow Song occurs to Shakespeare's Desdemona more spontaneously). As in the play, the song is interrupted by the wind. Perhaps this detail encouraged Rossini to the very effective arousal of one of his patented musical storms as Otello's passions are whipped to fury and murder (his memorable comic *temporale,* of course, occurs in *Il Barbiere*).

After singing the last, poignantly unembellished strophe of the song, Desdemona gives Emilia a "bacio estremo," and the maid leaves. This served to delete the powerful Shakespearean passage in which the two women mull over adultery and Emilia has her splendid proto-feminist attack on men and their double standard: "Let husbands know / Their wives have sense like them. . . ." Perhaps this was because Emilia was confined irrevocably to her comprimaria status, but more likely because the speech, for an Italian opera audience, was still at least a century ahead of its time. Instead, and all too predictably, Desdemona is given a beautifully simple *preghiera* (in the play she vainly asks Othello for time to "say one prayer"). Still, there is a terrible irony of the kind that follows Desdemona throughout the play in the fact that the singer prays that her "dearly beloved come to console" her (the Verdi–Boito prayer lacks this ironic touch).

She draws the curtains, lies down, and soon Otello enters from a secret door with a light and a dagger. The long recitative for him here is one of Rossini's finest; and, though without most of Shakespeare's sublime turns of phrase, it captures the essentials of the situation. There are his huge swings from fury to transfixed contemplation of "that face, on which Nature impressed its fairest print." There is also, for the first time in the opera, a suggestion from Otello that his "temuto aspetto" ("repulsive appearance") must have driven Desdemona from him. His "Ahi! my arm still trembles! cruel hesitation!" reflects Desdemona's observation in the play, "Some bloody passion shakes your very frame!" Also, his "It is the cause, it is the cause, my soul" appears in the libretto's "Eccone la cagion. . . ." And he then extinguishes his lamp (= "Put out the light, and then put out the light . . ."). Like Shakespeare's hero, Berio's hero tries to implicate Heaven in his quasi-judicial act of revenge: lightning flashes, and he sings, "Ah! amid the lightning heaven reveals her crime to me . . . and invites me to complete my vengeance."

The recitative reaches a furious climax; then Desdemona begins the duet proper with a flurry of big vocal gestures and divisions to these words: "Withhold not the blow . . . strike at this heart! Give vent to your evil rage, I shall die dauntless." As in the play, she does not go gently to her death. Otello has an answering strophe, and a few more interchanges raise passions to the limit. Gossett discovered evidence in the manuscript that Rossini had, according to his usual fashion, composed a slow middle section for the duet at this point. But, apparently recognizing this was not the place for leisurely warbling in parallel thirds, he happily pasted over it and hastened directly into the *agitato* finale, at the height of which Otello stabs Desdemona and curses, "Mori, infedel!"

Shakespeare's denouement from this point lasts nearly 300 lines and contains many wonderful moments. Among them are Desdemona's stunning momentary revival, the harrowing exchange—

> *Othello:* Will you, I pray, demand that demi-devil
> Why he hath thus ensnar'd my soul and body?
> *Iago:* Demand me nothing; what you know, you know.
> From this time forth I never will speak word . . .

—and Othello's great exit aria, "Soft you; a word or two before you go. . . ." The Berio–Rossini ending is much more efficient, lasting but three and a half minutes. It is certainly not Shakespeare's ending, but one must grant the powerful pathos it achieves in its own way. One of Otello's friends enters to say that Rodrigo has survived Iago's machinations and that Iago has confessed his villainy (he does not reappear on stage). Then the Doge, Desdemona's father, and Rodrigo cheerfully rush in, apprised of how Otello has been wronged. It is a fine moment for acting rather than singing when the Doge announces the Senate is reinstating Otello, Rodrigo relinquishes Desdemona to him, and Elmiro gives him her hand in marriage. His last speech is a response to this ghastly happy news: "The hand of your daughter . . . Yes . . . I must be united with her. Behold, punished she shall have me." He stabs himself, and ten seconds later the last chord sounds.

In the midst of his pages on *Otello*, Stendhal paused to make this general (and very accurate) prophecy: "whoever the young composer may be who is ultimately destined to eclipse Rossini, the great secret of his triumph will be found to lie in the *simplicity* of his style." The opera's third act does indeed show the way of the future. There is in it a clear premonition of the relaxation of formula and the increasingly dramatic deployment of vocal virtuosity that was to be achieved first by Bellini and then, most spectacularly, by Verdi. Trying to capture what was "different" about Rossini's opera, Stendhal wrote that it was "less caressing, less enchanting to the senses, but more stimulating, and (perhaps) more dynamic." The long arc of progress that began with the stimulating musical dynamism of act 3 was to reach a kind of fulfillment, appropriately enough, with the premiere of Verdi's next-to-last opera in 1887.

32

Hamlet's Outrageous Fortune

Late in his life, Hector Berlioz published an anthology of miscellaneous essays which he called "musical studies, adorations, whims, and criticisms." One of his whims was a music critic's scornful paraphrase of *Hamlet*'s most famous speech that is truly worthy of Shakespeare's black-tempered hero:

> To be or not to be, that is the question. Whether 'tis nobler in the mind to suffer wretched operas, ridiculous concerts, mediocre virtuosos, crazed composers . . . or take arms against a sea of troubles, and by opposing end them? To die, to sleep—no more. And by a sleep to say we end the torture of our ears, the suffering of the heart and mind, and the thousand natural shocks that the exercize of criticism imposes upon our intelligence and common sense. 'Tis a consummation devoutly to be wished. To die, to sleep—to sleep, perchance to have a nightmare—aye, there's the rub. For in that sleep of death what dreams may come when we have shuffled off this mortal coil: what mad theories will we have to entertain, what discordant scores will we have to hear, what imbeciles will we have to praise, what outrages will we see inflicted upon masterpieces, what absurdities will we hear pronounced, what wind-mills will we see taken for giants?

Toward the end of his *jeu d'esprit,* Berlioz wonders bitterly who would bear all these ills when "they could so easily be ended with a flask of chloroform or a revolver?" And he ends with one final wicked thrust: "But soft, here comes the young singer Ophelia, carrying a score and feigning a smile."

Since the first *Hamlet* opera appeared in 1789, several weary, flat, stale, and unprofitable adaptations have seen the light of day. Most of the names associated with these dozen or so versions of the quintessential Shakespearean—nay, English-language—play have sunk justly into oblivion: Buzzola, Hignard, Kalnins, Mareczek, Stadtfeld, Zafred. A few other names, however, do raise our interest. For instance, Felice Romani, the highly successful librettist for Bellini's *Norma,* Donizetti's *Elisir d'amore,* and numerous other bel canto operas. He collaborated on an *Amleto* with the then-prominent composer Saverio Mercadante; it premiered at Milan in 1822. Unfortunately, Romani took *Hamlet* to be merely a variation on the plot of the *Oresteia,* and so his libretto is all too reminiscent of Aeschylus' Orestes (= Hamlet), Clytemnestra (= Gertrude), and Aegisthus (= Claudius). Also, the role of Hamlet was sung by a woman.

One remarkable collaboration, while not in fact producing a *Hamlet* opera that survived, is worth pausing to study as a "prophecy" of magnificent Shakespearean music-drama to come. This is the *Amleto* composed by a twenty-five-year-old named Franco Faccio, which enjoyed a successful premiere (marred, to be sure, by the usual strife of cliques) at Genoa in 1865. Thereafter, Faccio's extraordinary career as a theatrical conductor blossomed, and in 1869 he began his long association with Verdi. This eventually led him to be chosen to conduct the first performance of the revised *Boccanegra,* the first Italian *Aida,* and the world premiere of *Otello.* Dean's perusal of Faccio's score led him to praise its theatricality but also to say his melodies were "too weak to support the weight placed upon them, especially in the ensembles." What makes this "lyric tragedy in four acts" notable is that its libretto was the first ever attempted by young Arrigo Boito, who over twenty years later would produce the text for Verdi's *Otello* and, several years after that, for *Falstaff.*

Given the constricting conventions of number-opera dramaturgy he had to work within, Boito has received high praise for this libretto. William Ashbrook sums up: "it is an extraordinary libretto for its period and, as the work of a twenty-one-year-old poet, full of promise. The vividness and musicality of the verse, though somewhat marred by eccentricity, are clearly the work of a sensitive, responsible artist." And Boito's chief biographer ventured: "that Boito wrote, in *Amleto,* a libretto raised to the dignity of art seems to me indisputable."* The libretto is, as any Shakespeare-based libretto must be, a striking feat of condensation. The original eighteen named characters and several comprimarios are pared down to a manageable nine roles, and the play's twenty scenes are reduced to eight. The vocal casting is on the whole predictable (Amleto a tenor, Ofelia a primo soprano, Geltrude [*sic*] a primo mezzo, Claudio a primo baritono, Polonio a basso), and one is also forced to admire how imaginatively Boito exploited Shakespeare for certain stock operatic events. For instance, in the first scene Claudius and the Queen are given a "pious drinking song" ("un pio brindisi") suitable for a recently deceased king! Ofelia's later distresses provide an obvious opening for a big two-part aria and then a "mad scene" complete with her onstage demise. Ofelia is turned into the sort of earnest, devout heroine that Italian opera of the time doted on: her first words, to her brother, are "Let me pray." Needless to say, a plot that springs into action with a Ghost saying, "Vendetta io vo" ("I want vengeance") was heaven-sent for Italian operatic treatment. It is also a nice touch that Boito's Ghost sings in the appropriately infernal *terza rima* of Dante.

And yet . . . each scene has a clear Shakespearean focus (each also carries an English epigraph like "How now, a rat?" or "Frailty, thy name is woman!"). And a respectable number of the play's supremely famous lines are managed to telling effect. The opening, at least, of the "To be, or not to be" soliloquy is well handled ("Essere o non essere! codesta la tesi ell' è . . ."), and "Get thee to a nunnery" ("Fatti monachella") becomes the refrain for an Amleto–Ofelia duet. "The

Mousetrap" ("La Trappola") is loyal, too, though of course the itinerant players become singers. Polonius meets his death precisely as Shakespeare writes it, with Amleto exulting "Un topo . . . un topo . . . un topo," and Amleto eventually "lugs his guts" away. Laertes—made by Shakespeare into a caricature of vendetta obsession—was of course grist for the Italian operatic mill: Boito fills him with "l'ira del cor." Though inevitably much of Shakespeare's longest, most spectacularly verbose play had to be cut, Boito was able to salvage much of both the graveyard scene and even the climactic scene in the "armament hall," with Amleto, held up by Orazio, uttering the curtain line, "Ed or t'aspetto, O morte!" ("And now, O death, I await you!"). While the libretto is short even by standards of the time and thus devastating to much of what makes *Hamlet* such an ambiguous, engrossing experience, Boito can still be applauded for presenting a recognizable skeleton of the play.

That master ironist Berlioz (and thus the most Hamlet-like of all composers) would have savored several ironies of his own funeral solemnities, which took place three days after his death on 8 March 1869. The most striking occurred when two tame horses drawing his hearse suddenly took it into their minds to burst through the cortège and take the composer through the gates of Montmartre cemetery all by himself. A minor irony lay in the fact that one of Berlioz's pallbearers was Ambroise Thomas, composer of the lone *Hamlet* opera that was enormously successful at its premiere—and that has managed, just barely, to keep a toehold in the repertory. Its premiere at the Opéra was on 9 March 1868, and it was reprised sixty-three times in its first season. Berlioz left no record of a public response to his old acquaintance's opera, if indeed he had the nerve, desire, or energy to attend a performance at all (he was either outside Paris or very ill for most of his last year).

I think it is safe, however, to imagine him walking out of the Opéra muttering something about there being more things in *Hamlet* than were dreamt of in the stage philosophy of Thomas and his librettists. We do know Berlioz viewed Thomas' music dimly, as appealing to a shallow audience and treading water in music history. In an 1856 letter, for instance, he described the Netherlands' monarch as "this dilettante king who only recognizes two composers, Donizetti and Thomas." And Berlioz' history of ferocious attacks on all adulterations of Shakespeare is surely another clear indication that (unless friendship with the composer might have made him bottle his acid) it would have sunk beneath contempt. As, in fact, we have seen that it did in the eyes of that other deep-dyed Shakespearean, Verdi. He perused the libretto and said, "It is impossible to have done worse. Poor Shakespeare!"

What kind of beast is the Thomas *Hamlet*? Critical reaction to several recent revivals has largely been pleasurable surprise.* Andrew Porter offers gentle understatement: "It is not exactly Shakespearean in spirit." Dean, on the other hand, summarizes more bluntly: "Thomas' opera . . . is a compound of the impressive and the deplorable." There are, certainly, many occasions for shocked eyebrow-raising (even hilarity) in the libretto fashioned by Michel Carré and Jules Barbier, but—as has happened several times elsewhere in this study—it becomes increasingly difficult to sustain a tone of righteous indignation about "improvements" on Shakespeare when we begin exploring some of the specific historical circumstances that generated them. Carré and Barbier were to become extremely successful librettists, with Gounod's *Faust, Roméo et Juliette,* and Offenbach's *Contes d'Hoffmann* among their numerous credits. They were brilliant perceivers of their composers' and their audience's tastes, and obviously reworked the original play with their own, un-Shakespearean goals firmly in mind. These, I think, deserve some respect, if not our unreserved approval.

At the outset, though, it is necessary to say quite simply: *Hamlet* may be the least suitable of all Shakespeare's plays to the exigencies of musical dramaturgy. By virtue of its sheer length, the percentage of lines that must inevitably be "lost in translation" into a libretto (where, in Polonius' words, brevity is the soul of wit) will be very high. The "good" Quarto contains over 3,700 lines, the Folio over 3,900. By comparison, turning *Macbeth*'s mere 2,100 lines or *A Midsummer Night's Dream*'s 2,175 lines into an opera of manageable length would seem like child's play.

Also spelling doom, one would think, is the fact that *Hamlet* is Shakespeare's most word-bound and thought-bound—in short, his densest—play. The greatest challenge, obviously, is the title role, which Hazlitt observed is "not . . . marked by strength of passion or will, but by refinement of thought and feeling." Acting the role, Hazlitt adds, "is like the attempt to embody a shadow." Conveying passion, not refinement of thought, is opera's forte, and so one can well imagine why the shadow-life of the Prince's mercurial mind has proved of meager interest to librettists and composers. One visitor to the Thomas premiere, in the course of a very favorable review, accepted this fact of operatic life philosophically, as perhaps we should be prepared to do. Paul Bernard wrote that he admired how "perfectly" the libretto captured the play's "same passionate fever, the implacable destiny dogging its victims, the death-knell ringing amid festivities, and above all the conviction which marches toward its goal without faltering." But he then adds, "Only the philosophical element is less developed, as was bound to be the case. One cannot conduct a discussion with melodies, and the musical aethetic must come to a standstill where the mists of thought commence."

Still, we can permit ourselves some astonishment at how much of Hamlet's famed interior life is jettisoned. The librettists' lack of interest in the time-honored image of Hamlet alone at center stage with his thoughts is strikingly shown in the lone soliloquy that survives in the opera, a shell of its former self. Literally translated from the French, it runs:

> To be or not to be! . . . O mystery!
> To die . . . to sleep . . . to sleep!
> Ah, if it were permitted to come to you.
> To break the bond that ties me to the earth!
> But then? . . . What is that unknown land
> From which no voyager has ever returned?
> To be or not to be! . . . O mystery!
> To die! . . . To sleep! . . . To sleep!

O mystery! O mystery!
To die! . . . To sleep! . . . Perchance to dream!

To add insult to injury, the score indicates that this entire passage—a haunting adagio—can be cut! One other affecting solo moment for Hamlet in the last act (the equivalent of his mortuary philosophizing with the gravediggers and Horatio) is also marked in the score as expendable.

Several non-Shakespearean aspects of the score can be explained by the fact that this was a French Grand Opera, inevitably entailing a five-act division, a huge opening crowd scene designed to rivet attention, and several action-stopping preludes or entr'actes to cover changes of elaborate scenery. Then there is the huge ballet divertissement, a cheerful, bouncy "Fête du printemps," placed by Opéra tradition in the fourth act. It is performed by local villagers who seem blissfully unaware that something is rotten in the state of Denmark, and it features a "pas des chasseurs," "valse mazurka," and "scène de bouquet" (this ballet takes up about fifteen percent of the score). It is easy to see why Verdi despised it: "*Hamlet* and dance tunes!! What a cacophony! Poor Shakespeare!"

Other interesting changes, large and small, are worth noting. Though trimming the cast to eleven characters was surely necessary, one is shocked that batty, devious-minded Polonius, who has almost 400 lines in the play, is shrunk to an infinitesimal twenty-three *words*. What is more, he is in these few words made an accomplice to the regicide (thus motivating Hamlet's rejection of Ophélie) and survives at the opera's end. Nineteenth-century opera's weakness for drinking songs is displayed in the "Chanson Bachique" with which Hamlet entertains the traveling singers—a show-stopper and the opera's best-known number. The opera's two gravediggers also lapse from Shakespeare's witty undertaker's humor into praise of wine. In order to make way for a large amount of non-Shakespearean matter, some drastic excisions were inevitable. Hamlet's "To be, or not to be" speech, for instance, is interrupted, not by Ophélie, but by the King entering and trying to pray (in the play over 500 lines separate these events).

And then there is the matter of the denouement, which, as Bernard suggested in his review, owed more to an Alexandre Dumas and Paul Meurice version of *Hamlet:* Hamlet slays Claudius, survives to become king, and banishes his mother to a convent (Laertes also survives). Thomas later concocted another ending in which Hamlet commits suicide, and the London Records version provides a compromise between the two that comes closer to the spirit of the original. It is hard to blame the librettists for the hash they make of the climax, for Shakespeare's two long last scenes are indeed dramatically verbose. And the sight of so many dead bodies always skirts dangerously near the risible. One is almost inclined to treat cordially any attempt to simplify and end the evening sooner. Bernard showed such indulgence after describing the happy ending in his review: "All this is less logical, but at least the final butchery doesn't exist any more, and, *ma foi,* for the Opéra stage I don't have the courage to regret this very much."

The most striking difference between play and opera is the prominence in the latter of the part of the romantic female role. Shakespeare's typical care to avoid demanding too much from his boy-actors is very clear, for Ophelia speaks less than five percent of the play's lines. As often happens in his plays, a very potent and memorable female role turns out to be a small one as well. But the virtuosic female executant loomed enormously large in nineteenth-century opera, and so Thomas and his librettists—being especially wedded to themes of romantic love-plighting and star-crossed suffering—systematically enlarged the love affair of their Hamlet and Ophélie.

The "problem" of this love affair as Shakespeare presents it is notorious. For there seems to be little love lost during the entire action. Surely, Shakespeare never wrote a speech harder for an audience to swallow than the one for Hamlet toward play's end: "I lov'd Ophelia. Forty thousand brothers / Could not with all their quantity of love / Make up my sum" (5.1.269–271). If so, Hamlet has indulged in "tough love" with a vengeance. Carré and Barbier, however, were intent upon taking Hamlet seriously and inventing a sentimental love interest for their opera. Perhaps with Ophelia's assertion that Hamlet "hath . . . of late made many tenders / Of his affection to me" (1.3.99–100) in mind, they created a big love duet for the first act, in which Hamlet sounds for all the world like Romeo: "Ah! Doubt that the light illumines, doubt that the sun shines by day, doubt that the sky vaults the earth, but never doubt my love!" When alone in the cemetery, Hamlet sings not of Alexander's stopping a beer-barrel, but of "O chère Ophélie! Pauvre enfant. . . ." The prominence of a lyrical Hamlet-as-romantic-lover, incidentally, is one reason why Hamlet-as-ironist-and-satirist is present so seldom in the opera.

Shakespeare created the role of Ophelia for a talented boy who must have had an agreeable singing voice. For Ophelia's mad scenes (4.5.20–73, 155–201) are interspersed with nearly fifty lines of song texts. There is thus a certain loyalty to Shakespearean *method* in the fact that Thomas worked his Ophélie's mad scene into a frenzy of coloratura acrobatics because he had the remarkable young Swedish soprano Christine Nilsson at his disposal. If giving vocal virtuosity free and spectacular reign is one essence of Shakespearean dramaturgy, then the extravaganza—ranking in fame with Lucia's Mad Scene and Lakmé's Bell Song— that takes up, along with the ballet, the entire fourth act is scarcely an indulgent excrescence but a thoroughly Shakespearean theatrical coup. Indeed, the scene of Ophélie's death derives from what is in effect Gertrudes's only verse "aria" in *Hamlet,* thus bringing onstage one of the few simply "beautiful" moments in this grim play:

> Her clothes spread wide,
> And mermaid-like awhile they bore her up,
> Which time she chaunted snatches of old lauds,
> As one incapable of her own distress,
> Or like a creature native and indued
> Unto that element. But long it could not be
> Till that her garments, heavy with their drink,
> Pull'd the poor wretch from her melodious lay
> To muddy death.   (4.7.175–183)

The last few pages of this scene, especially the evanescent final high B, are as gorgeous a musical translation of a Shakespearean passage as the repertory affords.

The scene certainly had the desired effect on the Opéra audience: "On opening night the scene of Ophélie's death electrified the entire house," wrote Bernard, and he expressed some concern that the "overwhelming success" of this one scene might initially make everything else in the opera pale. And as in any superlative Shakespeare performance, heroic execution was given its due: "Mlle. Nilsson, up to now a mere singer, is now an idol. One worried beforehand at the effect of the huge hall upon her fresh voice, but this charming stream has become an immense river. Rarely have I heard a voice fill that vast hall so marvelously: it is a crystal fountain of dazzling limpidness." Shakespeare's exploitation of a boy perhaps thirteen years old for his vocal dexterity may seem modest by comparison, but he shows the same clear focus on "performance" situations displayed in Thomas' huge mad scene.

Similarly, though the "real" Prince may seem to be but fitfully present in the opera (for instance, when La Reine says "what somber fire sparkles in his eyes!"), it can be said that Thomas, Carré, and Barbier were loyal to the performative heroism at the heart of Shakespeare's play. Just as the playwright conceived of the title role as a thrilling vehicle through which Richard Burbage could display his manifold talents, so, too, was the opera conceived with the remarkable virtuosity of the baritone Jean-Baptiste Faure well in mind. (Original plans for a tenor Hamlet were dropped when Faure became available.) Indeed, Bernard's description of Faure's heroic *tour de force* is cast in exactly the sort of praise that any admirable performance of Shakespeare's Hamlet will elicit:

> By right of seniority and because of the importance of his role, I will speak of Faure first. The creation of Hamlet will be accounted an epoch in his artistic life. He acted magnificently, he sang irreproachably. His voice, of a purity without equal, can be heard in effects of the most wonderful sweetness, and his diction is so careful that it is impossible to miss the slightest syllable. What a singer and what an actor! From one end of this long score to the other he succeeds with a phrase here, a word there, a gesture there. The role is overwhelming! Always on stage for four acts, where is there an athlete, other than he, capable of sustaining such a heavy task?

That Thomas' opera will not be revived in the future without such a superior baritone to champion it (and a fine coloratura soprano to second the motion) is thus another mark of its loyalty to the Shakespearean ethos. No sensible producer decides to cook up a *Hamlet* without first catching his Prince.

A few instances in which the opera seems to get right the "motive and cue for passion" also deserve notice. The rampart scene in which the Ghost first appears (and the bass sings seventy-five consecutive D's) is eerily effective and has been praised on all sides. Bernard was much taken with it: "Everything here is combined with a rare beauty; when the curtain falls one is astonished at the immense sensation one has experienced . . . the shadow of Gluck could well have appeared there." The *prière* was a staple of the time, so the King's failed attempt to pray was easily accommodated. And the "Murder of Gonzago," felicitously conceived (the saxophone makes its operatic debut here), happily satisfies the need for a big mid-opera crowd scene.

Verdi singled out but one scene in the libretto for praise: "well-managed, theatrical, and suitable to music." It is surely no coincidence that this scene—the duet for the Queen and Hamlet in act 3—is the only one that extensively and with approximate fidelity follows a scene from the play. This is the tremendous "closet" scene (3.4.1–218). It is the only scene that looms about as large as in the play: it takes up about thirteen minutes or seven percent of the running time, while Shakespeare's scene amounts to about six percent of the action. What Verdi must have approved of was the fact that this duet was not conceived as a big "decoration" with a conventional slow-and-fast-section structure, but rather as an asymmetrical, rhythmically supple vehicle for displaying the mother's and son's roller-coaster emotions (there are twelve tempo changes). The scene is a tantalizing indication that Messrs. Thomas, Carré, and Barbier might have produced something considerably more Shakespearean, if they had not been so heavily weighed down by the excess baggage of grand-opera convention. As it is, the play and opera exist in two distinct theatrical galaxies. They are, to borrow Hamlet's phrase, as different as a hawk and a handsaw.

33

Several Fat Knights

There are many more or less sacred cows in the Shakespearean canon: *The Merry Wives of Windsor* is not one of them. It has always taken a decided back seat to *Midsummer Night's Dream, Much Ado About Nothing, Twelfth Night*—even to the blithe comic fare of *The Taming of the Shrew, Love's Labour's Lost,* and *Comedy of Errors*. It is also a Shakespearean oddity: it has by far the highest percentage of prose lines (eighty-eight percent), and—like Verdi's *La Traviata*—it is the only work he set in his own time. Finally, it is his only play wholly preoccupied with rural English society. And then there is the tantalizing legend of its creation: *The Merry Wives* may perhaps never live down the probably apocryphal story that Shakespeare, at Queen Elizabeth's request, unwillingly resurrected Sir John after the death so affectingly described by the Hostess in *Henry V*.

The suggestion, often ventured, is that the play's weaknesses are due to the author's inability—given a fortnight's notice—to revive his comic inspiration for a character he had killed off: thrusting him into a hackneyed "ageing lover" farce was all he could muster. Demeaning the charismatic "hill of flesh" of the *Henry IV* plays was thus accomplished by Shakespeare himself . . . which makes it rather difficult to carp about the ways librettists and composers have played fast and loose with *The Merry Wives* to achieve their comic-opera rewards. Indeed, if John Dennis could "improve" the play by making sweet, amiable Fenton—who has but three percent of its lines—into the principle character, as he did for a 1702 London production under the title *The Comical Gallant*, then subsequent adapters of the comedy for the musical stage should have little to fear! And there is at least one good hint that Shakespeare may have condoned radical liberties taken with his trivial but ebullient farce. In 1602 an edition of the play appeared that ran about 1,600 lines; this was a drastic, often confused truncation of the full 2,700 lines eventually printed in the 1623 Folio. Several scholars who have analyzed the two versions have speculated that the shorter one may have been concocted by Shakespeare's company for performances for more plebeian audiences after the (alleged) royal or aristocratic premiere: much "caviar" is excized and it is much easier on the legs of the groundlings.* A play that could be cut by forty percent and survive could scarcely be called "well made."

Happily, admirers of the play are able to get past the huge come-down of Falstaff from the histories and its stock battle (much loved by opera librettists) between the old roué and the young lover for the soubrette; they delight in this most splendidly and thoroughly "local" of Shakespeare's plays, with its authentic picture of Elizabethan village life. Corollary to this is the enormous fun, as Mistress Quickly puts it, of the comic "abusing of God's patience and the King's English." The Arden editor—thinking of Quickly, Parson Evans (who makes Welsh "fritters of English"), young Page, Nym, Pistol, Dr. Caius, and the Host—writes of "one of the most astonishing galleries of perpetrators of verbal fun that even Shakespeare ever put in a play." It would of course be unreasonable to expect Continental musical adapters to show much interest in its "roast beef of Old England" glories, and it can be no surprise that the only *Merry Wives* opera to retain all the juicy comprimario roles is by the one composer who set Shakespeare's own words, Ralph Vaughan Williams. In almost too-populous consequence, his *Sir John in Love* score lists twenty singing roles. Only one librettist, Verdi's Boito, appears to have made a determined effort to mimic Shakespeare's verbal fun in his own language. Though, it was in this regard that Shaw ventured his one minuscule complaint about Verdi's "immensely vivacious and interesting" comedy: "Boito's libretto is excellent as far as a condensation of Shakespeare, except that he has not appreciated the great stage effectiveness of Falstaff's description to Ford of his misadventures in the basket, with its climaxes, dear to old Shakespearean actors, of 'Think of that, Master Brook.'"

Clearly, the main operatic interest of *The Merry Wives of Windsor* was bound to be its highly imitable plot-furniture and stock characters, not its inimitable ear for perversions of the King's—or, to be accurate, Queen's—English. This granted, it is easier for the Bardolatrous purist to remain calm in the face of radical cutting and pasting. Left with what is merely an exacerbated (and repetitious) farce plot, it is hard to work up annoyance, for instance, that Boito made Nannetta the daughter of Ford (not Page) and Dr. Caius (not Slender) the favored suitor . . . or to fuss that he thought it better for Nannetta than Quickly to impersonate the Queen of the Fairies. Indeed, composers and librettists sometimes risk dramaturgical disaster precisely when they follow Shakespeare too slavishly.

This happens, for example, in *Falstaff; ovvero, Le tre burle,* by history's most infamous operatic also-ran, Antonio Salieri. Though the opera was successful at its 1799 premiere in Vienna, it drags on at considerable length (a 1969 edition runs over 450 pages). This is obviously because, as the title indicates, all three of the practical jokes on Falstaff are retained: the basket dumping, the fat-woman-of-Brainford disguise, and the midnight rendezvous at Herne's Oak. Dean calls the work "superficial . . . in the flimsy Neapolitan style that Mozart's genius has rendered intolerable"; and one reason for its decimation of the play's cast was clearly to borrow stage time for the three *burle* ("jokes"). Thus, Anne and Fenton vanish, while Quickly disappears in favor of a confidante for the soprano prima donna, Alice Ford. Almost all the colorful minor characters vanish, too; even

Slender, who has been called "perhaps the finest comic portrayal in the play, and certainly one of the finest Shakespeare ever painted on a small canvas." Salieri's librettist, Carlo Franceschi, did, however, take Slender's name, give him a wife, and substitute them for Master and Mistress Page!

The subversive effects on Shakespeare's play of vocal star-casting are also obvious in Salieri's opera. Master Ford is the first tenor and so is given several high-profile bravura arias during the action, only one deriving from the play's text. The symmetry of opera casts in Salieri's time makes it inevitable that Mistress Ford be the prima donna. Her role burgeons, too, particularly with a long scene in which, disguised as a German maid, she delivers a love letter to Falstaff herself. (The mishmash of German and Italian surely delighted the first audiences at Vienna's famed Kärntnertor Theater.) It climaxes with the exit aria, "Oh, die Männer kenn ich schön!" ("Oh, I know men well!") Mr. and Mrs. Slender shrink to secondary status, as their casting—mezzo and bass—would suggest. Casting Falstaff as a baritone and giving him one bass servant, Bardolfo, curiously gives their interactions the flavor of Mozart's Giovanni and Leporello now and then, especially when Bardolfo dips into satiric patter-song.

Still and all, a remarkable number of the scenes in the opera unfold with extensive fidelity to *The Merry Wives*. Notable are "Herr Broch's" interview with Falstaff and the latter's marvelous boasting aria, "Nell' impero di Cupido sono un Cesare, un Achille." The basket scene is a near approximation, too, with such added touches as Mrs. Ford's fearing Falstaff won't fit in the huge receptacle and Mrs. Slender's wondering, "Is your lover an elephant?" The fat-woman travesty is remarkably close to Shakespeare (though she becomes the aunt of Ford's cook rather than of his maid), as is the entire Herne's Oak scene, where Falstaff is pinched "in time to the music" and the fairies perform a "danza orgiastica" around him. Without the Fenton–Anne–Slender–Caius subplot to unravel, Salieri is able to manage a much more efficient denouement, but he did not improve on the rather flat effect, in Shakespeare, of Falstaff's "I do begin to perceive that I am made an ass." His baritone sings, "I assure you, I swear to you, I already recognize my error!" Then he and the entire cast sing in unison, "Mai più!" ("Never again!") It remained for Nicolai, Verdi, and Vaughan Williams to recognize the value of closing with a flourishing gesture addressing the audience, as Shakespeare did on several occasions (not, however, in *The Merry Wives*).

When Otto Nicolai was first approached about a Shakespearean opera, he responded: "Zu Shakespeare passt nur Mozart!" ("Only Mozart is worthy of Shakespeare!") Fortunately, his resolve weakened. For two supremely happy characteristics of his *Lustigen Weiber von Windsor* happen to be precisely Mozartean: consistently appealing melodic invention and graceful ease of transition. (Both are perfectly demonstrated in the "komische fantastische" opera's lovely overture.) Perhaps the overlooked Shakespearean opera most worthy of revival and a continuing place in the repertory, *Die Weiber* also achieves a remarkably comprehensive fidelity to its source, given the musical fashions of the Biedermeier period it is steeped in (it premiered at Berlin in 1849, just a few months before Nicolai's death at age 39).

Nicolai's librettist, Hermann Mosenthal, also retained all three tricks on Falstaff, but he was able to streamline their preparation and execution so skillfully that he was able to retain Fenton, Anne, and her two foolish suitors . . . though Quickly disappears. There is also, in the vocal writing, a comfortable willingness to take stock characters at face value. The sweet tenor writing for Fenton, especially in his lovely aria "Horch, die Lerche," perfectly suits the thumbnail description he gets from Shakespeare: "He has eyes of youth; he writes verses, he speaks holiday, he smells of April and May." The imperious and vibrant wives, Frau Fluth (Mistress Ford, a soprano) and Frau Reich (Page, a mezzo), were of course transformed into the imperious coloraturists necessary for any major operatic venture of the time. Herr Reich, eager for his daughter to make a remunerative marriage, is given the same simpleminded bourgeois character that Beethoven gives to his jailer Rocco in *Fidelio*.

In fact, one might say the cozy, warmhearted, semirural ambience of Windsor—where life seems untouched by the commercial hugger-mugger of the Big City or the fretting ambitions of the Court—is also finely recreated in Nicolai's opera. The sleazier, darker side of Falstaff's character and of the Garter Inn that Boito managed to convey by interpolating passages from the *Henry IV* plays is nowhere in evidence here. (This darker side is perhaps what made Eleonora Duse tell Boito that his comedy struck her as melancholy.*) The spirits of Nicolai's merry wives are lighter and less vindictive than those in Verdi's opera, as one can easily tell from the almost–Johann Straussian lilt of their several numbers and their credo: "Cheerfulness and good spirits are the spice of life, and forgiveness is fun." If, as has been suggested, the one theme of the play is "the old one of the natural superiority of plain 'honesty' or virtue to the sophisticated or sophistical arts of the gallant or courtier," then the almost exactly equal measures of wit and good-natured sentiment blended into *Die lustigen Weiber* capture the mood of Shakespeare's Windsor.

Nicolai's opera is faithful to *The Merry Wives* in a third important way, one that may explain the disappointment one is bound to feel about his Falstaff. Dean asserts that "comparison with the humanity of Verdi's portrait inevitably reduces [Nicolai's knight] to a conventional buffo bass." This is of course so, especially when it is recalled that his one big solo in the whole work is purely conventional: a drinking song with which he leads a tippling competition. Its text—

> Als Büblein klein an der Mutter Brust,
> Hopp heissa bei Regen und Wind,
> Da war der Sekt schon meine Lust,
> Denn der Regen, der regnet jeglichen Tag . . .

—is roughly adapted from the Clown's song at the end of *Twelfth Night:*

> When that I was and a little tiny boy,
> With hey ho, the wind and the rain,
> A foolish thing was but a toy,
> For the rain it raineth every day.

Working solely with the Falstaff of the play, Mosenthal and Nicolai were bound to produce a purely *buffo* character: Shakespeare gave them nothing more. Their refusal to inflate the role into a clear star turn assisted by several secondary vocalists, as Verdi and Boito did, also made it possible to remain faithful to the ensemble nature of *The Merry Wives*. The fact that Falstaff, the Fluths, the Reichs, Anna, and Fenton all have substantial moments in the vocal spotlight (Frau Fluth, admittedly, is the most challenging role) loyally reflects the play's roughly democratic line distribution: Falstaff (16 percent), Mistress Ford (7 percent), Ford (11 percent), Mistress Page (14 percent), Quickly (9 percent), Evans (8 percent). As Shakespeare notably failed to do in the history plays (Falstaff has over 25 percent of the lines in *2 Henry IV*), in *The Merry Wives* he kept his voluble comic knight in his place. So did Nicolai and Mosenthal. Thus, they did not have to title their opera *Die lustigen Weiber von Windsor und der dicke Hans, Le Vieux Coquet,* or *Sir John in Love,* as other composers have. In fact, they alone retained Shakespeare's title exactly.

Ralph Vaughan Williams offered his musical version of *The Merry Wives of Windsor* in 1929 with the hope "that it may be possible to consider that even Verdi's masterpiece does not exhaust all the possibilities of Shakespeare's genius." *Sir John in Love*—a charming, inventively conceived, and utterly English work— proves that Vaughan Williams' hope was justified, even though the work has been unable to break Verdi's stranglehold on the active repertory. Indeed, seventeen years passed before it received its first (and, to one Vaughan Williams expert, "not very satisfactory") professional production. As with the Nicolai, it is a shame the opera is not staged every now and then, for it boasts much attractive music in the patented Vaughan Williams veins of lush, string-dominated romanticism and gay, folkloristic simplicity. And its dramaturgy is generally acute and efficient.

As a "reading" of *The Merry Wives,* however, this opera presents a fascinating schizophrenia: in several respects the score is remarkably loyal to Shakespeare's genius, in others the composer's genius manifestly got the upper hand. These latter "digressions" from Shakespeare should be dealt with at the outset. Most notable, perhaps, is the curiosity of one of the most exuberantly romantic, big-gestured composers of the early twentieth century being attracted to so un-prepossessingly prosaic a play (only 12 percent of its lines are in verse). The verbal ethos of *The Merry Wives* is colloquial, casual, everyday, not at all artificial: there are practically no invitations to lyric levitation, even in the few snatches of poetry allotted to the young lovers. But the composer did not let this stop him from haling in several lyric "events" at the slightest Shakespearean hint. For instance, Ford's brusque, witty apology for his jealous rage near play's end—

> Pardon me, wife, henceforth do what thou wilt.
> I rather will suspect the sun with cold
> Than thee with wantonness

—is expanded into a gorgeous, warmly felt set piece marked *andante piacevole*. And Falstaff's lines uttered in lascivious anticipation—"Go thy ways, old Jack. I'll make more of thy old body than I have done"—are not turned into comic swagger (as in Verdi), but rather glow to a sonorous, string-tutti *andante*. Thus, one's aural memory of the opera as a whole is one of frequent bursts of lyric warmth and fellow-feeling . . . which seems rather far from the boisterous, frenetic pace of the rustic bourgeois comedy that Shakespeare actually wrote. Very apt, therefore, is the conclusion of A. E. F. Dickinson in his study of the composer: *"Sir John . . .* has an impressive and congenial breadth, which compensates for its lack of dramatic tension."

Vaughan Williams' view of W'.ndsor life, in consequence, is idyllic. One can almost feel in its aura of "rural paradise lost" the anti-urban, anti–industrial-age bias of the composer (a bias steadfastly maintained by his successor, Benjamin Britten). This is also a subversion of the original play. For *Merry Wives* was above all a *satire* of rural life written for the delight of a sophisticated London audience. Though Shakespeare's satire of country life is more cutting in the Gloucestershire scenes of *2 Henry IV,* it is also a primary ingredient here: he knew his citified audience would enjoy condescending to a cast made up almost entirely of comic butts from, as we would say, "the sticks." Besides, life in the country in Elizabethan times was grim and exceedingly uncomfortable: it would never have occurred to the playwright to foist upon his own public such a roseate idyll as Vaughan Williams gives us. The composer's strenuously cheery view of country living is perhaps most obviously displayed in an "interlude" he added several years after completing the opera. This is a completely un-Shakespearean scene—having almost the feel of a masque—that takes its hint from a line describing Fenton as smelling "April and May" and offers a fragrant, hymeneal "Maying" chorus celebrating the love of Anne and Fenton.

Shakespeare's comic hero also suffers somewhat from the composer's urge to romanticize his source, as one might guess from his decision to throw out his original title, *The Fat Knight.* It is a sign of this transformation that we occasionally find ourselves almost hoping the unsinkable old boy might succeed. Vaughan Williams—who prepared his own libretto—helped his hero along by sanitizing his character at certain points. His financial motive for approaching the merry wives is softened, for instance, by giving to Bardolph his greedy line, "She is my region in Guiana, all gold and bounty." When he exits in all his courting finery, too, Vaughan Williams gives him a very impressive *maestoso,* and in the brief courting scene with Mistress Ford he is allowed a lovely refrain to her rendition of "Greensleeves" and a charming song of his own. At play's end, Ford's reconciliatory lines saying the two youngsters should be allowed to marry—

> Stand not amazed; here is no remedy.
> In love, the heavens themselves do guide the state.

—are given instead to Falstaff. And he, like Verdi's hero, is allowed to step forward to address the audience with universal words of wisdom (borrowed by the composer from an Elizabethan song). He ends with a couplet that is not that far removed from the Verdian "tutto nel mondo è burla": "There is underneath the sun / Nothing in true earnest done." There is certainly nothing in this opera of the callous, self-centered, boorish sponge of Shakespeare's *Henry IV*.

Given all this, we cannot be surprised that Dean finds the opera "disappointing . . . partly owing to the clumsiness of the libretto, partly because it treats the plot as a slow comedy of rustic life instead of a sophisticated farce." One is seldom inclined to differ with Dean, but here he has perhaps come a cropper. A few of his minor objections can, I think, be daffed aside fairly easily. He says that "Falstaff receives too little attention and the lovers too much," but, as we have seen, Falstaff does not loom as large in the play as Boito would have us think. As long as the optional "interlude" featuring the lovers is left out, their presence—while certainly greater than in *Falstaff*—is hardly oppressive. (Dean grants that their music is "beautiful in itself.") Dean knows the performers of the forest scene are "yokels dressed up," but—fondly recalling the glorious music of Nicolai's and Verdi's fairies—he feels it is a mistake "to deprive them of their magic." The rub, of course, is that Shakespeare makes no effort to evoke genuine fairy magic here (as he did in *Midsummer Night's Dream*); his intent was to reap satiric fun from rural superstition and folk-ceremonial hijinks.

Dean's main objection to *Sir John*—and presumably the source of the libretto's "clumsiness"—is its most distinctive stylistic feature: the interpolation of Elizabethan lyrics and tunes from songbooks and other plays of the period whenever the need for an aria or chorus occurs. Here Vaughan Williams was following the lead of his friend and fellow deep-dyed folklorist, Gustav Holst. Holst's hour-long "musical interlude" *At the Boar's Head* (1925), based on Falstaff's *Henry IV* tavern scenes, had made use of numerous old English melodies.* Dean objects that the "folky idiom of the music" is not "really suited to the play, which needs a quicksilver deftness if it is not to become tedious." Possibly. But it may be that one Falstaffian opera of quicksilver deftness—Verdi's—is enough, and that in fact the richly contemporary ethos of *The Merry Wives of Windsor* can best be captured by recourse to contemporary musical material. Frank Howes states the case for this view succinctly: "In Shakespeare's day music and life were at one; popular music reflected popular life; for every situation there is a song; and so the composer can cap his own music with a tune that belongs to the people of the play, to the period of the play, and to the topicality of the play." In fact, it is easy to picture Shakespeare and his contemporaries enjoying this opera more than any other— Shakespearean or otherwise—in the repertory. (Britten's music, I suspect, would have made them acutely uncomfortable.)

The other obvious defense of the technique is that it is thoroughly Shakespearean. Though there is but one in-set song in *Merry Wives* (sung as the fairies pinch Falstaff), there are countless instances of his stopping the action for a discrete musical event. And it is a mark of the composer's ingenuity that his interpolations, by and large, blend naturally into the flow of both Shakespeare's action and his music. A few examples must suffice. At the end of the opera's first act, Ford's jealousy aria is interrupted by Mistresses Ford and Page merrily singing the song that ends *Love's Labour's Lost*. They, but not Ford, are unaware of the irony of one of its stanzas:

> The cuckoo then on every tree
> Mocks married men; for thus sings he:
> "Cuckoo, cuckoo, cuckoo"—O word of fear,
> Unpleasing to a married ear!

They leave lightheartedly, arm in arm, and Vaughan Williams then lets Ford have the last three words of his soliloquy (and the last three of act 1): "Cuckold, cuckold, cuckold!" Another marvelously apt interpolation comes after the two wives rise to high dudgeon over Falstaff's identical letters to them. Vaughan Williams' Mistress Quickly enters and, like her Shakespearean counterpart, exerts a worldly-wise, calming influence by leading them into a trio rendition, in lilting waltz time, of a highly appropriate song from *Much Ado*:

> Sigh no more, ladies, sigh no more,
> Men were deceivers ever,
> One foot in sea, and one on shore,
> To one thing constant never . . .

And what is a Falstaff opera without a drinking song? Unfortunately, Shakespeare did not provide one for him in *The Merry Wives*. So the composer foraged shrewdly and borrowed this one from Justice Silence in *2 Henry IV* for his opera's only tavern scene. As usual, the play's actual lines are neatly used by Vaughan Williams to lead into the interpolation;

Falstaff: Go, brew me a pottle of sack finely.
Bardolph: With eggs, sir?
Falstaff: Simple of itself; I'll no pullet-sperm in my brewage.

> "A cup of wine that's brisk and fine,
> And drink unto thee, leman mine,
> And a merry heart lives long-a.
> Do me right, and dub me knight,
> Samingo."

These have all been Shakespearean interpolations, but others from various Elizabethan authors such as Fletcher, Jonson, and Middleton are also acutely emplaced. Falstaff is left alone to "compose a love-song," and Vaughan Williams inserts Jonson's dulcet song, "O that joy so soon should waste!"—which almost convinces us that Falstaff must have been quite a charmer in his youth. Later, Ford's jealous rage incites the witty host to lead a choral outburst of the famous, coarse ditty "Peg a' Ramsey" in *Twelfth Night*:

> When I was a bachelor, I lived a merry life,
> But now I am a married man and troubled with a wife.

And who would complain that Vaughan Williams has Mistress Ford prank her seductive pose for the libidinous knight with the time's most notable "hit" song:

> Alas, my love, you do me wrong,
> To cast me off so discourteously,
> And I have loved you so long
> Delighting in your constancy.
>   Greensleeves is all my joy . . .

After all, Shakespeare's only two references to the song occur in _Merry Wives_. The second one is in fact given to Falstaff to sing when Mistress Ford approaches him in the forest and he invokes: "Let the sky rain potatoes [potatoes were then thought to be aphrodisiac]; let it thunder to the tune of 'Greensleeves'. . . ." Falstaff then responds with equal courtly formality, performing (_appassionato,_ of course) his own interpolation: part of Philip Sidney's second song in his _Astrophil and Stella_ sonnet sequence, "Have I caught my heavenly jewel?" Here, too, a Shakespearean hint justifies the event, for the play's Falstaff may indeed have had the song in his repertory. His first amorous assault is: "'Have I caught thee, my heavenly jewel?' Why, now let me die, for I have liv'd long enough. This is the period of my ambition. O this blessed hour!" Vaughan Williams merely inserts three of Sidney's seven stanzas and then continues the speech, in the process happily evoking the stately and elaborate artificiality of all Petrarchan–Elizabethan love poetry.

It would be unfair, in conclusion, not to mention several other obvious ways the composer was faithful to the letter and the spirit of _The Merry Wives of Windsor_. Most obviously, the libretto is largely constituted from literal transcription of the play. In his preface, the composer says that only "a few unimportant remarks (e.g., 'Here comes Master Ford') are my own." And he consistently succeeds in discovering music even in Shakespeare's prose. As a setting of Shakespeare's own words, _Sir John in Love_ has only Britten's _Midsummer Night's Dream_ for competition in the entire Shakespearean-opera repertory. As already noted, he entirely retains Shakespeare's large cast, though Robin and young Page become nonsinging roles, and his only major cut is Falstaff's escape as the fat woman of Brainford. A. E. F. Dickinson has summarized oddly about the opera: "Seldom do individual episodes grip the senses in the Russian way. . . . One often longs for a more musically abandoned episode." But surely these are mystifying, not to say misguided, complaints to make against a composer so eager to achieve something in the _English_ way, a way that has not been strong on "abandon."

Verdi's _Falstaff_ has figured often in this study and has been commented on extensively by others elsewhere.* In all this discussion I have taken its pre-eminence for granted. By way of a coda reminding us of its splendors, I would like to draw attention to just one small but illuminating way Boito and Verdi displayed their fidelity to _The Merry Wives_. This is their treatment of the two lovers. In Shakespeare's play they play a small but important part. Though their lines amount to but four percent of the total, they are the spokespersons for genuine, romantic love. The playwright underscores this by doling out the play's very few blank verse lines largely to them. Fenton's lapses are few and probably intentional . . . for example, when he is unromantically bribing Quickly to help him win Anne (1.4.134–165). Their several speeches, scattered fleetingly through the action, are aural oases among all the earthy colloquialism the play chiefly revels in. These verse lines contrast their vernal smell, as the Host puts it, "of April and May" with the autumnal hubris of Falstaff and all the other older, wiser (but duller) adults.

Boito was particularly taken with the lovers and clearly sensed the importance of their tiny lyric blossomings. He wrote to Verdi, "Their love pleases me and makes the whole comedy fresher and more solid." In consequence, the four brief appearances of Nannetta and Fenton in the opera, totaling about seven minutes or five percent of the running time, are almost exactly in proportion with those of the play's lovers. And the aural effect of romantic oases is, of course, even more emphatically achieved when Verdi relaxes from the skittering busyness of his orchestra into the plangent, ethereal lyricism of their signature music (mostly in A-flat and in lulling triple meter). As did Shakespeare's blank verse, these fleeting moments ratify—in a sense, aerate—the action and give us some emotional perspective on all the relentless *buffo* botheration.

34

Coda: A *Dream* of the Bard

It is well known that the French baritone Victor Maurel created the Verdian roles of Jago and Falstaff in 1887 and 1893. It is less well known that shortly before this—in Paris in 1886—Maurel had, rather prophetically, impersonated Shakespeare himself. Just the year before, Ambroise Thomas had recast the baritone title role of his internationally successful *Hamlet* for a tenor, and this encouraged him to resurrect an *opéra comique* on another Shakespearean subject that he had composed over thirty-five years earlier. In this process he rewrote the principal tenor role for Maurel. This three-act comedy is misleadingly but (once one discovers the plot) very appropriately titled *Le Songe d'une nuit d'été*, or *The Summer Night's Dream*. It had received its premiere in 1850; an Italian version followed, and in 1866 it was performed in New York City (a French–English libretto printed for the occasion is extant).

Though it has slumbered in oblivion these many decades, *Le Songe* is worth looking at as a coda to this book, not least because its score suggests a work of considerable gaiety and cleverness—not to say zaniness—of invention. But furthermore, in spite of its occasional *boulevardier* humor, this one opera that actually brings the Bard himself on stage touches unwittingly on several significant truths about his life and art. An adventurous impresario could do much worse among the huge repertory of arcane "lost" operas than give this one a revival, though in certain respects the challenge would be daunting. The long spoken passages (all, I think, amenable to some cutting) would require much thespian verve. But the real trick would be to cast successfully the three principal roles, which demand the "startlingly virtuoso coloratura" singing that *Grove's Dictionary of Music and Musicians* tells us typified Thomas' early stage pieces.

And yet . . . what famous and engaging characters these three would be. The heroine is none other than Queen Elizabeth; she is naturally cast as a commanding *première chanteuse à roulades*. Thomas requires of his soprano a range from low B-flat to high E-flat, and he sends her through an obstacle course of the sort that is familiar from his most famous aria, "Je suis Titania" in *Mignon*. With her lady-in-waiting, Olivia (paired with a certain Latimer to provide the love interest), and others she also sings several elaborate bel canto duets with the usual cascades of parallel thirds. Shakespeare, the *premier ténor,* is given a more lyrical, bittersweet vocal identity. The final principal is—of all people—Shakespeare's "huge hill of flesh" himself, Sir John Falstaff. He is placed exactly equidistant from his usual Shakespearean habitats of Windsor and central London, becoming the General Keeper of Richmond Park. His role is cast for a *première basse chantante*—the eminent Charles Battaille created it—and it brilliantly evokes the jolly knight's extrovert charisma with a plethora of roulades, trill-work, and high D's and E's.

The plot's somewhat surreal mixture of fantasy, history, and reminiscences from several plays produces numerous ironies, some intended and others not. The action begins at a Thames-side tavern with, naturally, a drinking song for Falstaff, then with preparations for the arrival of Shakespeare and his company. These include a "Grand défilé [procession] culinaire comique" and a huge pyramid-shaped dessert borne on a pallet by several servants. The stage clears as the Queen and Olivia enter in disguise; the former speaks of her desire to meet this new playwright, who will one day be "le première poëte de l'Angleterre."

In good keeping with the character of the real-life Elizabeth, Thomas' heroine is very deft, perceptive, and manipulative. Falstaff swaggers in, boasting of the Queen's best game, which he keeps for himself. She chides him for

> Toutes vos bravades
> Vos fanfaronnades
> Vos rodomontades . . .

and makes fun of the ageing lover—as do Mistresses Ford and Page in *Merry Wives*—when he outrageously tries to court the two women. As for her great poet, the Queen appears to have very good reason to think his dissolute friends "threaten to destroy his high intelligence," for he also has a boisterous drinking song at his first appearance (a premonition of the prince's famous drinking song in Thomas' *Hamlet?*). Shakespeare leaves the stage briefly, then returns half-drunk for an interview with the disguised monarch.

She has done her homework. She knows, correctly, that at age eighteen he married a woman of twenty-six. But, amazingly, she also knows that Anne died two years later (in reality she died seven years after her husband, at age sixty-seven). It appears not to have been a happy two years, for Shakespeare says of her demise with a smile, "We owe it to the gods not to forget their gifts." It becomes clear that he is married instead to the bottle and is mired in the demimonde of, as the Queen condescends to put it, "the sailors and drunkards of Blackfriars." After he collapses insensible at her feet, she vows to become his "ange protecteur" and has him removed to Richmond Park.

Act 2 begins with a ballad for Falstaff, fearful at midnight of "le chasseur maudit"—a recollection of the "hideous and dreadful manner" of Herne the Hunter in *Merry Wives*. The theatrical coup of this act, though, is what might be called Shakespeare's detoxification-by-coloratura treatment for alcohol abuse. While still in semi-delirium, he is treated to a spectacular vocalise of arpeggios from offstage; then Elizabeth appears to him, elaborately costumed, in what he thinks must be a dream. She launches into some spellbinding coloratura pyrotechnics. He is transfixed and at first mistakes the radiant woman for his own Juliet, but she informs him that she is his genius and encourages him to reassert the glories of his art.

The third act takes place in a richly furnished gallery at Whitehall Palace and begins with a solo for the Queen on the theme of "The brow that bears the crown is also laden with care," a poignant reminiscence of King Henry's "What infinite heart's ease / Must kings neglect, that private men enjoy" from *Henry V*. Shakespeare is summoned and enters the presence. The Queen, "with a mysterious smile, and complete mistress of herself," convinces Shakespeare that his experience in Richmond Park was a dream. He is left to puzzle over his dream—perhaps a reminiscence of Bottom expounding his strange dream of the ass's head. Distraught, however, that he cannot recapture the splendor of his dream, Shakespeare becomes suicidal. The Queen returns to find she must finally disclose—after swearing him to secrecy—that it was indeed England's monarch who visited him the night before. The opera ends with her urging him to "live on in order to illustrate my reign," and with Falstaff, in a curiously affecting touch, suggesting that the playwright make of him "un héros de comédie." Affecting, because it seems likely that Falstaff was modeled on one (or many) of the life-of-the-tavern profligates that the real Shakespeare must have rubbed shoulders with in the vicinity of the Globe Theatre. There is thus a Pirandello-esque touch in this character in search of an author who boasts, early in the opera, ". . . me, without whom [Shakespeare] could not exist! me, whom he has surnamed his shadow!"

As I suggested above, this charming fantasy hits unintentionally on some important aspects of Shakespeare's life and art. Several have to do with creative habits. *Grove's* describes Thomas as a "thoroughly eclectic" composer, and that phrase is perfectly in the spirit of the playwright's own if-it-works-use-it modus operandi. The dependence on "startlingly virtuoso" execution and heroically skilled executants, we have already seen, is also one of the givens of Shakespearean dramaturgy. Likewise Shakespearean was Thomas' obvious tailoring (and re-tailoring) of his principal roles to specific individual talents.

Other verities—unimagined by Thomas and his librettists Rosier and de Leuven—resonate with a broader irony. Shakespeare *did* become the supreme artistic "illustrator" of the Elizabethan reign, though he was active only in its last decade or so. And Elizabeth—along with her aristocratic coterie—*was* in a real sense an "ange protecteur" for Shakespeare and his company. Without the protection of her royal patents, beginning in 1583, the theatrical profession might well have been dismantled by London's city fathers and the growing anti-theatrical Puritan sector of the population. Also, the worry of the opera's Queen over the bad influence of lowbrows on Shakespeare's "high intelligence" points to a tension that is constant in most of his plays. This is the tension between intellectual sophistication on one hand (Hamlet's "caviar," which sails over the head of the "general" public) and his shamelessly broad appeals to the more vulgar elements of his Globe audience on the other. Happily, I think it can be said, not even at the very end, with *The Tempest,* did Shakespeare show any inclination to resolve this tension. Why circumscribe one's paying audience so arbitrarily?

There is perhaps even a ring of truth in the fun had at Anne Hathaway's expense.* After all, Shakespeare seems to have left Anne permanently in Stratford throughout his London years, and many have observed that he writes more shrewdly and eloquently of fast women—notably the Dark Lady and Cleopatra—and the promiscuous "expense of spirit in a waste of shame" (Sonnet 129) than is entirely compatible with perfect marital fidelity. This might be expected of one whose livelihood was earned in the midst of London's most notorious brothel district. Shakespeare's attitude toward marriage in the opera—never again!—may thus be on the mark. For not once in all his plays did it occur to the real Shakespeare to portray a happy one.

Notes

Introduction: The Great Shakespearean Orange

xi   "nothing but death . . ." *Bernard Shaw: Dramatic Opinions and Essays,* ed. James Huneker (1906), II, pp. 141, 276; *Our Theatre in the Nineties* (1931), III, p. 80.

Top Bard . . . *The Dyer's Hand* (1962), p. 182.

"I can well believe . . ." David Magarshack, *Turgenev: A Life* (1954), p. 370; "to have reached . . ." Letter to Humbert Ferrand, 8 February 1865, *Berlioz: A Selection from His Letters,* ed. Humphrey Searle (1973), p. 194.

"The truth is . . ." *Bernard Shaw: The Great Composers,* ed. Louis Crompton (1978), p. 224. Shaw always spelled his archrival's name thus. Shakespeare, like most Elizabethans, appears to have been of Mark Twain's mind—that it shows a lack of imagination to spell a word only one way—for no two of his six extant signatures are alike.

xii   colorful circumstances . . . The following quotations are from these writers, respectively: Richard Brautigan, *The Revenge of the Lawn* (1971), p. 9; Stephen Holden, review, *New York Times,* 20 September 1987, p. 86; David Leavitt, review, *New York Times Book Review,* 19 April 1987, p. 29; James Wolcott, *New York Review of Books,* 13 August 1981, p. 51; V. S. Pritchett, review, *The New Yorker,* 12 December 1987, p. 133; John Updike, review, *New York Times Book Review,* 23 February 1986, p. 3.

xiii   "would all be bores . . ." "Some Reflections on Music and Opera," in *The Essence of Opera,* ed. Ulrich Weisstein (1964), p. 356. This essay first appeared in *The Partisan Review* 19 (1952).

"a grand cartoon . . ." "Learning from Disney and Dickens," *New York Times Book Review,* 30 January 1983, p. 3.

"With rare exceptions . . ." *The Operas of Verdi* (1981), III, p. vii.

"Neither [Shakespeare] . . ." and "If you want . . ." Letter to Stella Patrick Campbell, 13 January 1921, in *Collected Letters 1911–1925,* ed. Dan H. Laurence (1985), pp. 706–7.

xv   "I know it is . . ." *Collected Works* (1934), p. 515.

xvi   "Utility is . . ." *Prefaces to Shakespeare,* fourth series (1945), p. 212.

"concern to a surer . . ." Letter to Giuseppe Piroli, 7 February 1871, ed. A. Luzio *Carteggi Verdiani* (1935–1947), III, p. 79.

"When the situation . . ." Letter to Antonio Ghislanzoni, 13 September 1870, in Hans Busch, *Verdi's Aïda* (1978), p. 67.

"Shakespeare has always . . ." F. D'Arcais, "Rassegna Musicale," *Nuovo Antologia,* September–October 1879, p. 160.

xvii "Only a few . . ." *Les Écrits de Paul Dukas sur la musique* (1948), pp. 214–15.

"planes away . . ." "Shakespeare and Opera," in *Shakespeare in Music,* ed. Phyllis Hartnoll (1966), p. 89. See also Dean's "Shakespeare in the Opera House," *Shakespeare Quarterly* 18 (1965), pp. 75–93.

"Harriet, like M. Bovary . . ." *Where Angels Fear to Tread* (1947), pp. 120–21.

"an orthodoxy of . . ." Roger Savage, "The Shakespeare–Purcell *Fairy Queen:* A Defence and Recommendation," *Early Music* 1 (1973), p. 203.

"Is it true . . ." Letter to Léon Escudier, 14 January 1868, in Franco Abbiati, *Giuseppe Verdi* (1959), III, p. 161 ("Quale stonazione! Povero Shakespeare!"; *stonazione* is singing or playing out of tune); "It is impossible . . ." Letter to Escudier, 12 March 1868, in Franco Abbiati, *Giuseppe Verdi* (1959), III, p. 164.

"The title of the work . . ." *Shaw's Music,* ed. Dan. H. Laurence (1981), II, p. 133.

"Bitter disappointment . . ." *Memoirs,* trans. David Cairns (1975), p. 161.

xviii "One should never . . ." *Debussy on Music,* ed. François Lesure and Richard Smith (1977), p. 99.

"*Much Ado* without . . ." *Music of Three Seasons: 1974–1977* (1978), p. 483.

"more Shakespearean than . . ." Graham Bradshaw, "A Shakespearean Perspective: Verdi and Boito as Translators," in *Giuseppe Verdi: Falstaff,* ed. James Hepokoski (1983), p. 152; "Verdi and Boito . . ." Peter Conrad, *Romantic Opera and Literary Form* (1977), p. 5.

"As when a Tree's . . ." and "I found the whole . . ." Brian Vickers, *Shakespeare: The Critical Heritage 1693–1733* (1974), I, pp. 78, 344.

xix "Modern producers . . ." *Prefaces to Shakespeare,* second series (1930), p. 25.

"Withered be . . ." H. N. Hudson, *Lectures on Shakespeare* (1848), II. pp. 277–78.

*"Not once did he bother . . . My recent discovery of the probable source for the plot of *The Tempest* is reported in "*The Tempest* and *Primaleon:* A New Source," *Shakespeare Quarterly* 37 (1986), 423–39.

"They have been . . ." *Byron's Letters and Journals,* ed. Leslie Marchand (1976), VI, p. 18.

xx "Elizabethan art . . ." *Elizabethan Essays* (1934), p. 15.

*"of a mingled yarn . . ." The full quotation is: "The web of our life is of a mingled yarn, good and ill together: our virtues would be proud, if our faults whipt them not, and our crimes would despair, if they were not cherish'd by our virtues" (*All's Well That Ends Well,* 4.3.71–74). If one passage could be said to epitomize Shakespeare's presentation (and perception) of human nature, my choice would be this one. La Rochefoucauld perhaps paraphrased its wisdom more succinctly: "Virtue would not go so far did not vanity keep her company."

"I know the world . . ." Letter to Gilbert Duprez, February 1865, in *I Copialettere,* ed. G. Cesari and A. Luzio (1913), p. 455.

"In the first few days . . ." Letter to Verdi, 1 August 1889, in *Carteggi Verdiani,* II, p. 147.

xxi *"It is not possible . . ." *À Travers chants* (1862; rpt. 1970), p. 317. During preparations for *Otello,* Boito made a similar remark to Verdi: "We have desired to retouch the Perfection [of *Othello*] and have destroyed it. This is the reasoning of the Critic. But an opera is not a drama; our art lives by elements unknown to spoken tragedy." And he consoles Verdi by adding, "music is the most free of the arts, has a logic of its own more fleet and more free than the logic of spoken thought—and much more eloquent" (letter of 18 October 1880, *Carteggio Verdi-Boito,* ed. M. Medici and M. Conati (1978), I, p. 5.

* "The fate of Shakespeare . . ." *Romantic Opera and Literary Form,* pp. 69, 178. Dean chooses a gentler, more apt image to make Conrad's general point: "music and drama are notoriously unaccommodating bedfellows" ("Shakespeare in the Opera House," *Shakespeare Quarterly* 18 [1965], p. 75).

"it is a mistake . . ." Michael Goldman, "Acting Values and Shakespearean Meaning: Some Suggestions," *Mosaic* 10 (1977), p. 58.

xxii   "the performers . . ." and "never stop urging . . ." In *Verdi's Macbeth: A Sourcebook,* ed. Andrew Porter and David Rosen (1984), pp. 29–30.

Musiconia . . . *The Complete Singer-Actor: Training for Music Theater* (1985), p. 3.

"In Praise of Musique . . ." Richard Barnfield, *Poems in Divers Humors* (1598), sig. E2ʳ.

1. A Passage to Opera

4   "O brother . . ." *Titus Andronicus,* 3.1.214–33. In this passage I have reproduced the punctuation of the Folio.

5   "is an imitation . . ." In *The Essence of Opera,* ed. Ulrich Weisstein (1964), p. 356.

6   "I have long since . . ." *Shaw's Music* (1980), II, p. 133. ✓

* "being what I call . . ." Letter to Archibald Henderson, 3 January 1905, in *Collected Letters 1898–1910,* ed. Dan H. Laurence (1985), p. 482. Shaw was referring to W. E. Henley. In an undated letter to Vladimir Tchertkoff (August 1905), Shaw tweaked the Top Bard himself: "I have striven hard to open English eyes to the emptiness of Shakespear's philosophy, to the superficiality and second-handedness of his morality, to his weakness and incoherence as a thinker, to his snobbery, his vulgar prejudices, his ignorance, his disqualifications of all sorts for the philosophic eminence claimed for him" (ibid., p. 551).

"Tested by the brain . . ." *Our Theatres in the Nineties* (1931), III, pp. 155–56.

"For the most part . . ." *Dramatic Opinions and Essays,* ed. James Huneker (1906), I, p. 193.

7   "The ear is . . ." *Our Theatres in the Nineties,* I, p. 26. For other Shavian commentary in this vein, see Charles Haywood, "George Bernard Shaw on Shakespearian Music and the Actor," *Shakespeare Quarterly* 20 (1969), pp. 417–26.

* "It consisted of two . . ." *The Tenth Muse: A Historical Study of the Opera Libretto* (1970), p. 80. An example of a simile aria ("Se cresce un torrente") occurs in Vivaldi's *Orlando Furioso;* it runs, in translation: "When a stream swells to a troubled flood and breaks its banks, proudly its waters pour into the fields knowing no restraint. So great is the joy I feel . . ."

"very, very important . . ." *Playing Shakespeare* (1984), pp. 108–10.

8   "The problem . . ." Ibid., pp. 90–91.

"You've got . . ." Yveta Graff, "Tosca Talks," *Opera News,* 18 January 1986, p. 19.

9   "it's so easy . . ." *Playing Shakespeare,* p. 86 (emphasis added).

2. Things Like Truths

10   "The Dramatist . . ." "On Liszt's Symphonic Poems," *Prose Works,* trans. W. A. Ellis (1895), III, p. 249.

"the new naturalism . . ." Howard Felperin, *Shakespearean Representation* (1977), p. 60.

11   "All stage dialogue . . ." *The Life of Drama* (1964), p. 85.

"American Shakespeare . . ." Daniel Seltzer, "Acting Shakespeare Now," *Mosaic* 10 (1977), p. 61.

"The fear of being . . ." *Shakespeare Without Tears* (1942), p. 77.

"That we are all . . ." *The Life of Drama,* pp. 196, 205. "Every good dramatist," writes Kenneth Muir, "especially the poetic dramatist, eschews the methods of naturalism" (*Shakespeare the Professional* [1973], p. 12).

"I never really . . ." In *Players in Shakespeare,* ed. Philip Brockbank (1985), p. 62.

"I find more . . ." In *Playing Shakespeare,* p. 19.

"truth of reality . . ." Ibid., p. 209.

"an epitome . . ." *Hazlitt on Theatre,* ed. William Archer (1895; new ed. 1957), p. 133.

"Truth, *cher ami* . . ." *The Fall* (1957), p. 101.

12   "Tragedy is something . . ." In *Music at Night* (1931), p. 12. Huxley's quotation is from Wordsworth's sonnet "To Toussaint L'Ouverture."

"If a perfume . . ." Quoted as an epigraph to W. H. Auden's essay "Cav and Pag," in *The Dyer's Hand* (1967), p. 475.

"What is essential . . ." Quoted in John Fraser, "*The Tempest* Revisited," *The Critical Review* 11 (Melbourne, 1968), p. 76.

"poet never . . ." Prologue, *Epicoene, Works,* ed. C. H. Herford and P. Simpson (1925–1952), V, p. 164.

"To be deceived . . ." *In Praise of Folly,* trans. Clarence Miller (1979), p. 71.

* "On the stage . . ." Chapter x ("Quaedam eo vero qua falsa"), Book II, *Soliloquies,* in *Patrologiae cursus completus,* XXXII, column 893 (this trans. by Donald Howard). Gore Vidal made something like Augustine's point when he observed, "There are certain truths so true that they are practically unbelievable" (*The Second American Revolution* [1982], p. 182). Richard Howard praises the power of "incredible" art in the last stanza of his splendid poem, "Giuseppe Verdi" (*The New Republic,* 6 October 1979, p. 32):

> We listen to you and applaud not because
>   we believe you (we do not),
> we applaud you for making life credible.

This compliment, obviously, can be paid to Shakespeare as well.

13   "Critics may talk . . ." Richard Hurd, *Letters on Chivalry and Romance* (1762), p. 103.

* "when we speake . . ." *The Arte of English Poesie* (1589; reprint 1970), p. 202. Maynard Mack calls all of Shakespeare's heroes, except Hamlet, "hyperbolists" in his essay "Energy and Detachment in Shakespeare's Plays," in *Essays on Shakespeare and Elizabethan Drama in Honor of Hardin Craig,* ed. Richard Hosley (1962), p. 287.

"Handel's and Mozart's . . ." In William Irvine, *The Universe of GBS* (1949), p. 13.

* "freedom from . . ." "Mr. Shaw on Mr. Shaw," *New York Times,* 12 June 1927, section vii, p. 1. G. Wilson Knight makes this point in less colorful fashion in *The Wheel of Fire* (1949): "We should not look for perfect verisimilitude to life, but rather see each play as an expanded metaphor, by means of which the original vision has been projected into forms roughly correspondent with actuality.

. . . The persons, ultimately, are not human at all, but purely symbols of a poetic vision" (p. 16).

"To copy reality . . ." Letter to Clarina Maffei, 20 October 1876, *I Copialettere,* ed. A. Luzio (1913), p. 624.

14   "one of the worst . . ." *Opera as Drama* (1988 ed.), pp. 124–26.

"From a musical . . ." Letter to Giulio Ricordi, 20 November 1880, *I Copialettere,* p. 559.

"Have no doubt . . ." Letter of 22 August 1870, *Verdi's Aida,* ed. Hans Busch (1978), p. 55.

"I have always . . ." Letter of 28 September 1870, *Verdi's Aida,* p. 69. It should be added that Verdi could still share Kerman's disgust at the "usual" *cabaletta:* "Ah, these damned *cabalettas* that always have the same form and that all resemble each other . . ." (letter to Ghislanzoni, 27 October 1870, in Busch, p. 88).

15   "the improbable . . ." *Music and Politics* (1982), p. 107.

"I don't like . . ." "Die Ägyptische Helena," *Prosa* (1955), IV, pp. 457–60. Another translation of this passage is included in Weisstein, *The Essence of Opera* (1964), pp. 311–13.

3. Worlds of Sound

17   Balk has amusingly . . . *Performing Power* (1985), pp. 306, 309.

"His tongue . . ." *The Empty Space* (1968), p. 110.

"But no one . . ." *The Wreck of the Thresher and Other Poems* (1964), p. 32.

18   "I think that Shakespeare . . ." "On Acting Shakespeare," *Shakespeare Quarterly* 32 (1982), p. 140.

"We can take comfort . . ." In Barton, *Playing Shakespeare* (1984), p. 16.

"With the art . . ." *Prefaces to Shakespeare,* second series (1930), p. 209.

"Sit in a full . . ." *Characters, Complete Works,* ed. F. L. Lucas (1927), IV, pp. 42–43.

19   "Language shows . . ." "Explorata: or Discoveries," in *The Complete Poems,* ed. George Parfitt (1975), p. 435.

*"It is probable etc. . . ." J. L. Styan, *Shakespeare's Stagecraft* (1967), pp. 75–76, 157, 163. In *Shakespeare's Art of Orchestration* (1984), Jean Howard writes similarly of "shifts in tone and tempo," "aural shifts and contrasts," "changes of key," and "crescendo / decrescendo" patterns. Wesley Balk broadens the musical comparison far beyond Shakespeare when, in *The Complete Singer-Actor* (1985), he refers to the "essentially musical language demands of Shakespearean, Restoration, Georgian, and Greek drama" (p. 28).

"Consorts many . . ." *The Whore of Babylon* (1607), sig. A2ᵛ.

"Beauty to th' Eye . . ." "A Short Discourse of the English Stage," in *Critical Essays of the Seventeenth Century,* ed. J. E. Spingarn (1957), II, p. 95.

"he must be . . ." "Preface," *The Fairy Queen* (1692).

"The voice of . . ." *Apology for the Life of Mr. Colley Cibber* (1740), p. 62. Cibber's point is echoed in Betterton's *History of the English Stage* (1741): "in all good Speech there is a sort of Music, with respect to its Measure, Time and Tune" (p. 46).

"the Tones of his . . ." Quoted in Alan Downer, "Nature to Advantage Dressed: Eighteenth Century Acting," *PMLA* 58 (1943), p. 1023.

20   "Shakespeare has . . ." J.B., "On the Art of Acting," *The Spectator* 2 (1829), p. 811.

"The text of a play . . ." *Prefaces to Shakespeare,* first series (1927), p. xv.

"a man in such . . ." Letter to his father, 26 September 1781, quoted in William Mann, *The Operas of Mozart* (1977), p. 294.

"Verbal relish *etc.* . . ." *Playing Shakespeare,* pp. 47, 48, 54, 65–67.

21   "trivial sentiments . . ." *Preface to the Plays of William Shakespeare* (1765), in Brian Vickers, *Shakespeare: The Critical Heritage 1765–1774* (1979), p. 67.

"Many . . . of the Tragical . . ." In *Shakespeare: The Critical Heritage 1693–1733,* ed. Brian Vickers (1974), pp. 25–26. It is no wonder that Macaulay called Rymer "the worst critic who has ever lived."

22   "[It] is a lyric . . ." *Prefaces to Shakespeare,* second series (1930), p. 1.

"all this toil . . ." In *Shakespeare: The Critical Heritage 1765–1774,* p. 152.

"Shut our minds . . ." *Prefaces to Shakespeare,* second series, pp. 19, 64.

"Mr. Keith has . . ." *The Dramatic Event* (1956), p. 131.

23   "you have to know . . ." In *Actors Talk About Acting,* ed. Lewis Funke and John Booth (1961), p. 32.

"This, in analysis . . ." *Prefaces to Shakespeare,* second series, p. 186.

*"This is not good sense . . ." *Our Theatres in the Nineties* (1931), III, pp. 80–81. Granville-Barker likewise often emphasizes the musicality of Shakespeare's writing: "To the last . . . he would write an occasional passage of word-music with a minimum of meaning to it (but of maximum emotional value, it will be found, to the character that has to speak it)" (*Prefaces to Shakespeare,* first series [1927], p. 8).

24   "Obviously you need . . ." Quoted in Barton, *Playing Shakespeare,* p. 195.

"The words do not . . ." *Our Theatres in the Nineties,* III, p. 155.

"Note how actual . . ." *Prefaces to Shakespeare,* second series, pp. 185–86.

"inherent, powerful . . ." Quoted in Frank Walker, *The Man Verdi* (1962), p. 489.

"Before I answer . . ." *Our Theatres in the Nineties,* III, p. 339.

25   *"Not at all . . ." *The Operas of Verdi* (1979), II, pp. 374–75. Richard Strauss wrote to Hofmannsthal about the loss of words in performance: "It's unbelievable, a great pity but true, how little of the text can be caught by opera-goers no matter what pains the composer takes" (letter of 20 April 1914, in *A Working Friendship* [1961], p. 192).

4. Sweet Smoke of Rhetoric

26   * *Love's Labour's Lost* unfolds . . . The play has been tranformed into opera, most notably by Nicolas Nabokov, to a libretto by W. H. Auden and Chester Kallman. The premiere took place at Brussels in 1973; its German title emerged as *Verlor'ne Liebes Müh.*

27   "from its broad . . ." Arthur Quiller-Couch, *Shakespeare's Workmanship* (1918; reprint 1937), p. 42.

"The 'modern' theater . . ." Basil Langton, letter to the editor, *New York Times,* 15 March 1987, p. H22.

"move predominantly . . ." Clifford Leech, "The Acting of Shakespeare and Marlowe," *G. F. Reynolds Lectures* (1963), p. 36.

Shaw reminisced . . . "The Old Acting and the New," *The Saturday Review,* 14 December 1895, in *Our Theatres in the Nineties* (1931), I, p. 285.

28   "The aim is total . . ." Quoted in Benedict Nightingale, "Decoding an Enigma's Genius," *New York Times,* 8 November 1987, p. H35.

*"a typically British . . ." Frank Rich, "Stage: *Breaking the Code*," *New York Times,* 16 November 1987, p. C15. Walter Kerr, similarly impressed by his virtuosity, wrote of Jacobi that his "performance is a wide-open one, not readily topped. When he bites his fingernails to show his edginess, he doesn't do it furtively as most self-conscious people do. He does it as though he were demonstrating the proper way to bite one's nails before a very large classroom. Or a very large theater. . . . He is a well-equiped ringmaster, and we are watching a *tour de force*" ("When Actors Beat the Odds," *New York Times,* 3 January 1988, p. H25).

"By doing Shakespeare . . ." *The Diaries of Peter Hall,* ed. John Goodwin (1983), pp. 314–15.

"The actual filming . . ." *The Dramatic Event* (1956), p. 147.

29   *unnatural excursions . . . Shaw made just this point about the rhetorical "excursions" that are essential to Shakespeare in a boisterous letter to William Archer: "Unless you can make your actors and actresses, in lovely dresses and lovely tights, go off suddenly into dances of words and rhymes like children at play you cannot put Shakespear on the stage. If you cut them out, there is no Shakespear: nothing but [Dickens'] Mr. Wopsle and the Melancholy Dane" (*Collected Letters 1911–1925* [1985], p. 611).

"My God, my God . . ." "Nineteenth Expostulation," *Devotions upon Emergent Occasions* (1624; 1923 ed. by John Sparrow), p. 113.

30   "for my plays bear . . ." Letter to Thomas Demetrius O'Bolger, 16 February 1916, *Collected Letters 1911–1925,* p. 375.

"figures and how . . ." From Part III, "Of Ornament"; Chapter X, "A division of figures, and how they serve in exornation of language," in *The Arte of English Poesie* (1589; facsimile reprint 1970), p. 171. A more recent tour of rhetorical figures is Arthur Quinn's *Figures of Speech: Sixty Ways to Turn a Phrase* (1983).

"the chiefe grace . . ." *Arte,* p. 95; "your Cadences," p. 93; "your figures rhetori call," pp. 206–7; "our speech is made," pp. 206–7; "speech drawn out," p. 186; "the figure that worketh," p. 208; "is when we make," p. 208; "Coocko-spell," p. 210.

31   "the cadence . . ." Quoted in Leonard Ratner, *Classic Music* (1980), p. 33.

"To be persuasive . . ." Ibid., p. 31.

brief definitions . . . Ibid., p. 91.

32   examples of poetic feet . . . Quoted in ibid., p. 71.

"O eloquent . . ." *History of the World, The Works* (1829; reprint 1965), VII, pp. 900–901.

34   One writer . . . See Stephen Booth, *An Essay on Shakespeare's Sonnets* (1969), pp. 72–75.

"When I do read . . ." The student was Kimberly Compare.

35   three "Modified Sonnets" . . . *A Swim off the Rocks* (1976), p. 20.

"Sonnets can be . . ." *Playing Shakespeare,* p. 103.

36   "not a Plant . . ." *An Apology for the Life of Colley Cibber,* ed. B. R. Fone (1968), p. 210.

"I announce a life . . ." *The Poetry and Prose,* ed. Louis Untermeyer (1949), p. 446.

5. The Story of O

37   "name to conjure . . ." All quotations from *The Poetry and Prose,* ed. Louis
      Untermeyer (1949), pp. 600–601.
      *"all the Italian . . ." These reminiscences would support his assertion elsewhere
      as to what kind of opera influenced him: "The experts and musicians of my
      present friends claim that the new Wagner and his pieces belong far more truly to
      me, and I to them. . . . But I was fed and bred under the Italian dispensation,
      and absorbed it, and doubtless show it" (*The Collected Writings of Walt Whitman*
      [1964], II, p. 694). In these reminiscences he also notes that "the opening of the
      Secession war" came on 13 April 1861: "I had been to the opera in Fourteenth
      Street that night" (*Poetry and Prose,* p. 602).

38   "Out of the cradle . . ." Elise Tompkins published four poems in *The New Republic*
      (27 September 1980) titled "Four Lectures by Robert Lowell." In the one devoted
      to "Out of the cradle" are these lines: ". . . Most operatic thing he ever wrote / A
      tour de force . . . highly / Organized musically . . ."
      "dared use anything . . ." Quoted in *The Poetry and Prose,* p. xxiv.

39   "is always capable . . ." Frank Kermode, introduction to *Timon of Athens,* in *The
      Riverside Shakespeare,* p. 1443.
      "The figure of . . ." *Arte,* p. 221.

41   "We may say . . ." *The Mutual Flame* (1955), p. 77. See my discussion of Nietzsche's
      principles, Thomas Mann's "Tod in Venedig," and Britten's *Death in Venice* in
      Literature as Opera (1977), pp. 334–44.
      "a tendency in our . . ." *Playing Shakespeare,* p. 54.
      "You know what . . ." In *Actors on Acting,* ed. Toby Cole (1970), p. 413.

42   "pays the price . . ." *Interpretations of Poetry and Religion* (1957), p. 167.
      "A great actor . . ." Derek Granger, quoted in *Time,* 15 November 1982, p. 80.
      *"heroic nudity . . ." "Literature is always calling in the doctor for consultation
      and confession and always giving evasions and swathing suppressions in place of
      that 'heroic nudity' on which only a genuine diagnosis of serious cases can be
      built" (*Poetry and Prose,* p. 519).
      "was equally overpowering . . ." *Brief Chronicles* (1943), p. 175.
      "was a far more English . . ." In *Writers at Work,* fourth series, ed. George
      Plimpton (1976), pp. 136–37.
      "is musical, but . . ." *Shaw's Music,* III, p. 161.
      "a dainty, pleading . . ." *Our Theatres in the Nineties,* II, p. 120.
      "The part of Othello . . ." *Brief Chronicles,* p. 303.

43   "Marivaux, not Shakespeare . . ." *The Titans* (1957), p. 62; this is a study of three
      generations of the Dumas family.
      "There is something majestic . . ." *Where Angels Fear to Tread,* p. 118.
      "a vulgar beast . . ." *Shaw's Music,* I, p. 660.
      "the bravest actor *etc.* . . ." These several quotations are from *Olivier,* ed. Logan
      Gourlay (1973), pp. 21, 55, 189, 193, 107.

44   "Attention to detail . . ." Quoted in ibid., p. 52.
      *"I never could . . ." George Eliot, *Middlemarch* (ed. G. Haight, 1956), p. 48.
      Boswell tells a charming anecdote that places Samuel Johnson in Mr. Casaubon's
      company. In his entry for 23 September 1777, when Johnson was 68, he reports:
      "Johnson desired to have 'Let ambition fire thy mind' played over again, and
      appeared to give a patient attention to it; though he owned to me that he was very

insensible to the power of musick. I told him, that it affected me to such a degree, as often to agitate my nerves painfully, producing in my mind alternate sensations of pathetick dejection, so that I was ready to shed tears; and of daring resolution, so that I was inclined to rush into the thickest part of the battle. 'Sir (said he), I should never hear it, if it made me such a fool'" (*Life of Johnson* [1953], p. 874).

"I must sing . . ." *Poetry and Prose,* p. 289.

"Almost the first thing . . ." *Shakespeare Without Tears* (1942), p. 79.

*"Implied emotion . . ." Bernard Beckerman, *Shakespeare at the Globe: 1599–1609* (1966), pp. 153–55. Erich Auerbach also emphasizes the "dynamic immediacy" of Shakespearean style in *Mimesis* (1953), p. 325. W. H. Auden makes a very similar judgment about operatic expression, quoted in the text on p. 106.

"general swell . . ." *Hazlitt on Theatre,* ed. William Archer (1895; reprint 1957), pp. 102–3.

"the big speeches . . ." *Brief Chronicles,* pp. 191, 22–23.

45 "She never minces . . ." *Leigh Hunt's Dramatic Criticism: 1808–1831,* ed. L. H. and C. W. Houtchens (1950), pp. 267–68.

"the forgotten heroic . . ." Introduction to Lillah McCarthy, *Myself and My Friends* (1934), p. 4.

"there is much more . . ." "Mr. Shaw on Mr. Shaw," *New York Times,* 12 June 1927, sec. vii, p. 1.

the four principals . . . "Shaw's Rules for Directors," *Theatre Arts* 33 (1949), p. 7.

"There is no time . . ." Letter to John Barrymore, 22 February 1905, in *Collected Letters 1911–1925* (1985), p. 903.

"built to an heroic . . ." *Prefaces to Shakespeare,* fourth series (1945), pp. 219–20.

titled "Vocalism" . . . *Poetry and Prose,* pp. 360–61.

6. Passion Play

49 *"Posse of fat . . ." "Opera," 22 February 1918, in *H. L. Mencken on Music,* ed. Louis Cheslock (1961), pp. 101–4. For Mencken, Wagner was an exemplary mountebank: "Greater men, lacking [Wagner's] touch of the quack, have failed where he succeeded—Beethoven, Schubert, Schumann, Brahms, Bach, Haydn. Not one of them produced genuinely successful opera. . . . Schubert wrote more actual music every morning between 10 o'clock and lunch time than the average opera composer produces in 150 years, yet he always came a cropper in the opera house."

"that Fire, Impetuosity . . ." *Some Account of the Life of . . . Shakespeare,* in *Shakespeare: The Critical Heritage 1693–1733,* ed. Brian Vickers (1974), II, p. 191.

50 "Brahms kept . . ." *The Young Mencken,* ed. Carl Bode (1973), p. 544.

"that Shakespeare individualizes . . ." *Shakespeare at the Globe* (1966), pp. 151, 156.

"How passionate . . ." *Prefaces to Shakespeare,* first series (1927), p. 16.

53 "You may here . . ." James Shirley, "To the Reader," *Comedies and Tragedies by Francis Beaumont and John Fletcher* (1647), sig. A3ᵛ.

"an individual . . ." *Shakespeare at the Globe,* pp. 145–46.

54 "The spectator's enjoyment . . ." George Stone, *David Garrick* (1979), pp. 31–32.

"a characteristic quality . . ." Earle Ernst, *The Kabuki Theatre* (1956), pp. 173–74, 199, 76.

"the dramatic passions . . ." *An Essay on the Art of Acting* (1753), *Works* (1753–54), IV, p. 357.

55 "a good *Taste* . . ." *Letters Concerning Taste,* p. 3.

the *Affektenlehre* . . . For a fuller discussion of the *Affektenlehre* and eighteenth-century theories of the passions, see my *Literature as Opera* (1977), pp. 38–41.

I wrote several years ago . . . See *Literature as Opera,* p. 36.

"*Ah Shaespeare* . . ." In Franco Abbiati, *Giuseppe Verdi* (1959), III, p. 566.

"the last master . . ." "The Naïveté of Verdi," in *The Verdi Companion,* ed. Weaver and Chusid (1979), pp. 10–12.

"Shakespeare stands . . ." "La perfezione morale è la perfezione dell'arte, e che percio Shakespeare sovrasta agli altri, perchè è più morale" (*Tutte le opere,* ed. Mario Martelli [1973], II, p. 1645).

56 "the intensity and . . ." *As They Liked It: An Essay on Shakespeare and Morality* (1947), pp. 6, 8, 21. Harbage's remarks seem especially suitable to the greatest operas of Mozart and Verdi.

"like Victor Hugo's . . ." *Bernard Shaw: The Great Composers,* ed. Louis Crompton (1978), p. 220.

plays were being "rediscovered" . . . Between 1829 and 1834 six different translations of Shakespeare and eight different plays appeared in Italy, the first complete (prose) Shakespeare appearing in 1839. See David Kimbell, "The Young Verdi and Shakespeare," *Proceedings of the Royal Musical Association* 101 (1974–75), pp. 59–73.

"A movement analogous . . ." "A History of Romanticism," in *The Complete Works,* ed. F. C. de Sumichrast (1903), XVI, p. 16.

"He loved to present . . ." Ibid., p. 34.

57 "There is hardly . . ." "Aspects of Verdi's Dramaturgy," in *The Verdi Companion,* pp. 136–37.

"It would be a difficult . . ." "A History of Romanticism," p. 149.

"The golden age . . ." In *The Essence of Opera,* ed. Weisstein (1964), p. 360.

"Only he holds . . ." Preface, *November Boughs* (1888), in *Poetry and Prose,* p. 513.

58 "The Dead Tenor" Ibid., p. 458.

"To a Certain Cantatrice" Ibid., p. 82.

7. Simple Savor

59 "Verdi, when he is . . ." *Bernard Shaw: The Great Composers,* p. 222.

60 "The holy fool . . ." Caryl Emerson, *Boris Godunov: Transpositions of a Russian Theme* (1986), p. 206.

61 *"Blow me about . . ." *Othello,* 5.2.279–82. Granville-Barker notes a few other memorable examples of simple savor: "It is interesting to compare this whole passage [in *Othello*] with that in which Lear apostrophizes the dead Cordelia. We have the same intimate simplicity of phrase. Death, and the death of one so dear, is no matter for rhetoric. And even for Cleopatra, attired in all her splendor, Charmian finds the simplest of terms: 'a lass unparalleled' . . . How far we are in all these from Romeo's grief . . . !" (*Prefaces to Shakespeare,* fourth series [1945], p. 142).

62 "Brevity and sublimity . . ." Letter to Francesco Piave, 22 September 1846, *Verdi's Macbeth: A Sourcebook,* ed. Andrew Porter, David Rosen (1984), p. 10.

"The tenor has . . ." Letter to Ranzanici, 17 March 1847, *Verdi's Macbeth,* p. 55.

"a very short death . . ." Letter of January 1847, *Verdi's Macbeth,* p. 37.

* Several draft versions . . . All *Re Lear* quotations are from my "Verdi's *King Lear* Project," *Nineteenth Century Music* 9 (1985), 83–101. See the further discussion of *Re Lear* in Chapter 23.

63   "In the duet . . ." Letter of 28 September 1870, *Verdi's Aida* (1978), p. 68.

64   "Here you wanted . . ." Letter to Ghislanzoni, 30 September 1870, *Verdi's Aida*, p. 71.

    *"I would not . . ." Letter to Ricordi, 10 July 1870, *Verdi's Aida*, p. 31. Verdi had to repeat himself to Ghislanzoni a month later: "It seems to me that the *theatrical word* is missing, or if it is there, it is buried under the rhyme or under the verse and so doesn't jump out as neatly and plainly as it should" (letter of 14 August 1870, p. 47). Three days later, to Ghislanzoni again: "But then, when the action warms up, it seems to me that the *theatrical word* is missing. I don't know if I make myself clear when I say '*the theatrical word*,' but I mean the word that clarifies and presents the situation neatly and plainly" (p. 50).

8. These Fierce Moments

65   "An image of articulateness . . ." *Earth Walk* (1970), p. 14.

    "Arabella must . . ." *A Working Friendship* (1974), pp. 534, 528 (original emphasis).

66   "dramatic thought . . ." *The Mutual Flame* (1955), p. 77.

    The *dithyramb* . . . Nietzsche made much of the dithyramb in his analysis of ancient and modern lyric drama; a dithyrambic orgy figures in the climax of Britten's *Death in Venice,* based on Thomas Mann's Nietzsche-impregnated novella. See my *Literature as Opera,* pp. 334–44.

67   "Elizabethan drama . . ." Granville-Barker, *Prefaces to Shakespeare,* first series (1927), p. 8.

    *"episodic intensification . . ." *Character Problems in Shakespeare's Plays* (1922), pp. 112–13. Schücking quotes Goethe, in a letter of 18 April 1827, saying that Shakespeare "on every occasion makes his characters say what is effective, right, and appropriate to the situation, without troubling overmuch to reflect whether the words may not possibly come into apparent conflict with some other passage" (p. 119).

68   "the most important . . ." *The Structure of Shakespearean Scenes* (1981), p. 210.

    "Elizabethans in general . . ." A. R. Humphreys, Arden edition, *Much Ado About Nothing* (1981), p. 51.

    "Within the framework . . ." *Shakespeare at the Globe,* pp. 56–57.

    "never pinned . . ." *Prefaces to Shakespeare,* first series (1927), pp. xxxi–xxxii.

    "The architectural . . ." *Shakespeare at the Globe,* p. 45.

69   *"the evidence points . . ." *Shakespeare at the Globe,* p. 132. In *Action Is Eloquence: Shakespeare's Language of Gesture* (1984), David Bevington summarizes Elizabethan acting style in different terms: "Elizabethan acting was at once ceremonious in a conventional framework, romantic in dramatizing high passions, and epic in its scope and elevation of tone" (pp. 71–72).

    "It is probable . . ." *Shakespeare's Stagecraft* (1967), p. 75.

    *"There are very few . . ." *Playing Shakespeare,* p. 94. J. L. Styan confirms this view: "'Soliloquy' is a late seventeenth- or eighteenth-century literary concept,

and 'speaking to oneself' was not a device that Shakespeare or the Elizabethan stage would have recognized" ("Stage Space and the Shakespeare Experience," in *Shakespeare and the Sense of Performance,* ed. Marvin and Ruth Martin [1989], p. 198).

"he delivered the speech . . ." *Brief Chronicles* (1943), p. 195.

70   "When he played . . ." "Sing That Shakespeare," *New York Times,* 31 May 1970, p. D9.

"coactors in the scene . . ." *The Antipodes* (1640), sig. D3^r, E^v.

"When he doth hold . . ." J. Cocke, "A Common Player" (1615), quoted in E. K. Chambers, *The Elizabethan Stage* (1923; ed. 1967), IV, p. 256.

"Behold me in my . . ." *Shaw's Music* (1981), I, p. 157.

"Opera taught me . . ." "The Play of Ideas," *New Statesman and Nation,* new series 39 (1950), p. 511.

"twinkled with crotchets . . ." Quoted in *Collected Letters 1898–1910* (1985), pp. 390–91.

"The normal procedure . . ." *Shakespearean Production* (1964), p. 278.

"Shaw creates a play . . ." *On Theatre,* trans. John Willett (1964), p. 11.

71   "Drama was exposure . . ."; "plunges deep and reaches high . . ."; "roughness of texture . . ."; "the greatest of rough . . ."; "Grand Opera . . ." *The Empty Space* (1968), pp. 36, 88, 68, 17. Warming to his subject, Brook further remarks, "Opera is a nightmare of vast feuds over tiny details; of surrealist anecdotes that all turn round the same assertion: nothing needs to change. Everything in opera must change, but in opera change is blocked" (p. 17).

9. Soliloquy, Set Speech, Aria I

73   * star vehicles . . . For further discussion of star vehicles, see Chapter 17.

74   "Each great Shakespearean . . ." Michael Goldman, *Acting and Action in Shakespearean Tragedy* (1985), pp. 13–14.

75   "the 'great' characters . . ." *The Life of Drama* (1964), pp. 68–69.

"The problem is how . . ." *Playing Shakespeare,* p. 90.

"Here a beautiful voice . . ." *Shakespeare Without Tears* (1942), p. 83.

"The actor who . . ." and "It is useless . . ." *Disposizione Scenica* (1887), pp. 37, 108.

10. Soliloquy, Set Speech, Aria II

83   "cheapen mythology . . ." *Romantic Opera and Literary Form* (1977), p. 48.

85   "the architectonic superiority . . ." Beckerman, *Shakespeare at the Globe* (1966), p. 45.

86   "Shakespearean drama . . ." Ibid., p. 106.

89   "very powerful situations . . ." That is, "posizioni potentissime, varietà, brio, patetico." Letter to Antonio Somma, 22 April 1853, *Re Lear e Ballo in Maschera: Lettere di Giuseppe Verdi ad Antonio Somma* (1902), p. 46.

"It is not the plot . . ." "Aspects of Verdi's Dramaturgy,' in *The Verdi Companion,* ed. Weaver and Chusid (1979), p. 141.

11. *The M-Word*

90   "Such an old-fashioned . . ." Shaw's remarks were appended in a postscript to a
     review of orchestral concerts in *The Star,* 31 January 1890, in *Shaw's Music* (1981),
     I, p. 911.

     "that shabby little shocker . . ." *Opera as Drama,* p. 205.

     "I have some cousins . . ." Letter of 10 January 1911 (in the possession of the
     Truman Library). Truman was 27 at the time.

     "I am an Irishman . . ." Letter to Viola Tree, 29 November–5 December 1911, in
     Collected Letters 1911–1925, p. 61.

91   "A good libretto . . ." "Some Reflections on Music and Opera," in *The Essence of
     Opera,* ed. Ulrich Weisstein (1964), pp. 357–58.

     "My idea of heaven . . ." "Immortality," *The Complete Writings* (1929), II,
     p. 832.

     "merely functioning as . . ." Frits Noske, *"Otello:* Drama Through Structure," in
     Essays on Music for Charles Warren Fox, ed. Jerold Graue (1979), p. 47.

     "a vulgar melodramatist . . ." Peter Brooks' précis of the view of Martin Turnell in
     *The Melodramatic Imagination: Balzac, Henry James, Melodrama, and the Mode of
     Excess* (1976), p. 3.

     "A melodramatic actor . . ." George H. Lewes, *On Actors and the Art of Acting*
     (1875), p. 15.

     "The tragedian . . ." *Nicholas Nickleby,* chapter 29.

     "Massive and concrete . . ." *Great Expectations,* chapter 31.

92   "Essentially, melodrama . . ." Michael Booth, *English Melodrama* (1965), pp. 14,
     38.

     "Puccini's operas . . ." "Intellectual Content," in *The Art of Judging Music* (1969),
     p. 300.

     "All his mature tragedies . . ." Arden *Macbeth* (1962), p. lix.

     "Is not *Hamlet* . . ." *Oxford Lectures on Poetry* (1926), p. 283.

     "the most splendidly . . ." *Our Theatres in the Nineties,* III, p. 313.

     "the half-dozen big . . ." Ibid., p. 332.

     "Time and again . . ." *The Life of Drama* (1964), pp. 177–78.

     "this tinkerer . . ." *On Acting* (1986), p. 79.

     "Oh, *Le Roi s'amuse* . . ." Letter to Francesco Piave, 8 May 1850, in Franco Abbiati,
     Giuseppi Verdi (1959), II, p. 62.

93   "What we have . . ." and following quotations: *The Life of Drama,* pp. 177, 196, 28,
     178, 205, 216, 272, 213, respectively.

     "mostly about the police . . ." "A Word More," *Anglo-Saxon Review,* March 1901,
     in *Bernard Shaw: The Great Composers* (1978), p. 220.

     "A good melodrama . . ." Letter to Ellen Terry, 26 March 1896, *Ellen Terry and
     Bernard Shaw* (1931), p. 21.

     Instead, he was won . . . See Martin Meisel's discussion in *Shaw and the Nine-
     teenth Century Theater* (1963), p. 222.

     "this thing . . ." Letter to Ellen Terry, 30 November 1896, *Ellen Terry and Bernard
     Shaw,* p. 97.

94   "a simple and sincere . . ." *Our Theatres in the Nineties,* I, p. 98.

     "I missed the big . . ." *Bernard Shaw and Mrs. Patrick Campbell: Their Cor-
     respondence,* ed. Alan Dent (1952), p. 211.

     *Here is a sampling . . . The following quotations are, respectively, from *The

Melodramatic Imagination, pp. ix, 4, 40, 41, 13, 40. Another important attempt to describe and understand melodrama dispassionately is Robert Heilman's *Tragedy and Melodrama: Versions of Experience* (1968). Seeking to use "melodrama as a neutral descriptive term, not as a pejorative" one (p. 75), Heilman asserts that "within the category, plays may range from contemptible to distinguished." Heilman concludes his study by observing, "Tragedy is not always called for; there are occasions when the melodramatic response is right and necessary" (p. 300). Shakespeare's plays and most operas consistently demonstrate the justness of Heilman's conclusion.

95   "melodrama died . . ." *The Life of Drama,* p. 210.

"the greatest, most substantially . . ." *Selected Prose* (1952), p. 269; "They set against . . ." *The Melodramatic Imagination,* p. 198.

"Emerson says *manners* . . ." "The Perfect Human Voice," in *Prose Works,* ed. Floyd Stovall (1964), II, p. 674.

"intense, excessive . . ." *The Melodramatic Imagination,* p. 3.

96   *"singularly dandified . . ." "Emerson's Books (The Shadows of Them)," in *The Poetry and Prose,* p. 887. Whitman's sketch of Emerson perhaps captures the essence of the anti-melodramatic sensibility (in the same essay he accuses Emerson of not really knowing or feeling the power of Shakespeare's poetry): "He is best as a critic, or diagnoser. Not passion or imagination or warp or weakness, or any pronounced cause of speciality, dominates him. (I know the fires, emotions, love, egotisms, glow deep, perennial, as in all New Englanders— but the façade hides them well—they give no sign.)"

"My dear boy . . ." *Memoirs,* trans. David Cairns (1975), pp. 123–24.

*"true enthusiast" . . . "long brooding silences . . ." François Guizot and Ernest Legouvé, respectively, quoted in the *Memoirs,* p. 521. One Shakespearean hero has always struck me as particularly like Berlioz: the "rough, unswayable, free" Coriolanus. Coriolanus's witheringly cynical views of his countrymen are especially reminiscent of Berlioz' professional experiences throughout his life with the Parisian musical establishment. See the discussion of Berlioz and *Much Ado* in Chapter 26.

97   "instance of the mischief . . ." *Shaw's Music* (1981), II, p. 190.

"This dumb girl"; "a monstrous, pie-bald . . ." *Prose Works* (1893), II, pp. 57, 94.

98   "I should like . . ." Letter of 2 August 1928; Hofmannsthal's reply came in a letter of 5 August 1928 (*A Working Friendship,* p. 499).

"a stunning of . . ." *Prose Works* V, p. 133.

"I value a worthy . . ." *Characters* (1615), in *The Complete Works* (1937), IV, p. 43.

"Tut, I can counterfeit . . ." *Richard III,* 3.3.5–9.

"he rolled out . . ." Richard Cumberland, *Memoirs* (1807), I, p. 80.

99   "His whole action . . ." *Memoirs* (1799), II, p. 265.

"Some dozen lines . . ." *Works* (1844), I, p. 37.

"It is . . . nothing *melo*dramatic . . ." Letter of 12 January 1821, in *Byron's Letters and Journals,* ed. Leslie Marchand (1978), VIII, p. 23.

*"all kinds of unnatural . . ." "*The Gladiator*—Mr. Forest—Acting," *Brooklyn Eagle,* 26 December 1846. Whitman refers to the "loud mouthed ranting style— the tearing of everything to shivers—which is so much the ambition of some of our players." Hamlet's "robustious" player was still alive and well.

"neither on Stilts . . ." *Characters* (1615), in *The Complete Works,* IV, p. 42.

"[A]cted very well . . ." *Macready's Reminiscences, and Selections from his Diaries* (1875), p. 3.

"shout cautiously . . ." *Collected Letters 1898–1910,* p. 211.

100   "The actor's need . . ." *Playing Shakespeare,* p. 147.

"The possibility of . . ." *The Life of Drama,* p. 175.

"requires extreme mechanical . . ." *Shaw's Music,* II, p. 36. Andrew Porter made the same point about Meyerbeer in his review of a concert performance of *Robert le Diable:* "More than any other composer, perhaps, Meyerbeer is dependent on what performers make of him" (*The New Yorker,* 7 March 1988, p. 118).

"We knew that Sir Laurence . . ." Quoted in *Olivier,* ed. Logan Gourlay (1973), p. 189.

"There is no denying . . ." Review in the *New York Herald Tribune,* 22 December 1945, in *The Art of Judging Music* (1948), pp. 113–14.

12. The Nature of Melodrama

102   "I could not help . . ." *Emerson in His Journals,* ed. Joel Porte (1982), p. 99.

103   "a Bloody Farce . . ." "A Short View of Tragedy" (1693), in *Shakespeare: The Critical Heritage 1693–1733,* ed. Brian Vickers (1974), II, pp. 27–8, 54.

"its noble savage . . ." *Our Theatres in the Nineties,* III, p. 332.

"Almost all the Faults . . ." *Shakespeare: The Critical Heritage,* II, p. 86. Surely the most magisterial riposte to criticism of Shakespeare was penned by Hazlitt: "People would not trouble their heads about Shakespear, if he had given them no pleasure, or cry him to the skies, if he had not first raised them there. The world are not grateful *for nothing*" (*Complete Works,* XII, p. 187).

"Melodrama typically offers . . ." *Tragedy and Melodrama: Versions of Experience* (1968), p. 95.

105   "Melodramatic good . . ." *The Melodramatic Imagination* (1976), p. 16.

"An alternative expression . . ." *The Life of Drama* (1964), p. 41.

"A female who . . ." Letter to Ricordi, 22 April 1887, quoted in Franco Abbiati, *Giuseppe Verdi* (1959), IV, p. 332.

"there never was . . ." Letter to Mrs. Patrick Campbell, 13 January 1921, in *Collected Letters 1911–1925,* p. 707.

106   *"Opera . . . cannot . . ." "Some Reflections on Music and Opera" (1952), in *The Essence of Opera* (1964), pp. 356–57. Virgil Thomson ventures a similar generalization about opera: "The best opera composers have usually avoided, in writing for the lyric stage, any duality of allusion that might weaken the impact of the expressive content" ("Intellectual Content," in *The Art of Judging Music* [1969], p. 299).

". . . tears or throwing things . . ." *A Swim off the Rocks* (1976), p. 15.

107   "noble exaltation" and following quotations . . . *Longinus on the Sublime,* trans. G. M. Grube (1957), pp. 12–13, 19–21. The actual author of the treatise is unknown; it appears to have been written in the first or second century A.D. It became very popular in the eighteenth century and inspired Pope's "Essay on Criticism."

"You can't be too stagey . . ." *Collected Letters 1898–1910,* p. 391.

"One virtue of . . ." *Opera as Drama* (1988), p. 26.

"the more novelty . . ." Letter to Cesare De Sanctis, 29 March 1851, in *Carteggi Verdiani,* I, pp. 4–5.

108   "fine theatrical effects . . ." Letter to Cammarano, 4 April 1851, quoted in Abbiati, II, pp. 122–23

*"throws light on . . ." M. R. Ridley, Arden *Othello* (1958), p. lxx. In the related incredibility involved in Othello's sudden rush to jealous judgment, Granville-Barker also sees the domination of the bold effect: "That we may rather feel with Othello in his suffering than despise him for the folly of it, *we* are speeded through time as unwittingly as he, and left little more chance for reflection. . . . Most unconscionable treatment of time truly, had time any independent rights! But effect is all" (*Prefaces to Shakespeare*, fourth series [1945], p. 35).

"tends to sweep . . ." Liner note, p. 5, London Records *Il Trovatore* (#90-5250).

"the fact that . . ." *The Operas of Verdi* (1979), II, p. 267.

"no nice discrimination . . ." and "This play is . . ." Vickers, *Shakespeare: The Critical Heritage*, V, p. 144.

109 "Points of view . . ." *Prefaces to Shakespeare*, fourth series (1945), p. 171. Granville-Barker makes the same point in reverse when he refers to more realistic modern drama, "with its scenic illusion, its quiet acting, its gains in subtlety and loss in power" (*Prefaces to Shakespeare*, second series [1930], p. 248).

"otherwise we . . ." Letter of 5 August 1928, *A Working Friendship*, p. 501.

"a climactic moment . . ." *The Melodramatic Imagination*, p. 4.

"spontaneous and uninhibited . . ." *The Life of Drama*, p. 216.

"Implied emotion . . ." *Shakespeare at the Globe*, p. 153. Echoing Beckerman's assertion is Brooks' observation that in melodrama "emotions are given a full acting-out" (*The Melodramatic Imagination*, p. 41).

"there operated a spontaneous . . ." *Shakespeare at the Globe*, pp. 215–16.

"It is delusive . . ." *The Melodramatic Imagination*, p. 35.

110 *"It was from Handel . . ." *Shaw's Music* (1981), III, p. 640. Shaw's boast was thoroughly justified. Rodin said of him, "M. Shaw ne parle très bien; mais il s'exprime avec une telle violence qu'il s'impose" (quoted in *Collected Letters 1898–1910*, p. 618). Shaw had occasion to express his credo in a letter to a feckless would-be biographer (a professor of English at the University of Pennsylvania) who feared he would not be "just" to Shaw. Shaw wrote, "who are you that you should be just? That is mere American childishness. Write boldly according to your own bent: say what you WANT to say and not what you think you ought to say or what is right or just or any such arid nonsense" (*Collected Letters 1898–1910*, p. 516).

"appears suddenly . . ." *Longinus on the Sublime*, p. 4.

"His plays are . . ." In *George Bernard Shaw, A Critical Survey*, ed. Louis Kronenberger (1953), p. 156.

"Even at its most extrovert . . ." *The Operas of Verdi*, II, pp. 45–46.

"asked for a new mode . . ." *Shaw and the Nineteenth Century Theater* (1963), pp. 97, 117.

"require a special technique . . ." "Mr. Shaw on Mr. Shaw," *New York Times*, 12 June 1927, sec. vii, p. l.

111 "There are too many . . ." *On Acting* (1986), p. 104.

"pure melodrama . . ." *Our Theatres in the Nineties*, III, pp. 154–55.

"cater to longings . . ." *Tragedy and Melodrama*, pp. 87, 168–69.

"constantly reaching . . ." *The Melodramatic Imagination*, p. 35.

"swept off their feet . . ." "Some Reflections," p. 358.

"something in himself . . ." Quoted in *Olivier*, p. 27. Melodrama's "typical figures are hyperbole, antithesis, and oxymoron," writes Brooks (p. 40).

"at the very first rehearsal . . ." *On Acting*, p. 107.

"You do not play . . ." Letter of 18 December 1919, *Collected Letters 1911–1925*, p. 650.

112 "[*Ariadne*] is about . . ." Letter of July 1911, in *A Working Friendship*, p. 94. Longinus makes a similar distinction in urging that the first source of sublimity is "natural high-mindedness"; it cannot be produced, he adds, by "men whose thoughts and concerns are mean and petty throughout life" (*Longinus on the Sublime*, pp. 11–12).

*"Part of my quest . . ." "Stories That Break into Song: Choosing Plots for Opera," *The Kenyon Review* 4 (1982), p. 77. Or, it should be added, very *low* values: Eaton based an early opera "on Ma Barker, the huge (Wagnerian?) leader of a midwestern mob. . . . She seemed to me to have the superhuman (or subhuman) characteristics of a serious operatic figure" (pp. 76–77).

113 "But the moment . . ." *Emerson in His Journals*, ed. Joel Porte (1982), p. 99. In a brief peroration to his attempt to raise the estimation of melodrama, Heilman urges: "Melodrama is not indissolubly bound to the meretriciousness which, in our unconscious but habitual debasing of forms, we take to be its normal state. It is not meretricious when it portrays recognizable evil—as in Renaissance revengers or [Shelley's] Count Cenci or Brecht's Nazis or Koestler's Gletkin—or plausible courage and fidelity, as in the Talbots of *Henry VI*" (*Tragedy and Melodrama*, pp. 288–89).

13. Kaleidoscope; or, Ringing the Changes

114 "To use two hours . . ." *The Empty Space* (1968), p. 31.

115 *"enabled the dramatist . . ." Ibid., p. 86. In *Shakespeare's Stagecraft* (1967) J. L. Styan refers to the Shakespearean soliloquizer "alone upon his thousand square feet of platform" (p. 72). The Globe stage is estimated to have been 41 feet wide and 24 to 29 feet deep; see Irwin Smith, *Shakespeare's Globe Playhouse* (1956). Most Broadway theaters have stages between 37 and 39 feet wide and 28 to 35 feet deep.

"as though I had . . ." Letter to Modest Tchaikovsky, 20 August 1876, *Life and Letters of P. I. Tchaikovsky*, trans. Rosa Newmarch (1906), p. 184. Cosima Wagner displayed a flawless grasp of the obvious when, a few months after the premiere of *Parsifal*, she confided to her diary, "R. does not care for brevity" (*Cosima Wagner's Diaries*, ed. Martin Gregor-Dellin [1980], II, p. 923). In the vein of Tchaikovsky's remark is Noël Coward on *Falstaff*: "Lovely opera, but too long" (*Diaries*, ed. Graham Payn [1982], p. 561).

"No unbiased listener . . ." Letter of 23 October 1923, *A Working Friendship*, p. 372.

"high and fresh variety . . ." *An Apology for the Life of Colley Cibber* (1740), ed. B. R. Fone (1968), p. 55.

"clarity of statement . . ." *Prefaces to Shakespeare*, second series (1930), pp. 116–17.

"aural shifts . . ." *Shakespeare's Art of Orchestration* (1984), pp. 2, 24, 42, 81, 32.

116 "I turn over a score . . ." *Shaw on Theater* (1958), p. 144.

"Contrast, ringing" and "violent switches . . ." *Playing Shakespeare*, p. 21.

"If I were to offer . . ." Ibid., p. 55.

117 "I find that the opera . . ." Letter to Antonio Somma, 22 April 1853, in A. Pascolato, *Re Lear e Ballo in Maschera: Lettere di Giuseppe Verdi ad Antonio Somma* (1902), p. 46.

"the *great tyrant* . . ." Letter to Giulio Ricordi, 17 December 1870, in *Verdi's Aida*, ed. Hans Busch (1978), p. 115.

"What is certain . . ." Letter to Somma, 7 April 1856, in Pascolato, p. 78.

118   "I am aware . . ." "A' chi legge," *Madrigali Guerrieri, et Amorosi*, in *Tutte le Opere*, ed. G. F. Malipiero (1929), VIII, n.p.; "Variety wisely . . ." *Correspondance inédite* (1879), p. 113.

"moments of lyrical . . ." *Prose Works*, V, p. 305.

"What makes the charm . . ." Letter of 22 December 1927, in *A Working Friendship,* p. 462.

"with your most exquisite . . ." Letter to Verdi, 30 April 1853, translated from microfilm copy in the Verdi Institute, New York University. For a fuller exploration of the *Re Lear* project, see Chapter 23.

119   * about 2,500 lines . . . This is Alfred Harbage's estimate in "Elizabethan Acting," *PMLA* 44 (1939), p. 704. Harbage's guess about the length of acting versions of Renaissance plays is perhaps corroborated by the publication in 1623 of a 3,300-line version of John Webster's *Duchess of Malfi*; its title-page promises that it is based on "The perfect and exact Copy, with diverse things Printed, that the length of the Play would not beare in the Presentment."

seventy-five feet from . . . Marvin Rosenberg suggests this distance in "Elizabethan Actors: Men or Marionettes?" *PMLA* 69 (1954), p. 920.

* 2½, hours, or 155 minutes . . . The Arden editor of *Richard III* (1981), Antony Hammond, guesses that this 3,600-line play, if performed without cuts, would have taken "at least two and three-quarter hours, and perhaps half an hour longer." His explanation for this estimate is pertinent to my own hypothesis: "The shorter time is calculated on an average rate of not more than 200 words a minute, and a performance rate of not more than 22 lines a minute. It would be hard to keep up such a rate: if one slowed down by a fifth (17 lines a minute) it would take 211 minutes to speak the dialogue. Most modern performances *go much slower still*" (p. 66—emphasis added). A highly successful 1988 production of *Richard III* at the Guthrie Theater in a version that contained about 2,550 lines ran 195 minutes (excluding intermission time), which yields a rate of 13 lines per minute.

But surely we can assume that *Richard III* was originally cut for performance. Many of the 200 or so lines from the Folio that do not appear in the early quartos (derived, probably, from acting versions of the play) constitute very plausible cuts. The Guthrie Theater production mentioned above was based on earlier stage versions by Colley Cibber, Henry Irving, and Tyrone Guthrie. It cut 1,100 lines (and added fifty from *3 Henry VI*). On the propriety of cuts, McKellen is blunt: "uncut productions . . . are ridiculous. The plays were always cut" ("Ian McKellen on Acting Shakespeare," *Shakespeare Quarterly* 33 [1982], p. 141).

Andrew Gurr recently addressed the question of performance time again in *Playgoing in Shakespeare's London* (1987) and concluded that the "likely duration of most performances was nearer three hours" (p. 33). Gurr pertinently quotes an October 1594 promise from the sponsor of Shakespeare's company, the Lord Chamberlain, to the Lord Mayor that performances "will now begin at two, & have don betwene fower and five."

McKellen has . . . "You can do so much more, and more effectively, in a small house. I suppose the optimum size is a theatre seating about 400 people" (*op. cit.,* p. 140).

120   * the typical Shakespearean play . . . Here are the statistics for each of the plays.

Excluded are three collaborative plays (*Pericles, Henry VIII,* and *The Two Noble Kinsmen*) and a likely unfinished one (*Timon of Athens*). The plays are listed roughly in the order in which they were written; given in parentheses are the total number of lines (taken from F. E. Halliday, *A Shakespeare Companion 1564–1964* [1964], p. 516), the number of scenes over fifteen lines long (scene divisions are those of the *Riverside Shakespeare*), the running time in minutes, and the average length of the play's scenes in minutes:

Comedy of Errors (1,778, 11, 99, 9); *Love's Labour's Lost* (2,789, 8, 155, 19); *1 Henry VI* (2,677, 26, 149, 5³/₄); *2 Henry VI* (3,162, 22, 176, 8); *3 Henry VI* (2,904, 28, 161, 5³/₄); *Richard III* (3,619, 23, 201, 8³/₄); *Titus Andronicus* (2,523, 14, 140, 10); *The Taming of the Shrew* (2,649, 14, 147, 10¹/₂); *The Two Gentlemen of Verona* (2,294, 17, 127, 7¹/₂); *Romeo and Juliet* (3,052, 20, 170, 8¹/₂); *Richard II* (2,756, 18, 153, 8¹/₂); *A Midsummer Night's Dream* (2,174, 9, 120, 13); *King John* (2,570, 16, 142, 8³/₄); *The Merchant of Venice* (2,660, 20, 148, 7¹/₂); *1 Henry IV* (3,176, 18, 176, 9³/₄); *2 Henry IV* (3,446, 19, 191, 10); *Henry V* (3,380, 28, 188, 6³/₄); *Much Ado About Nothing* (2,826, 19, 157, 8¹/₄); *Julius Caesar* (2,478, 16, 138, 8¹/₂); *As You Like It* (2,587, 19, 144, 7¹/₂); *Twelfth Night* (2,690, 18, 149, 8¹/₄); *Hamlet* (3,931, 20, 218, 11); *The Merry Wives of Windsor* (3,018, 18, 167, 9¹/₄); *Troilus and Cressida* (3,496, 23, 194, 8¹/₂); *All's Well that Ends Well* (2,966, 22, 165, 7¹/₂); *Othello* (3,316, 13, 184, 14); *Measure for Measure* (2,821, 15, 157, 10¹/₂); *King Lear* (3,334, 25, 185, 7¹/₂); *Macbeth* (2,108, 27, 117, 4¹/₂); *Antony and Cleopatra* (3,063, 36, 170, 4³/₄); *Coriolanus* (3,410, 27, 189, 7); *Cymbeline* (3,339, 27, 185, 6³/₄); *The Winter's Tale* (3,075, 15, 171, 11¹/₂); *The Tempest* (2,064, 8, 115, 14¹/₄).

"roughness of texture . . ." *The Empty Space,* p. 88.

121 "an over-all duration . . ." Letter of 1 November 1923, *A Working Friendship,* p. 374.

122 * the average number . . . In parentheses are given the number of discrete musical entities or "numbers", the running time of the opera, and the average length of these entities in minutes:

Aida (21, 145, 7); *A Midsummer Night's Dream* (18, 143, 8); *Barbiere di Siviglia* (20, 160, 8); *La Bohème* (14, 107, 7¹/₂); *Madama Butterfly* (17, 130, 7¹/₂); *Carmen* (26, 150, 6); *Così fan tutte* (31, 182, 6); *Don Giovanni* (24, 164, 7); *Elektra* (11, 98, 9); *Falstaff* (22, 132, 6); *Faust* (19, 192, 10); *Die Fledermaus* (15, 144, 9¹/₂); *Der fliegende Holländer* (15, 140, 9); *La Gioconda* (21, 163, 8); *Les Contes d'Hoffmann* (21, 158, 7¹/₂); *Lohengrin* (21, 215, 10); *Lucia di Lammermoor* (16, 125, 8); *Die lustigen Weiber von Windsor* (17, 143, 8¹/₂); *Macbeth* (16, 130, 8); *Manon* (23, 185, 8); *Manon Lescaut* (13, 113, 8¹/₂); *Norma* (16, 155, 10); *Le Nozze di Figaro* (28, 173, 6); *Eugene Onegin* (22, 150, 7); Rossini's *Otello* (14, 124, 9); Verdi's *Otello* (19, 134, 7); *Rigoletto* (21, 122, 6); *Roméo et Juliette* (22, 165, 7¹/₂); *Samson et Dalila* (18, 122, 7); *Tannhäuser* (17, 170, 10); *Tosca* (16, 118, 7); *La Traviata* (19, 128, 7); *Il Trovatore* (23, 145, 6¹/₂); *Turandot* (12, 117, 9¹/₂); *Die Zauberflöte* (21, 155, 7).

These figures yield an average of 18¹/₄ scenes or "numbers" per action, with an average scene length of 7¹/₄ minutes. It will be noticed that the sample includes no baroque operas, which I believe are a special case. Decidedly un-Shakespearean in their catering to a wealthy, leisured, and *seated* audience and in their dependency on extremely elaborate stage machinery and special effects, these operas some-times sprawled to five or six hours and typically consisted of 45 scenes or, on average, 42 arias (these figures are from Robert Freeman's aptly titled *Opera Without Drama: Currents of Change in Italian Opera, 1675–1725* [1981], pp. 17–20). Doubtless audiences today would respond to an *uncut* baroque opera rather the way Sir Isaac Newton did to Handel's *Radamisto* in 1720: "The first Act he heard with pleasure, the 2d stretch'd his patience, at the 3d he ran away." The average length of baroque numbers, however, is not unusual: a recent recording of Handel's *Giulio Cesare* runs 243 minutes, with its 40 numbers averaging six minutes in length.

14. The Pyramid Game

124   "a man [one] would leap . . ." *The Complete Poems,* ed. George Parfitt (1975), p. 381.

125   "Stick to my plays . . ." *Collected Letters 1898–1910,* pp. 215–16. To a young actress he wrote in 1921, "I don't know whether you are a musician. If not, you don't know Mozart; and if you don't know Mozart you will never understand my technique" (*Collected Letters 1911–1925,* p. 754.).

134   *a point of boasting . . . "I am so good a proficient in one quarter of an hour, that I can drink with any tinker in his own language during my life" (*1 Henry IV,* 2.4.17–20).

137   *"O, you are" . . . The Folio prints this speech as prose; some editors arrange it as imperfect poetry.

15. Operatic Bard, Bardic Opera

138   "A musician only . . ." "Richard Reorchestrated," *The Star,* 23 March 1889, in *Shaw's Music* (1981), I, pp. 586–87. The prefix in Shaw's title is worth emphasizing.
Only two operas . . . One *Riccardo III* is by Luigi Canepa (Milan, 1879), the other by Gaston Salvayre (St. Petersburg, 1883). For a census of the Shakespearean operatic repertory, see Winton Dean, "Shakespeare in the Opera House," *Shakespeare Survey* 18 (1965), pp. 75–93.

139   Heilman . . . grants . . . *Tragedy and Melodrama* (1968), pp. 177–79, 290.
"powerful, singular . . ." Letter to Escudier, 20 August 1856, in *Rivista Musicale Italiana* 35 (1928), p. 22.
"most Shakespearean opera . . ." "Aspects of Verdi's Dramaturgy," in *The Verdi Companion,* ed. William Weaver and Martin Chusid (1979), p. 135.
"portrayal of humanity . . ." *The Operas of Verdi* (1979), II, p. 431.
"Oh, *Le Roi s'amuse* . . ." Letter to Piave, 8 May 1850, in Abbiati, *Giuseppe Verdi* (1959), II, p. 62.

140   *most loved of Verdi's . . . See Budden, *The Operas of Verdi,* II, p. 67. Shaw, in 1889, called *Trovatore* "the most popular opera of the nineteenth century until Gounod's *Faust* [1859] supplanted it" (*London Music in 1888–1889,* Vol. XXXIII of *The Works of Bernard Shaw* [1938], p. 367).

141   Shaw does call it . . . "*Trovatore* and *The Huguenots,*" *The World,* 4 June 1890, *Shaw's Music,* III, p. 587.

142   *two hundred breathtaking lines . . . In *Shakespeare Our Contemporary* (1964) Jan Kott calls this thoroughly incredible scene "one of the greatest scenes written by Shakespeare, and one of the greatest ever written" (p. 37).

146   "attractiveness lies . . ." Antony Hammond, *Richard III* (Arden ed. 1981), p. 105.

147   "A melody, based . . ." Letter of 26 October 1870, in *Verdi's Aida,* ed. Busch (1978), p. 86.

"the principal role . . ." Letter to Piave, 17 April 1853, in Abbiati, II, p. 241. The role, he added, is "finer and more dramatic and more original than [Leonora]. If I were a prima donna (a fine thing that would be!), I would always rather sing the part of the gypsy." It is possible Verdi was playfully using reverse psychology on a soprano—Barbieri-Nini—who was expressing unhappiness with the role of Leonora; he was suggesting that she sing Azucena instead.

*"What is vital . . ." Letter of 13 July 1928, *A Working Friendship* (1974), p. 486. Likewise, Bernard Beckerman may be describing this genius for achieving a unique *tinta* when he observes that "every element in a great Shakespearean play—character, structure, speech—individually and collectively, is brought into an artistic unity through a structural and poetic expression of an unseen referent at its center" (*Shakespeare at the Globe*, p. 62).

"a coherent masterpiece . . ." *The Operas of Verdi*, II, p. 112.

148   "the darkest . . ." William Weaver, "The Grim Richness of *Il Trovatore*," program essay for London Records OSA 13124, p. 5.

"furiously pessimistic . . ." M. Owen Lee, "Elemental, Furious, Wholly True," *The Opera Quarterly* 1 (1983), p. 8.

"People say . . ." Letter to Clarina Maffei, 29 January 1853, in *I Copialettere*, ed. A. Luzio (1913), p. 532.

"world full of murder . . ." *Shakespeare Our Contemporary*, p. 38.

"the solemnity we . . ." *Shakespeare's History Plays* (1944), p. 200.

"my pieces are not . . ." *Collected Letters 1911–1925*, p. 566.

Shaw amusingly . . . "The vulgar realism of sitting down is ten times more impossible for the Count Di Luna than for the Venus of Milo" (*London Music in 1888–1889*, p. 368).

"The element of ritual . . ." Hammond, *Richard III* (1981), pp. 72–73.

150   "If in opera . . ." Letter of 4 April 1851, Abbiati, II, p. 122.

151   "the continual hurry . . ." Johnson was referring to *Antony and Cleopatra* (in *Shakespeare: The Critical Heritage 1765–1774*, ed. Brian Vickers [1979], V, p. 148).

"If it allowed . . ." *London Music in 1888-1889*, pp. 367–68.

"it would be a good . . ." Letter of 4 April 1851, Abbiati, II, p. 123.

152   * Shakespeare, too . . . "On the whole, Elizabethan drama was very casual about motivating anything" (Bert States, *Great Reckonings in Little Rooms: On the Phenomenology of Theater* [1985], p. 193).

153   *a half dozen memorable . . . Hamlet's ghostly father promises to unfold a tale that will make "Thy knotted and combined locks to part, / And each particular hair to stand on end, / Like quills upon the fearful porpentine"; and Macbeth begins his tragic career by yielding to a "suggestion / Whose horrid image doth unfix my hair"; and at the end he recalls when his "fell hair" would "rouse and stir / As life were in't." See also *Julius Caesar*, 4.3.280; *The Tempest*, 1.2.213; and *2 Henry VI*, 3.2.171, 318.

"the one thing . . ." *Shaw's Music*, II, p. 549.

"unbounded and . . ." *London Music in 1888–1889*, p. 369.

16. Heroic Utterance; or, Be Not Too Tame

157   "Now a subscription night . . . "*Trovatore* and *The Huguenots*," *The World*, 4 June 1890, in *Shaw's Music* (1981), II, p. 78.

"If you ask . . ." "A Word More," *Anglo-Saxon Review,* March 1901, in *Bernard Shaw: The Great Composers,* ed. Louis Crompton (1978), p. 220.

158 "sudden achievement . . ." *The Actor's Freedom: Toward a Theory of Drama* (1975), pp. 95, 58.

"chiefest part . . ." C. C. Stopes, *Burbage and Shakespeare's Stage* (1913), p. 121.

"Shakespeare and Burbage . . ." *On Acting* (1986), p. 101.

159 "hold forth . . ." *Dramatic Opinions and Essays,* ed. James Huneker (1906), II. p. 362.

*did mean something special . . . Shakespeare was much more likely to use *glass* to signify what we now call a mirror. Explaining the phrase, "to hold, as 'twere, the mirror up to nature," Virgil Whitaker has written: "This sentence has been taken to imply a doctrine of verisimilitude, or even something approaching the 'slice of life' sought by modern writers. It means, in fact, the precise opposite, as an examination of Shakespeare's terms will show. 'Mirror' denotes for him an exemplar, a perfect pattern, whether good or of evil. . . . This is indeed the common Elizabethan meaning, preserved in the title of *A Mirror for Magistrates*" (*The Mirror Up to Nature* [1965], p. 90).

"did the companies of Burbage . . ." "Elizabethan Acting," *PMLA* 54 (1939), p. 689.

"simply by formal . . ." and "formal acting . . ." "Elizabethan Acting," pp. 690, 705. For a discussion of Elizabethan acting with an anti-formalist thrust, see Marvin Rosenberg, "Elizabethan Acting: Men or Marionettes?" *PMLA* 69 (1954), pp. 915–27.

"Elizabethan acting . . ." *Action Is Eloquence* . . . (1984), pp. 71–72.

160 "non-personal . . ." and "The hard parts . . ." "Elizabethan Acting," pp. 918, 926. John Barton makes the same point when he speaks of Shakespeare's "violent switches between naturalistic and heightened language" (*Playing Shakespeare,* p. 21).

"I would suggest" and "The Elizabethan actor . . ." "Shakespeare's Actors," *Review of English Studies,* new series 1 (1950), pp. 203–4.

"wantonizing Stage-gestures . . ." I. H., *This World's Folly* (1615), sig. B2ʳ.

"she is compelled . . ." Review, London *Times,* 2 December 1817, in *The Complete Works,* ed. P. P. Howe (1967), XVIII, p. 265.

"thumping his breast . . ." Charles Macklin, *Memoirs* (1799), II, p. 265.

161 "tranquilized benevolence . . ." Quoted in Diana Rigg's *No Turn Unstoned* (1982), p. 26.

"ambition and pretension . . ." *The Empty Space* (1968), p. 36.

"if one could find . . ." Quoted in *Actors Talk About Acting,* ed. Lewis Funke, John Booth (1961), p. 35.

"Larry went for . . ." Quoted in *Olivier,* ed. Logan Gourlay (1973), p. 27.

*"The part of Ellie . . ." *Collected Letters 1911–1925,* p. 743. It is thus no shock to find an old Shaw hand complaining, "even though a new generation of theater is now discovering Shaw, we don't have too many actors who can handle the language and act in a broad presentational operatic style *truthfully*" (Basil Langton, "In Praise of Shaw's Theatricality," *New York Times,* 15 March 1987, p. H22).

"Emotion has . . ." *Playing Shakespeare,* pp. 110, 15, 67.

"Mountaineers who . . ." *Brief Chronicles* (1943), p. 204 (this was written ten years before Sir Edmund Hillary's ascent).

"it is like a mountain . . ." Quoted in Charles Marowitz, "*Lear* Log," *Encore* 10

(1963), p. 22. John Barton speaks of Shakespeare's verse as presenting to some actors "a mountain to be climbed . . ." (*Playing Shakespeare*, p. 25).

"this terrifying . . ." Quoted in Rasponi, *The Last Prima Donnas* (1985), p. 192.

162 "actor who should . . ." *Shaw's Music*, I, p. 329.

"I am a highly . . ." Quoted in *The Last Prima Donnas*, p. 569.

*"unscrupulously to emphasize . . ." *Bernard Shaw: The Great Composers*, pp. 225–26. Shaw was by no means the first to make this point. One review of the premiere of *Macbeth* in 1847 warned Varesi, the baritone singing the title role, "if he performs Macbeth or other works of that sort often, his voice will not last two years. Today's manner of writing, so unhappily sponsored by Verdi, leaves but few years of existence to even the best-constituted and apparently soundest voices" (*Verdi's Macbeth: A Sourcebook*, ed. Andrew Porter, David Rosen [1984], p. 379).

"The anxiety . . ." *The New Republic*, 15 October 1977, p. 44.

163 "sempre pppppppianissimo . . ." Manuscript full score, fol. 28ᵛ.

"Betterton kept . . ." Antony Aston, *A Brief Supplement to Colley Cibber, esq.,* quoted in Alan Downer, "Nature to Advantage Dressed: Eighteenth Century Acting," *PMLA* 58 (1943), p. 1006.

164 "We seem so far . . ." Interview with the author.

Claire Bloom . . . Conversation with the author.

"Mr. Booth's Attitudes . . ." Quoted in Alan Downer, "Nature to Advantage Dressed: Eighteenth Century Acting" *PMLA* 58 (1943), p. 1006.

"the very still-life . . ." *A View of the Stage*, in *The Complete Works*, ed. P. P. Howe (1930), V, p. 304.

"Of her attitudes . . ." Quoted in Alan Downer, "Players and Painted Stage: Nineteenth Century Acting," *PMLA* 61 (1946), p. 528.

"If you are fond . . ." *Byron's Letters and Journals*, ed. Leslie Marchand (1975), IV, p. 216.

"where upon the stage . . ." J.B., "On the Art of Acting," *The Spectator* 2 (1829), p. 811.

*"it wasn't easy . . ." Quoted in *Playing Shakespeare*, p. 16. McKellen gives an example of a Shakespearean "aria" being "released" in a smaller theater: "It's surprising how doing a play in a small theater can release it in some way. When I played Henry V, I worked in a very small theater and we had no army. So I imagined that the army was in the audience and I knelt down at the front of the stage and whispered 'Once more unto the breach . . .' I was able to get just as much passion into that and bravado and patriotism by whispering as I could by shouting. In fact I think I got more because it was more real" (p. 185). One can, of course, think of numerous great operatic scenes and arias that could have a similar effect in a similar venue; indeed, many of the operas we constantly meet on the biggest stages were originally intended for drastically smaller ones. The consequences of this disparity are discussed in Chapter 20.

"We are so often . . ." *Playing Shakespeare*, p. 185.

165 "Exalting the solitary . . ." *Prefaces to Shakespeare*, second series (1930), p. 216.

"from isolation . . ." *Shakespeare and the Energies of Drama* (1972), p. 33.

166 Agate offered . . . *Brief Chronicles* (1943), p. 203.

"Keep in mind . . ." Letter to Ricordi, 24 May 1871, *Verdi's Aida*, p. 163.

*"So-called *polished singing* . . ." Letter to Ricordi, 10 July 1871, *Verdi's Aida*, p. 183 (Verdi's italics). When Verdi wrote to the director of the Paris Opéra about

casting Amneris and Amonasro for *Aida's* 1880 premiere there, he had occasion to give his crucial phrase a French translation: "it is necessary, above all and before everything else, that they be full of *spirit, spirit,* and still more *spirit* [*tout de l'âme, de l'âme, et encore de l'âme*]: do you remember Waldmann and Pando-lfini . . . two singers! But they also had the *devil inside them* [*le Diable au corps*]! That is what is needed for these two roles. Without this no success is possible!" (letter to Auguste Vaucorbeil, 4 January 1880, in Ursula Gunther, "Documents inconnus concernant les relations de Verdi avec L'Opéra de Paris," in *Atti del terzo Congresso internazionale di studi Verdiani* [1974], p. 582). Later, Verdi was to insist that the singer impersonating Alice Ford in *Falstaff* should possess the "greatest vivacity" and "deve avere il diavolo adosso" (letter of 12 June 1892, in Abbiati, *Giuseppe Verdi,* IV, p. 443).

"largeness of imaginative . . ." Madeleine Doran, *The Endeavors of Art* (1954), p. 52.

"dangerous freedom . . ." Michael Goldman, *The Actor's Freedom* (1975), pp. 55–56.

"dynamic throbbing . . ." *Mimesis* (1946), p. 328.

"a heightened Pathos . . ." *Prose Works* (1893), V, p. 150.

"superhuman" qualities . . . "Stories That Break into Song: Choosing Plots for Opera," *The Kenyon Review* 4 (1982), p. 77.

"Only one thing . . ." "Some Reflections on Music and Opera," in *The Essence of Opera,* ed. Weisstein (1964), p. 358.

167 * "Beautiful voice . . ." The anecdote is reported in Giuseppi Adami, *Giulio Ricordi e i suoi musicisti* (1933), pp. 51–52. Thus Shaw, witheringly, on Signor Fancelli: "He is not an actor, but what he does is done in earnest, and he sustains comparison only with men who cannot act and who are not in earnest" (*Shaw's Music,* I, p. 146).

"a mere expletive . . ." *A General View of the Stage* (1759), p. 155.

Mrs. Normer . . . See Vera Brodsky Lawrence, *Strong on Music* (1988). Other witty travesty titles were *The Cat's in the Larder* (*La Gazza Ladra*), *Fried Shots* (*Der Freischütz*), *The Roof Scrambler* (*La Sonnambula*), *Sam Parr with the Red Coarse Hair* (*Zampa, or the Red Corsair*).

* "one who seldom . . ." *Hazlitt on Theatre,* ed. William Archer (1895; new ed. 1957), pp. 57–58. Among the shouts of disapproval to be heard at Kabuki performances is one that seems apt to Hazlitt's sketch: *daikon,* which refers to a large, white, common, cheap, and rather tasteless vegetable of the radish family (Earle Ernst, *The Kabuki Theatre* [1956], p. 83). The *daikon* is also well known on the opera stage.

168 "Fire, Impetuosity . . ." *Some Account of the Life of . . . Shakespeare* (1709), in *Shakespeare: The Critical Heritage 1693–1733,* II, p. 191; "fire, spirit. . . ." The original Italian is "fuoco, anima, nerbo, ed entusiasmo" (letter to Escudier, 30 March 1872, in Abbiati, *Giuseppe Verdi,* III, pp. 567–68).

"I went back . . ." Introduction to Lillah McCarthy, *Myself and Friends* (1934), p. 4.

"bond of stage method . . ." *The Saturday Review* 99 (1905), p. 169.

"a strong personality . . ." *The Music Theater of Walter Felsenstein,* ed. Peter Fuchs (1975), p. 16.

"First, it is not . . ." *Hazlitt on Theatre,* p. 51.

169 "I personally had . . ." *Brief Chronicles,* p. 199.

"best Cleopatra . . ." Ibid., p. 175. Agate was writing in 1925, a dozen years after Achurch retired.

"Miss Janet Achurch . . ." *Our Theatres in the Nineties,* III, p. 81 (review of 20 March 1897). A melancholy footnote to this rave review is Shaw's remark in a letter three years later: "Janet has now lost all power of doing anything but her own particular *Io son io* . . ." (letter to William Archer, 8 July 1900, *Collected Letters 1898–1910,* p. 176).

* Shakespearean and operatic . . . A soprano very much in this extraordinary tradition is Leonie Rysanek, according to Peter Conrad, who wrote, apropos her Sieglinde and Kostelnička in Janáček's *Jenufa* in New York City in March 1988: "Leonie Rysanek personifies opera: hers is the dramatic impulse raised to its highest pitch by music . . . her characters are always in extremis, and her voice is equal to their anguish or exultation. . . . She has always been wild onstage, with her interpolated screams and scenery-chewing histrionic abandon. Behind her you can sense the precedent of those Dionysian women from whose madness Nietzsche derived the art of opera" (*Opera News,* June 1988, p. 38).

It is poignant for me that, while writing this chapter, the death of the tenor James McCracken was announced. For his Otello in a San Francisco Opera performance of the early 1960s (partnered by Tito Gobbi's Jago) was my first experience of a performer "driven by the devil." In the *Village Voice* of 17 May 1988 (p. 94), Leighton Kerner remembered McCracken in an essay titled, aptly enough, "Heroic Madness." Because several of Kerner's observations are pertinent to a chapter on heroic utterance (and because the memory of that Otello is indelible), I would like to quote them at length:

> In a field where safe-and-sane performance has become a blight of bland-ness, James McCracken carried on an older tradition of going for broke, vocally and histrionically. His Otello, Don Jose, and Tannhäuser, to take the conspicuous examples, ennobled the desperation and insanity of the charac-ters. Other justly famous tenors, working with many of the McCracken roles, make an orthodoxy of keeping a tight rein on their acting as well as their singing. Calculation, for them, is primary, and they're convinced something should always be held in reserve. A very sane and career-wise principle, that, and the longevity of their time at the top proves their good sense.
>
> But opera is a mad game, or at least it should be. Unless we're dealing with comedy, and sometimes even then, the heroes and heroines of this medium are more than merely players; they are insanely in love, insanely jealous, insanely patriotic, insanely religious, and/or insanely criminal. The source of all this craziness lies in the literature on which so much of the standard repertory is based, and the fuel that keeps it going is the singing.
>
> McCracken never failed to realize that in performance. The credo put him firmly in the midcentury tradition of Maria Callas, although he was spared her premature vocal deterioration, and in the earlier line of Martinelli, and, by all accounts, Tamagno, the first singer of Verdi's Moor. Like Callas in several tragic roles and like Beverly Sills as Elizabeth in *Roberto Devereux,* McCracken took on his most arduous parts as if they were fires by which to be tried. He knew the risks of Otello's perilous entrance-music (*"Esultate!"*) and of dozens of other points in that opera's four acts, but each succeeding time I heard him in the role revealed new, unexpected dangers over-come. . . . [He was] one rugged and noble individualist whom opera could hardly spare.

"I was stunned . . ." "On Novelty and Familiarity," in *The Complete Works* (1967), XII, p. 301.

170 "It was as if . . ." *Eleonora Duse* (1927), p. 151.

"The event remains . . ." and "spun from within like a spider's web . . ." John Coveney, "The Lehmann Mystique," *Opera News* (12 March 1988), p. 14; Vincent Sheean, quoted in the same article, p. 14.

"a case apart . . ." Lanfranco Rasponi, *The Last Prima Donnas* (1985), p. 441.

*"I then heard . . ." *Praeterita,* Vol. XXXV, *Complete Works* (1903–1912), p. 175. Ruskin appears to have had his very bad nights at the opera, too. In a letter to Lady Burne-Jones in 1882 he leaves the distinct impression that *Die Meistersinger* did not entirely appeal to him: "Of all the *bête,* clumsy, blundering, boggling, baboon-blooded stuff I ever saw on a human stage, that thing last night beat—as far as the story and acting went—and of all the affected, sapless, soulless, beginningless, endless, topless, bottomless, topsiturviest, tuneless, scrannelpipiest—tongs and boniest—doggerel of sounds I ever endured the deadliness of, that eternity of nothing was the deadliest. . . ." The next evening, however, brought a *Don Giovanni* (with Patti) that was "one feast of glorious sound for three hours" (*Complete Works,* XXXVII, p. 402).

171 "The 'Sempre libera' . . ." *Mawrdew Czgowchwz* (1972), pp. 26–27.

"The way Melba sang . . ." *Mary Garden's Story* (1951), pp. 93–94.

172 "It soared, a bird . . ." *Ulysses* (1961), pp. 275–76.

17. Star Quality; or, Risky Business

173 "film is the director's . . ." *On Acting* (1986), p. 246.

"he was glorious . . ." Letter to William Harness, 15 December 1811, *Byron's Letters and Journals,* ed. Leslie Marchand (1973), II, p. 149.

"if there is to be . . ." "To Revitalize Opera" (1938), *Roger Sessions on Music* (1979), p. 141.

"the total elimination . . ." Quoted by Benedict Nightingale in "Decoding an Enigma's Genius," *New York Times,* 8 November 1987, p. H35.

174 "stagecraft concentrates . . ." Granville-Barker, *Prefaces to Shakespeare,* first series (1927), p. xxiv. Pertinently, he continues, "We think now of the plays themselves; their first public knew them by their acting; and the development of the actor's art from the agilities and funniments of the clown, and from formal repetition or round-mouthed rhetoric . . . was a factor in the drama's triumph that we now too often ignore."

"consists in making . . ." Michael Goldman, *Acting and Action in Shakespearean Tragedy* (1985), p. 71.

"an actor's performance . . ." Ibid., p. 3.

"those who went . . ." Coleridge, *Shakespearean Criticism,* ed. Thomas Raysor (1930), II, p. 97.

"a great player like Alleyn . . ." *The Elizabethan Popular Theatre* (1982), p. 96.

"one of the lightning careers . . ." *The Actor's Freedom* (1975), p. 58.

175 T. W. Baldwin speculates . . . *The Organization and Personnel of the Shakespearean Company* (1927), p. 238.

*The seventeen other single-star vehicles (that is, plays with the star taking more than 20 percent of the lines) are *Two Gentlemen of Verona, Richard III, Taming of the Shrew, Richard II, 2 Henry IV, Henry V, Julius Caesar, As You Like It, Hamlet,*

Measure for Measure, Timon of Athens, King Lear, Macbeth, Coriolanus, Pericles, Winter's Tale, and *The Tempest*. The five two-star vehicles are *King John, 1 Henry IV, Romeo and Juliet, Othello,* and *Antony and Cleopatra*. Nine plays can fairly be called ensemble pieces, with three or more characters given substantial roles (*Comedy of Errors, 2 Henry VI* and *3 Henry VI, Midsummer Night's Dream, Much Ado About Nothing, Twelfth Night, Merry Wives of Windsor, All's Well That Ends Well, Cymbeline*).

* single most important . . . In the course of acknowledging his own creative debt to fine actors, Shaw offered this pertinent praise of Burbage: "The actor also may enlarge the scope of the drama by displaying powers not previously discovered by the author. If the best available actors are only Horatios, the authors will have to leave Hamlet out, and be content with Horatios for heroes. Some of the difference between Shakespear's Orlandos and Bassanios and Bertrams and his Hamlets and Macbeths must have been due not only to his development as a dramatic poet, but to the development of Burbage as an actor. Playwrights do not write for ideal actors when their livelihood is at stake: if they did, they would write parts for heroes with twenty arms like an Indian god" (*Shaw on Shakespeare,* ed. Edwin Wilson [1961], p. 251).

"self-evident . . ." *Lectures on Dramatic Art and Literature* (1809; Eng. trans. John Black, 1894), p. 454; "From the time of Burbage . . ." *Conceptions of Shakespeare* (1966), p. 39. "Leading" ceased to be synonymous with "Shakespearean," of course, with the advent of film and, later, television cameras, which demand essentially un-Shakespearean techniques of projection.

Burbage bestrode . . . Richard Flecknoe, in his "Short Discourse of the English Stage" (1664), wrote of Burbage as "the admired example of the age . . ." (Stopes, *Burbage and Shakespeare's Stage,* p. 122).

"emphatically not . . ." *Dramatic Character in the English Romantic Age* (1970), p. 7.

176   "Which is your *Burbage* . . ." *Bartholomew Fayre, Works,* VI, pp. 119–20.
177   "an egomaniac's delight . . ." Anthony Hammond, *Richard III* (1981), p. 69.
       "to a considerable extent . . ." Clifford Leech, "Shakespeare, Cibber, and the Tudor Myth," in *Shakespearean Essays,* ed. Alwin Thaler (1964), p. 79.
       "noble art . . ." *Ulysses* (1961), p. 589.
       "fundamental truths . . ." Hofmannsthal, letter to Strauss, 27 June 1928, in *A Working Friendship* (1974), p. 482.
       "*Pygmalion* is essentially . . ." *Collected Letters 1911–1925,* p. 146.
       "One actor in . . ." "What is Acting?" in *The Dramatic Event* (1956), p. 51.
178   "his coactors . . ." *The Antipodes,* sig. D3^r.
       "over-act prodigiously . . ." Induction to *The Staple of News* (1631), *Works,* VI, p. 280.
       "speak warmly . . ." Thomas Davies, Garrick's first biographer, quoted in Diana Rigg, *No Turn Unstoned* (1982), p. 186.
       "the virtuoso supreme . . ." Simon Williams, *German Actors of the Eighteenth and Nineteenth Centuries* (1985), p. 98.
       "The operatic stage . . ." Review in *The Star,* 18 April 1890, in *Shaw's Music,* II, √ p. 32.
179   "art of making . . ." "Duse and Bernhardt" (1895), in *Our Theatres in the Nineties* (1931), I, p. 158.
       "Janet has now . . ." Letter to William Archer, 8 July 1900, in *Collected Letters 1898– 1910,* p. 176.

*prudence and "normality" . . . Maynard Mack writes of Hamlet as "partly an *eiron* figure . . . and his language shows it. The other heroes, hyperbolists, use the language of the *alazon*" ("Energy and Detachment in Shakespeare's Plays," *Essays on Shakespeare and Elizabethan Drama,* ed. Richard Hosley [1962], p. 287). For a broader discussion of the *alazon* and *eiron,* see Northrop Frye, *The Anatomy of Criticism* (1957), pp. 36–37, 226.

180   "Do you not see . . ." *The Poetry and Prose,* ed. Louis Untermeyer (1949), p. 365.

"Caesar and Pompey . . ." *An Apology for the Life of Colley Cibber* (1740; 1968 ed. B. R. Fone), p. 227.

"Young singers . . ." Interview in Yveta Graff, "Tosca Talks," *Opera News,* 18 January 1986, p. 23.

"Jacobi is always . . ." Frank Rich, review, *New York Times,* 22 December 1988, p. C11 (Rich's emphasis).

"the actor seems . . ." *Great Reckonings in Little Rooms: On the Phenomenology of Theater* (1985), pp. 161–62.

181   "As acting itself . . ." *The Actor's Freedom* (1975), p. 55.

"His playing disdained . . ." London *Times,* 12 July 1989. Anthony Holden wrote for the *Sunday Times,* "The quality I most admired about Olivier was his penchant for risk, thrillingly mirrored in the sense of danger he conveyed on the stage" (16 July 1989).

"It seems tew be . . ." Josh Billings [Henry Wheeler Shaw], *His Works, Complete* (1888), pp. 114–15.

182   *"Courage is needed . . ." "Il faut du courage pour être romantique, car il faut hasarder." Stendhal [Marie Henry Beyle], "Racine et Shakspeare" (1823), *Œuvres Complètes* (1954), XVI, p. 28. Vladimir Horowitz made precisely the same point in assessing his own executive style: "I am a nineteenth-century Romantic. I am the last. I take terrible risks. Because my playing is very clear, when I make a mistake you hear it. But the score is not a bible, and I am never afraid to dare. The music is behind those dots. You search for it, and that is what is meant by the grand manner" (obituary, *New York Times,* 6 November 1989, p. B9).

*"Part of the epic scale . . ." *Peter Hall's Diaries,* ed. John Goodwin (1983), p. 356 (emphasis added). Wesley Balk, in *The Complete Singer-Actor: Training for Music Theater* (1985), makes a similar point: "In a great performance we feel that the performer is totally committed to daring her or his limits" (p. 39).

"there are too many . . ." *On Acting,* p. 104.

"If you want . . ." This and the following remarks by singers of the role of Tosca are from Yveta Graff, "Tosca Talks," *Opera News,* 18 January 1986, pp. 19–22.

183   "it is always possible . . ." "Introduction to the Italian Theater," in *The Genius of Italian Theater,* ed. Eric Bentley (1964), p. 15.

"In every great performance . . ." *The Complete Singer-Actor,* (1985), p. 40.

"As usual . . . he raises . . ." *Observer,* 21 August 1955, quoted in *Olivier* (1973), p. 189.

"Each time you . . ." Quoted in *Playing Shakespeare,* p. 112.

"Virtuosity, in theater . . ." *Great Reckonings in Little Rooms,* pp. 119–20.

"It is the destiny . . ." Interview in Rasponi, *The Last Prima Donnas,* p. 157.

184   "Being a masterpiece . . ." *Bulletin of the Société Internationale de Musique,* 1 February 1914, in *Debussy on Music,* ed. François Lesure (1977), p. 307.

"Shakespeare allows . . ." *Shakespeare Without Tears* (1942), p. 103.

"the 'great' characters . . ." *The Life of Drama* (1964), pp. 68–69.

"Bulwer's old title . . ." *Brief Chronicles* (1943), p. 280.

185 "We have each . . ." Quoted in ibid., p. 239.

"every good performance . . ." Ibid., p. 228. Cibber, for instance, praised the Italian divo Farinelli (Carlo Broschi) because he was able to "exquisitely surprise us" (*Apology,* p. 210).

"there is nothing . . ." "Tosca Talks," *Opera News,* 18 January 1986, p. 22.

"If you gave all . . ." Conrad Osborne, "From the Top," *Opera News,* 4 February 1984, p. 52.

"I *hate* performers . . ." *Shaw's Music,* III, p. 224.

"When she sang . . ." *The Last Prima Donnas,* p. 499. Mafalda Favero spoke similarly of her own Butterfly: "After a performance of Cio-Cio-San I was literally undone, and it took me a couple of days to regain my composure and strength. But that was the only way I could approach a role—unsparing and dedicated. I am fully aware that Butterfly shortened my career by at least five years" (p. 516).

186 "worst sins . . ." *Bernard Shaw: The Great Composers,* ed. Louis Crompton (1978), p. 214. Shaw's discussion (12 April 1893) was prompted by his first look at a *Falstaff* score.

"there is no lack . . ." C. Mellini, *La Rivista,* 27 February 1847, in Porter and Rosen, *Verdi's Macbeth* (1984), p. 371.

*"to kindle art . . ." *Shaw's Music,* III, p. 209. Granville-Barker's attempt to capture the magical ingredient came down to a milder word: style. Explaining what is necessary to do *Love's Labour's Lost* justice, he wrote: "The whole play, first and last, demands style. A vexingly indefinable thing, a hackneyed abracadabra of a word! One should apologize for bringing it into such a practical discussion as this pretends to be. . . . But the theatre must deal in magic sometimes" (*Prefaces to Shakespeare,* first series [1927], p. 14). The classic expression of Granville-Barker's point, of course, occurs in Alexander Pope's *Essay on Criticism,* where "nameless grace" is given priority:

> Some beauties yet no precepts can declare,
> For there's a happiness as well as care.
> Music resembles poetry, in each
> Are nameless graces which no methods teach,
> And which a master hand alone can reach.

187 * magnification, and escalation . . . I first made this observation in Chapter 1 of *Literature as Opera* (1977).

* the novel ends . . . Some of the first reviewers of *Henderson the Rain King* used phrases that suggest its "operatic" spirit remarkably well. One spoke of Bellow's "heroes, yea-sayers, and visionaries," another of the novel's "exceptional amalgam of vehement forces," another of its "emotional abundance" and its "profound sufferers," another of "this purest distillation of . . . affirmation." The composer Leon Kirchner produced an opera from the novel: *Lily.* The New York City Opera premiere in 1977 proved unsuccessful.

18. Technique and the Gemini Factor

188 "One must try . . ." Berlioz was expressing his difficulty composing *Les Troyens:* "Another hurdle in my path is that the feelings to be expressed move me too much. . . . That is what held me up so long in the *Romeo and Juliet adagio*"

(letter to Princess Sayn-Wittgenstein, 12 August 1856, *Hector Berlioz: A Selection from His Letters,* ed. Humphrey Searle [1973], p. 150).

"In the audience . . ." *The Idea of the Actor* (1984), p. 3.

*"what we see him . . ." *The Complete Works,* ed. F. L. Lucas (1927), IV, p. 43 (emphasis added). Michael Goldman recently reiterated Webster's paradox in different terms: "All acting roles have a quality we may call iconic—they give the impression of a fixed or masklike definition. We feel we are watching a figure that, although animated, is yet a type or effigy. It's through the interplay between the iconic and the animate, between mask and face, that drama is able to deploy some of the uncanniness associated with acting itself. So we may think of the actor's task as both projecting an icon and filling it with life" ("Performer and Role in Marlowe and Shakespeare," in *Shakespeare and the Sense of Performance,* ed. Marvin and Ruth Martin [1989], p. 93).

*"In acting . . ." Quoted in *Actors on Acting,* ed. Toby Cole (1970), p. 581. Sam Goldwyn gave Belasco's point a characteristic twist when he opined, "The most important thing in acting is honesty; once you learn to fake that, you're in!"

189 "Johnson, indeed . . ." *The Life of Johnson* (1953), p. 1252. John Philip Kemble (1757–1823) flourished in the late eighteenth century.

"inner honesty . . ." *The Complete Singer-Actor* (1985), p. 26.

"non-technical . . ." and "the idea . . ." *Performing Power* (1985), pp. 55, 109.

"provides a field day . . ." *New York Times,* 18 September 1987, p. C3.

190 *"Fool or Monarch . . ." Robert Lloyd, *The Actor. A Poetical Epistle* (1760), p. 4 (original emphasis). This sentimental notion of the actor "becoming" the character is perhaps the most time-honored cliché of hype in all theatrical criticism. Incredibly, for instance, Flecknoe would have us think that Burbage was a thorough Stanislavskian. In his "Discourse" he describes Burbage as "a delightful *Proteus,* so wholly transforming himself into his Part, and putting off himself with his Cloathes, as he never (not so much as in the Tyring-house) assum'd himself again until the Play was done" (*Critical Essays of the Seventeenth Century,* ed. J. E. Spingarn [1909], II, p. 95).

"I live every part . . ." Quoted in Yveta Graff, "Tosca Talks," *Opera News,* 18 January 1986, p. 21.

"control the floodtide . . ." Quoted in Barton, *Playing Shakespeare,* p. 147.

"the whole technique *etc.* . . ." *The Kabuki Theatre,* pp. 188, 192, 195, 290.

"that the actor *etc.* . . ." *Brecht on Theatre,* trans. John Willett (1964), pp. 194, 58, 136.

*"The cardinal principle . . ." *The Idea of the Actor,* p. 164. John Fuegi, in an essay "Meditation on Mimesis: The Case of Brecht," in *Drama and Mimesis* (Vol. 3 of *Themes in Drama,* 1980), arrives at a similar conclusion: Brecht's is "a style where the actor does not 'become the part' but instead suggests or presents the part to an audience while retaining his own character as a human being who is an actor" (p. 111).

191 "We shall do . . ." "The Acting of Marlowe and Shakespeare," *G. F. Reynolds Lectures* (1963), pp. 36–37.

"the theatrical self-consciousness . . ." *The Idea of the Actor,* p. 11.

"was never more modern . . ." Fuegi, *op. cit.,* p. 110.

"the histrionic dynamics . . ." *The Idea of the Actor,* p. 10.

"*Hamlet* is the most interesting . . ." *The Actor's Freedom,* p. 148.

192 "One could really . . ." "Über die liesse sich freilich eine interessante Oper schreiben" (act 1, scene 3).

"Give your sense . . ." Letter of 24 July 1911, in *A Working Friendship* (1974), p. 100.

193 "Any striking . . ." *Acting and Action in Shakespearean Tragedy* (1985), p. 79.

"The problem is how . . ." *Playing Shakespeare,* p. 90.

194 "All artists . . ." *Disposizione Scenica,* p. 4.

* "Always, even when the whirlwind . . ." The Italian: "Sempre, anche quando è più violento il turbine della passione, dovete conservare il vostro senno ed essere padroni di voi medesimi." Curiously, Hamlet's "Be not too tame neither, but let your discretion be your tutor" is not included by Boito. One is inclined to wonder, whimsically, if this is because under-acting was not a problem on the nineteenth-century opera stage.

"the language of the play . . ." *Playing Shakespeare,* p. 147.

195 "Hotspur's pain is . . ." Ibid., pp. 135–36.

"The moral . . ." Ibid., p. 145.

"Say to the audience . . ." Ibid., pp. 127–28 (Barton's emphasis).

196 "It's delicious . . ." Ibid., p. 127.

"he should play them . . ." Ibid., p. 126.

"the choric function . . ." Ibid., p. 137.

"Shakespeare very often . . ." Ibid.

"we can be pretty sure . . ." Ibid., pp. 138–39.

197 the sweat and the heaving . . . One of the incidental pleasures of opera video taped live for television—though to many viewers the effect is off-putting—is that we can see at close range how physically demanding the effort of vocal projection is.

* "This kind of acting . . ." Michael Hattaway, *Elizabethan Popular Theatre* (1982), p. 96. One is inclined to think this intimacy between actor at the forestage and audience was not an absolutely silent one but was volubly enhanced by the spectators. The eighteenth-century audience, according to Worthen, carried on this tradition in a way that opera audiences have done since commercial performances began: "The brassy theatricality of the spectators' behavior revealed their sympathetic involvement with one another. Rather than deflating the dramatic spectacle, the performer on the near side of the stage seemed to broaden it, to form part of the evening's entertainment" (*The Idea of the Actor,* p. 83). Several other aspects of legitimate theater in the age of Garrick evoke the nature of operatic performances. Worthen notes that it provided "opportunities to compare the talents of various actors in the same parts" and was "attended by an experienced and discriminating public." Pointing—"a device emphasizing technique rather than meaning"—was common, as was the habit of "detaching a famous speech from the action of the play and delivering it directly to the audience, repeating it if necessary" (pp. 71–72).

Bert States, in *Great Reckonings in Little Rooms,* posits three modes of acting: the Self-expressive Mode (I/ACTOR), the Collaborative Mode (YOU/AUDIENCE), and the Representational Mode (HE/ACTOR). Shakespearean and operatic dramaturgy would seem to depend principally upon a complex intertwining of the first two Modes.

198 "directly at his fellow's . . ." Quoted in E. K. Chambers, *The Elizabethan Stage* (1967), IV, p. 256.

"shows less regard . . ." Alan Downer, "Nature to Advantage Dressed: Eighteenth-Century Acting," *PMLA* 58 (1943), p. 1018.

"a magician . . ." *The Empty Space* (1968), pp. 110–111.

"could permit both rant . . ." *Shakespeare's Stagecraft* (1967), p. 37.

"the great English tragedians . . ." *The Art of Acting* (1926), pp. 28–29. William Archer made fun of the wide vocal range in his satiric essay, "The Fashionable Tragedian" (1877), referring to the "trick of alternating between *basso profondo* and falsetto, like a ventriloquist imitating a conversation between the Giant Cormoran and Jack the Giant-killer" (p. 7).

"I didn't think . . ." *Actors on Acting,* ed. Toby Cole (1970), p. 414.

"move along the full . . ." *The Complete Singer-Actor,* p. 41.

199   "the plays of Shakespeare . . ." Quoted in *Actors on Acting,* p. 421.

"were great *elocutionists* . . ." *Conceptions of Shakespeare* (1966), p. 50.

The Art of Singing . . . The full title: *Andreas Ornithoparcus His Micrologus, or Introduction: Containing the Art of Singing.* The Latin original was first published in 1517; the English translation appeared in 1609; all quotations are from pp. 88–90.

"the music is always . . ." Interview with the author.

"In general our singers . . ." "I nostri cantanti non sanno fare in generale che la voce grossa; non hanno elasticità di voce, ne sillibazione chiara e facile, e mancano d'accento e di fiato" (letter to Ricordi, 13 June 1892, in Abbiati, *Giuseppe Verdi,* IV, p. 444).

200   "If you forget . . ." *Playing Shakespeare,* p. 46.

"Actors and singers . . ." *Shaw's Music* (1981), I, p. 896.

"not to make bug-eyes . . ." *Disposizione Scenica* (1887), p. 5.

"his perfect intelligence . . ." *The Saturday Review* 99 (1905), p. 170.

201   "the prima donna . . ." John Percival, "Shakespeare at Stratford," *Plays and Players,* June 1963, pp. 34–35. "Rhetoric?" Percival adds, "There is little in Tom Fleming's bookish Prospero. A snappy asperity enlivens his tone at times, but no color or extravagance or persuasion" (p. 34). Is there a Shakespearean protagonist who can exist on stage without executive extravagance?

202   "John Wood's . . ." Robert Smallwood, "Shakespeare at Stratford-on-Avon, 1988," *Shakespeare Quarterly* 40 (1989), p. 85.

"A director often . . ." Quoted in *Playing Shakespeare,* p. 18.

"Shakespeare's language . . ." Ibid., pp. 146–47.

203   *yin*–and–*yang* coexistence . . . One of the most sensible interpretations of Hamlet's advice to the players, Eric Bentley's, also strikes a note of typically Shakespearean balance: "Dramatic, like all other art, necessarily involves both imitation and selection, nature and artifice, truth and beauty. . . . To interpret Hamlet's advice correctly we would need to know whether Shakespeare's contemporaries were leaning too far toward the natural or toward the artificial; Hamlet's was against the unbalance, whichever it was. Natural and artificial are not names of rival styles in acting; they are the names for lack of style in acting. Acting is both natural and artificial, yet to the extent that it comes off, to the extent that it is good acting, it is not notably either" (*The Dramatic Event* [1956], pp. 78–79).

Maynard Mack sought . . . "Engagement and Detachment in Shakespeare's Plays," in *Essays on Shakespeare and Elizabethan Drama,* ed. Richard Hosley (1962), p. 275. Clark's observation occurs in *The Nude in Art* (1956), p. 130.

*since Shakespeare evinced . . . On Shakespeare's simultaneous engagement and detachment, Alvin Kernan writes, "In the end the theater seems to have been paradoxical for him: at one and the same time, only a transitory illusion and an image of trancendent reality, a trick and a vision, mere entertainment and a means of directing life to meaningful ends" (*The Playwright as Magician* [1979], p. 158).

* "The first thing . . ." *Eleonora Duse* (1926), pp. 154–55. Thus, Bernhardt might be viewed, like Hamlet, as a "type" of the Shakespearean or operatic virtuoso *performer*. Indeed, Symons' effusive description of "The Genius of Eleonora Duse," on the other hand, serves nicely as a sketch of the "type" of the heroic Shakespearean or operatic *protagonist*: "Duse is the symbol of relentless fate, of destiny, remorseless, inevitable, taking us unawares. . . . There was an ardent flame in Duse's blood. . . . [She was] the artist of her own soul and it is her strength of will . . . which makes her what she is: a great impersonal force rushing towards the light; looking to every form of life for sustenance, for inspiration; seeming always to live in every nerve and brain-cell with a life which is unslackening" (p. 157).

19. The Right Voices

204   "All waits . . ." From "Vocalism," *The Poetry and Prose,* ed. Louis Untermeyer (1949), p. 361.

"Who would condescend . . ." *Medwin's Conversations with Byron,* ed. Ernest Lovell (1966), p. 93.

"short, sad . . ." and "the Romantics were . . ." Timothy Webb, "The Romantic Poet and the Stage: A Short, Sad History," in *Romantic Theatre,* ed. Richard Cave (1986), p. 10.

"this close analysis . . ." *Prefaces to Shakespeare,* second series (1930), p. 220 (emphasis added).

205   "It was the happiness . . ." "A Short Discourse of the Stage," in *Critical Essays of the Seventeenth Century,* ed. J. E. Spingarn, II, pp. 94–95.

"The powers of . . ." *The Actor's Freedom: Toward a Theory of Drama* (1975), p. 100.

"When someone says . . ." Letter of 26 October 1871, *Verdi's Aida,* ed. Hans Busch (1978), p. 244.

206   "The membership . . ." "Shakespeare and his Actors: Some Remarks on *King Lear,*" *Proceedings of the Comparative Literature Symposium* 12 (1981), p. 184.

"The performers of . . ." *Discovering Shakespeare* (1981), p. 31.

207   "Years he numbered . . ." "Epitaph on S. P., a Child of Q. El. Chapel," *Complete Poems,* ed. George Parfitt (1975), p. 81.

208   * This information . . . We know the two actors who played these roles because, in one edition, several speech prefixes for one scene (4.2) were inadvertently given to the actors rather than the characters (the compositor was presumably reading from a prompt copy).

* highly comic duo . . . The editor of *Much Ado About Nothing* thinks that Kempe's "cherished abilities" were "a major consideration" as Shakespeare created Dogberry, and quotes from a play of 1601 a character's declaration that Kempe's face "would be good for a foolish Mayor or a foolish justice of peace" (Arden ed. [1981], A. R. Humphreys, p. 23). W. A. Armstrong, in "Actors and Theatres," writes: "As for Dogberry . . . the bumbling discursiveness of the characterization seems designed to accommodate . . . such digressions, by-play, and improvisations as Kempe may have brought to the role. That Shakespeare's clowns were shaped to fit the actors who played them seems beyond question" (*Shakespeare Survey* 7 [1964], p. 195). David Wiles discusses Kempe's roles and his influence on Shakespeare at length in *Shakespeare's Clown* (1987).

209 * Ringler suggests . . . "Shakespeare and His Actors," pp. 186–93. The six interior characters are: legitimate son of Gloucester, naked Bedlam beggar, the beggar clothed as a poor "peasant," a stranger to his father at the Dover cliff, a "Zomerzet" speaker when he kills Oswald, and mysterious armored champion when he fights his brother Edmund. Ringler notes that Armin had, in about 1598, written a play called *The Two Maids of More-Clack*, in which he himself had impersonated an adult man who—like Tom of Bedlam—is mentally impaired. On the matter of the Fool as a boy actor, see Huntington Brown, "Lear's Fool: A Boy, Not a Man," *Essays in Criticism* 13 (1963), pp. 164–71.

"he has abounding energy . . ." *Prefaces to Shakespeare,* fourth series (1945), p. 155.

"One of his arts . . ." *Disposizione Scenica* (1887), p. 5.

* a quicksilver gaiety . . . One wonders if the very odd little scene—often cut (as it was in the BBC version)—in which Iago plays the entertaining jester to Desdemona as they await Othello's arrival in Cyprus is a reminiscence of Armin's normal "clown" assignment. Besides, can one possibly imagine an actor who had shone as Touchstone or Feste being wasted on the play's *real* Clown . . . the dreariest in all Shakespeare and a role only a paltry thirty lines long?

* one particular "line" . . . On the subject of Shakespeare writing to type, Baldwin is categorical: "Shakespeare cut his plays to fit the actors of his company even in details of age and physical appearance, fashioning for each principal actor a suitable principal part" (*The Organization and Personnel of the Shakespearean Company* [1927], p. 282). Ringler rejects this view, saying that Baldwin "goes contrary to all that we know of repertory companies, whose actors, rather than being limited to certain prescribed roles, pride themselves on their virtuosity and strive to act in as great a variety of roles as possible" (*op. cit.,* p. 187). The reality of Shakespeare's company must surely have lain somewhere in between: some actors were indeed versatile, others (and good actors they might be, too) were not. These were happy to play to type, as many actors on stage and screen do today.

210 "Then fear not . . ." This epigram is printed among poems ascribed to Jonson, *Works,* VIII, p. 439.

211 "thinks of himself . . ." *As They Liked It* (1947), p. 52.

"an audience not aristocratic . . ." H. J. Oliver, Arden *Merry Wives of Windsor* (1971), p. xxx. This version, extant in the 1602 quarto, runs about 1,600 lines, compared to the 3,000 lines of the Folio version.

to appeal to the vanity . . . See Henry Paul, *The Royal Play of Macbeth* (1950).

212 "I have substituted . . ." Letter to Ricordi, 12 November 1871, in *I Copialettere,* ed. A Luzio (1913), p. 676 (Verdi's emphasis).

* "He was not an artist . . ." *The Operas of Verdi* (1981), III, pp. 441–42. Likewise, it is extremely hard to imagine Shakespeare or Verdi making a confession at the end of their careers such as Wagner did to King Ludwig II: "I have had to surrender all my works to a kind of audience and theatrical practice which I recognize to be deeply immoral" (letter of 28 September 1880, *Selected Letters,* ed. S. Spencer, B. Millington [1987], p. 903).

"Varesi is . . ." Letter to Alessandro Lanari, 19 August 1846, in Porter and Rosen, *Verdi's Macbeth: A Sourcebook* (1984), p. 5. All quotations from *Macbeth* documents will be from this volume.

"Either there'll be . . ." Letter of 3 September 1846, p. 8.

"the first time Verdi . . ." "Verdi and the Contemporary Italian Operatic Scene,"
The Verdi Companion (1979), p. 93.

213   "I saw how much . . ." Letter of 31 January 1847, pp. 39–40.

"I do not well recall . . ." Letter of 2 January 1847, p. 29.

"a company of . . ." All *Re Lear* letters are quoted from my "Verdi's *King Lear*
Project," in *Nineteenth Century Music* 9 (1985), pp. 99–100.

214   "I have a part . . ." Letter of 26 October 1861, *Carteggi Verdiani*, ed. A. Luzio
(1935), II, p. 62.

"since no one will . . ." Letter to Tito Ricordi, April 1863, Abbiati, *Giuseppe Verdi*,
II, p. 732.

*"Don't forget that . . ." Letter of 15 December 1868, Abbiati, *Giuseppe Verdi*, III,
p. 235. Verdi's adamant insistence on "the right voices" is displayed in what
happened when he arrived in St. Petersburg to oversee rehearsals for the
premiere. The soprano Emma Lagrua was unable to perform and no satisfactory
replacement could be found. Verdi's wife wrote to Count Arrivabene to describe
the upshot:

> [T]he news I'm about to give you will make your eyes, mouth and ears all open
> at once. But though you may exclaim in all the keys of your basso profondo
> voice 'Ha!' 'He!' and 'Ha!' the news is true nonetheless. Verdi will not give his
> new opera in Petersburg—this year. Alas! Singers' voices are as fragile
> as . . . (I'll leave you to complete the phrase) and Mme Lagrua's voice, to her
> and Verdi's misfortune, is an appalling example. . . . [quoted in Budden, III,
> p. 434].

The premiere took place a year later.

*"La Fabbri with her . . ." and "this is a comedy . . ." Letter to Giulio Ricordi,
13 June 1892, Abbiati, *Giuseppe Verdi*, IV, p. 442–43. Only the casting of Falstaff
himself caused no discussion, everyone being convinced Maurel was perfect for
the role. Verdi said, in a conversation with Italo Pizzi, "The opera is finished, but
I am having great difficulty in finding singers. There are many characters in
Falstaff and not one is secondary." As he often did, Verdi haled in Shakespeare to
underscore his point: "None of Shakespeare's characters is secondary. All of them
are of equal importance, even those who only have a few words to say. They all
have their own individuality, their own very distinctive characteristics." (*Encounters with Verdi*, ed. Marcello Conati [1984], p. 349).

215   Hepokoski has studied . . . "Verdi, Giuseppina Pasqua, and the Composition
of *Falstaff*," *Nineteenth Century Music* 3 (1980), pp. 239–50 (quoted passage on
p. 240).

"So it seems that even . . ." *The Operas of Verdi*, III, p. 433.

"I shall give the part" and following quotations . . . *A Working Friendship* (1974),
pp. 241–43 (original emphasis).

216   "I have sacrificed . . ." Letter to his father, 26 September 1781, *The Letters of Mozart
and His Family*, ed. Emily Anderson (1966), p. 769.

217   "In his Neapolitan . . ." New Grove *Masters of Italian Opera* (1983), p. 34.

"extremely intuitive . . ." Quoted in Ara Guzelimian, "Knight of Bel Canto,"
Opera News, July 1989, p. 12.

"The extension . . ." and "You have merely . . ." Quoted in Herbert Weinstock, *Vincenzo Bellini: His Life and His Operas* (1971), p. 99.

"particularly responsive . . ." New Grove *Masters of Italian Opera,* p. 123.

218 "built upon vigor . . ." *Prefaces to Shakespeare,* first series (1927), p. xxvi.

"Bel canto is not . . ." *Pure Contraption* (1974), p. 48.

219 "a tremendous success . . ." *Reminiscences of the Opera* (1864; reprint, 1976), pp. 279–81.

20. Staging Shakespeare and Opera; or, Big Tiny Little

220 "There was a time . . ." *Shaw's Music* (1981), II, p. 151.

221 "the *Royaltie* of *Sight* . . ." Henry Wotton, *The Elements of Architecture* (1614; reprint, ed. Frederick Hard, 1968), p. 4.

222 "we must have faith . . ." "The Acting of Marlowe and Shakespeare," *G. F. Reynolds Lectures* (1963), pp. 36–37.

* "Feminine charm . . ." and "Shakespeare . . . has left . . ." *Prefaces to Shakespeare,* first series (1927), p. xxix–xxx. He adds, "His studies of women seem often to be begun from some spiritual paces beyond the point at which a modern dramatist leaves off. Curious that not a little of the praise lavished upon the beauty and truth of them—mainly by women—may be due to their having been written to be played by boys!"

223 "See if you can persuade . . ." Letter of 21 September 1871, *Verdi's Aida,* ed. Busch (1978), p. 223. Verdi's qualm was expressed in a letter of 7 September 1871: "Of course our prima donnas will be very beautiful; but what if, later on, there happened to be one who wasn't? The audience might joke about it, and that would displease me because that moment is too important" (p. 219). In the end, Ghislanzoni's phrasing was retained.

"She was not . . ." Rasponi, *The Last Prima Donnas* (1985), p. 417.

"The poetry of . . ." *Dramatic Opinions and Essays* (1906), II, p. 363.

"If there is one way . . ." and "it is interesting . . ." *Great Reckonings in Little Rooms: On the Phenomenology of Theater* (1985), pp. 57–60.

224 "Such was the exquisite . . ." *Hymenaei* (1606), *Works,* VII, p. 229.

225 The culmination . . . My *Shakespeare and the Courtly Aesthetic* (1980) is devoted to exploring Shakespeare's response to the Stuart Court's artistic revolution, focusing in particular upon *The Tempest.* For further consideration of the masque, see Stephen Orgel and Roy Strong, *Inigo Jones: The Theatre of the Stuart Court* (1973).

"gentle expressions . . ." "Hee was (indeed) honest, and of an open, and free nature, had an excellent *Phantsie* [imagination], brave notions, and gentle expressions" (*Works,* VIII, p. 584).

226 "O Showes, Showes . . ." Ibid., pp. 403–4.

227 "The difference betwixt . . ." "A Short Discourse of the English Stage," in *Critical Essays of the Seventeenth Century,* ed. J. E. Spingarn, II, pp. 94–95.

"The Elizabethan drama . . ." *Prefaces to Shakespeare,* first series, pp. xvii–xviii.

228 "Prospect of Grottos . . ." and following quotations, in Henry Purcell, *The Fairy Queen,* in *The Works of Henry Purcell* (1968), XII, pp. xiiiff. For some of my information about Purcell's opera I am obliged to Roger Savage, "The Shakespeare–Purcell *Fairy Queen:* A Defence and Recommendation," *Early Music* I (1973), pp. 203–21.

* "great designs . . ." *The Spectator,* ed. Donald Bond (1965), I, p. 24. Handel's first London opera, *Rinaldo,* had received its premiere on 24 February 1711 at the Haymarket Theater. Addison here pretty much describes the state of Baroque

opera throughout Europe. For an accurate, witty satire on the staging of Italian opera of this period, see Piero Weiss, "Pier Jacopo Martello on Opera (1715): An Annotated Translation," *Musical Quarterly* 66 (1980), pp. 378–403.

"a good acting play . . ." *Medwin's Conversations with Byron,* ed. Ernest Lovell (1966), p. 120

229 "The spirit was evaporated . . ." *Complete Works,* ed. P. P. Howe (1930), V, p. 275.

"The first inconvenience . . ." *Essays on Chivalry, Romance, and the Drama* (n.d.), p. 224.

"As the nineteenth century . . ." "Pictorial Acting and Ellen Terry," in *Shakespeare and the Victorian Stage,* ed. Richard Foulkes (1986), p. 82.

"heavy upholstery . . ." Margaret Webster, *Shakespeare Without Tears* (1942), p. 13.

"I have never seen . . ." Quoted in Diana Rigg, *No Turn Unstoned* (1982), p. 155.

"Drastic methods . . ." *Evenings with the Orchestra,* ed. Jacques Barzun (1956), p. 109.

230 "For a whole century . . ." Letter of 25 October 1917, *Collected Letters 1911–1925,* p. 512 (Shaw's emphasis).

*"My opera is a drama . . ." Conversation with A. De Lauzières-Thémines in March–April 1886, in *Encounters with Verdi,* ed. M. Conati (1984), p. 204. Perhaps Verdi was making a similar point of smaller being better in his remarks about his second-largest opera (*Don Carlo* is surely the Verdi canon's heavyweight): "I have never had the fortune of seeing one of my operas done well at La Scala. Even *Aida* was performed better in a small provincial city, Parma, than in Milan" (letter to Ricordi, 21 February 1881, *Verdi's Aida,* p. 425).

*"You talk to me . . ." Letter of 13 June 1892, in Abbiati, *Giuseppe Verdi,* IV, p. 442. Conversely, when Verdi came to consider the production of his version of *Macbeth*—Shakespeare's first play to move in the direction of spectacular effects under the Stuarts—he insisted on the importance of machinery. To Alessandro Lanari, who commissioned the opera, he wrote, "the things that need special care in this work are: *Chorus and Machinery*" (15 October 1846, *Verdi's Macbeth,* p. 11).

231 "more an idyllic picture . . ." *Shaw's Music,* II, p. 633.

*"Be more sparing . . ." Letter of 21 September 1921, *Letters of Giacomo Puccini,* ed. G. Adami (1931), p. 278. The fact that, of the great composers, Puccini is the one whose works are as a whole the most amenable to drop-dead visual panoply may be explained by Mosco Carner's observation: "Puccini's gifts as a man of the theater include one that many playwrights might envy and that only a few opera composers have possessed. This is what the French call *l'optique du théâtre, an eye for the purely visual effect of a scene*" (New Grove *Masters of Italian Opera* [1983], p. 328).

"Realism, by its very nature . . ." *The Fashionable Tragedian* (1877), p. 21.

"It is possible that . . ." *Prefaces to Shakespeare,* first series (1927), p. xxxi.

"The order is reversed . . ." Letter to Antonio Gallo, 17 August 1869, *I Copialettere,* p. 620.

232 "not to overvalue . . ." *Shakespeare at the Globe* (1966), p. 213.

"Nowhere does the Deadly *etc.* . . ." *The Empty Space* (1968), pp. 10, 17, 11.

"greatest freedoms . . ." Ibid., p. 86.

"vigorous movement of . . ." *Shakespeare at the Globe,* p. 106.

"If in opera . . ." Letter to Cammarano, 4 April 1851, Abbiati, *Giuseppe Verdi,* II, p. 123.

*"an opera may . . ." "Opera," *New York Evening Mail,* 22 February 1918, in *Mencken on Music,* ed Louis Cheslock (1961), p. 103. Addison was of Mencken's mind. His satire on opera in *The Spectator* #5 begins, "An Opera may be allowed to be extravagantly lavish in its Decorations, as its only Design is to gratify the Senses, and keep up an indolent Attention in the Audience" (I, pp. 22–23). Addison's epigraph for this essay is still an apt one for productions that overwhelm the work itself. It is from Horace's *Ars Poetica:* "Spectatum admissi risum teneatis?" ("Admitted to the sight, would you not laugh?")

21. Squeezing the Orange

237 * what I hope for . . . The reader may notice that one important opera—Verdi's *Macbeth*—is not addressed. An entire chapter (pp. 179–215) of my *Literature as Opera* is devoted to this revolutionary work.

238 *"What I am about . . ." Letter of 10 May 1886 (Boito's emphasis), *Carteggio Verdi–Boito,* ed. M. Medici and M. Conati (1978), I, p. 104. Winton Dean wittily seconded Boito's point when he criticized a composer who approached Shakespeare more as a slave than a free man: "Instead of taking Shakespeare by the scruff of the neck, Collingwood treats him with such respect that we are tempted to concentrate on the words and wish the music would not keep interfering" (quoted by Andrew Porter in "Translating Shakespeare's Operas," *Opera* 31 [1980], p. 753).

"An opera is . . ." Letter of 18 October 1880, *Carteggio Verdi–Boito,* I, p. 5.

"adherence to . . . In *Shakespeare: The Critical Heritage,* V, p. 146.

"It is hoped . . ." *The Tempest, an Opera taken from Shakespear* (1756; reprint, 1969), sig. A4ᵛ.

239 "If you wish to set . . ." Conversation with Étienne Destranges, in *Encounters with Verdi,* ed. M. Conati (1984), p. 214.

22. Incredible Credo?

240 "Shakespeare's very complicated . . ." *Opera as Drama* (1988), p. 135; "a romantic satanist . . ." *Romantic Opera and Literary Form* (1977), p. 67; "a born villain . . ." "*Otello:* Drama through Structure," in *Essays on Music for Charles Warren Fox,* ed. Jerald Graue (1979), p. 27; "slightly embarrassing . . ." "Departing from Shakespeare," *Times Literary Supplement,* 30 January 1987, p. 114; "one of the most . . ." *Operas of Verdi* (1981), III, p. 318; "a profession . . ." Katherine Bergeron, "How to Avoid Believing (While Reading Iago's Credo)," in *Reading Opera,* ed. Arthur Groos and Roger Parker (1988), p. 190; "acute discomfort . . ." James Hepokoski, "Boito and F.-V. Hugo's 'Magnificent Translation': A Study of the Genesis of the *Otello* Libretto," in *Reading Opera,* p. 55.

   NOTE: in this essay "Iago" will denote Shakespeare's character; "Jago" (the score's spelling), Verdi's.

"Most beautiful, this Credo . . ." Letter of 3 May 1884, *Carteggio Verdi–Boito,* I, p. 76.

241 "simplification of Iago's . . ." *Opera as Drama,* p. 136.

"spiritual void . . ." *Operas of Verdi,* III, p. 317.

*bogus reasons . . . Robert Heilman systematically lays bare the speciousness of Iago's announced motives in *Magic in the Web* (1956), pp. 25 ff.

"the motive for . . ." *Brief Chronicles* (1943), p. 295.

242 *The original Italian (*Carteggi Verdiani*, ed. A. Luzio [1935], II, p. 110) is:

> Tesa è l'insidia—ho in man le frodi,
> Ti gonfia, Invidia—che mi corrodi!
> Fiele ch'io stillo—sul mio sentier,
> Furente assillo—del mio pensier.
>
> L'idea qui regna—salda, segreta (*Toccandosi la fronte*)
> E all'opra assegna—cammino e meta.
> Sì fiera imprenta—vi pose il mal
> Che non la sventa—forza mortal.
>
> D'Otello il fato—io guido, io nomo,
> Son scellerato—perchè son uomo,
> Perchè ho la scoria—dell'odio in cor,
> Mentr'ei di gloria—vive e d'amor.
>
> Ma il paziente—tempo matura
> Assiduamente—gioia e sventura.
> Trionfa o truce—mio genio altier!
> Or pianga il Duce—rida l'Alfier!!

243 "The principle cause . . ." Quoted by Hepokoski, *op, cit.,* pp. 55–59.

"Jago is Envy . . ." *Disposizione Scenica* (1887), p. 4.

244 "I have been . . ." Letter written after 26 April 1884, *Carteggio Verdi–Boito,* I, p. 74.

*a *Re Lear* libretto . . . I discuss this project, never brought to completion, in Chapter 23.

"He is not a repulsive . . ." Letter of 8 January 1855, *Re Lear e Ballo in Maschera: Lettere di Giuseppe Verdi ad Antonio Somma,* ed. A. Pascolato (1902), p. 71.

"Extremely good!" Letter to Domenico Morelli, 7 January 1880, *Copialettere,* ed. G. Cesari and A. Luzio (1913), p. 693.

245 "distracted, nonchalant . . ." Letter to Morelli, 24 September 1881, *Copialettere,* p. 318.

"The coarsest . . ." *Disposizione Scenica,* p. 5.

*"the accomplished hypocrite . . ." *Hazlitt on Theatre,* ed. William Archer (1895), p. 16; "not in the least picturesque . . ." "On Mr. Fechter's Acting," *Atlantic Monthly,* August 1869; "I played Iago . . ." *On Acting* (1986), p. 101. Margaret Webster similarly chided actors who "have adopted the sinister mien of a typical Italianate villain to an extent which would cause any sensible housewife to hide the silver spoons the moment he crossed the threshold" (*Shakespeare Without Tears* [1942], p. 234). Verdi made precisely this point to his friend Morelli: "a squat, malicious figure will make everyone suspicious and won't fool anyone!" (letter of 24 September 1884, *Copialettere,* p. 318).

"variety of colors . . ." Letter of 4 January 1855, in Pascolato, *op. cit.,* pp. 68–69. A few days later Verdi called the first version "too pretentious" and "too much seeking after effect" (letter of 8 January, p. 70).

"unbalances the Shakespearean . . ." *Romantic Opera and Literary Form,* p. 66.

246   "A curious thing!" Letter to Boito, 9 September 1886, *Carteggio Verdi–Boito,* I, p. 116; "Iago cannot . . ." Letter to Ricordi, 11 November 1886, Abbiati, *G. Verdi,* IV, p. 299; "Just as Iago . . ." Letter to Ricordi, 11 May 1887, ibid., p. 337.

"I have said . . ." Letter of 8 January 1855, in Pascolato, *op. cit.,* p. 71.

*Hepokoski suggests (*op. cit.,* p. 57) this desire to tone down the lyricism of Jago's role may owe something to another passage (unmarked by Boito) in Hugo's commentary:

> Morally [Iago] has the hypocrisy of Tartuffe. Intellectually he has the skepticism of Don Juan. He lacks only supernatural power to be Mephistopheles. Poetically—for Iago sometimes improvises—he never produced nor could have produced anything but epigrams. *Lyricism is thus denied him,* as is faith, and for him the sublime is only the neighbor of the ridiculous. (emphasis added)

"an excellent good fellow . . ." *Complete Works* (1930), IV, p. 17.

247   "Iago's statement . . ." *The Dyer's Hand* (1962), p. 257 (emphasis added).

"This Iago is . . ." Letter to Domenico Morelli, 7 January 1880, *Copialettere,* p. 694.

*Christian context . . . Heilman lavishly supports his statement that the world of Othello "is emphatically Christian, as the language abundantly shows" (*op. cit.,* p. 41).

248   "Othello is like . . ." Letter to Verdi, 18 October 1880, *Carteggio Verdi–Boito,* I, p. 4.

"Of Iago's opinions . . ." *Prefaces to Shakespeare,* fourth series (1945), p. 158.

249   "a conglomeration . . ." *Operas of Verdi,* III, p. 359.

"a remarkable . . ." *Prefaces to Shakespeare,* fourth series, pp. 154–55.

"Which way I fly . . ." *Paradise Lost:* Book IV, ll. 75, 109–10; Book IX, ll. 129–30.

"There is too much . . ." Review of 22 July 1891, in *Shaw: The Great Composers,* ed. Louis Crompton (1978), p. 212.

23. Two Lears

251   *not frightened . . . Actually, Winton Dean lists in his census seven prior—and forgotten—operas based on the play ("Shakespeare in the Opera House," p. 89).

thoughtful synopsis . . . Letter of 28 February 1850, *Copialettere,* pp. 478 ff. It also appears in Charles Osborne's *Letters of Giuseppe Verdi* (1971), pp. 152–56.

252   *"essentially innocents . . ." "Verdi the Man and Dramatist," *Nineteenth Century Music* 2 and 3 (1978–79), pp. 110–42, 214–30. I examine the *Re Lear* libretto and related correspondence in detail in "Verdi's *King Lear* Project," *Nineteenth Century Music* 9 (1985), pp. 83–101.

253   "The scene . . ." Quoted by Vincent Godefroy in *The Dramatic Art of Giuseppe Verdi* (1977), II, p. 341.

*"stormily positive . . ." Andrew Porter was less than entranced, finding "disappointingly little music in the piece." The opera, he adds, "it seems has missed the point of the play" (*The New Yorker,* 27 July 1981, p. 71).

"Transforming a . . ." Deutsche Grammophon recording liner notes, p. 50 (my
  translation).
"on the whole harder . . ." Ibid.
*wild variance . . . Porter noted in his review that the King seems senile and
  sleepy at the beginning, not "Shakespeare's great image of authority" (p. 71).
255   "the isolation . . ." Deutsche Grammophon recording liner notes, p. 51.
256   "Three weeks I . . ." Ibid., p. 54.

24. The Measure of Wagner

258   "may well be . . ." George Steiner, *The New Yorker*, 3 October 1988, p. 102.
"The hour of . . ." "Die schwache Stunde kommt fur jeden; da wird er dumm und
  läszt mit sich reden" (act 3, scene 3).
*"the mightiest Poet . . ." "The Art-Work of the Future" (1849), *Prose Works*,
  trans. W. A. Ellis (1892), I, p. 141; "applying Beethovenian . . ." Ibid., V, p. 149.
  He also observed, "Beethoven is the very counterpart of Shakespeare even in his
  attitude toward the formal laws of his art, his fulfilling abrogation of them" (V,
  p. 109).
"merely to make . . ." Quoted in C. F. Glasenapp, *The Life of Richard Wagner*
  (1876; Eng. trans. by W. A. Ellis, 1900), I, p. 93.
"nothing but a . . ." and "The plan was . . ." *Prose Works*, I, p. 4.
259   *"least successful . . ." *Richard Wagner: His Life, His Work, His Century* (1983),
  p. 73. Wagner had recently been bowled over by Wilhelmine Schröder-Devrient's
  Leipzig performances as Romeo in Bellini's *I Capuleti e i Montecchi*. Years later he
  wrote of *Liebesverbot* that "even the gentle Sicilian Bellini constituted a factor in
  this composition" (*My Life* [1936], p. 102). The melody Wagner gave Isabella to
  plead for her brother's life is pure Bellini:

Wagner moved the opera's site from Vienna to Palermo in homage to Bellini.
"to a better self . . ." Michael Balling, *Das Liebesverbot* (Edition Breitkopf). Two
  "private" recordings of the work are extant. A highlights disc of 1963 conducted
  by Meinhard von Zallinger appeared on the Golden Age of Opera label
  (EJS-273); a two-record set derived from a 1964 Vienna performance conducted
  by Robert Heger appeared on the Roger Franck label (W301).
"a first glance . . ." "Shakespeare and Opera," pp. 138–39. Dean's further assess
  ment of the music and dramaturgy of the opera is worth consulting. An *Opera
  News* correspondent, reporting on the West Coast premiere of *Das Liebesverbot*,
  concluded that "the score sounds like giddy outtakes from Nicolai, Lortzing, and
  Donizetti, with an occasional pre-echo of Gilbert and Sullivan" (July 1990, p. 36).
"the object possesses . . ." Quotations from Schiller's "On Sentimental and Naïve
  Poetry" are from Berlin's "The Naïveté of Verdi," in *The Verdi Companion* (1979),
  pp. 4, 10.
260   "at its peril . . ." *Drama, Stage, and Audience* (1975), p. 31.
"in general a want . . ." *The Characters of Shakespear's Plays* (1817), *Complete Works*
  (1967), IV, p. 345.

"most theological . . ." Louise Schleiner, "Providential Improvisation in *Measure for Measure*," *PMLA* 97 (1982), p. 227; "most specifically Christian . . ." Virgil Whitaker, *Shakespeare's Use of Learning* (1953), p. 215.

261 "wonderful summer trip . . ." All Wagnerian citations here and following are taken from *My Life*, pp. 101–2, 140–47. I have on several occasions altered the translation after consulting the original German.

"a vivacious celebration . . ." Burnett James, *Wagner and the Romantic Disaster* (1983), p. 32.

"made active interveners . . ." *Life of Richard Wagner*, I, p. 183.

262 "I would wish . . ." *The Anatomy of Abuses*, ed. F. Furnivall (1876), I, p. 99.

* got their way . . . In 1650 the Puritans enacted a law providing that when "any married woman shall . . . be carnally known by any man other than her husband (except in Case of Ravishment) . . . every person, as well the man as the woman offending therein . . . shall suffer death as in case of Felony, without benefit of clergy" (*Acts and Ordinances of the Interregnum, 1642–1660*, ed. C. H. Firth and R. S. Rait [1911], II, p. 388).

263 "the comic ebullience . . ." J. W. Lever, Arden ed. (1965), p. xciii.

264 "They are all . . ." Quoted in Gregor-Dellin, *op. cit.*, p. 74. Wagner's epithet is "Scheisskerle." When Wagner tried to persuade the director of the Leipzig theater to mount his opera by assigning the role of Mariana to the director's debutante daughter, the sober-minded director informed him that, even in the unlikely event that the Leipzig magistrates approved of the venture, he could not possibly allow his daughter to perform in such an unsavory work!

25. Taming The Shrew

265 "the most sublime . . ." *New York Times Book Review*, 6 December 1981, p. 68; "the loves of . . ." Letter to Léon Escudier, 30 June 1865, in J. Prod'homme, "Lettres inédites de G. Verdi à Léon Escudier," *Rivista Musicale* (1928), p. 191.

"I have never . . ." In *Shakespeare: The Critical Heritage*, V, pp. 146, 124, 135.

266 "degrading . . . might have passed . . ." *Shaw on Shakespeare*, ed. Edwin Wilson (1961), p. 187.

"has embarrassed generations . . ." John Bean, "Comic Structure and the Human ization of Kate in *The Taming of the Shrew*," in *The Woman's Part: Feminist Criticism of Shakespeare*, ed. Carolyn Lenz *et al.* (1980), p. 67; "self-extinguishing . . ." Margie Burns, "The Ending of *The Shrew*," *Shakespeare Studies* 18 (1986), p. 41; "superficial tidiness . . ." Valerie Wayne, "Refashioning the Shrew," *Shakespeare Studies* 17 (1985), p. 174; "some vestiges . . ." Bean, *op. cit.*, p. 74; "Shakespeare's comic . . ." Wayne, *op. cit.*, p. 174.

267 * sentimental self-indulgence . . . In his Arden edition of the play (1981) Brian Morris summarizes: "The 'obedience' speech, taken as a whole, is completely in accord with normal Elizabethan opinion on the rights and status of wives" (p. 146). Morris quotes the "impeccably orthodox and familiar" homily "Of the State of Matrimonie" that was read periodically in all English parishes: "For the woman is a weake creature, not indued with like strength and constancy of minde, therefore they bee the sooner disquieted, and they bee the more prone to all weake affections and dispositions of minde, more then men bee, and lighter they bee, and more vaine in their fantasies and opinions."

"surely they are . . ." *A Godly form of Householde Governemente* (1598), p. 211.

268 "A good woman . . ." Juan Vives, *The Instruction of a Christen Woman* (1592), sig. U2ᵛ. The last of several sixteenth-century editions of his work appeared the year before Shakespeare wrote *The Shrew*.

"marriage as a rich . . ." David Daniell, "The Good Marriage of Katharine and Petruchio," *Shakespeare Survey* 37 (1984), pp. 29–30.

"the wager has . . ." Peter Saccio, "Shrewd and Kindly Farce," *Shakespeare Survey* 37 (1984), p. 39.

Margaret Webster . . . *Shakespeare without Tears* (1942), p. 142; "the play is . . ." *The Meaning of Shakespeare* (1951), I, pp. 68–69.

"spirit remains . . ." "*The Taming of the Shrew*: Shakespeare's Mirror of Marriage," *Modern Language Studies* 5 (1975), pp. 98–99.

269 *Der Widerspenstigen* . . . A piano–vocal score of the opera was published, and there is a full-length recording (1952) conducted by Karl Elmensdorff (Urania URLP 221).

"the greatest comic . . ." *Shaw's Music* (1981), I, p. 451; "Goetz's *Taming* . . ." I, p. 613; "such masterpieces . . ." III, p. 762.

"the bite of . . ." "Shakespeare and Opera," p. 128.

"romantic love . . ." Brian Morris, *op. cit.*, p. 142.

"was too much like . . ." *Chapters of Opera* (1911), p. 247.

270 "capital show . . ." Howard Taubman, *New York Times*, 14 April 1958, p. 22. A full-length recording made by the Kansas City Opera appeared in 1972 (CRI SD272).

271 *"unquestionably the best . . ." *Op. cit.*, p. 130. The opera's score has been published. Dean refers to a recording, but I have not been able to locate one.

26. Some Ado about Berlioz

272 *"Who might . . ." *The Memoirs of Hector Berlioz*, trans. David Cairns (1975), p. 515. Many have concluded, along with Jacques Barzun, that "the parallel between Shakespeare and Berlioz is hard to resist" (*Berlioz and the Romantic Century* [1969], p. 222). Dean writes glowingly (*op. cit.*, p. 133): "Of all composers he was the most naturally adapted to produce a great Shakespeare opera, not only on account of his genius, literary as well as musical, but through some affinity of temperament. Whenever he touched Shakespeare . . . he created something of outstanding quality."

"music is one . . ." Letter of 10 July 1848, *Hector Berlioz: A Selection from His Letters* (hereafter *Letters*), ed. Humphrey Searle (1973), p. 107.

"Shakespeare is the supreme . . ." Quoted in *Memoirs*, p. 563; "It is much harder . . ." *Memoirs*, p. 97; "Shakespeare, the great . . ." Letter to Humbert Ferrand, 10 November 1864, *Letters*, p. 192.

*And it is doubtful . . . Barzun summarizes (*op. cit.*, p. 220): "In his writings [Berlioz] alludes to or quotes some one hundred and fifty passages drawn from twenty-two out of the thirty-four plays."

"only glimpse . . ." *Memoirs*, p. 97.

273 "I am quite . . ." Letter to Princess Sayn-Wittgenstein, 3 September 1856, *Letters*, p. 150.

"I saw *Hamlet* . . ." Letter of 26 May 1844, *Letters*, p. 104; "miserable poetaster etc. . . ." *Memoirs*, p. 92.

*"Toad, swollen . . ." *Memoirs*, p. 251. Berlioz naturally despised that most eminent of anti-Shakespeareans: "Voltaire told France that Shakespeare was a

Huron, a drunken Iroquois; and France believed Voltaire. . . . In France today, should Voltaire return and again express his opinion, no matter how great a Voltaire he was, is, and will remain, people would laugh in his face. Some that I know might even do worse" (*Evenings with the Orchestra,* trans. Jacques Barzun [1973], pp. 182–83).

"You fool! . . ." and "*Hamlet* and . . ." Ibid., p. 92.

"it is beautiful . . ." Letter to Sayn-Wittgenstein, 12 August 1856, *Letters,* p. 149.

"admirably arranged . . ." Letter to Ferrand, 15/16 May 1834, *Letters,* p. 57. See my chapter on Berlioz and *Benvenuto Cellini* in *Literature as Opera,* pp. 151–77.

"Shakespeare! . . ." *Memoirs,* p. 463; "which begins . . ." Letter to Sayn-Wittgenstein, 19 November 1863, *Letters,* p. 186; "For you, morons . . ." *Memoirs,* p. 473; "yesterday I . . ." Letter of 21 September 1862, *Letters,* p. 179.

274   "greatest symphony . . ." and "conceived of a vast . . ." *Memoirs,* pp. 97, 484.

*based on the general "plan" . . . Cairns writes (*Memoirs,* p. 563) that *Les Troyens* "represents the supreme fruit of that seminal discovery [of Shakespeare by Berlioz] in 1827 . . . the whole work shows at every turn an enrichment of classical *tragédie lyrique* by Shakespearean wealth of poetic (musical) language and irony of contrast." *Don Carlo* is sometimes suggested as Verdi's "Shakespearean" historical opera, but its ethos seems to me more Schillerian and Romantic.

"my infernal . . ." Letter to Sayn-Wittgenstein, 22 July 1862, quoted in New Berlioz Edition of *Béatrice et Bénédict* (1980), p. xiii; "excruciating pain . . ." Letter to his son Louis, 10 August 1862, *Letters,* p. 177.

275   "I have only . . ." Letter of 21 November 1860, *Letters,* p. 168; "I am getting . . ." Letter of 6 July 1861, *Letters,* p. 172.

"a shocking bad . . ." Letter to Ellen Terry, 3 June 1903, *Shaw on Shakespeare,* p. 156. Shaw thought *All's Well* Shakespeare's "only really interesting comedy" (letter to William Archer, 10 January 1922, *Collected Letters 1911–1925* [1985], p. 758).

"the play's . . ." A. R. Humphreys, Arden *Much Ado About Nothing* (1981), p. 69.

"miserable specimen . . ." M. R. Ridley, *Shakespeare's Plays: A Commentary* (1936), p. 106; "least amiable . . ." Alfred Harbage, *As They Liked It* (1947), p. 192.

"It is gay . . ." Letter to Peter Cornelius, 29 November 1860, in Peter Cornelius, *Literarische Werke* (1905), II, p. 756.

276   *"There is a kind . . ." Berlioz' translation of this passage, for spoken dialogue, is: "Il y a entre elle et le seigneur Bénédict une guerre d'epigrammes, et ils ne se rencontrent jamais qu'ils ne s'engage entre eux une escarmouche d'esprit."

"a caprice . . ." Letter to Sayn-Wittgenstein, 22 July 1862, *Letters,* p. 176.

278   "that of strategic . . ." Humphreys, *op. cit.,* p. 72.

"Since those days . . ." *Memoirs,* p. 162.

279   "I am going . . ." Letter to Joseph d'Ortigue, 19 January 1833, *Correspondance Générale* (1972), II, p. 68.

"radiant beauty . . ." *Memoirs,* p. 461.

*"Benedict the married . . ." It should be added that Berlioz almost immediately thrust himself into the yoke again, marrying his mistress of twelve years, Marie Recio (by most accounts a very disagreeble woman).

*"nothing about . . ." Humphreys, *op. cit.,* pp. 26, 72. Fine as its music is, it is doubtful that *Béatrice et Bénédict* will ever make more than the occasional concert-version appearance. In English-speaking countries, especially, it will seem odd to experience the spoken scenes in French, and it is hard to imagine the music sung in English. Furthermore, it is very difficult to perform. After the premiere, in

fact, Berlioz made this surprising but plausible assertion: "This little work is much harder to perform than *Les Troyens* because it has *humor*" (letter to Sayn-Wittgenstein, 21 September 1862, *Briefe von Hector Berlioz an die Fürstin Carolyne Sayn-Wittgenstein* [1903], p. 126).

"The predominant . . ." *Memoirs*, p. 478.

27. The Tempest *Tossed*

280 *"Dean discovered . . ." Dean examines several of these mostly obscure operas in often amusing detail (*op. cit.*, pp. 105–15).

"I now know . . ." Joseph Desaymard, *Emmanuel Chabrier d'après ses lettres* . . . (1934), p. 119. This letter also figures in my introduction to *Literature as Opera*, pp. 1–3.

281 "Let magick . . ." *The Tempest, an Opera taken from Shakespear,* (1756; reprint 1969), sig. G2^r.

282 "it seems probable . . ." *Reminiscences of the Opera* (1864; reprint 1976), pp. 277–78. Lumley's first choice for his *Tempest* project was Mendelssohn: "Of all modern musicians, no one had been so well fitted for the task of clothing the ideas of Shakespeare with appropriate music as the immortal composer of *A Midsummer Night's Dream.*"

A brief look at the Scribe . . . *Œuvres complètes* (1876), Vol. 47.

283 *bull by the horns . . . Dean's conclusion: "Such a literal and unadorned setting, apart from reducing the pace to a crawl, sets up an impossible challenge to the poetry. *Sprechgesang* and a *parlando* delivery . . . only confirm that music is superfluous. Nor does Martin recreate the lyrical and magic elements of the plot in musical terms; the love of Ferdinand and Miranda remains cold and Prospero becomes a droning bore" (pp. 114–15).

Verdi jotted down . . . Plate xi, *Copialettere*, p. 423.

284 "Gotter has written . . ." Quoted in Alfred Einstein, "Mozart and Shakespeare's *The Tempest,*" *Essays in Music* (1956), p. 197. The following quotations of Gotter are from Werner Deetjen, "*Der Strum* als Operntext bearbeitet von Einsiedel und Gotter," *Shakespeare Jahrbuch* 64 (1928), pp. 82–84.

285 Edward Dent observes . . . "Emanuel Schikaneder," *Music and Letters* 37 (1956), pp. 15–18.

"singularly remote . . ." Dean, *op. cit.*, p. 108. Johann Reichardt's *Geisterinsel* opera was published in an edition by Thomas Bauman in 1986.

"a magic opera . . ." Einstein, *op. cit.*, p. 201.

"*Die Zauberflöte* is . . ." *Emanuel Schikaneder: Ein Beitrag zur Geschichte des deutschen Theaters* (1951), p. 220.

286 "*Die* Zauberflöte . . ." *Leigh Hunt's Dramatic Criticism: 1808–1831,* ed. L. H. and C. W. Houtchens (1950), pp. 217–18.

28. Britten's Dream

287 "He's not a composer . . ." Quoted in Ethan Mordden, *Opera Anecdotes* (1985), p. 257.

"Everything but . . ." Robert Craft, *Stravinsky: Chronicle of a Friendship* (1972),

p. 26. In fairness, Stravinsky's curt remark on Britten's opera should be noted: "it is a mistake to conclude each act with people going to sleep" (p. 258). When Stravinsky and Britten were jointly awarded the Erasmus Prize in 1965, Stravinsky wrote to his publisher, "How tactless these Dutch" (*Selected Correspondence* [1984], II, p. 376).

"great ingenuity . . ." Geoffrey Bullough, *Narrative and Dramatic Sources of Shakespeare* (1957), I, p. 367.

288 "almost formidable . . ." Peter Evans, *The Music of Benjamin Britten* (1979), p. 236; "the most successful . . ." Dean, *op. cit.*, p. 118. In addition to these discussions of *Dream*, see also the chapters on the opera in Eric White, *Benjamin Britten: His Life and Operas* (1983), and Patricia Howard, *The Operas of Benjamin Britten* (1969).

"All is dross . . ." *Brief Chronicles* (1943), p. 43.

*only one line . . . The one line, about the "sharp" Athenian law "Compelling thee [Hermia] to marry with Demetrius," was needed because Hermia's harsh father Egeus (along with most of the play's first scene) was cut from the opera. Words and phrases were also altered occasionally.

289 "the fairy denizens . . ." Program note for the fall 1988 *Dream* production of the Los Angeles Music Center Opera, p. 21.

290 *"The average operatic . . ." For a discussion of typical Shakespearean and operatic running times, see pp. 118–20 above.

29. Wherefore Romeo?

293 "As in other . . ." Dean, *op. cit.*, p. 152.

"What a subject . . ." *Memoirs*, p. 160.

"downright threadbare . . ." *Prose Works*, V, p. 141.

"It is a duet . . ." Reported by Étienne Destranges, *Encounters with Verdi*, ed. M. Conati (1984), p. 215.

*"And yet *Romeo!*" Conversation with Étienne Destranges, *Encounters with Verdi*, p. 214. Several years later Verdi felt the need to clarify his remarks about Gounod: "They claim I made a remark criticizing *Roméo et Juliette*. But it's not true. I only expressed, and continue to express, the opinion that, believing the Shakespeare play to be much fuller of pathos, if I wrote an opera on it (as was once my intention) *I would not have ended with a duet, as Gounod did*" (*Encounters with Verdi*, p. 273).

294 *Guelph versus . . . The Montecchi and Capelletti were real families who belonged to these opposing factions, but they were prominent in different cities— the Montecchi in Verona, the Capelletti in Cremona.

295 "pathetick strains . . ." From Johnson's 1765 edition, Yale *Works* (1968), VIII, p. 957. The quoted phrase is Dryden's.

296 "voluptuous and . . ." Dean, *op. cit.*, p. 151.

30. Barber's "Pair So Famous"

298 "If ever a Shakespearean . . ." *Brief Chronicles* (1943), p. 176.

"is not in its essence . . ." M. R. Ridley, Arden *Antony and Cleopatra* (1956), p. l; "It is, no doubt . . ." A. C. Bradley, *Oxford Lectures on Poetry* (1926), pp. 282–83.

*not proved attractive . . . Dean lists only four operas based on the play. The only one he comments on is G. F. Malipiero's *Antonio e Cleopatra* (1938), and then unkindly: "Malipiero's melodic self-denial is fatal . . . However sensitive the word-setting, recitative is not enough; it only hampers the greatest poetry" (*op. cit.*, p. 156). Malipiero created remarkable havoc with the plot, deleting *all* the Roman scenes and the rustic with his snakes and giving Octavius but a few words. Dryden, incidentally, deleted Caesar entirely in his adaptation of the play, *All for Love* (1677).

*Shakespeare, I believe . . . Shakespeare is blatantly anachronistic in his diction in this play; only two words in the whole play—*lictors* and *triumpherate*—are authentic to the supposed time of the action. I discuss the decidedly contemporary "feel" of this play in the Epilogue of *Shakespeare and the Poet's Life* (1990), pp. 196–203.

"feminine arts . . ." From Johnson's 1765 edition, *Works,* VIII, p. 873.

"little sympathy . . ." Letter to Léon Escudier, 30 June 1865, J. G. Prod'homme, "Lettres inédites de G. Verdi à Léon Escudier," *Rivista Musicale Italiana* 35 (1928), p. 190.

299   "Of all the subjects . . ." Letter to Princess Sayn-Wittgenstein, 5 November 1859, *Hector Berlioz: A Selection from His Letters,* ed. Humphrey Searle (1973), p. 162.

*read Boieldieu-esque . . . See the conversation between Berlioz and Boieldieu, apropos the Egyptian queen, quoted on p. 96.

"Swinburnean mélange . . ." Harold Schonberg, *New York Times,* 17 September 1966, p. 16.

"uninterrupted bore . . ." *Variety,* 21 September 1966.

"haircurlingly awful . . ." Donal Henahan, *New York Times,* 8 February 1975.

*more favorable critical response . . . Andrew Porter discusses the revised version in *Music of Three Seasons: 1974–1977* (1978), pp. 97–102.

300   *"Take, oh take . . ." The play was *Rollo, Duke of Normandy; or, The Bloody Brother* (dated about 1624). One wonders whether "thy frozen bosom" becomes "thy frozen blossom" in the opera out of oversight or prudery. The first stanza is well known for its appearance in *Measure for Measure.* Some scholars think the first stanza is Shakespearean, the second added by a later hand.

301   *for his bass-baritone . . . Antony is a bass-baritone in the 1966 score; in the 1976 score he is described as a baritone, though the Antony in the 1983 recording (New World Records, #322-24) is still a bass-baritone. It seems to me that favoring the higher voice would greatly improve the opera's effectiveness in performance.

302   "all that Rotary Club . . ." Quoted in liner note, New World recording.

"Perhaps there could . . ." Letter to Sayn-Wittgenstein, 13 December 1859, *A Selection from His Letters,* p. 164.

31. The Other Otello

306   "the marchese Berio . . ." etc. *Life of Rossini* (1824; Eng. trans. R. N. Coe, 1956), pp. 206–17.

*"very good . . ." Letter to Samuel Rogers, 3 March 1818, *Byron's Letters and Journals,* ed. Leslie Marchand (1976), VI, p. 18. Byron also sniffed that "the first Singer would not *black* his face." Incidentally, Otello is never referred to as a Moor in the opera, always as an "Africano"—perhaps to allow the libretto's entire

avoidance of the "race theme" and to allow the divo to avoid the distress of blackface.

307   *disappointed love for Desdemona . . . From the libretto: ". . . you despise me; I care no longer for your hand. . . . You have scorned me for a vile African. . . ." From Cinthio: "The wicked Ensign . . . fell ardently in love with Disdemona. . . . But she, whose every thought was for the Moor, never gave a thought to the Ensign. . . . The love which he had felt for the Lady now changed to the bitterest hate" (Bullough, *Narrative and Dramatic Sources of Shakespeare,* VII, pp. 243–44). Disdemona, by the way, is the only character Cinthio actually names.

"to hear a performance . . ." *Life of Rossini,* p. 222.

308   "is a pure farce plot . . ." "A Word More," *Anglo-Saxon Review,* March 1901, in *Shaw: The Great Composers,* ed. Louis Crompton (1978), p. 224.

310   *a tragic ending . . . Rossini, however, was not above concocting a happy ending when obliged to do so for the more delicate audiences of Rome. Gossett, in "The Operas of Rossini" (Ph.D. diss., Princeton, 1971), calls this "perhaps the most disreputable and cynical revision of Rossini's artistic life" (p. 321).

"is contrived . . ." *Life of Rossini,* p. 226.

"truly divine . . ." *Briefwechsel und Tagebücher,* ed. H. Becker (1960), I, p. 359.

"a masterpiece . . ." *Rossini: A Study in Tragi-Comedy* (1934), p. 66.

"if one were to choose . . ." Introduction, "Otello" (facsimile edition of the manuscript copy, 1979), n.p. Gossett's extended appreciation of this scene deserves consulting; it appears, in slightly revised form, as the recording's liner essay (Philips 6769-023).

"The last act . . ." Quoted in Toye, p. 151. Bernard Shaw wrote in 1892, "If we are to have any Rossini celebrations [of his centennial], the best opera for the purpose would be *Otello,* partly because comparison between it and Verdi's latest work would be interesting, and partly because it is one of the least obsolete of his operas" (*Works,* Vol. XXVII, p. 46).

"Rossini came . . ." New Grove *Masters of Italian Opera* (1983), p. 37.

*"Nessun maggior . . ." "There is no greater sadness than to remember time of happiness in the midst of misery" (*Inferno,* 5.121–23). In the same vein, Shakespeare worked up a speech out of Ovid for Prospero to utter in the last scene of *The Tempest* ("Ye elves of hills, brooks, standing lakes, and groves . . ." 5.1.33–50).

312   "whoever the young . . ." and "less caressing" . . . *Life of Rossini,* pp. 217, 117 original emphasis).

32. Hamlet's *Outrageous Fortune*

313   "To be or not to be . . ." *À Travers chants* (1862), pp. 330–31.

314   "too weak to support . . ." "Shakespeare and Opera," p. 165.

*"it is an extraordinary . . ." New Grove Dictionary essay; "that Boito wrote . . ." Piero Nardi, *Vita di Arrigo Boito* (1942), p. 107. Dean writes of this "superb libretto" that it is "admirable alike in language, construction, and handling of the play. Every step is clearly motivated, every character developed in action. As in *Otello,* Boito reconciles Shakespeare with operatic conventions without debasing him, and it is astonishing how little he needs to omit or

alter. . . . We can only regret that Verdi did not set this libretto" (pp. 165–66). In this same article Dean describes several other *Hamlet* operas.

315 "this dilettante king . . ." Letter to Theodore Ritter, 23 May 1856, in *Hector Berlioz: A Selection from his Letters,* ed. Humphrey Searle (1973), p. 146.

"It is impossible . . ." Verdi's response is quoted in full on p. xvii.

*Critical reaction . . . Several reviews are quoted in the program essay by Jeremy Commons and Peter Murray for the London recording (#410-184-1) featuring Joan Sutherland and Sherrill Milnes.

"It is not exactly . . ." "Translating Shakespeare's Operas," *Opera* 31 (1980), p. 642.

"Thomas' opera . . ." *Op. cit.,* p. 166.

316 "not . . . marked by . . ." *Hazlitt on Theatre,* pp. 10–11.

"same passionate fever . . ." This and all following quotations are from "Hamlet," *Revue et Gazette musicale de Paris,* 15 March 1868, pp. 81–84.

33. Several Fat Knights

322 *Several scholars . . . H. J. Oliver, the Arden editor of *The Merry Wives of Windsor* (1971), summarizes: "Behind the Quarto text, then, there would seem to be a version of *The Merry Wives* that was designed for an audience not aristocratic and not primarily intellectual, whereas the full Folio text has much that would appeal only to the more sophisticated" (p. xxx).

"one of the more astonishing . . ." Arden edition, p. lxxiv.

"immensely vivacious . . ." *Shaw's Music,* III, p. 219.

"superficial . . . in the flimsy . . ." "Shakespeare and Opera," *op. cit.,* p. 120.

323 "perhaps the finest . . ." Arden edition, p. lxxiii.

"Zu Shakespeare . . ." Quoted in Dietmar Holland's liner essay, DGG recording of *Die lustigen Weiber* #2740-159.

324 *Duse tell . . . After attending a Parisian *Falstaff,* Duse wrote wittily: "Let the God-Arrigo forgive me, but it seems to me a thing so . . . melancholy, that *Falstaff.* Scorn me utterly, but that's what it is!!" (quoted in Nardi, *Vita di Arrigo Boito,* p. 606).

"the old one . . ." Arden edition, p. lxvii.

"comparison with the . . ." *Op. cit.,* p. 123.

325 "that it may be . . ." Preface, *Sir John in Love* (1971 ed.).

"not very satisfying . . ." Frank Howes, *The Music of Ralph Vaughan Williams* (1954), p. 269.

326 "*Sir John* . . . has . . ." Dickinson, *Vaughan Williams* (1963), p. 271.

327 "disappointing . . ." Dean, *op. cit.,* pp. 126–27.

*Holst's hour-long . . . Holst lists thirty-six "old English melodies" he uses in his score; he adds, wittily, that it contains three original tunes that "are, I hope, my own." Imogen Holst, in *The Music of Gustav Holst* (1986 ed.), admits that her father's entry in the Falstaff competition was "condemned by nearly everyone who heard it" (p. 67).

"In Shakespeare's day . . ." Howes, *op. cit.,* p. 274.

329 "Seldom do . . ." Dickinson, *op. cit.,* pp. 267, 271.

*by others elsewhere . . . See in particular James Hepokoski, *Giuseppe Verdi: Falstaff* (1983) and the works cited in his bibliography (pp. 172–75). Especially incisive is Graham Bradshaw's essay, "Verdi and Boito as Translators," which serves as the epilogue in Hepokoski's volume.

330 "Their love pleases . . ." Letter of 12 July 1889, *Carteggio Verdi–Boito* (1978), I, p. 150. Thomas Bauman treats this subject in detail in "The Young Lovers in *Falstaff*," *Nineteenth Century Music* 9 (1985), pp. 62–69.

Coda: *A* Dream *of the Bard*

334 *at Anne Hathaway's expense . . . In a program essay for a 1910 production of *The Dark Lady of the Sonnets*, Shaw also had fun with Anne (*Shaw on Shakespeare*, pp. 247–48). In this *jeu d'esprit* Shakespeare's bust speaks to a partygoer dressed as Lady Macbeth:

> "You are another of my failures. I meant Lady Mac to be something really awful; but she turned into my wife. . . ."
> "Your wife! Anne Hathaway!! Was she like Lady Macbeth?"
> "Very," said Shakespear, with conviction. "If you notice, Lady Macbeth has only one consistent characteristic, which is, that she thinks everything her husband does is wrong and that she can do it better. If I'd ever murdered anybody she'd have bullied me for making a mess of it and gone upstairs to improve on it herself. Whenever we gave a party she apologized to the company for my behavior."

Stephen Dedalus speculates wickedly on the courtship of Will and Anne in *Ulysses* (p. 191): "He chose badly? He was chosen, it seems to me. If others have their will Ann hath a way. By cock, she was to blame. She put the comether on him, sweet and twentysix. The greyeyed goddess who bends over the boy Adonis, stooping to conquer, as prologue to the swelling act, is a boldfaced Stratford wench who tumbles in a cornfield a lover younger than herself."

Index